D0026641

ENZYME KINETICS AND MECHANISM

ENZYME KINETICS AND MECHANISM

Dr. Paul F. Cook

Grayce B. Kerr Centennial Professor of Biochemistry
Department of Chemistry and Biochemistry
University of Oklahoma

Dr. W.W. Cleland

M.J. Johnson Professor of Biochemistry
Department of Biochemistry
Enzyme Institute
University of Wisconsin

LONDON AND NEW YORK

Vice President	Denise Schanck
Senior Editor	Robert L. Rogers
Associate Editor	Summers Scholl
Senior Publisher UK	Jackie Harbor
Production Editor	Simon Hill
Copyeditor	Heather Whirlow Cammarn
Cover Designer	Lisa Langhoff
Typesetter	Keyword
Printer	Sheridan Books, Inc.

ISBN 0815341407
9780815341406

Library of Congress Cataloging-in-Publication Data

Cook, Paul F.
Enzyme Kinetics and Mechanism / Paul F. Cook, W.W. Cleland.
p. ; cm.
ISBN 978-0-8153-4140-6 (alk. paper)
1. Enzyme Kinetics. I. Cleland, W. W. (William Wallace), 1930- II. Title.
[DNLM: 1. Enzymes–metabolism. 2. Kinetics. QU 135 C771e 2007]
QP601.5.C65 2007
572′.744–dc22 2007000470

Published in 2007 by Taylor & Francis Group, LLC,
270 Madison Avenue, New York, NY 10016, USA and
2 Park Square, Milton Park, Abingdon, Oxon, OX14 4RN, UK.

Printed in the United States of America on acid-free paper.

10 9 8 7 6 5 4 3 2 1

We dedicate this book to

Karen and Sandy Cook

and to

Erica, Elsa, and Joan Cleland

CONTENTS

Preface xv

Note to Reader xx

Chapter 1 Nomenclature 1

Reaction Components 1
Kinetic Mechanism 3
Kinetic Constants 4
Microscopic Rate Constants 5
Macroscopic Rate Constants 7

Chapter 2 Introduction to Kinetics 9

Chemical Kinetics 9
First-Order Kinetics 9
Second-Order Kinetics 11
Pseudo-First-Order Kinetics 11
Saturation Kinetics 12
Temperature Dependence of Kinetic Parameters 12

Steady-State Enzyme Kinetics 13
Initial Rate 13
Steady State 14
The Michaelis–Menten/Briggs–Haldane Equation 14
Overall Rate Equation 15
Presentation of Initial Rate Data 16

Chapter 3 Enzyme Assays 19

Assays with Reduced Pyridine Nucleotides 21
Fixed Time Assays 24
Coupled Assays 25
Analysis of Time Courses 32

Chapter 4 Derivation of Initial Velocity Rate Equations and Data Processing 35

Derivation of Rate Equations 35
Algebraic Solution 36
King–Altman Method 37
Alternative Reaction Pathways 39
Conversion of the Initial Velocity Rate Equation to Kinetic Constants 40
Net Rate Constant Method 43
Rapid Equilibrium Approximation 46
Derivation of Equations for Isotope Exchange 52
Derivation of Equations for Isotope Effects 54
Shorthand Notation for Rate Constants 55
Data Processing 56

Chapter 5 Initial Velocity Studies in the Absence of Added Inhibitors 59

Uni Bi Enzyme Reactions 60
Bireactant Enzyme Reactions 68

Ordered Sequential Mechanisms 68
Equilibrium Ordered Mechanism 69
Steady-State Ordered Mechanism 72
Theorell–Chance Mechanism 73
Random Sequential Reactions 74
Dependence of K_m on Reactant Concentration 78
Ping Pong Reactions 79
Determination of V_{max} 83
Can Sequential and Ping Pong Mechanisms Resemble One Another? 85
Crossover Point Analysis 88
Terreactant Enzyme Mechanisms 90
No Constant Term—Ping Pong Mechanisms 90
Constant Term Present—Sequential Mechanisms 91
Methods for Telling Which Denominator Terms Are Missing 94
Haldane Relationships 98
Alternative Substrate Studies 101
Kinetics of Metal Ions 105
Cooperativity and Allosterism 108
Positive Cooperativity 108
Negative Cooperativity 110
The Hill Plot 112
Allosterism 112
Kinetic Mechanism of Regulation 114
Transmission of Allosteric Effects—Coupling of Active and Allosteric Sites 116
Practical Considerations 117
From Data to Interpretation 120

Chapter 6 Initial Velocity Studies: Presence of Added Inhibitors 121

Irreversible Step 122
Types of Inhibition 122
Competitive Inhibition 122

Noncompetitive Inhibition 124
Uncompetitive Inhibition 125
Product Inhibition 128
Uni Bi Reaction Mechanisms 128
Uni Bi Steady-State Ordered 128
Uni Bi Rapid Equilibrium Random 130
Rules for Predicting Product Inhibition Patterns 132
Bireactant Enzyme Mechanisms 134
Ordered Kinetic Mechanisms 134
Steady-State Ordered 134
Theorell–Chance 142
Equilibrium Ordered 143
Random Kinetic Mechanisms 144
Rapid Equilibrium Random with EBQ Dead-End Complex 144
Steady-State Random 147
Ping Pong Mechanisms 147
Classical (One-Site) Ping Pong Mechanism 147
Nonclassical (Two-Site) Ping Pong Mechanism 150
Alternate Product Inhibition 150
Dead-End Inhibition 153
Rules for Predicting Dead-End Inhibition Patterns 157
Combination of Inhibitor with More Than a Single Enzyme Form 161
Mixed Product and Dead-End Inhibition 161
Multiple Combinations of a Dead-End Inhibitor 166
Substrate Inhibition 173
Complete Substrate Inhibition 173
Partial Substrate Inhibition 182
Induced Substrate Inhibition 184
Alternate Substrate Inhibition 186
Ping Pong Mechanism 186
Sequential Mechanisms 187
Double Inhibition 191

Slow Binding and Tight Binding Inhibition 195
Slow Binding Inhibition 196
Effect of a Reversible Inhibitor on a Slow Binding Inhibitor 199
Tight Binding Inhibition 199
Slow Tight Binding Inhibition 203
Some Practical Considerations 204

Chapter 7 Pre-Steady-State and Relaxation Kinetics 205

Reactant Concentration in Excess of Enzyme Concentration 206
Irreversible First-order Reactions 206
Reversible First-order Reactions 206
Consecutive First-order Reactions 207
Parallel First-order Reactions 212
Burst in a Time Course 213
Reactant Concentration Comparable to Enzyme Concentration 216
Temperature Jump 218
Methods of Pre-Steady-State Analysis 220
Rapid Spectral Acquisition 220
Single Wavelength Methods 220
Stopped Flow 221
Rapid Quench 222
Relaxation Methods 223

Chapter 8 Isotopic Probes of Kinetic Mechanism 225

Isotopic Exchange 226
Isotopic Exchange at Equilibrium 227
Ping Pong Exchange Patterns 227
Exchange at Equilibrium in Sequential Mechanisms 230
Isotope Exchange Not at Equilibrium 234
Oversaturation and Iso Mechanisms 235
Countertransport of Label 238

Positional Isotopic Exchange (PIX) 240
Determination of Stickiness 243
Isotope Partitioning 244
Experimental 244
Data Analysis 246
Theory 246
Theoretical Limits 248
Variation of *V*/*K* with Viscosity 249

Chapter 9 Isotope Effects as a Probe of Mechanism 253

Types of Isotope Effects 254
Nomenclature 254
Measurement of Kinetic Isotope Effects 254
Direct Comparison of Initial Rates 254
Equilibrium Perturbation 257
Internal Competition 260
Remote Label Method 264
Types of Isotope Effects 267
Equilibrium Isotope Effects 267
Equations for Isotope Effects 271
Calculation of Dissociation Constant 273
Isotope Effects on More Than One Step 275
Determination of Intrinsic Isotope Effects 275
Northrop's Method 276
Multiple Isotope Effect Method 277
Multiple Isotope Effects in Stepwise Mechanisms 280
Intermediate Partitioning 280
Reactant Dependence of Isotope Effects 282
Substrate Dependence of Isotope Effects in Bireactant Sequential Mechanisms 282
Ordered Mechanisms 284
Random Mechanisms 285
Ping Pong Mechanisms 287
Substrate Dependence of Isotope Effects in Terreactant and Higher Order Mechanisms 289

Product Dependence of Isotope Effects 290
Ordered Kinetic Mechanisms 291
Random Kinetic Mechanisms 292
Ping Pong Kinetic Mechanisms 294
Isotope Effects as a Probe of Regulatory Mechanism 295
Isotope Effects as a Probe of Chemical Mechanism 297
Isotope Effects as a Probe of Transition-State Structure 306
Formate Dehydrogenase 307
Alcohol Dehydrogenase 308
Glutamate Mutase 309
Acyl and Phosphoryl Transfers 310
Phosphoryl Transfer 313
Glycosyl Transferases 318
Binding Isotope Effects 320
Transient-State Kinetic Isotope Effects 321

Chapter 10 pH Dependence of Kinetic Parameters and Isotope Effects 325

pH–Rate Studies 325
Single Buffers with Overlap 327
Mixed Buffer Systems 327
pH Dependence of Equilibrium Dissociation Constant 328
pH Dependence of V/K 336
Activity Lost at Low pH 336
Activity Lost at Low pH and pK_1 above pH 7.3 340
Activity Lost at High pH 343
Activity Lost at High pH and pK_1 Less Than 7.3 345
V/K Profile Decreases at both Low and High pH 346
V/K Profiles That Show Two pKs at Low or High pH 350
Identifying the Groups Seen in V/K Profiles 351
pH Dependence of V_{max} 354
V *Profile with a Sticky Substrate and Proton* 354
pH Dependence of V_{max} in a Ping Pong Mechanism 357

Metal Ion Binding 358
pK_i Profiles for Metal Ions 358
pH Dependence of Isotope Effects 359
pH-Dependent Step is Sensitive to Isotopic Substitution 359
Random Addition of Proton and Substrate to Enzyme 360
Dead-End Protonation of Enzyme 361
Dead-End Protonation of Enzyme and Enzyme–Reactant Complex 362
Dead-End Formation of Protonated Enzyme–Reactant Complex 363
^DV in a Ping Pong Mechanism 364
pH- and Isotope-Sensitive Steps Differ 364

Appendices

A1. King and Altman Patterns and Distribution Equations 368

A2. Rate Equations, Definitions of Kinetic and Inhibition Constants, Haldanes, Distribution Equations, and Rate Constant Calculations for a Number of Multireactant Mechanisms 375

Index 397

PREFACE

Elucidation of the mechanism of an enzyme is not possible without a study of the kinetics of the reaction, which looks at the enzyme in the act of catalyzing its reaction. This book, which should serve as a text for graduate students and a reference for the senior investigator, provides an in-depth coverage of steady-state enzyme kinetics directed toward a determination of the kinetic and chemical mechanism of an enzyme reaction. There are a number of books on the market devoted to the theory of enzyme kinetics. Why write another text? There are several reasons we undertook this task. First, there are a number of aspects of theory that are either not covered or covered in a cursory manner in other texts, for example, isotope effects and pH-rate profiles. Second, although most books currently available provide the equations applicable to a given mechanism and a graphical analysis of the equations presented, they do not provide the underlying reasons for a given initial velocity pattern. In other words, although useful to a point, they do not convey an understanding at the intuitive level, and this is left to the reader. We have attempted to address these issues in our book.

The book is presented in three parts, roughly organized according to the way data are collected to determine an enzyme's mechanism. The systematic approach developed by W.W. Cleland is used throughout. First, one determines the kinetic mechanism; that is, the order in which substrates combine with the enzyme and products leave. Second, one determines the relative rates of the steps in the reaction at high and low levels of substrates. Third, one uses pH profiles and isotope effects to obtain information on the chemical mechanism, including acid-base catalysis. And finally, one uses isotope effects to determine transition state structure.

The first third of the book is introductory, while the last two-thirds of the book are devoted to a determination of kinetic and chemical mechanisms, respectively. Of course, it is difficult to separate kinetic and chemical mechanism, since they are interdependent. Perhaps the simplest way to illustrate this is to consider a Ping-Pong or double displacement kinetic mechanism. Demonstration that an enzyme has a Ping-Pong kinetic mechanism indicates the chemistry likely occurs via a covalent intermediate. In addition, information on kinetic mechanism is obtained, for example, from the substrate dependence of isotope effects, which is considered in Chapter 9.

The first four chapters of the book are a necessary precursor to any in-depth discussion of theory. Chapter 1, devoted to nomenclature, sets the language used throughout the remainder of the book, while Chapter 2 provides an introduction to kinetics in general, and the basics of steady-state enzyme kinetics. Chapter 3 provides information on monitoring enzyme reactions, a necessary precursor to any kinetics study. Continuous and stopped-time assays are considered, as are coupled enzyme assays, with a how-to on optimizing coupled assays with respect to cost. Chapter 4 is devoted to derivation of rate equations and data fitting. Kinetics is mathematically based, and this chapter provides the background necessary to derive any rate equation. In addition to a consideration of the usual algebraic and King-Altman methods, the net rate constant method and methods for derivation of equations for isotope exchange and isotope effects are considered. The introductory material ends with a discussion of data fitting.

Chapters 5 and 6 are devoted to a complete consideration of initial velocity studies in the absence and presence of inhibitors. Although many of the topics presented in these chapters appear in other books on steady-state kinetics, the explanation underlying the theory does not. The latter provides the reader with a much more complete understanding of existing theory, and is particularly useful for the instructor. A how-to is provided when possible, as are practical considerations at the end of the chapters, and literature examples illustrating theory when they add to the overall coverage. Chapter 5 includes information on non-Michaelis-Menten behavior and allosteric enzymes with respect to determination of kinetic mechanism of regulation. Chapter 6 provides a comprehensive coverage of inhibition from product to dead-end to slow binding. It is arguably the most complete coverage in a kinetics text. Although the text emphasizes the steady-state approach, the essentials of pre-steady-state kinetics are also covered in Chapter 7. Chapter 8 looks at other than initial rate studies to examine kinetic, and in some cases chemical mechanisms of enzymes. In addition to isotope exchange at equilibrium, isotope exchange with the reaction not at equilibrium is covered, as well as oversaturation; PIX, isotope partitioning and effects of changing viscosity are considered, and examples of each are discussed in detail.

Chapters 9 and 10 are devoted to isotope effects and pH studies with an emphasis on chemical mechanism. The following topics are discussed in detail: the use of isotope effects to obtain information on the intrinsic isotope effect (which reflects transition-state structure), kinetic mechanism from the reactant dependence of isotope effects, binding isotope effects and transient isotope effects, and the use of isotope effects to determine transition-state structure. The chapter is replete with examples from the literature. Chapter 10 details how pH-rate profiles and the pH dependence of isotope effects can be used to obtain information on acid-base chemistry. Again, examples are used liberally to illustrate the theory.

Following the final chapter are two appendices. The first provides a number of worksheets to be used in conjunction with Chapter 4 in the derivation of rate equations using the King-Altman approach; the investigator can photocopy pages to aid in derivation of the equation. The second appendix includes rate equations for a number of kinetic mechanisms in terms of rate constants, and other details of the rate equations.

Acknowledgements

We thank Bryce Plapp, Frank Raushel, Vernon Anderson, Brian Fox, Gordon Hammes, and David Ballou for many helpful comments and discussion during the preparation of the manuscript. We also thank Babak Andi for help in preparation of figures. Finally, we wish to acknowledge all of the graduate students, postdoctorates, and others who have carried out experimental work in our labs, using the theory presented in this book.

The authors are grateful for a research grant to WWC from the National Institutes of Health (GM18938), and research grants to PFC from the National Institutes of Health (GM071417), and the National Science Foundation (MCB 0091207).

Personal Notes

Enzymes are fascinating molecules, and they catalyze all of the chemical reactions in a living cell, as well as ones outside cells such as digestion. Determining the mechanisms by which enzymes catalyze their reactions has occupied Paul Cook and me during our entire careers. When we started out, one had to prepare enzymes from their original source in lengthy procedures. There were only limited structures available, and mutagenesis had not yet been invented. Cloning and overproducing enzyme in bacteria had not been conceived of. So we used kinetics as a tool to determine first the kinetic mechanisms, and later the chemical mechanisms of a number of enzymes. Even today when X-ray and NMR methods provide structures and one can create mutants at will, the kinetics must be the ultimate arbiter of mechanism, as it looks directly at the reaction while it is taking place.

I started my career as an Assistant Professor of Biochemistry at the University of Wisconsin in 1959, and at first worked on the substrate specificity of phosphatidic acid synthesis (I had been hired as a lipid chemist, as I had done a postdoc with Gene Kennedy studying cerebroside synthesis). But I was teaching in a first-year graduate course in which enzyme kinetics was one topic. I discovered that theory for reactions with more than one substrate was not available, so I started playing with equations. This led to experimental studies of product inhibition of dehydrogenases and transaminases, as well as to three papers in BBA in which I laid out the theory for the kinetics of enzymatic reactions with two or more substrates or products. Because we were having more success with our kinetic studies, I dropped the lipid work, although the development of dithiothreitol as a reducing agent for disulfides was a product of that project.

Paul Cook came to my lab as a postdoc in 1976. Six papers in Biochemistry resulted from his work, which covered equilibrium isotope effects, pH profiles for creatine kinase and several other enzymes, and the use of primary and secondary isotope effects to determine kinetic and chemical mechanisms. Paul left in 1980, and has continued to work in mechanistic enzymology and use kinetics as a tool, being successively at LSU Medical Center at New Orleans, University of North Texas, and now the University of Oklahoma.

While Paul moved slowly north from New Orleans, I have stayed at Wisconsin, only moving from the Biochemistry building to the Enzyme Institute in 1989. The Institute was run by the Graduate School for 50 years until a rogue Dean transferred it to the Biochemistry Department several years ago. But my lab is still in the Enzyme Institute building.

W. W. "Mo" Cleland

I have had an enjoyable career to this point studying enzymes using steady-state kinetic techniques. The background I received early on was broad with degrees in biology and chemistry at the undergraduate level. The training I received in the fundamentals of scientific investigation in the laboratory of Randolph T. Wedding and in the pure logic and power of kinetic techniques in Mo's lab determined my career path.

I have never found an enzyme I did not enjoy studying, and have studied many from all reaction classes in the last 30-odd years. I have also been very much interested in the theoretical aspects of the kinetics of enzyme-catalyzed reactions and have made modest contributions through the years. We have constructed this text using the overall chronology of experiments pointed out in the first paragraph of the Preface, and a teaching style we both use.

This text has been many years in preparation, with a number of false starts. It was not until Mo and I agreed to co-author the text that I knew it would be finished. Since that time it has been an enjoyable enterprise. The book is designed as both a teaching and reference text; it will hopefully be useful in both capacities.

Paul F. Cook

NOTE TO READER

This text is organized according to the way experiments should be carried out to determine the mechanism of an enzyme. Thus, experiments directed toward the determination of kinetic mechanism precede those directed toward the study of chemical mechanism. When applicable, examples from the literature are used to support and demonstrate the theory. In all cases we have attempted to explain the meaning underlying the rate equations presented, so that one can explain data at the intuitive level.

In the text, references to the primary literature are given in a shorthand format with the journal abbreviated followed by the volume and the number of the first page; for example, *Biochemistry* (1984) *23*, 5471–5478 is given as (B *23*, 5471). A list of journals and the abbreviations used throughout the text is provided below. Appropriate books are provided as full citations.

AdvEnz	Advances in Enzymology and Related Areas of Molecular Biology
AB	Analytical Biochemistry
ARB	Annual Review of Biochemistry
ACIEE	Angewande Chemie, International Edition in English
ABB	Archives of Biochemistry and Biophysics
APP	Archiv für Experimentelle Pathologie und Pharmakologie

BBA	Biochimica et Biophysica Acta
BJ	Biochemical Journal
BZ	Biochemische Zeitschrift
B	Biochemistry
CJC	Canadian Journal of Chemistry
ClinChem	Clinical Chemistry
CRB	CRC Critical Reviews in Biochemistry
EJB	European Journal of Biochemistry
FASEB J	Federation of American Societies for Experimental Biology Journal
JACS	Journal of the American Chemical Society
JBC	Journal of Biological Chemistry
JCEd	Journal of Chemical Education
JCSCC	Journal of the Chemical Society, Chemical Communications
JMB	Journal of Molecular Biology
JPC	Journal of Physical Chemistry
ME	Methods in Enzymology
PrExpPur	Protein Expression and Purification
Z. Electrochem.	Zeitschrift für Electrochemie
ZN	Zeitschrift für Naturforschung

Other References

References in the text to the following books are given by the name of the first author and date, plus page numbers if appropriate.

Cook, P.F., Ed. *Enzyme Mechanism from Isotope Effects*, CRC Press, Boca Raton, FL, 1991.

Cleland W.W., O'Leary M.H., Northrop, D.B., Eds. *Isotope Effects on Enzyme-Catalyzed Reactions* [Proceedings of the 6th Steenbock Symposium], University Park Press, Baltimore, MD, 1977.

Eisenthal, R., and Danson, M.J. *Enzyme Assays, A Practical Approach*, IRL Press, New York, 1992.

Fersht, A. *Structure and Mechanism in Protein Science*, W.H. Freeman and Co., 1999.

Martell, A.E., and Smith, R.M. *Critical Stability Constants*, 5 volumes, Plenum Press, New York, 1974–1989.

1

NOMENCLATURE

An understanding of the language of steady-state kinetics is an important prerequisite to an understanding of the theory and experimental approaches employed by the kineticist. This chapter is devoted to a discussion of the nomenclature developed by Cleland (BBA 67, 104).

Reaction Components

Enzymes catalyze chemical reactions with specific reactants that are called substrates and products. Rather than referring to the actual name of the reactants, a convenient shorthand notation has been developed in which reactants, inhibitors, activators, and enzyme are replaced by letters. Substrates are designated by the letters A, B, C, and D in the order to which they add to enzyme, and products are designated P, Q, R, and S in the order that they are released from the enzyme surface. To illustrate, the alcohol dehydrogenase reaction is shown in equation 1-1, where A and B represent NAD^+ and ethanol, and P and Q represent acetaldehyde and NADH:

$$\text{NAD}^+ + CH_3CH_2OH \rightleftharpoons CH_3CHO + \text{NADH} + H^+ \quad (1\text{-}1)$$

NAD^+	Ethanol	Acetaldehyde	NADH
A	B	P	Q

A metal ion activator can be designated by M and the ionic charge or by the actual chemical symbol for the metal ion, for exampe, M^{2+} or Mg^{2+}, while other activators are designated by X, Y, and Z. The letters E, F, and G are reserved for enzyme forms as will be discussed further. The product of an enzyme-catalyzed reaction is a substrate in the opposite direction for a reversible reaction. Thus, the physiologic reaction direction (if known) is chosen for substrates, but it is always best to specify the reaction direction. Inhibitors are designated I and J.

The following two reactions schematically depict enzyme combining with substrate to generate product and the combination of enzyme and inhibitor, respectively:

$$\mathrm{E + A \rightleftharpoons (EA \rightleftharpoons EP) \rightleftharpoons E + P} \qquad (1\text{-}2)$$

$$\mathrm{E + I \rightleftharpoons EI} \qquad (1\text{-}3)$$

There are two kinds of enzyme forms called *stable* and *transitory*. An enzyme form is termed stable if, when isolated from the rest of the reaction components, it has a long half-life with respect to the assay time scale, and reactants add to it in a bimolecular step. Stable enzyme forms that occur in a single reaction cycle, that is, conversion of one molecule of reactant to product, are given the abbreviation E, F, or G. An enzyme form is termed transitory if, when isolated from the rest of the reaction components, its half-life is short relative to the assay time scale. The latter are usually enzyme–reactant complexes. In equations 1-2 and 1-3, E is a stable enzyme form while EA, EP, and EI are transitory enzyme forms, since they form and decompose rapidly. Consider the examples given in equations 1-4 and 1-5, where E and F are stable enzyme forms and EA, FP, FB, EQ, EAB, and EPQ are transitory enzyme forms. Differences between E and F will be discussed in detail later, but they are chemically and/or structurally different forms of the enzyme.

$$\mathrm{E \xrightleftharpoons[A]{A} (EA \rightleftharpoons FP) \xrightleftharpoons[P]{P} F \xrightleftharpoons[B]{B} (FB \rightleftharpoons EQ) \xrightleftharpoons[Q]{Q} E} \qquad (1\text{-}4)$$

$$\mathrm{E \xrightleftharpoons[A]{A} EA \xrightleftharpoons[B]{B} (EAB \rightleftharpoons EPQ) \xrightleftharpoons[P]{P} EQ \xrightleftharpoons[Q]{Q} E} \qquad (1\text{-}5)$$

There are two types of transitory enzyme complexes, those in which the active site is fully occupied by substrates and products and those in which it is partially filled. Transitory complexes in which the active site is completely filled and kinetically competent are called *central* complexes, while the others are termed *noncentral*. The transitory enzyme complexes are denoted by E and the bound reactant, for example, EA and EQ. Central complexes are sometimes additionally enclosed in parentheses, for example, (EAB) and (EPQ), or with both reactants in parentheses along with arrows to show the interconversion of EAB and EPQ. Complexes with one, two, or three reactants and/or products bound are called binary, ternary, and quaternary. In equation 1-4, EA, FP, FB, and EQ are binary central complexes, while in equation 1-5, EAB and EPQ are ternary central complexes. Noncentral complexes are not present in equation 1-4, while EA and EQ are noncentral transitory complexes in equation 1-5.

Kinetic Mechanism

Kinetic mechanisms, that is, the order of addition of reactants to and release of products from the enzyme active site, fall into two classes. Reactions in which all reactants must be bound to an enzyme before any reaction occurs are called sequential, while those in which a product is released between the addition of two substrates are called ping-pong.

A sequential kinetic mechanism is depicted in equation 1-5; both A and B are bound to E before conversion to P and Q takes place. For this type of reaction there is usually only a single stable enzyme form present, E. Sequential mechanisms can be termed either *ordered*, when there is an obligatory order of addition of reactants to an enzyme and/or release of products from enzyme, or *random*, when there is not. The term rapid equilibrium is used to indicate a special case in which the equilibrium between an enzyme–reactant complex and free reactant is established in a reaction cycle. The latter will occur when subsequent steps are slow compared to the rate of dissociation of reactant.

The example shown in equation 1-4 depicts a ping-pong kinetic mechanism in which P is released prior to the addition of B. Two stable enzyme forms are present in this type of reaction, and these enzyme forms can usually be isolated. In the ping-pong mechanism a reactant binds to the enzyme active site, undergoes a chemical transformation, and leaves a fragment (or part of the substrate) on the enzyme prior to dissociating as a product. The second reactant then binds, undergoes a second chemical transformation involving the fragment of the first substrate, and dissociates as the second product.

The reaction order for a given enzyme is obtained from the number of kinetically important reactants in both reaction directions. The terms Uni, Bi, Ter, and Quad are used for enzyme-catalyzed reactions with one, two, three, and four reactants in a given reaction direction. The examples used in equations 1-4 and 1-5 above have two reactants, A and B, and two products, P and Q, and are thus called Bi-Bi reactions. Most mechanisms can be described by stating the type of mechanism followed by the reaction order in forward and reverse directions, for example, ping-pong Bi-Bi or random Bi-Ter. For more complex ping-pong mechanisms, one may need to designate groups of substrates and products released. For example, a mechanism in which A and B add followed by the release of P prior to the addition of C, and these steps precede the release of Q and R, would be called a ping-pong Bi-Uni-Uni-Bi mechanism.

Transitory complexes may undergo an isomerization, as in equation 1-6:

$$E \rightleftharpoons EA \rightleftharpoons EA^* \rightleftharpoons EA^*B \qquad (1\text{-}6)$$

Such isomerizations do not change the algebraic form of the rate equation. In some instances, isomerizations in noncentral complexes can be detected by studying the quantitative relationships between kinetic constants, as will be discussed in subsequent chapters. If, however, a stable enzyme form isomerizes so that a conformational change is required before reactant can bind to reinitiate the cycle, a separate term will be present in the denominator of the rate equation. If such an isomerization is present, the mechanism is called an iso-mechanism. Thus, equation 1-7 represents an iso-mechanism:

$$E \rightleftharpoons E^* \rightleftharpoons EA^* \rightleftharpoons EP \rightleftharpoons E \qquad (1\text{-}7)$$

There is a term (or terms) in the denominator of the rate equation for every enzyme form. The isomerization of EP to EP* would give a more complex expression for the EP term, but the complexity would not be evident from simple initial velocity studies alone. The difference between isomerization of stable and transitory enzyme forms is that the former can be detected in appropriate inhibition studies, while the latter cannot.

Kinetic Constants

The letter K together with a subscript referring to a reactant represents the Michaelis constant for that reactant; for example, K_a, K_b, K_c, and K_p represent the Michaelis constant (K_m) values for substrates A, B, and C and for the product P.

The K_m is defined as the concentration of reactant that gives half the maximum rate at saturating concentrations of all other reactants and with products maintained at zero concentration.

The term K_i with an accompanying subscript referring to a reactant or inhibitor is reserved for reactant or inhibitor inhibition constants. Thus, K_{ia} represents the inhibition constant for A in its normal capacity—that is, forming a binary EA complex, equation 1-8—while K_{ip} and K_{iq} refer to product inhibition constants for P and Q. (The K_{ia} term may not be a simple dissociation constant and may contain the rate constants for an isomerization step.) The uppercase subscript in K_I is used for dead-end combinations of reactant or inhibitor; for example, formation of the EBQ complex will generate a denominator term containing K_{Ib}. The latter is an example of substrate inhibition.

$$\mathrm{E} \underset{k_2}{\overset{k_1\mathbf{A}}{\rightleftharpoons}} \mathrm{EA} \underset{k_4}{\overset{k_3\mathbf{B}}{\rightleftharpoons}} (\mathrm{EAB} \underset{k_4}{\overset{k_5}{\rightleftharpoons}} \mathrm{EPQ}) \underset{k_8\mathbf{P}}{\overset{k_7}{\rightleftharpoons}} \mathrm{EQ} \underset{k_{10}\mathbf{Q}}{\overset{k_9}{\rightleftharpoons}} \mathrm{E}, \quad \mathrm{EQ} \overset{K_{Ib}}{\rightleftharpoons} \mathrm{EQB} \tag{1-8}$$

Finally, the term K_{eq} represents the equilibrium constant. All constants written with uppercase K except K_{eq} have units of concentration. The units of K_{eq} depend on the number of reactants and products. With an equal number of reactants and products, K_{eq} is dimensionless; with two products and one substrate, it has units of concentration.

The terms V_1 and V_2 represent the maximum velocities of the enzyme-catalyzed reaction in the forward and reverse reactions, respectively. Units are given as concentration per time and most conveniently as millimolar per minute. If the concentration of enzyme active sites in millimolar is known ($\mathbf{E_t}$), then the turnover numbers in both reaction directions can be calculated in minutes^{-1} as V_1/E_t and V_2/E_t. These are equivalent to k_{cat}, which is usually expressed in units of seconds^{-1}. Note that throughout this book reactant and enzyme concentrations will be signified by letters in boldface type, such as $\mathbf{E_t}$ and $\mathbf{A}$.

Microscopic Rate Constants

Each of the individual steps in a reaction scheme will have associated rate constants, denoted by lowercase k. For example, in equation 1-8, k_1 and k_3 represent the rate constants for conversion of E to EA and EA to EAB, respectively. First-order processes will depend on the concentration of the enzyme form being transformed alone, while second-order processes will depend on the concentrations of the enzyme form and reactant. In equation 1-8, k_1 is a second-order rate constant for the reaction of E and A, while k_5 is a first-order rate constant for reaction of EAB. Thus k_3, k_8, and k_{10} are additional second-order rate constants, while k_2, k_4, k_6, k_7, and k_9 are first-order rate constants. Second-order rate constants have units that are the product of reciprocal concentration and reciprocal time (molar^{-1} seconds^{-1}), while first-order rate constants have units of reciprocal time (seconds^{-1}). By convention, odd-numbered subscripts denote forward rate processes, k_1, k_3, etc., while even numbers denote reverse rate processes, k_2, k_4, etc.

Thus far, we have been representing the reaction as a linear sequence of steps as depicted in equation 1-8 above. This becomes somewhat tedious for more complex mechanisms. A shorthand notation has been developed in which the enzyme is represented by a line, addition of reactants is denoted by an arrow drawn perpendicular to the line, and release of products is denoted by a perpendicular arrow pointing away from the line. Stable and transitory enzyme forms are written under the line before or after the appropriate arrows. Rate constants are then written for the forward direction, that is, the conversion of A and B to P and Q, on the left of the perpendicular arrows, while rate constants for the reverse reaction direction are written on the right of the perpendicular arrows. Examples of ping-pong Bi-Bi and random Bi-Bi reactions are given in equations 1-9 and 1-10:

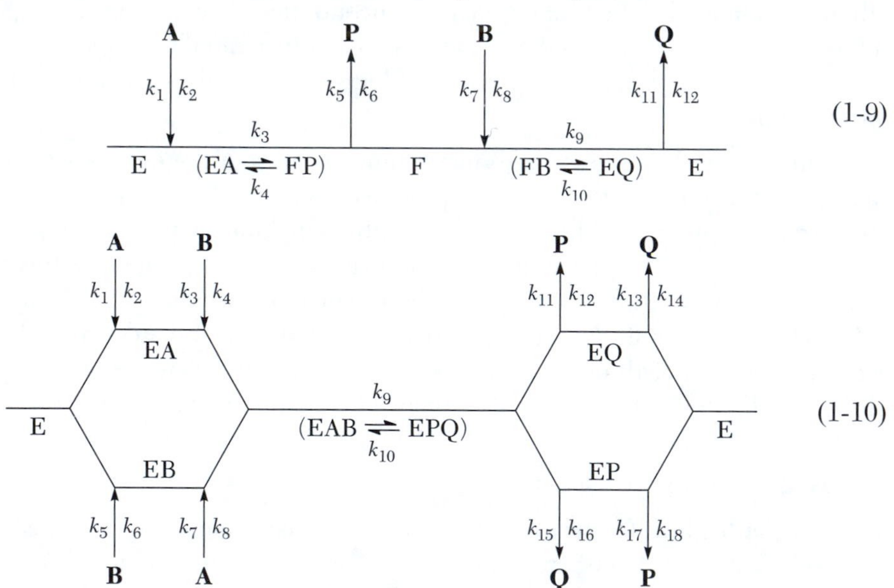

As indicated, rate constants for interconversion of central complexes are also given above and below the enzyme line.

Combination of dead-end inhibitors is represented by a double-headed arrow drawn from the complex with which the inhibitor combines to the enzyme complex formed, as shown in equation 1-11:

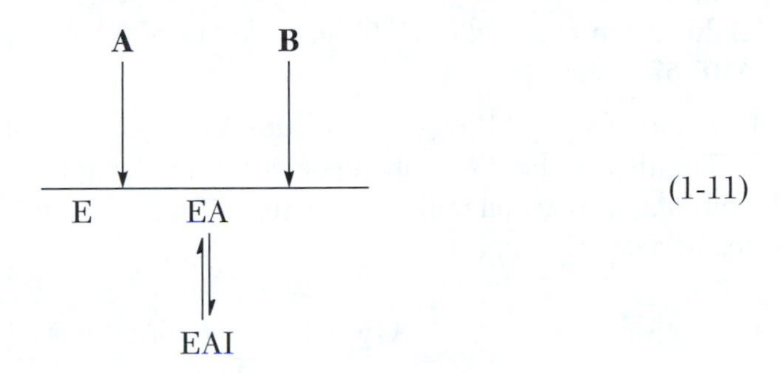

(1-11)

Macroscopic Rate Constants

When one substrate is varied at fixed levels of all others, the saturation curve for that substrate can usually be described by the Michaelis–Menten equation:

$$v = \frac{V_1\mathbf{A}}{K_a + \mathbf{A}} \tag{1-12}$$

There are two limits to equation 1-12. As **A** approaches zero, v becomes equal to $(V_1/K_a)\mathbf{A}$, so that V_1/K_a is a first-order rate constant. As **A** approaches infinity, v becomes equal to V_1, a zero-order rate constant. These constants are independent variables. The above constants are actually pseudo-first-order and pseudo-zero-order since they include the concentration of enzyme. Once corrected for enzyme concentration ($V_1/K_a\mathbf{E_t}$ and $V_1/\mathbf{E_t}$), they become second- and first-order rate constants, respectively. The constant $V_1/K_a\mathbf{E_t}$ can have values up to the diffusion-limited rate of combination of E and A. In equation 1-13, if k_3 is very fast with respect to k_2, the reactant A never has a chance to dissociate from EA and $V_1/K_a\mathbf{E_t}$ becomes equal to k_1, which is the diffusion-limited rate of formation of EA, that is, 10^7–10^9 M^{-1} s^{-1} (Fersht, 1999). Such enzyme-catalyzed reactions are said to be diffusion-limited. An example of a diffusion limited enzyme-catalyzed reaction is that catalyzed by triose-phosphate isomerase (BJ *129*, 301):

$$\mathrm{E} \underset{k_2}{\overset{k_1\mathbf{A}}{\rightleftharpoons}} \mathrm{EA} \xrightarrow{k_3} \mathrm{E} + \mathrm{P} \tag{1-13}$$

$V_1/\mathbf{E_t}$ has values from $< 1\ \mathrm{s}^{-1}$ to $10^7\ \mathrm{s}^{-1}$, with a value of about $0.15\ \mathrm{s}^{-1}$ reported for lysozyme (JBC *248*, 4786) and a value of $4 \times 10^7\ \mathrm{s}^{-1}$ reported for catalase (ABB *57*, 288).

The ratio of V_1 and V_1/K_a is the K_m for A or the steady-state dissociation constant for EA, that is, the dissociation constant under conditions in which the enzyme is catalyzing its reaction in the steady state. Thus, $K_a = (\mathbf{E}_{ss})\ (\mathbf{A}_{ss})/(\mathbf{EA}_{ss})$, where ss is steady state, or

$$K_a = \frac{\mathbf{A}(\sum \text{all enzyme forms as } \mathbf{A} \to 0)}{(\sum \text{all enzyme forms as } \mathbf{A} \to \infty)} \tag{1-14}$$

The K_a is also functional, providing an idea of the concentration range around which **A** must be varied to observe a change in rate. The constant may also give an indication of the intracellular concentration of A since many enzymes operate *in vivo* at a reaction concentration that is nonsaturating. The latter allows the maximum change in reaction flux as the concentration of nutrient is changed. K_a may be greater than, less than, or equal to the dissociation constant of A, K_{ia} (see Chapter 2).

2

INTRODUCTION TO KINETICS

In order to understand the kinetics of enzyme-catalyzed reactions, it is necessary to understand simple chemical kinetics (non-enzyme-catalyzed). This is important, at least at the level of first- and second-order reactions, for enzyme-catalyzed reactions are first-order approaches to equilibrium, and the overall reaction is second-order or more complex.

Chemical Kinetics

First-Order Kinetics

When the rate of a reaction is proportional to the concentration of the reactant, kinetics are called *first-order*. The mechanism shown in equation 2-1 is described by the rate law in equation 2-2:

$$\mathrm{A} \xrightarrow{k} \text{products} \tag{2-1}$$

$$v = -\frac{\mathrm{d}\mathbf{A}}{\mathrm{d}t} = k\mathbf{A} \tag{2-2}$$

The first-order rate constant k has dimensions of time^{-1} and is usually expressed as seconds^{-1} or minutes^{-1}. One way to determine the first-order rate constant is to measure the initial velocities of the reaction at different levels of **A** and plot the velocities versus **A**, which should give a line going through the origin. The slope is then k. A more usual analysis involves following the time course of a single reaction. If equation 2-2 is integrated, one obtains equation 2-3:

$$\mathbf{A} = \mathbf{A_0}e^{-kt} \tag{2-3}$$

or

$$\ln\left(\frac{\mathbf{A}}{\mathbf{A_0}}\right) = -kt \tag{2-4}$$

where $\mathbf{A_0}$ is the concentration of A at $t = 0$. A plot of ln ($\mathbf{A}/\mathbf{A_0}$) versus t allows k to be determined from the slope. Note that if $\log_{10}$ ($\mathbf{A}/\mathbf{A_0}$) is plotted versus t, the slope is $k/2.303$.

This analysis is valid for irreversible reactions that go to completion. For a reversible reaction:

$$\mathrm{A} \underset{k_2}{\overset{k_1}{\rightleftharpoons}} \mathrm{P} \qquad K_{\mathrm{eq}} = \frac{k_1}{k_2} \tag{2-5}$$

the rate equation with P originally absent is

$$v = -\frac{\mathrm{d}\mathbf{A}}{\mathrm{d}t} = k_1\mathbf{A_0} \tag{2-6}$$

A plot of initial velocities versus $\mathbf{A_0}$ has a slope of k_1 and goes through the origin. The overall rate equation for mechanism 2-5 is given in equation 2-7:

$$v = -\frac{\mathrm{d}\mathbf{A}}{\mathrm{d}t} = k_1\mathbf{A} - k_2\mathbf{P} = k_1\mathbf{A} - k_2(\mathbf{A_0} - \mathbf{A}) = (k_1 + k_2)\mathbf{A} - k_2\mathbf{A_0} \tag{2-7}$$

When equation 2-7 is integrated, equation 2-8 is obtained:

$$\mathbf{A} = \left(\frac{\mathbf{A_0}}{k_1 + k_2}\right)\left(k_2 + k_1 e^{-(k_1+k_2)t}\right) \tag{2-8}$$

or

$$-(k_1 + k_2)t = \ln\left(\frac{\mathbf{A} - \dfrac{k_2\mathbf{A_0}}{k_1 + k_2}}{\dfrac{k_1\mathbf{A_0}}{k_1 + k_2}}\right) = \ln\left(\frac{\mathbf{A} - \mathbf{A_{eq}}}{\mathbf{A_0} - \mathbf{A_{eq}}}\right) \tag{2-9}$$

where

$$\mathbf{A_{eq}} = \frac{k_2\mathbf{A_0}}{k_1 + k_2} \tag{2-10}$$

Thus one gets a linear plot from equation 2-8 only when one knows the value of $\mathbf{A_{eq}}$, although a fit of the time course data to equation 2-7 or 2-8 will provide estimates of the rate constants and $\mathbf{A_{eq}}$. Note that, in the exponent term, the first-order constant is the sum of k_1 and k_2 and not just k_1 alone.

Second-Order Kinetics

This occurs when two molecules react together, mechanism 2-11, and the rate is proportional to the concentrations of both, equation 2-12:

$$\mathrm{A} + \mathrm{B} \xrightarrow{k} \text{products} \tag{2-11}$$

$$v = -\frac{d\mathbf{A}}{dt} = -\frac{d\mathbf{B}}{dt} = k\mathbf{AB} \tag{2-12}$$

The second-order rate constant k has dimensions of concentration^{-1} time^{-1} and typically is expressed as molar^{-1} seconds^{-1}.

The initial velocity plotted against **A** at various **B** levels is a series of lines passing through the origin with a slope of $k\mathbf{B}$, and a replot of the slopes versus **B** then determines k. The integrated form of equation 2-12 is

$$\left(\frac{1}{\mathbf{B_0} - \mathbf{A_0}}\right)\ln\left(\frac{\mathbf{A_0B}}{\mathbf{AB_0}}\right) = kt \tag{2-13}$$

In the special case where $\mathbf{B_0} = \mathbf{A_0}$, or two molecules of A react with each other:

$$v = -\frac{d\mathbf{A}}{dt} = k\mathbf{A}^2 \tag{2-14}$$

which integrates to

$$\frac{1}{\mathbf{A}} - \frac{1}{\mathbf{A_0}} = kt \tag{2-15}$$

In the case of $\mathbf{B_0} = \mathbf{A_0}$, one uses the sum of **A** and **B** as **A** in equations 2-14 and 2-15.

Pseudo-First-Order Kinetics

Many bimolecular reactions involve the reaction of two molecules whose concentrations differ by several orders of magnitude. Then the concentration of the one present at a higher level is essentially constant during the reaction and the kinetics will appear first-order. The apparent first-order rate constant is the product of the true second-order rate constant and the concentration of the reactant in excess. This is normally the case for substrates combining with enzymes (see below), since the concentration of the enzyme is much lower than that of the substrate.

Saturation Kinetics

When a reactant is adsorbed on a surface or combines with a catalyst that is not used up and then reacts by a first-order process, the rate is

$$v = -\frac{d\mathbf{A}}{dt} = \frac{k\mathbf{A}}{K_d + \mathbf{A}} \tag{2-16}$$

where k is the first-order rate constant for reaction of the adsorbed reactant or the complex with the catalyst and K_d is the dissociation constant of the reactant. A plot of v versus $\mathbf{A}$ gives a rectangular hyperbola with a horizontal asymptote at high $\mathbf{A}$ of k and an initial slope of k/K_d. The kinetics at low $\mathbf{A}$ are first-order with k/K_d as the apparent rate constant, while at very high $\mathbf{A}$ the kinetics are zero-order (that is, the rate does not vary with $\mathbf{A}$). In between these extremes, the kinetics are of mixed order. Equation 2-16 can be linearized in several ways, but the simplest is to invert both sides as shown in equation 2-17:

$$\frac{1}{v} = \left(\frac{K_d}{k}\right)\left(\frac{1}{\mathbf{A}}\right) + \frac{1}{k} \tag{2-17}$$

A plot of $1/v$ versus $1/\mathbf{A}$ determines $1/k$ from the vertical intercept and K_d from the ratio of slope and vertical intercept.

Temperature Dependence of Kinetic Parameters

The basic equation for the temperature dependence of a unimolecular rate constant is

$$k = \left(\frac{\mathbf{k}T}{h}\right) e^{-\left(\frac{\Delta G^{\ddagger}}{RT}\right)} \tag{2-18}$$

where $\mathbf{k}$ and h are Boltzmann's and Planck's constants, k is the rate constant in seconds^{-1}, and T is the temperature in kelvins. Equation 2-16 can also be written

$$k = \left(\frac{\mathbf{k}T}{h}\right) e^{-\left(\frac{\Delta H^{\ddagger}}{RT}\right)} e^{\left(\frac{\Delta S^{\ddagger}}{R}\right)} \tag{2-19}$$

Dividing by T and taking the natural logarithms gives

$$\ln\left(\frac{k}{T}\right) = \ln\left(\frac{\boldsymbol{k}}{h}\right) - \frac{\Delta H^\ddagger}{RT} + \frac{\Delta S^\ddagger}{R} \tag{2-20}$$

The slope of a plot of ln (k/T) versus $1/T$ is $-\Delta H^\ddagger/R$, while the intercept is

$$\ln\left(\frac{\boldsymbol{k}}{h}\right) + \frac{\Delta S^\ddagger}{R} = 23.76 + \frac{\Delta S^\ddagger}{R} \tag{2-21}$$

The value of R is 1.987 cal mol^{-1} deg^{-1} with the activation parameters in calories per mole or calories per mole·degree. The most accurate activation parameter is $\Delta H^\ddagger$, since it is determined directly from the slope of the plot. The next most accurately known parameter is $\Delta G^\ddagger$, which is determined from equation 2-18 above. The uncertainty here derives from the accuracy of the assumed equation, which is only approximately correct. The least well-determined activation parameter is $\Delta S^\ddagger$, since the extrapolation to determine the intercept of the plot is long and the difference between the intercept and 23.76 will be small (a $\Delta S^\ddagger$ value of 10 cal/mol·deg adds just over 5 to the intercept).

Another way to plot the variation of rate with temperature is to plot ln (k) versus $1/T$ (the classical Arrhenius plot). In this case the slope is given by $(\Delta H^\ddagger + RT)/R$. This can be derived by evaluating $\partial(\ln T)/\partial(1/T) = -T$. This plot appears linear in the accessible temperature range, but actually it is curved and asymptotically approaches infinity at high temperatures. For accurate analysis of temperature studies, one should plot ln (k/T) versus $1/T$.

Steady-State Enzyme Kinetics

Steady-state kinetic studies are carried out in the seconds to minutes time scale, as opposed to pre-steady-state studies that make use of rapid kinetic techniques and operate in the millisecond to second time scale. There are a number of models for describing kinetic behavior, but generally it is assumed that an enzyme with more than one subunit binds one molecule at each active center and that there is no interaction between active centers. The case of interaction between sites will be considered later.

Initial Rate

Consider the mechanism depicted in equation 2-22.

$$\mathrm{E + A} \underset{k_2}{\overset{k_1}{\rightleftharpoons}} (\mathrm{EA} \underset{k_4}{\overset{k_3}{\rightleftharpoons}} \mathrm{EP}) \underset{k_6}{\overset{k_5}{\rightleftharpoons}} \mathrm{E + P} \tag{2-22}$$

Steady State

The enzyme-catalyzed reaction is assumed to be at steady state when the concentrations of intermediates along the reaction pathway, that is, **E**, **EA**, and **EP**, change much more slowly than the reactant concentration so that we can assume $d\mathbf{E}/dt = d(\mathbf{EA})/dt = d(\mathbf{EP})/dt = 0$. In addition, equilibria are assumed to be established rapidly, that is, in the mixing time prior to measurement of the rate. The steady state is achieved whenever $\mathbf{A_t} > \mathbf{E_t}$; under these conditions, $d\mathbf{A}/dt \gg d\mathbf{E}/dt$.

The steady-state rate is measured as close to time zero as possible such that little change in the concentration of added substrate occurs and the final and initial concentrations of substrate are the same. Conventionally, the rate measured when the reactant concentration has changed by $<10\%$ is usually an initial rate. However, for reversible reactions in which the equilibrium position favors the substrates, the initial velocity should be determined before the reaction has proceeded to 10% of its final position. It is assumed that the concentration of substrate is the final concentration in the reaction mixture prior to initiation of reaction. The initial rate is obtained as a function of the concentration of substrate as it is varied around its Michaelis constant (K_m). The K_m values for the majority of substrates of enzyme-catalyzed reactions are in the range 1 μM (10^{-6} M) to 1 mM (10^{-3} M), while the assay concentration of most enzymes is in the range 10^{-8}–10^{-9} M. Since $\mathbf{A_{total}} = \mathbf{A_{free}} + \mathbf{EA}$ for the mechanism in equation 2-22, and $\mathbf{A_{total}} \gg \mathbf{E_{total}}$, then $\mathbf{A_{total}} \sim \mathbf{A_{free}}$. The concentration of added reactant is used for steady-state studies, and it is normally assumed that the substrate concentration is known absolutely and has no error. (Operationally, it is important to determine the concentration of the substrate stock solution as accurately as possible to minimize error.)

But if one is studying a very slow mutant enzyme, it may no longer be true that $\mathbf{A_{total}} \gg \mathbf{E_{total}}$, and one must use more complicated rate equations (see Chapter 7).

The Michaelis–Menten/Briggs–Haldane Equation

The initial rate of a given reaction is described mathematically by a rate equation that describes the initial rate of the reaction under any given set of conditions. The simplest treatment of the steady-state rate equation was provided by Henri Michaelis and Maude Menten (BZ *49*, 533). These authors made the assumption that an enzyme-catalyzed reaction could be described by the mechanism given in equation 2-23:

$$\mathrm{E + A} \underset{k_2}{\overset{k_1}{\rightleftharpoons}} \mathrm{EA} \xrightarrow{k_3} \mathrm{E + P} \qquad (2\text{-}23)$$

In the above mechanism Michaelis and Menten further reasoned that the binding of substrates would be rapid compared to the subsequent chemical steps that converted substrate to product (in rapid equilibrium) and that at saturation by the substrate the rate would be equal to the microscopic rate constant k_3. The rate equation shows that the initial rate would be a saturable function of substrate concentration and retains their name. The equation is given below, and one means of deriving it is provided in Chapter 4.

$$v = \frac{k_3\mathbf{A}\mathbf{E_t}}{(k_2/k_1) + \mathbf{A}} = \frac{V_1\mathbf{A}}{K_a + \mathbf{A}} \qquad (2\text{-}24)$$

In equation 2-24, $k_3\mathbf{E_t}$ and k_2/k_1 were called the maximum rate (V_{max}) and the K_m, respectively. The K_m in this treatment is equal to the dissociation constant (K_d) for EA.

Briggs and Haldane (BJ *19*, 338) modified the treatment of Michaelis and Menten by suggesting there was no *a priori* reason that the binding of substrate should come to equilibrium. Rather, they assumed that a steady state was reached in which the concentration of enzyme forms was determined by the rates of formation and breakdown and did not change during the initial velocity phase. They reasoned that the chemical interconversion of substrate to product could be more rapid than the release of substrate from the enzyme active site. Thus, although the algebraic form of the rate equation did not differ from that of equation 2-24, the expression for the K_m became equal to $(k_2 + k_3)/k_1$. Thus, depending on the relative rates of the chemical steps and release of the substrate from the EA complex, the K_m may be equal to the dissociation constant or larger. In actuality the expression for K_m is more complex than that given above and can be less than K_d, as will be discussed below. However, the basic principles put forth over eight decades ago still apply to the treatment of steady-state kinetic data. The rectangular hyperbolic function given in equation 2-24 applies to any case in which the initial rate is measured at different concentrations of one substrate with all others maintained constant at any concentration.

Overall Rate Equation

The overall rate equation for an enzyme-catalyzed reaction is a complex expression in which the numerator and denominator provide different but equally important information. The overall rate equation derived from mechanism 2-22 is

$$v = \frac{V_1\mathbf{A} - V_2\left(\frac{K_a}{K_p}\right)\mathbf{P}}{K_a + \mathbf{A} + \left(\frac{K_a}{K_p}\right)\mathbf{P}} \qquad (2\text{-}25)$$

Mechanism 2-22 and equation 2-25 represent a Uni-Uni reaction such as that catalyzed by a racemase, isomerase, mutase, or a lyase such as fumarase. The kinetic parameters given in equation 2-25 are complex expressions of the microscopic rate constants given in mechanism 2-22 (see Chapter 4 for derivation of the rate equation). The numerator of equation 2-25 represents the thermodynamic driving force of the reaction and provides an indication as to the direction of the reaction for given concentrations of A and P. At equilibrium the rates of formation of A and P are equal, and thus the initial rate is equal to zero. Under these conditions, $V_1\mathbf{A_{eq}} = V_2(K_a/K_p)\mathbf{P_{eq}}$, and $\mathbf{P_{eq}}/\mathbf{A_{eq}}$, which equals K_{eq}, is equal to $(V_1/K_a)/(V_2/K_p)$. The relationship between K_{eq} and the measured kinetic parameters was described by Haldane and bears his name, the *Haldane relationship*.

The denominator of equation 2-25 represents a distribution of the enzyme species among all of the possible forms that could exist for mechanism 2-22. The denominator corrects for the slowdown of the reaction as a result of enzyme being potentially distributed among a number of enzyme forms depending on conditions of reactant concentration. Each of the terms in the denominator of the rate equation represents one or more forms of the enzyme. The limits of the equation with respect to reactant concentration allow assignment of each term in the denominator to a form (or forms) of the enzyme that exists along the reaction pathway in mechanism 2-22. The limit of the denominator of equation 2-25 as **A** and **P** tend to zero is K_a. The predominant form of the enzyme under these conditions is E, and thus K_a in the denominator represents E. The limit as either **A** or **P** tends to infinity with the other tending to zero will represent EA plus EP. (The distribution between the latter two cannot be determined unless the rate constants for interconversion of EA and EP are known.)

It is thus the denominator of the rate equation that provides information on the kinetic mechanism, while the numerator provides information on the thermodynamic driving force. The overall rate equation differs, depending on the kinetic mechanism of an enzyme-catalyzed reaction. One attempts to determine what terms exist in the denominator of the rate equation to elucidate the kinetic mechanism of an enzyme-catalyzed reaction. The determination of kinetic mechanism will be considered beginning in Chapter 5.

Presentation of Initial Rate Data

A determination of the terms present in the denominator of the rate equation is obtained by measuring the initial rate as a function of the concentration of one reactant, maintaining the level of all others fixed, and then repeating this at several additional fixed levels of the second reactant. Data are displayed graphically by use of the linear transform of the Michaelis–Menten equation devised by Lineweaver and Burk, equation 2-26. [There are other linear transforms of the Michaelis–Menten equation, for example, Eadie–Hofstee (v vs $v/\mathbf{A}$) and Hanes–Wolf ($\mathbf{A}/v$ vs $\mathbf{A}$), but the Lineweaver–Burk or double reciprocal plot is by far the most commonly used and as a result is the most easily recognizable by the majority of investigators in the field.]

$$\frac{1}{v} = \left(\frac{K_a}{V_1}\right)\left(\frac{1}{\mathbf{A}}\right) + \frac{1}{V_1} \tag{2-26}$$

The reciprocal of the initial velocity is plotted against the reciprocal of the reactant concentration. An example of a double reciprocal plot is shown in Figure 2-1.

The slope of the double reciprocal plot represents K_a/V_1, while the vertical intercept represents $1/V_1$. These are the reciprocals of the two independent kinetic constants V_1/K_a and V_1. These terms are the limits of the Michaelis–Menten equation as discussed in Chapter 1 and reflect the enzyme form that predominates when the reactant concentration is extrapolated to zero or infinite concentration, respectively. Thus, one should be able to read the initial velocity double reciprocal plots to determine the absence or presence of enzyme forms in the denominator of the rate equation. Note that the value of K_a is obtained from the abscissa intercept

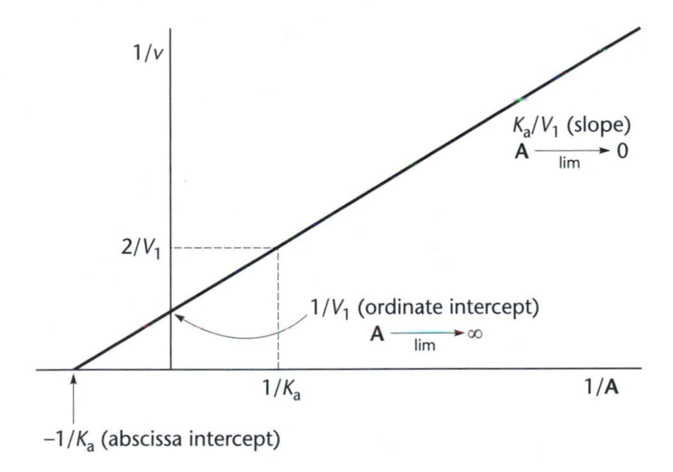

Figure 2-1. Lineweaver–Burk or double reciprocal plot of initial velocity and substrate concentration. The graphical estimation of kinetic constants is shown.

or the value of $1/\mathbf{A}$ when $1/v$ is extrapolated to zero. The K_a can also be obtained as required by the definition of K_a from the value of A at $v = 2/V_1$. For the plot shown in Figure 2-1 and a Uni-Uni reaction, the slope term represents free enzyme, while the intercept term represents EA plus EP. This method of interpreting initial velocity patterns will be used throughout the discussion of initial velocity data in the next section of the book.

3

ENZYME ASSAYS

The first requirement for studying an enzyme-catalyzed reaction via steady-state kinetics is a means of following the reaction. Parameters that must be taken into account for a good enzyme assay include sensitivity, interference from other components, ease of use, and cost. A more detailed consideration of assays can be found, for example, in the book by Eisenthal (1992).

An important criterion in choosing an assay is sensitivity. If the reactant concentration must be varied around 1 μM in order to see significant changes in rate with changes in substrate concentration, the monitor used must be sensitive enough to allow measurement of an initial rate. The absorbance of a 1 μM solution of NADH in a 1 cm path length cuvette is 0.006, so one needs a 10 cm path length cuvette to increase the sensitivity sufficiently so that accurate changes in absorbance could be measured under conditions where 10% of the NADH concentration changed, that is, 0.0006 for a 1 cm path length cell.

There are a number of different ways an enzyme reaction can be monitored, and these can be divided into continuous assays and fixed- or stopped-time assays. A continuous assay allows one to follow a reaction while it is occurring, that is, in real time. The advantages to the continuous assay are (1) initial velocities are readily measured, (2) errors are rapidly and easily detected, and (3) controls are easily carried out. The simplest of the continuous assays are ones in which a reactant or product of the enzymatic reaction being studied is colored so that its disappearance or appearance can be followed in a spectrophotometer as an indicator of reaction progress. There are a number of species that may serve as a means of monitoring reaction progress, such as fumarate (monitored at 240 nm), or acetyl-CoA (monitored at 232 nm). Many dehydrogenases use NAD^+ or $NADP^+$ as the oxidant, which are reduced to NADH or NADPH during the reaction. Both the oxidized and reduced forms absorb at 260 nm, but only the reduced forms have significant absorbance at 340 nm with a millimolar extinction coefficient of 6.22 for a 1 cm light path (Figure 3-1).

What is monitored for a given reaction depends on the properties of the reactants and techniques available. NADH can be monitored by either its absorbance at 340 nm, Figure 3-2, or its fluorescence, which gives higher sensitivity, at 465 nm, exciting at 340 nm. However, dust and dirt can interfere with fluorescence measurements, and greater care is needed with fluorescence assays than with those using UV–visible absorbance, although they are more sensitive. The disappearance of NADH can also be followed at 340 nm, although the upper limit of its concentration is about 400 µM ($\sim 2.4 A_{340}$) in 1 cm cuvettes (higher concentrations can be used in shorter path length cells or by monitoring the reaction at 366 nm, where the extinction coefficient is 3.1 $mM^{-1}cm^{-1}$). Circular dichroism is particularly useful for racemases, which interconvert stereoisomers that absorb opposite components of circularly polarized light and therefore have opposite signs in their CD spectra, but these assays are not very sensitive for most molecules. Changes in CD have been used to monitor the interconversion of D- and L-mandelate by mandelate racemase (AB *94*, 329) and to follow the proline

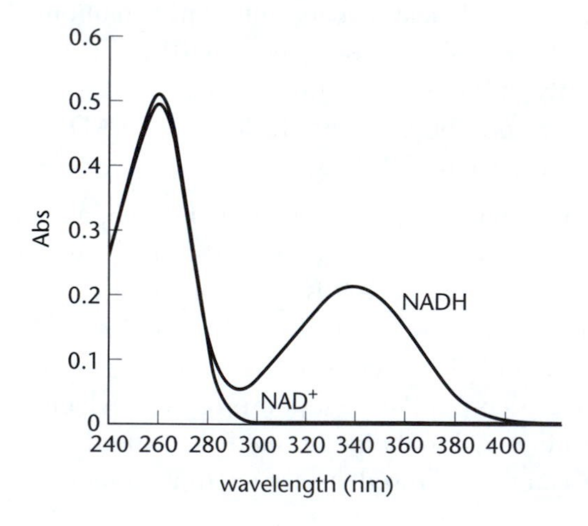

Figure 3-1. Absorbance spectra for NAD (λ_{max}, 260 nm) and NADH (λ_{max}, 260 nm, 340 nm).

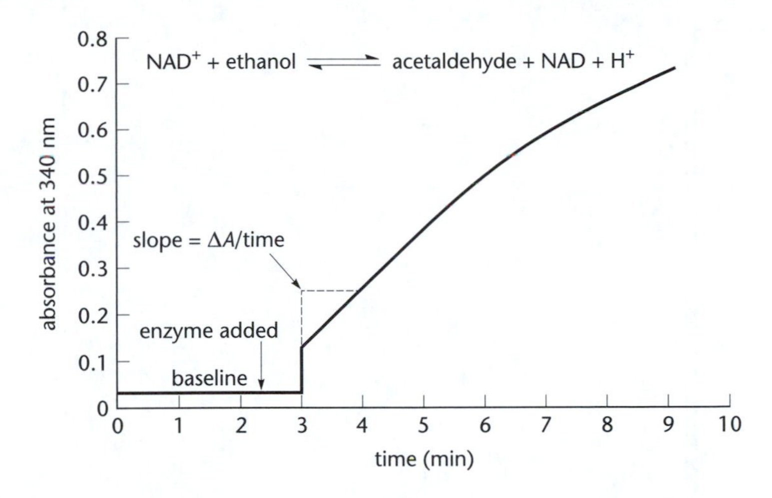

Figure 3-2. Time course for the alcohol dehydrogenase reaction.

racemase reaction (B *14*, 4515). Another possible assay uses a pH stat to measure protons produced or taken up during a reaction (APP 97, 242). These assays make use of automated titration of protons generated by use of calibrated base solutions as developed for dihydrofolate reductase (AB *152*, 1) or hydrogenase (AB *131*, 525), or can make use of a chromophoric pH indicator such as *p*-nitrophenol (ClinChem *17*, 214). Ion-selective electrodes can also be utilized to monitor changes in the concentration of an ion with time; sulfide concentration changes with time have been monitored with the sulfide electrode (PrExpPur *1*, 70). The oxygen electrode can be used to follow oxygen uptake in a closed system, and NMR can be used to follow slow enzymatic reactions.

Assays with Reduced Pyridine Nucleotides

The maximal wavelength of NADH absorbance is well separated from the absorbance of protein (280 nm), buffers, etc., and thus little interference occurs from other species in the reaction mixture. An enzyme assay is prepared in a cuvette with a fixed path length with buffered reactants, activators, or inhibitors; the cuvette is placed in a spectrophotometer set at 340 nm, linked to a strip chart recorder or computer; and a baseline is recorded in the absence of enzyme. Enzyme is then added, and the increase or decrease in absorbance with time is recorded. An assay for the alcohol dehydrogenase reaction is shown in Figure 3-2.

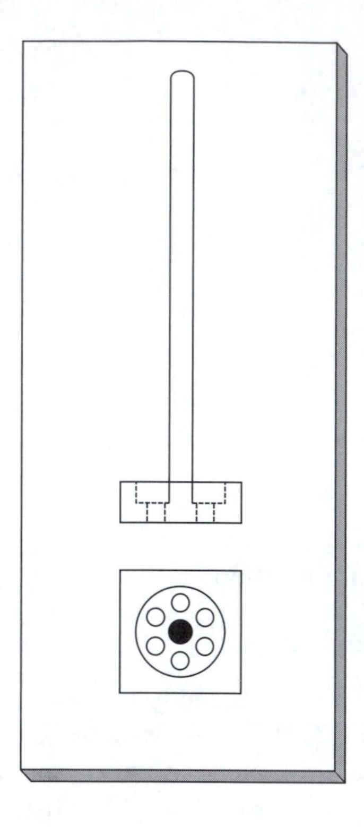

Figure 3-3. Adder–mixer for a 1 cm square cuvette. Side and top views are shown. The dark circle in the middle of the top view is the handle, while the clear circles represent holes; the depth of the cup is indicated in the side view as the distance from the bottom of the handle to the top of the base.

The initial rate is the slope of the time course extrapolated to time zero. The rate should be corrected by subtracting the rate of a minus-enzyme control. The slope of the time course has units of Δabsorbance/minute, and must be converted to millimolar per minute by division by the millimolar extinction coefficient of the reduced nucleotide and the path length used. The rate is proportional to enzyme concentration, which is most often given as units per milliliter (a unit is the amount of enzyme required to convert 1 μmol of reactant to product in 1 min at a given pH and temperature). The relationship between measured rates and the $V/\mathbf{E_t}$ or k_{cat} of an enzyme was discussed in Chapter 2.

When the above assays are carried out, the enzyme is best added with an adder–mixer, which is a plastic plate sized to fit the cuvette that has a well with small holes in the base and a handle (Figure 3-3). An aliquot of enzyme (50 μL, for example) is placed in the well, and the adder–mixer is pushed up and down in the cuvette 3–4 times to mix the contents without breaking the surface of the liquid (to prevent formation of bubbles). The chart motor is started (or the computer is prompted) with one hand when the adder–mixer is placed in the cuvette, and when the mixer is withdrawn and the cover is placed on the cuvette compartment, recording of absorbance is initiated. This leads to a gap in the initial recording corresponding to the 3–4 s it takes to mix the enzyme into the cuvette and close the cover. The initial velocity is obtained by extrapolating the recorded trace back to the mixing time and taking the tangent to the curve. Even when a computer is added to store the data, the time course should be printed out so one can evaluate the data and eliminate artifacts.

Several artifacts may require care in order to measure initial velocities in a spectrophotometric assay. First, there may be a lag in achieving a steady-state rate, especially when a coupled assay is used (see page 25). One must ignore the lag and extrapolate the time course back to the mixing time. Ships curves[1] are very useful for doing this so that a tangent at zero time can be obtained. If a computer is used to store and process data, be sure to delete the portion of the curve containing the lag before initial velocities are calculated. Unless the substrate concentration is very high, the time course of the reaction will be curved as the substrate is used up and the products accumulate, so measurement of initial velocities requires careful evaluation of the time course.

Several parameters can be varied to make it easier to determine initial velocities. Reducing the enzyme level and the chart speed together will reduce the initial portion of the curve that is missing as a result of mixing and reduce the effects of coupling lags (see page 25). Increasing the full-scale sensitivity of the spectrophotometer and reducing enzyme level will allow recording of the earlier portion of the time course and thus minimize the curvature caused by substrate depletion. Increasing the path length of the cuvette, along with reducing enzyme levels, has the same effect. It should be clear that by balancing these various parameters both sensitivity and the time of assay can be adjusted to permit accurate determination of initial velocities. Many bad kinetic data result from failure to pay attention to these problems and/or from allowing a computer to calculate an incorrect initial velocity from a time course that is too curved, shows a lag, or contains some other perturbation.

Other artifacts that may distort initial velocities are bubbles, dust particles, voltage surges, or inadequate mixing. Figure 3-4 shows these effects. Being careful with the adder–mixer will prevent bubbles, while centrifuging solutions (particularly the enzyme solution) will remove dust or other particles. Proper use of the adder–mixer should ensure good mixing, but when time courses look like those in Figure 3-4A,B, the best recourse is to discard the assay and try again. The effect

[1]Plastic templates with various compound curved shapes used by naval engineers to draw the shape of a ship's hull, look for them where engineering or architectural supplies are sold.

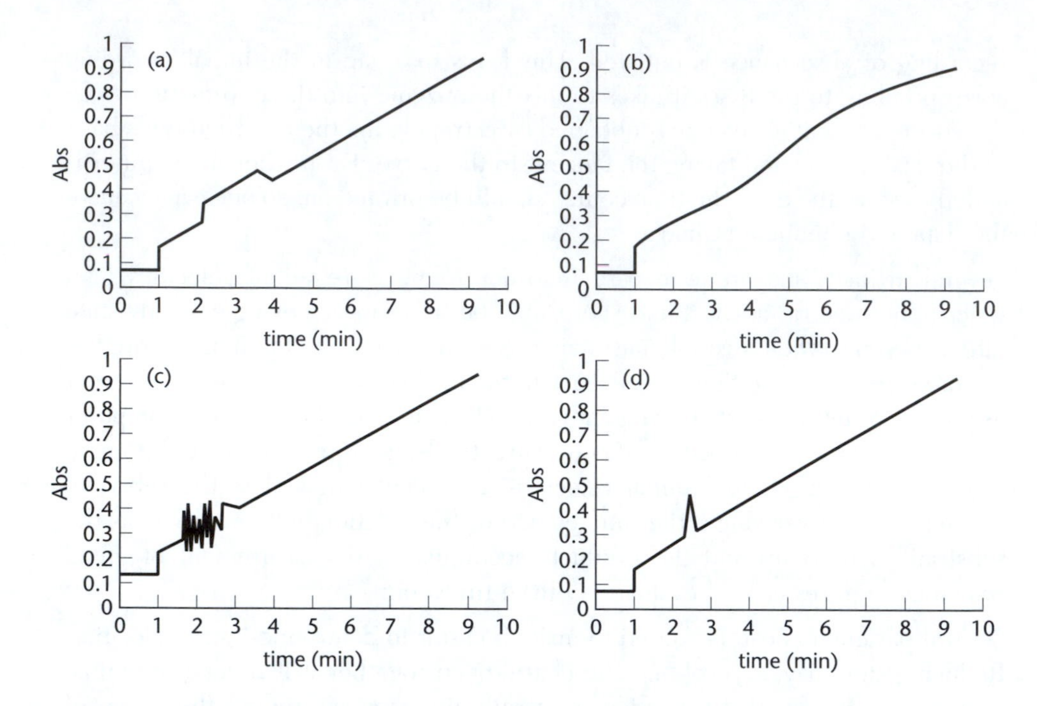

Figure 3-4. Common artifacts observed in spectrophotometric assay. (A) A bubble passes through the light path. (B) The reaction solution was not thoroughly mixed. (C) Dust passing through the light path. (D) A voltage surge.

of a single bubble, dust particle, or voltage surge can be corrected by proper use of ships curves and your eye, but this is not easily possible by computer. This is why it is essential to print out the time course and observe it visually. If possible, the first derivative (slope) of the time course can be determined by a computer fit of the time course.

Fixed Time Assays

Many reactions either do not have a convenient probe for continuous monitoring or the probe is not easily adapted to the reaction, not sensitive enough, or perhaps suffers from interference from other reaction components. For some reactions the conversion of a radioactive substrate to product must be used to follow the reaction. In these cases, a fixed-time assay must be utilized, requiring that a reaction mixture be aliquotted at different times after initiation of reaction and that the reaction be terminated by a means such as acid, base, or organic solvent that will not destroy the reactant to be measured. A reactant is then quantified by a variety of techniques, in some cases requiring chromatographic isolation.

The proper way to run a fixed time assay is to make up a reaction mixture containing all components except the one used to start the reaction (normally, but not necessarily, the enzyme) and then add the final component. As soon thereafter as practical (3–4 s), take an aliquot of the reaction and quench it at the same time as the time record is begun. One minute later quench a second aliquot, and do this at 1-min intervals until four aliquots are obtained. For example, use a 1 mL reaction mixture and remove 0.2 mL aliquots, so that 80% of the reaction mixture is utilized.

The time between aliquots can be longer if less enzyme is used, but it is too difficult to time samples when the reaction is run faster. The four aliquots should cover no more than 10–15% of the reaction in order to determine initial velocity. Note that the first aliquot is taken *after* mixing and is not a zero time control. This way the first aliquot contains not only the product formed in the 3–4 s after mixing but also the product present in the reaction prior to mixing. The aliquots are analyzed and a plot of product versus time used to determine initial velocity.

Contrast this procedure with the typical fixed time assay used in the past, where a zero time control was taken before mixing and only one time point was utilized. With this procedure there is no way to know whether the reaction was still linear when product formation was measured. If one is running a reciprocal plot for an enzymatic reaction and taking points after 5 min the reaction may still be linear with higher substrate concentrations but may have reached equilibrium and stopped with lower ones. A control reaction without enzyme must also be run and the rate from the minus-enzyme control must be subtracted from the rate in the presence of enzyme.

Coupled Assays

Fixed time assays are laborious, and thus it pays to find other ways to assay reactions that do not involve color changes. Such reactions can often be coupled to one or more reactions (enzymatic or nonenzymatic) that result in the utilization or production of a colored compound. A coupled assay can be depicted schematically (B 7, 2782; BBA 293, 552) as in equation 3-1.

$$\mathrm{S} \xrightarrow{v_0} \mathrm{Int} \xrightarrow[\text{Enzyme 2}]{V/K_{\mathrm{Int}}} \mathrm{P} \tag{3-1}$$

In equation 3-1, S and Int (intermediate in the overall coupled assay) are the substrate and product of the enzyme reaction for which an initial rate, v_0, is to be measured. The product Int is the substrate of enzyme 2, the coupling enzyme that produces a colored product P. The steady-state level of Int in the assay must be less than its K_{Int} (K_m) for enzyme 2 and its product inhibition constant for enzyme 1, since it is necessary for Int to be converted to P as fast as it is produced. Thus, in order to monitor v_0 for enzyme 1, the activity of enzyme 2 must be sufficiently greater than that of enzyme 1 so that it can utilize very low concentrations of its substrate Int. The rate constant for an enzyme-catalyzed reaction under conditions where reactant concentration is limiting is V_{max}/K_m (or V/K), the pseudo-first-order rate constant for conversion of reactant to product. In mechanism 3-1, V_{max} is for enzyme 2 and K_{Int} is the Michaelis constant for Int (see Chapter 2). The steady-state rate of conversion of S to Int, if an irreversible (or practically irreversible) reaction for enzyme 2 under the above conditions is assumed, is then

$$\frac{d\mathbf{P}}{dt} = v_0 = \left(\frac{V}{K_{Int}}\right)\mathbf{Int} \tag{3-2}$$

The complete rate equation for P formation in mechanism 3-1, however, is

$$\frac{d\mathbf{P}}{dt} = v_0(1 - e^{-(V/K_{Int})t}) \tag{3-3}$$

and a plot of the integrated form of equation 3-3 is shown in Figure 3-5. In this figure the rate increases from zero at zero time to a constant steady-state rate.

The change in rate at early time results from the buildup of Int to a concentration sufficient for (V/K_{Int})**Int** to equal v_0; that is, the rate is limited only by the conversion of S to Int. Extrapolation of the linear steady-state region of the time course to zero time gives a negative intercept, $\Delta\mathbf{P}$, which corresponds to the steady-state level of Int, which is $v_0/(V/K_{Int})$. From the integrated rate equation it can be shown that the lag time τ is given by K_{Int}/V. If there is more than one coupling enzyme, the lag is given by

$$\tau = \sum_i \left(\frac{K_i}{V_i}\right) \tag{3-4}$$

where i represents the number of coupling enzymes utilized, K_i is the K_m for the intermediate used by a given coupling enzyme, and V_i is the activity of the coupling enzyme in units per milliliter.

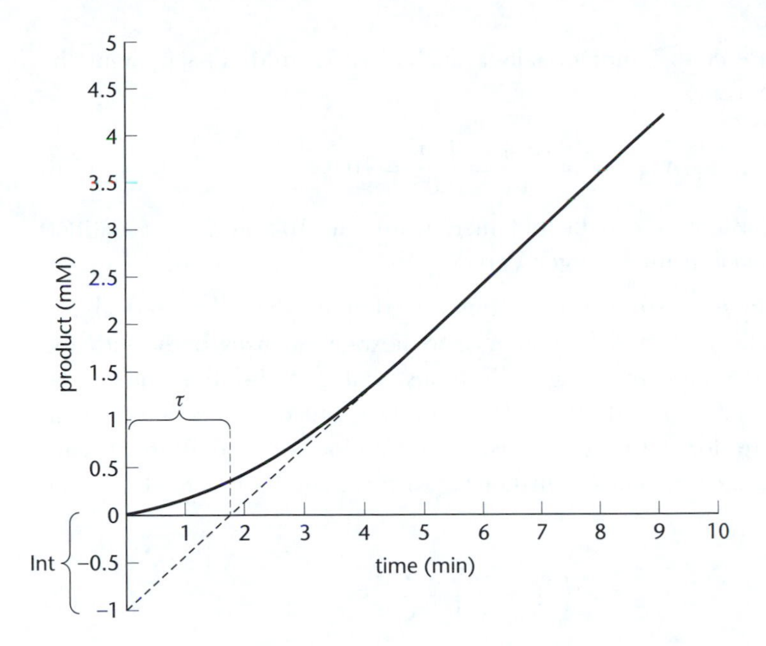

Figure 3-5. Time course of a coupled assay. τ is the lag time, and **Int** is the steady-state level of intermediate(s).

As is evident from equation 3-4, the lag time, τ, can be adjusted by varying the concentration of the coupling enzyme or enzymes. Calculation of the length of time required to reach the steady-state rate in terms of the lag time can be done as follows. First, V/K_{Int} is substituted in equation 3-3 by $1/\tau$ to give equation 3-5:

$$\frac{d\mathbf{P}}{dt} = v_0(1 - e^{-t/\tau}) \tag{3-5}$$

The rate will reach 95% of v_0 when $(1 - e^{-t/\tau})$ is equal to 0.95. Solving this equation gives $t = 3\tau$, so that for a lag of 3 s one must wait 9 s to reach 95% of the steady-state rate. If one desires 99% of v_0, $(1 - e^{-t/\tau})$ is set equal to 0.99, and now $t = 4.6\tau$, so one must wait 14 s. Depending on the availability of coupling enzymes, cost, etc., one may decide to accept a longer lag, and this can easily be done by decreasing the amount of the first enzyme and the chart speed of the recorder so that little reaction has occurred before the lag is over. As an example, consider the hexokinase reaction in which the product glucose-6-phosphate (G6P) is coupled with glucose-6-phosphate dehydrogenase (G6PDH) to produce the color of NADPH at 340 nm. If a lag of 3 s (0.05 min) is desired, that is a wait of 9–14 s to reach steady state, and given that the K_m for G6P for G6PDH from

Leuconostoc mesenteroides (commercially available) is 0.5 mM, V_{G6PDH} must be 10 units/mL in the assay.

$$V_{G6PDH} = \frac{K_{G6P}}{\tau} = \frac{0.5}{0.05} = 10 \tag{3-6}$$

(Data for specific enzymes can be obtained from the 10-volume set entitled *The Enzyme Handbook* from Springer-Verlag.)

This particular coupled system is in fact more complex than depicted above. Glucose is a mixture of α- and β-anomers and hexokinase uses both, with the *V*/*K* value for the α-anomer being 1.17 times that for the β-anomer. The G6P formed is thus 40% α and 60% β (EJB *36*, 68), but only the β-anomer of G6P is a substrate for G6PDH. Thus, mutarotation of α-G6P to β-G6P has to take place and contributes a further lag to this system. The total lag time is thus

$$\tau = \left(\frac{K_{\beta\text{-G6P}}}{V_{G6PDH}}\right) + \frac{\alpha}{k} \tag{3-7}$$

where α is 0.40 in this case and k is the rate of mutarotation of α to β, 3.8 min^{-1} (ME *41*, 488). This gives a mutarotation lag of 0.0105 min or 6.3 s, which must be added to that resulting from the coupling enzyme. With a 3 s coupling lag, this means that it will take 28 s to reach 95% of the steady-state rate. The amount of hexokinase must therefore be kept low enough that the initial velocity can still be measured after half a minute.

The cost of the coupling enzyme per assay (cost per unit multiplied by the number of units) is easily determined in the above case since there is only a single coupling enzyme. If there are two or more coupling enzymes, a more complex cost analysis is required (AB 99, 141). First the lag is given by

$$\tau = \frac{K_a}{V_a} + \frac{K_b}{V_b} \tag{3-8}$$

where a and b refer to the two coupling enzymes. The cost is then

$$\text{cost} = x_a V_a + x_b V_b \tag{3-9}$$

where cost is per milliliter of assay mixture and x_a and x_b are the cost per unit for the two enzymes. Solution for V_a in equation 3-8 gives

$$V_a = \frac{K_a V_b}{\tau V_b - K_b} \tag{3-10}$$

and substitution into equation 3-9 gives

$$\text{cost} = \frac{x_a K_a V_b}{\tau V_b - K_b} + x_b V_b \tag{3-11}$$

Differentiating cost with respect to V_b and setting this derivative to zero gives the units per milliliter of enzyme b that minimizes cost.

$$V_b = \left(\frac{K_b}{\tau}\right)\left(1 + \left[\frac{x_a K_a}{x_b K_b}\right]^{1/2}\right) \tag{3-12}$$

Substitution of this value of V_b into equation 3-10 gives the level of enzyme a needed to minimize cost:

$$V_a = \left(\frac{K_a}{\tau}\right)\left(1 + \left[\frac{x_b K_b}{x_a K_a}\right]^{1/2}\right) \tag{3-13}$$

With three or more coupling enzymes (as when the formation of AMP is followed by use of adenylate kinase, pyruvate kinase, and lactate dehydrogenase as coupling enzymes), the same approach produces a series of general equations for minimizing cost:

$$V_i = \left(\frac{K_i}{\tau}\right)\left(1 + \sum_{j(j\neq i)} \left[\frac{x_j K_j}{x_i K_i}\right]^{1/2}\right) \tag{3-14}$$

where V_i, K_i, and x_i are the units per milliliter required, the apparent Michaelis constant in millimolar and the cost per unit of the coupling enzyme being considered, K_j and x_j are the values for another coupling enzyme, and the summation is carried out for all of the other coupling enzymes.

An example of a cost analysis for the coupling enzyme assay of a protein kinase is given below. The equations are

$$\begin{aligned}
\text{peptide} + \text{MgATP} &\xrightarrow{\text{protein kinase}} \text{peptide-P} + \text{MgADP} \\
\text{PEP} + \text{MgADP} &\xrightarrow{\text{pyruvate kinase}} \text{pyruvate} + \text{MgATP} \\
\text{pyruvate} + \text{NADH} &\xrightarrow{\text{lactate dehydrogenase}} \text{lactate} + \text{NAD}^+
\end{aligned} \tag{3-15}$$

The data needed to calculate the units needed to minimize cost are $\tau = 3\ \text{s} = 0.05$ min; for pyruvate kinase, $K_{\text{MgADP}} = 0.3$ mM and cost = 0.08 ¢/unit; for lactate dehydrogenase, $K_{\text{pyruvate}} = 0.14$ mM and cost = 0.045 ¢/unit. By application of equations 3-12 and 3-13, for pyruvate kinase

$$V = \left(\frac{0.3}{0.05}\right)\left(1 + \left[\frac{(0.045)(0.14)}{(0.08)(0.3)}\right]^{1/2}\right) = 9.07\ \text{units/mL; cost} = 0.73\ ¢/mL$$

and for lactate dehydrogenase

$$V = \left(\frac{0.14}{0.05}\right)\left(1 + \left[\frac{(0.08)(0.3)}{(0.045)(0.14)}\right]^{1/2}\right) = 8.27\ \text{units/mL; cost} = 0.37\ ¢/\text{mL}$$

Thus nearly equal levels of the two enzymes will minimize cost, despite the fact that one costs twice as much as the other.

In all coupled assays it is usual to use saturating, or at least sufficiently high levels of the other substrates for each coupling enzyme so that their concentrations do not limit the rate as the assay proceeds. Thus, in the pyruvate kinase/lactate dehydrogenase coupled system for analysis of MgADP produced by a kinase, the levels of PEP and NADH should be at least 200 μM. Higher levels of both are feasible, but the level of NADH is limited to about 400 μM because this level gives $2.4A_{340}$ and this is near the upper limit of the adherence of most instruments to Beer's law. The level of NADH used can also be increased by using shorter path length cuvettes or by using spacers in the cuvette that reduce the path length. Spacers that reduce the path length of a 1 cm cuvette to 0.5, 0.2, or 0.1 cm are available.

After cost has been optimized, it is necessary to calculate how fast a v_0 the assay will measure without piling up unacceptable levels of intermediates. The level of an intermediate in the steady state is given by $v_0/(V/K_{\text{Int}})$, and the value of v_0 that makes this level equal to $0.02K_{\text{Int}}$ is probably an upper limit for a practical assay. Thus, $v_0 = 0.02V$ for each coupling enzyme. For the protein kinase example discussed above, the values of v_0 giving this limit are 1.8 mM/min for pyruvate kinase and 1.65 mM/min for lactate dehydrogenase. The level of protein kinase must thus be kept low enough not to exceed 1.65 mM/min. To have the assay handle faster rates, the levels of the coupling enzymes would have to be correspondingly increased.

Table 3-1 includes some commonly used coupled assays involving two or more enzymes for determining specific products.

Table 3-1. Enzymes Commonly Utilized in Coupled Enzyme Assays

Molecule followed	Coupling enzyme	Reactions	Assay for
MgADP	Pyruvate kinase	MgADP + PEP → MgATP + pyruvate	Kinases/ATPases
	lactate DH[a]	pyruvate + NADH → lactate + NAD^+	
MgATP	hexokinase	MgATP + glucose → MgADP + glucose-6-P	kinases
	glucose-6-P DH	glucose-6-P + $NADP^+$ → 6-phosphogluconate + NADPH	(creatine kinase)
MgATP	phosphoglycerate kinase	MgATP + 3-P-glycerate → MgADP + 1,3-bis-P-glycerate	
	glyceraldehyde-3-P DH[b]	1,3-bis-P-glycerate + NADH → glyceraldehyde 3-P + NAD^+ + P_i	
	triose Isomerase	glyceraldehyde 3-P → dihydroxyacetone-P	
	α-glycerophosphate DH	dihydroxyacetone-P + NADH → glycerol-P + NAD^+	
AMP	adenylate kinase	AMP + MgATP → MgADP + ADP	pyrophosphorylytic
	pyruvate kinase	MgADP + PEP → MgATP + pyruvate	enzymes
	lactate DH	pyruvate + NADH → lactate + NAD^+	
P_i	phosphorylase	P_i + $glycogen_n$ → glucose 1-P + $glycogen_{n-1}$	phosphatases
	phosphoglucomutase	glucose 1-P → glucose 6-P	ATPases
	glucose-6-P DH	glucose 6-P + $NADP^+$ → 6-phosphogluconate + NADPH	
PP_i	pyrophosphatase	$MgPP_i$ → Mg^{2-} + 2 P_i	pyrophosphorylytic
	phosphate as above		enzymes
PP_i	PP_i-F6P kinase	$MgPP_i$ + fructose 6-P → fructose 1,6-bis-P + Mg^{2+} + P_i	pyrophosphorylytic
	aldolase	fructose 1,6-bis-P → glyceraldehyde 3-P + dihydroxyacetone-P	enzymes
	triose isomerase	glyceraldehyde 3-P → dihydroxyacetone-P	
	α-glycerolphosphate DH	dihydroxyacetone-P + NADH → α-glycerol 3-P + NAD^+	
CO_2 (HCO_3^-)	PEP carboxylase	PEP + HCO_3^- → oxaloacetate + P_i	CO_2
	malate DH	oxaloacetate + NADH → malate + NAD^+	

[a]DH = dehydrogenase.

[b]The sensitivity of this assay can be increased by adding triosephosphate isomerase and α-glycerolphosphate DH.

As suggested above, coupled assays do not require use of enzyme and can also make use of a chemical couple as shown in the following example.

$$\begin{array}{l} \text{L-serine} + \text{acetyl-CoA} \xrightarrow{\text{serine acetyltransferase}} O\text{-acetyl-L-serine} + \text{CoASH} \\ \text{CoASH} + \text{DTNB} \rightarrow \text{CoAS–SNB} + \text{TNB} \end{array} \tag{3-16}$$

where DTNB is Ellman's reagent [5,5′-dithiobis(2-nitrobenzoate)] and CoAS–SNB is the mixed disulfide between CoASH and 5-thio-2-nitrobenzoate (TNB). The ε_{412} for TNB is 13,600 $M^{-1}cm^{-1}$. By the same reasoning as for enzyme-coupled reaction, $\tau = 1/k[\text{DTNB}]$, where k is the second-order rate constant for the disulfide exchange reaction between CoASH and DTNB. One must determine the second-order rate constant, k, by measuring the observed rate as a function of concentration of one reactant at several levels of another. Then the [DTNB] required to give a lag of 0.05 min can be calculated as $1/(k\tau)$.

Analysis of Time Courses

While most kinetic studies involve measurement of initial velocities, some workers prefer following the time course of the enzymatic reaction, and there are times when this may be preferable. The usual rate equation for an enzyme-catalyzed reaction when the concentration of only one substrate is varied is

$$-\frac{d\mathbf{A}}{dt} = \frac{V\mathbf{A}}{K + \mathbf{A}} = \frac{d\mathbf{P}}{dt} = \frac{V(\mathbf{A_0} - \mathbf{P})}{K + \mathbf{A_0} - \mathbf{P}} \tag{3-17}$$

This integrates to

$$t = \frac{\mathbf{P}}{V} - \left(\frac{K}{V}\right)\ln\left(1 - \frac{\mathbf{P}}{\mathbf{A_0}}\right) \tag{3-18}$$

which can be put in a linear form:

$$\frac{\mathbf{P}}{t} = V + \frac{K\left[\ln\left(1 - \frac{\mathbf{P}}{\mathbf{A_0}}\right)\right]}{t} \tag{3-19}$$

so that a plot of $\mathbf{P}/t$ versus $[\ln(1 - \mathbf{P}/\mathbf{A_0})]/t$ allows determination of K from the slope and V from the vertical intercept. The horizontal intercept gives V/K. Since $\mathbf{A_0} = \mathbf{A} + \mathbf{P}$, one can write equation 3-19 in terms of $\mathbf{A}$, rather than $\mathbf{P}$, as

$$\frac{\mathbf{A_0} - \mathbf{A}}{t} = V - \frac{K\left[\ln\left(\frac{\mathbf{A_0}}{\mathbf{A}}\right)\right]}{t} \tag{3-20}$$

and now a plot of $(\mathbf{A_0} - \mathbf{A})/t$ versus $[\ln (\mathbf{A_0}/\mathbf{A})]/t$ allows determination of V, K, and V/K.

For a more extensive description of the equations involved in analyzing time courses, see ME *63*, 159.

The advantage of this method is that less enzyme and substrate are required, important if these are expensive and in short supply. If there are two or more substrates for the reaction, the levels of the others must be kept at least 20 times the level of the one being followed or be kept at a saturating level ($20K_m$) during the entire reaction. The disadvantage of the method is that it does not readily permit determination of all of the kinetic constants when there is more than one substrate. Inhibition patterns for dead-end inhibitors can be determined from time courses at different levels of inhibitor, but product inhibition causes the equations to become more complex and the analysis more difficult. Other problems with following time courses include pH changes and enzyme instability. As a result this approach is seldom used. For its use in the study of lactate dehydrogenase, however, see JBC *244*, 1285. Further examples are given in ME *63*, 159 and ME *249*, 61.

Statistical analysis of time courses is tricky, since all measurements from a single time course are highly correlated with each other so that a least-squares fit to any of the above equations will give incorrectly small standard errors for K and V. Duggleby has given a full description of the way statistical analysis of time courses must be carried out (ME *249*, 61).

4

DERIVATION OF INITIAL VELOCITY RATE EQUATIONS AND DATA PROCESSING

The language of kinetics is the rate law, and it is thus important to consider derivation of rate equations. In this chapter several methods for the derivation of rate equations will be presented and discussed in terms of ease of use. Finally, at the end of the chapter, the statistics of data fitting will be discussed.

Derivation of Rate Equations

When the initial velocity equation is required for a two-substrate (Bi) reaction, it is immaterial how many products are formed. Thus, for the purpose of deriving the initial rate equation for an Ordered Bi Bi reaction, the mechanism is illustrated in equation 4-1:

$$\mathrm{E} \underset{k_2}{\overset{k_1\mathbf{A}}{\rightleftharpoons}} \mathrm{EA} \underset{k_4}{\overset{k_3\mathbf{B}}{\rightleftharpoons}} \mathrm{EAB} \xrightarrow{k_5} \mathrm{E} + \text{products} \tag{4-1}$$

where k_5 is the net rate constant reflecting the chemical steps and release of products. There are a number of procedures one can use, and several will be illustrated.

Algebraic Solution

Under steady-state conditions, the following expression can be written for the change in **E** with time:

$$\frac{d\mathbf{E}}{dt} = k_2(\mathbf{EA}) - k_1\mathbf{A} + k_5(\mathbf{EAB}) \tag{4-2}$$

Based on the steady-state assumption, the following expressions can be written for the change in the concentration of the intermediate enzyme species with time:

$$\frac{d\mathbf{EA}}{dt} = k_1(\mathbf{A})(\mathbf{E}) - (k_2 + k_3\mathbf{B})(\mathbf{EA}) + k_4(\mathbf{EAB}) = 0 \tag{4-3}$$

$$\frac{d\mathbf{EAB}}{dt} = k_3(\mathbf{B})(\mathbf{EA}) - (k_4 + k_5)(\mathbf{EAB}) = 0 \tag{4-4}$$

A conservation equation for enzyme can be written as follows:

$$\mathbf{E_t} = \mathbf{E} + \mathbf{EA} + \mathbf{EAB} \tag{4-5}$$

Solving equations 4-3 and 4-4 for **E** and **EA**, respectively, gives

$$\mathbf{E} = \frac{(k_2 + k_3\mathbf{B})(\mathbf{EA}) - k_4(\mathbf{EAB})}{k_1\mathbf{A}} \tag{4-6}$$

$$\mathbf{EA} = \frac{(k_4 + k_5)(\mathbf{EAB})}{k_3\mathbf{B}} \tag{4-7}$$

Substituting the expression for **EA** from equation 4-7 into equation 4-6 gives

$$\mathbf{E} = \frac{\dfrac{(k_2 + k_3\mathbf{B})(k_4 + k_5)(\mathbf{EAB})}{k_3\mathbf{B}} - k_4(\mathbf{EAB})}{k_1\mathbf{A}} \tag{4-8}$$

Substituting equations 4-7 and 4-8 into the conservation equation, equation 4-5, and solving for **EAB** gives

$$\mathbf{EAB} = \frac{k_1k_3\mathbf{ABE_t}}{k_2(k_4 + k_5) + k_1\mathbf{A}(k_4 + k_5) + k_3k_5\mathbf{B} + k_1k_3\mathbf{AB}} \tag{4-9}$$

The initial rate may be defined at any step along the reaction pathway, but the derivation is simpler if one uses an irreversible step. Thus

$$v = k_5(\mathbf{EAB}) \tag{4-10}$$

and substituting for **EAB** from equation 4-9 gives the final initial rate equation in terms of the rate constants defined in mechanism 4-1:

$$v = \frac{k_1 k_3 k_5 \mathbf{ABE_t}}{k_2(k_4 + k_5) + k_1(k_4 + k_5)\mathbf{A} + k_3 k_5 \mathbf{B} + k_1 k_3 \mathbf{AB}} \tag{4-11}$$

By considering that $k_5\mathbf{E_t} = V_1$; $k_2/k_1 = K_{ia}$; $(k_4 + k_5)/k_3 = K_b$; and $k_5/k_1 = K_a$, then

$$v = \frac{V\mathbf{AB}}{K_{ia}K_b + K_a\mathbf{B} + K_b\mathbf{A} + \mathbf{AB}} \tag{4-12}$$

The above approach can be used to derive the rate expression for simple mechanisms such as that shown in mechanism 4-1, but it becomes tedious for more complex mechanisms.

King–Altman Method

An algorithm method for the derivation of rate equations was developed by King and Altman (JPC *60*, 1375) and further discussed by Cleland (BBA *67*, 104). This method is still widely used for complex, especially branched mechanisms, and this will be illustrated for the following fully reversible reaction mechanism:

$$\mathrm{E} \underset{k_2}{\overset{k_1\mathbf{A}}{\rightleftharpoons}} \mathrm{EA} \underset{k_4}{\overset{k_3}{\rightleftharpoons}} \mathrm{EP} \underset{k_6\mathbf{P}}{\overset{k_5}{\rightleftharpoons}} \mathrm{E} \tag{4-13}$$

The different forms of enzyme (E, EA, and EP) are now arranged in a convenient geometrical array. Arrows are drawn showing the permissible interconversions and each arrow is labeled with (a) the proper rate constant and (b) the concentration factor for a substrate or product (if one is involved in that step). Two arrows are associated with reversible steps. For the above mechanism, the basic geometric figure is in the form of a triangle, with the rate constant in one reaction direction on one side and that in the opposite reaction direction on the opposite side, as in equation 4-14:

EA
$k_1\mathbf{A}$ k_2
E k_4 k_3 (4-14)
$k_6\mathbf{P}$ k_5
EP

The next step is to determine and write down all the possible patterns that connect all forms of enzyme and that contain one less line than there are enzyme forms. (The patterns must not contain closed loops.) The total number of patterns is given by the following expression

$$\text{no. of patterns} = \frac{m!}{(n-1)!(m-n+1)!} \tag{4-15}$$

where m is number of interconversion steps (lines in the basic figure) and n is the number of enzyme forms. For the above mechanism, the number of patterns is equal to 3!/2!1! or 3, and the geometric patterns are given below.

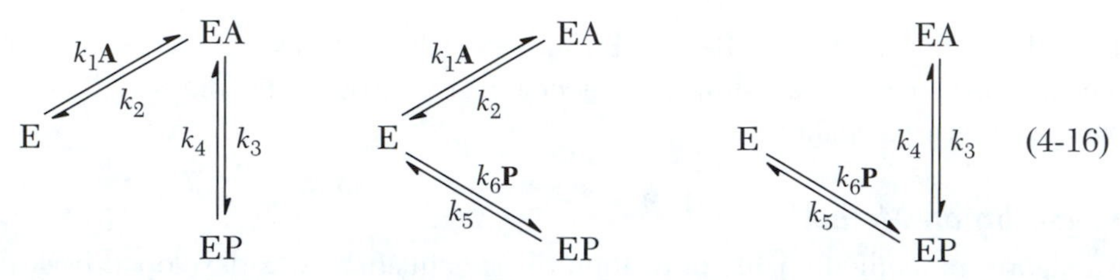

The proportion of a particular form of enzyme as a function of the total enzyme concentration is given by the following general equation. (The equations for a given mechanism are collectively called the distribution equations.)

$$\frac{\mathbf{E_i}}{\mathbf{E_t}} = \frac{\sum \text{one term from each pattern}}{\Delta} \tag{4-17}$$

where $\mathbf{E_i}$ represents the ith enzyme form, and Δ is a denominator that is the same for all equations and is equal to the sum of all the numerator terms. For a completely reversible reaction, there are as many terms in the numerator as there are patterns. In the case of an initial rate equation, some of the terms will be zero, since the release of product is irreversible under the conditions used. Thus an expression such as $k_6\mathbf{P}$ is zero if $\mathbf{P}=0$. Each numerator term is determined from its corresponding pattern as follows:

(a) Follow along a path in the basic figure that is concerned with the conversion of enzyme forms other than the one that is being considered.

(b) Multiply together the rate constants (and the concentration factors) associated with the path followed.

(c) Repeat the procedure for each of the patterns in the basic figure and sum the products.

For the present mechanism, the distribution equations would be

$$\frac{\mathbf{E}}{\mathbf{E_t}} = \frac{k_2k_4 + k_2k_5 + k_3k_5}{\Delta} \tag{4-18}$$

$$\frac{\mathbf{EA}}{\mathbf{E_t}} = \frac{k_1k_4\mathbf{A} + k_1k_5\mathbf{A} + k_4k_6\mathbf{P}}{\Delta} \tag{4-19}$$

$$\frac{\mathbf{EP}}{\mathbf{E_t}} = \frac{k_1k_3\mathbf{A} + k_2k_6\mathbf{P} + k_3k_6\mathbf{P}}{\Delta} \tag{4-20}$$

Note that the denominator (Δ) is the same for each equation. When the three equations are added, the left-hand terms add up to one and thus this denominator is merely the sum of all the numerator terms on the right, as can be derived from equation 4-5. By multiplying each of the equations by $\mathbf{E_t}$, the same expressions for **E**, **EA**, and **EP** are obtained as when the algebraic methods discussed above is used.

The velocity of the reaction is given by equation 4-21:

$$v = k_5(\mathbf{EP}) - k_6(\mathbf{P})(\mathbf{E}) \tag{4-21}$$

Substitution of the expressions for **EP** and **E** into the above equation gives

$$v = \frac{(k_1k_3k_5\mathbf{A} - k_2k_4k_6\mathbf{P})\mathbf{E_t}}{\Delta} \tag{4-22}$$

and the complete initial velocity equation in terms of microscopic rate constants is given in equation 4-23:

$$v = \frac{(k_1k_3k_5\mathbf{A} - k_2k_4k_6\mathbf{P})\mathbf{E_t}}{k_2k_5 + k_2k_4 + k_3k_5 + k_1(k_3 + k_4 + k_5)\mathbf{A} + k_6(k_2 + k_3 + k_4)\mathbf{P}} \tag{4-23}$$

Alternative Reaction Pathways

When a reaction can proceed via alternative pathways, loops are introduced into the basic King and Altman diagram and the use of the relationship $m!/(n - 1)!(m - n + 1)!$ overestimates the number of patterns. Thus, for the basic figure below, which one would use to derive the initial velocity equation for a random bireactant mechanism with E, EA, EB, and EAB as enzyme forms

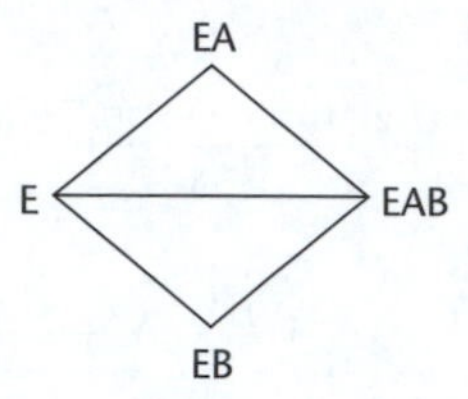

$n = 4$ and $m = 5$ so that the calculated number of patterns 5!/3!2! is 10. But the total of 10 includes the two three-sided loops shown below, which must be eliminated.

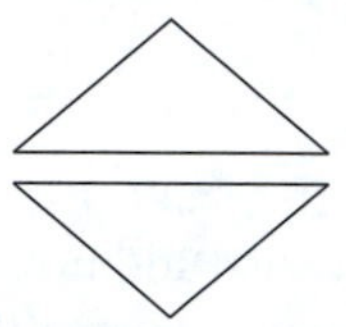

Therefore the total number of patterns is eight, and each pattern must be generated by eliminating two lines from the initial diagram.

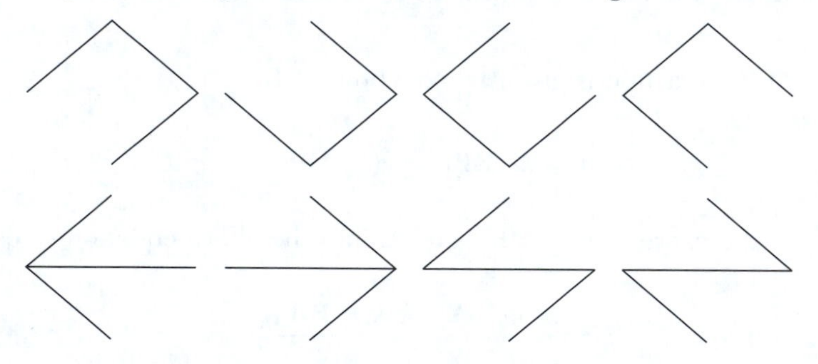

Any possible loop of r sides will occur $(m - r)!/(n - 1 - r)!(m - n + 1)!$ times so that the total number of patterns must be reduced by this number. Each of the above three-sided loops will occur 2!/0!2! times, or once, and thus 2 is subtracted from 10. Additional King–Altman patterns and distribution equations are provided in Appendix 1.

Conversion of the Initial Velocity Rate Equation to Kinetic Constants

The above initial velocity rate equation, equation 4-23, may be written in coefficient form as follows (BBA 67, 104):

$$v = \frac{(\text{num}_1)\mathbf{A} - (\text{num}_2)\mathbf{P}}{(\text{constant}) + (\text{coef A})\mathbf{A} + (\text{coef P})\mathbf{P}} \tag{4-24}$$

It should be noted that the equation has this general form irrespective of the number of intermediate enzyme complexes, although the definitions of the num, coef, and constant terms will differ. Equation 4-24 in coefficient form may be converted to one containing kinetic constants by multiplying each term in the numerator and denominator of equation 4-24 by $\text{num}_2/(\text{coef A})(\text{coef P})$. Thus, equation 4-25 is produced:

$$v = \frac{\dfrac{(\text{num}_1)(\text{num}_2)\mathbf{A}}{(\text{coef A})(\text{coef P})} - \dfrac{(\text{num}_2)(\text{num}_2)\mathbf{P}}{(\text{coef A})(\text{coef P})}}{\dfrac{(\text{constant})(\text{num}_2)}{(\text{coef A})(\text{coef P})} + \dfrac{(\text{num}_2)(\text{coef A})\mathbf{A}}{(\text{coef A})(\text{coef P})} + \dfrac{(\text{num}_2)(\text{coef P})\mathbf{P}}{(\text{coef A})(\text{coef P})}} \tag{4-25}$$

Kinetic parameters are defined in terms of kinetic constants as follows:

$$K_\text{a} = \frac{(\text{constant})}{\text{coef A}} = \frac{k_2k_5 + k_2k_4 + k_3k_5}{k_1(k_3 + k_4 + k_5)} \tag{4-26}$$

$$K_\text{p} = \frac{(\text{constant})}{\text{coef P}} = \frac{k_2k_5 + k_2k_4 + k_3k_5}{k_6(k_2 + k_3 + k_4)} \tag{4-27}$$

$$V_1 = \frac{(\text{num}_1)}{\text{coef P}} = \frac{k_3k_5\mathbf{E_t}}{k_3 + k_4 + k_5} \tag{4-28}$$

$$V_2 = \frac{(\text{num}_2)}{\text{coef P}} = \frac{k_2k_4\mathbf{E_t}}{k_2 + k_3 + k_4} \tag{4-29}$$

$$K_\text{eq} = \frac{(\text{num}_1)}{(\text{num}_2)} = \frac{k_1k_3k_5}{k_2k_4k_6} \tag{4-30}$$

By use of the above definitions, equation 4-25 is converted to equation 4-31:

$$v = \frac{V_1V_2\mathbf{A} - \dfrac{(\text{num}_2)V_2\mathbf{P}}{(\text{coef A})}}{V_2K_\text{a} + V_2\mathbf{A} + \dfrac{(\text{num}_2)\mathbf{P}}{(\text{coef A})}} \tag{4-31}$$

The remaining terms are defined by multiplying the **P** terms in numerator and denominator by $(\text{num}_1)/(\text{num}_1)$, giving equation 4-32:

$$v = \frac{V_1V_2\mathbf{A} - \dfrac{V_1V_2\mathbf{P}}{K_\text{eq}}}{V_2K_\text{a} + V_2\mathbf{A} + \dfrac{V_1\mathbf{P}}{K_\text{eq}}} \tag{4-32}$$

and since K_{eq} is V_1K_p/V_2K_a for mechanism 4-13 (see Chapter 5, page 98, Haldane Relationships), equation 4-32 can be rewritten as equation 4-33:

$$v = \frac{V_1\mathbf{A} - \dfrac{V_2K_a\mathbf{P}}{K_p}}{K_a + \mathbf{A} + \dfrac{K_a\mathbf{P}}{K_p}} \tag{4-33}$$

When initial velocities are determined by varying **A** at $\mathbf{P} = 0$ or by varying **P** at $\mathbf{A} = 0$, equations 4-34 and 4-35 are obtained as expected for a unireactant mechanism (Chapter 2):

$$v = \frac{V_1\mathbf{A}}{K_a + \mathbf{A}} \tag{4-34}$$

$$v = \frac{V_2\mathbf{A}}{K_p + \mathbf{P}} \tag{4-35}$$

The same general procedure is used for steady-state reactions involving more than one substrate. In contrast to reactions with one substrate, it is necessary to define more constants than just one Michaelis constant for each reactant, two maximum velocities, and the equilibrium constant. The additional constants are called inhibition constants, since they are often product inhibition constants for the reactant when it is a product in the opposite reaction direction. The inhibition constants are written as K_{ia}, K_{ib}, K_{ip}, and K_{iq} for an Ordered Bi Bi reaction and are defined in equation 4-36:

$$K_{ib} = \frac{\text{(coef PQ)}}{\text{(coef BPQ)}} \qquad K_{ia} = \frac{\text{constant}}{\text{(coef A)}} = \frac{\text{(coef P)}}{\text{(coef AP)}}$$
$$K_{ip} = \frac{\text{(coef AB)}}{\text{(coef ABP)}} \qquad K_{iq} = \frac{\text{constant}}{\text{(coef Q)}} = \frac{\text{(coef B)}}{\text{(coef BQ)}} \tag{4-36}$$

The Michaelis and inhibition constants are defined by ratios of denominator term coefficients. The ratio must be such that when letters corresponding to reactants are cancelled, the reactant associated with the kinetic constant will remain in the denominator. The ratios (const)/(coef A), (coef B)/(coef AB), and (coef P)/(coef AP) are all suitable definitions for a kinetic constant associated with A since cancellation of letters leaves 1/A in each case. However, the ratio for the Michaelis constants must have the same denominator term as that used in defining the maximum velocity, the coefficient that includes the letters of all reactants. Thus for the reaction of A and B to give P and Q, K_a is (coef B)/(coef AB) and either of the other two ratios provided above may be used as the definition of K_{ia}. For a reversible Bi Bi reaction, each term in the numerator and denominator of the rate equation in coefficient form is multiplied by (num_2)/(coef AB)(coef PQ). Rate equations for a number of mechanisms are provided in Appendix II.

When a stable enzyme form isomerizes, it is necessary to define two additional iso-inhibition constants for each isomerizing form, written as K_{iia}, K_{iiq}, etc. Additional terms then occur in the denominator of the rate equation. For example, in mechanism 4-37, conversion of A to P leaves the enzyme in a form F that has to isomerize back to E before A can add, and an **AP** term is added to the rate equation. For mechanism 4-37 the iso-inhibition constants are defined as K_{iia} = (coef P)/(coef AP) and K_{iip} = (coef A)/(coef AP). For any mechanism, the definition of a K_{ii} for a given reactant is the ratio of the denominator term coefficient used to define the maximum velocity for formation of this reactant, divided by the coefficient of the denominator term including this reactant as well. For an Iso Ordered Bi Bi mechanism, for example, K_{iia} = (coef PQ)/(coef APQ) and K_{iiq} = (coef AB)/(coef ABQ).

$$\mathrm{E} \underset{k_2}{\overset{k_1\mathbf{A}}{\rightleftharpoons}} \mathrm{EA} \underset{k_4}{\overset{k_3}{\rightleftharpoons}} \mathrm{EP} \underset{k_6\mathbf{P}}{\overset{k_5}{\rightleftharpoons}} \mathrm{F} \underset{k_8}{\overset{k_7}{\rightleftharpoons}} \mathrm{E} \qquad (4\text{-}37)$$

The rate equation for mechanism 4-37 is given below.

$$v = \frac{V_1V_2\mathbf{A} - \dfrac{V_1V_2\mathbf{P}}{K_{eq}}}{V_2K_a + V_2\mathbf{A} + \dfrac{V_1\mathbf{P}}{K_{eq}} + \dfrac{V_2\mathbf{AP}}{K_{iip}}} \qquad (4\text{-}38)$$

It differs from equation 4-32 in the **AP** term in the denominator, which could also be written $(V_1/K_{eq})\mathbf{AP}/K_{iia}$. For additional information on iso-mechanisms see ME *249*, 111.

Net Rate Constant Method

Another method that can be used for rapid derivation of rate equations for mechanisms with an irreversible step and with a linear or simple branched pathway makes use of partitioning of enzyme forms (B *14*, 3220). Consider the following mechanism:

$$\mathrm{E} \underset{k_2}{\overset{k_1\mathbf{A}}{\rightleftharpoons}} \mathrm{EA} \underset{k_4}{\overset{k_3}{\rightleftharpoons}} \mathrm{EP} \xrightarrow{k_5} \mathrm{E} + \mathrm{P} \qquad (4\text{-}39)$$

A net rate constant can be defined for each step along the reaction path. The net rate constant represents the same rate as if the step were irreversible and thus must include a consideration of all steps up to a committed (irreversible) step, so it also takes into account partitioning of the intermediate generated. The prime indicates a net rate constant. Mechanism 4-39 can be rewritten as mechanism 4-40 in terms of net rate constants.

$$\mathrm{E} \xrightarrow{k_1'} \mathrm{EA} \xrightarrow{k_3'} \mathrm{EP} \xrightarrow{k_5'} \mathrm{E} + \mathrm{P} \tag{4-40}$$

One begins the derivation at the last irreversible step and works backward, defining each of the net rate constants. If the step is irreversible, the net rate constant, k', will be equal to the rate constant, k, and therefore $k_5' = k_5$. The rate constant k_3', however, must take into account the partitioning of the EP complex toward formation of P, and k_1' must take into account partitioning of EA and EP. The net rate constants are then determined as follows:

$$k_3' = k_3\left(\frac{k_5}{k_4 + k_5}\right) = \frac{k_3 k_5}{k_4 + k_5} \tag{4-41}$$

$$k_1' = k_1\mathbf{A}\left(\frac{k_3'}{k_3' + k_2}\right) = \frac{k_1 k_3 k_5 \mathbf{A}}{k_2 k_4 + k_2 k_5 + k_3 k_5} \tag{4-42}$$

The amount of each enzyme form is inversely proportional to the net rate constant going from it toward product. The sum of these will be equal to $\mathbf{E_t}$ based on conservation of mass. The distribution equations can thus be written as in equations 4-43 to 4-45:

$$\frac{\mathbf{E}}{\mathbf{E_t}} = \frac{1/k_1'}{1/k_1' + 1/k_3' + 1/k_5'} \tag{4-43}$$

$$\frac{\mathbf{EA}}{\mathbf{E_t}} = \frac{1/k_3'}{1/k_1' + 1/k_3' + 1/k_5'} \tag{4-44}$$

$$\frac{\mathbf{EP}}{\mathbf{E_t}} = \frac{1/k_5'}{1/k_1' + 1/k_3' + 1/k_5'} \tag{4-45}$$

The initial rate (v) is then equal to any net rate constant multiplied by the amount of enzyme in that form, for example, $v = k_1'\mathbf{E}$. Substituting for **E** from above gives

$$v = k_1'\left[\frac{1/k_1'}{1/k_1' + 1/k_3' + 1/k_5'}\right]\mathbf{E}_t = \frac{\mathbf{E}_t}{1/k_1' + 1/k_3' + 1/k_5'} \quad (4\text{-}46)$$

The rate is therefore the reciprocal of the sums of the reciprocals of all net rate constants times the total enzyme concentration. Substituting the expressions for each of the net rate constants from equations 4-41 and 4-42 into equation 4-46 ($k_5' = k_5$), one gets equation 4-47:

$$v = \frac{\mathbf{E}_t}{\dfrac{1}{\dfrac{k_1k_3k_5\mathbf{A}}{k_2k_4 + k_2k_5 + k_3k_5}} + \dfrac{1}{\dfrac{k_3k_5}{k_4 + k_5}} + \dfrac{1}{k_5}} = \frac{k_1k_3k_5\mathbf{A}\mathbf{E}_t}{k_2k_4 + k_2k_5 + k_3k_5 + k_1\mathbf{A}(k_3 + k_4 + k_5)} \quad (4\text{-}47)$$

Expressions for V and V/K are easily obtained by use of net rate constants. Consider again the expression for v in equation 4-47. The V/K for A reflects the rate at limiting **A**. The important term(s) in the denominator will thus be the one(s) containing **A**. In the above example, k_1' is the only net rate constant containing **A**, and since the relationship is reciprocal, the term will predominate as **A** approaches zero. Thus,

$$\frac{V}{K_a}\mathbf{A} = \frac{\mathbf{E}_t}{(1/k_1')} = k_1'\mathbf{E}_t \text{ and } \frac{V}{K_a\mathbf{E}_t} = \frac{k_1k_3k_5}{k_2k_4 + k_2k_5 + k_3k_5} \quad (4\text{-}48)$$

V, on the other hand, contains all terms that do not contain **A**. As **A** approaches infinity, $1/k_1'$ goes to zero, and

$$\frac{V}{\mathbf{E}_t} = \frac{1}{(1/k_3' + 1/k_5')} = \frac{k_3k_5}{k_3 + k_4 + k_5} \quad (4\text{-}49)$$

The net rate constant method can also be used for simple branched pathways, but for complex pathways it is easier to use the algorithm method developed by King and Altman. Consider the following scheme for a mechanism with a branched pathway:

$$\mathrm{E} \underset{k_2}{\overset{k_1\mathbf{A}}{\rightleftharpoons}} \mathrm{EA} \underset{k_4}{\overset{k_3\mathbf{B}}{\rightleftharpoons}} \mathrm{EAB} \underset{k_6}{\overset{k_5}{\rightleftharpoons}} \mathrm{EPQ} \begin{array}{c} \xrightarrow{k_7} \mathrm{EQ} \xrightarrow{k_9} \\ \xrightarrow[k_{11}]{} \mathrm{EP} \xrightarrow[k_{13}]{} \end{array} \mathrm{E} \tag{4-50}$$

The value of $k_5{}'$ is derived in the usual way, that is, $k_5(k_7 + k_{11})/(k_6 + k_7 + k_{11})$, and $k_3{}'$ and $k_1{}'$ are derived by the procedure outlined above. But the net rate constant for conversion of EPQ to E must take into account the values of k_9 and k_{13} and the levels of EQ and EP. If the lower path did not exist, the value of $k_{7,9}$ for the upper path would be $1/(1/k_7 + 1/k_9)$, where $1/k_7$ represents the amount of enzyme in EPQ and $1/k_9$ the amount in EQ. To allow for the amount of the EP form, however, we must add $(k_{11}/k_7)(1/k_{13})$ to the denominator so that we get

$$\text{app } k_{7,9} = \frac{1}{\dfrac{1}{k_7} + \dfrac{1}{k_9} + \left(\dfrac{k_{11}}{k_7 k_{13}}\right)} = \frac{k_7}{1 + \dfrac{k_7}{k_9} + \dfrac{k_{11}}{k_{13}}} \tag{4-51}$$

By a similar logic, we get

$$\text{app } k_{11,13} = \frac{k_{11}}{1 + \dfrac{k_7}{k_9} + \dfrac{k_{11}}{k_{13}}} \tag{4-52}$$

for flux through the bottom pathway. Adding these together gives

$$k_{7,11} = \frac{k_7 + k_{11}}{1 + \dfrac{k_7}{k_9} + \dfrac{k_{11}}{k_{13}}} \tag{4-53}$$

as the net rate constant for conversion of EPQ to E. The overall rate equation can then be obtained as $\mathbf{E_t}$ divided by the sum of the reciprocals of k_1', k_3', k_5' and $k_{7,11}'$. If one of the products is present, the equations are more complex, and the reader should consult B *14*, 3220 for a more complete discussion.

Rapid Equilibrium Approximation

When one or more segments of a mechanism can be assumed at equilibrium in the steady state, the approximation of Cha can be utilized to simplify the derivation (JBC *243*, 820). The rapid equilibrium Bi Bi kinetic mechanism will be used as an example. In this mechanism, all steps are considered in rapid equilbrium, with the exception of those involving the interconversion of the central complexes, EAB and EPQ. The mechanism may be illustrated as shown below.

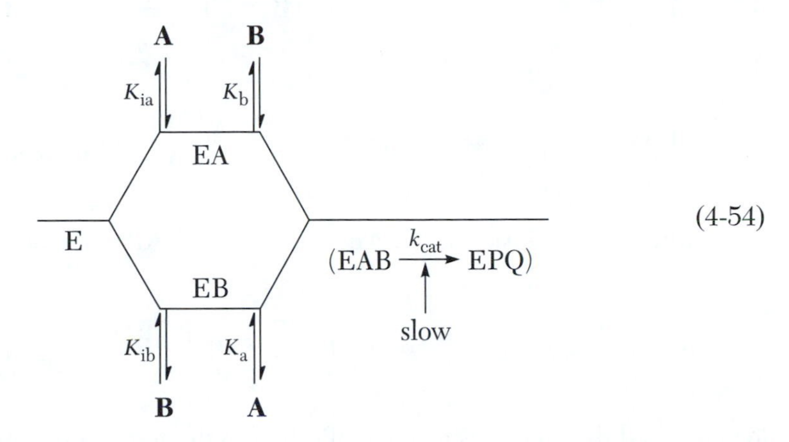

(4-54)

For mechanism 4-54 all the kinetic constants are dissociation constants. K_{ia} and K_{ib} represent dissociation constants for dissociation of EA and EB complexes, respectively, while the Michaelis constants, K_a and K_b, represent dissociation constants for EAB dissociating to give EB plus A and EA plus B, respectively.

The above dissociation constants are defined as follows:

$$K_{ia} = \frac{\mathbf{(E)(A)}}{\mathbf{(EA)}} \quad K_{ib} = \frac{\mathbf{(E)(B)}}{\mathbf{(EB)}} \quad K_a = \frac{\mathbf{(EB)(A)}}{\mathbf{(EAB)}} \quad K_b = \frac{\mathbf{(EA)(B)}}{\mathbf{(EAB)}} \tag{4-55}$$

The initial rate can be defined as in equation 4-56, and thus the concentration of each enzyme form is expressed in terms of **EAB**:

$$v = k_{cat}\mathbf{EAB} \tag{4-56}$$

It is necessary to use only three of the above four equilibria. The concentration of **EAB** is independent of the pathway by which it is formed and hence $K_{ia}K_b = K_aK_{ib}$. Only three of the four kinetic constants are independent. Using the relationships for K_{ia}, K_a and K_b one obtains the following:

$$\mathbf{(EB)} = \frac{K_a\mathbf{(EAB)}}{\mathbf{A}} \quad \mathbf{(EA)} = \frac{K_b\mathbf{(EAB)}}{\mathbf{B}} \quad \mathbf{(E)} = \frac{K_{ia}\mathbf{(EA)}}{\mathbf{A}} = \frac{K_{ia}K_b\mathbf{(EAB)}}{\mathbf{AB}} \tag{4-57}$$

Since the conservation equation states that $\mathbf{E_t}$ is equal to the sum of the concentrations of all enzyme species

$$\mathbf{E_t} = \mathbf{E} + \mathbf{EA} + \mathbf{EB} + \mathbf{EAB} = \left(\frac{K_{ia}K_b}{\mathbf{AB}} + \frac{K_b}{\mathbf{B}} + \frac{K_a}{\mathbf{A}} + 1\right)(\mathbf{EAB}) \tag{4-58}$$

and

$$\mathbf{EAB} = \frac{\mathbf{E_t}}{\dfrac{K_{ia}K_b}{\mathbf{AB}} + \dfrac{K_b}{\mathbf{B}} + \dfrac{K_a}{\mathbf{A}} + 1} = \frac{\mathbf{ABE_t}}{K_{ia}K_b + K_b\mathbf{A} + K_a\mathbf{B} + \mathbf{AB}} \tag{4-59}$$

Substituting the expression for $(\mathbf{EAB})$ into $v = k_{cat}(\mathbf{EAB})$, one obtains

$$V = \frac{V_1\mathbf{AB}}{K_{ia}K_b + K_b\mathbf{A} + K_a\mathbf{B} + \mathbf{AB}} \tag{4-60}$$

where $V_1 = k_{cat}\mathbf{E_t}$.

Cha extended the above, using a modification of the method of King and Altman that can be used when portions of a mechanism are considered in rapid equilibrium. Consider the following mechanism:

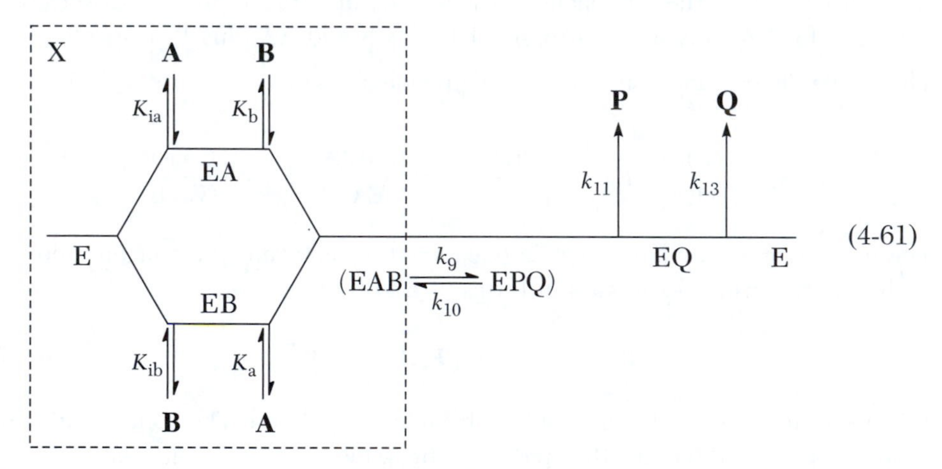

(4-61)

where everything in the box is considered in rapid equilibrium. The fraction of any enzyme form in the box is given by

$$f_i = \frac{\mathbf{E_i}}{\sum_{i=1}^{n} \mathbf{E_i}} \tag{4-62}$$

By use of the equilibria in equation 4-57

$$f_\mathrm{E} = \frac{\mathbf{E}}{\mathbf{E}+\mathbf{EA}+\mathbf{EB}+\mathbf{EAB}} = \frac{1}{1+\dfrac{\mathbf{A}}{K_\mathrm{ia}}+\dfrac{\mathbf{B}}{K_\mathrm{ib}}+\dfrac{\mathbf{AB}}{K_\mathrm{ia}K_\mathrm{b}}} \qquad (4\text{-}63)$$

Thus it follows that

$$f_\mathrm{EA} = \frac{\mathbf{EA}}{D} = \frac{\mathbf{A}}{K_\mathrm{ia}D} \quad f_\mathrm{EB} = \frac{\mathbf{EB}}{D} = \frac{\mathbf{B}}{K_\mathrm{ib}D} \quad f_\mathrm{EAB} = \frac{\mathbf{EAB}}{D} = \frac{\mathbf{AB}}{K_\mathrm{ia}K_\mathrm{b}D} \qquad (4\text{-}64)$$

where D is the denominator of equation 4-63. One then uses the King–Altman algorithm method as shown below:

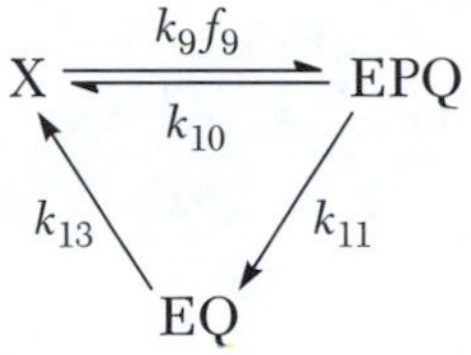

$$\mathrm{X} \underset{k_{10}}{\overset{k_9 f_9}{\rightleftharpoons}} \mathrm{EPQ} \xrightarrow{k_{11}} \mathrm{EQ} \xrightarrow{k_{13}} \mathrm{X} \qquad (4\text{-}65)$$

where f_9 is f_EAB, equation 4-64 (the net rate constant method can also be used). The above diagram gives the following patterns:

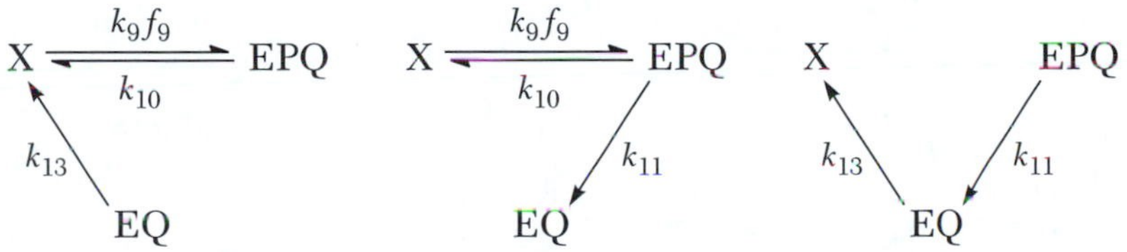

$$\mathrm{X} \underset{k_{10}}{\overset{k_9 f_9}{\rightleftharpoons}} \mathrm{EPQ},\ \mathrm{EQ} \xrightarrow{k_{13}} \mathrm{X} \qquad \mathrm{X} \underset{k_{10}}{\overset{k_9 f_9}{\rightleftharpoons}} \mathrm{EPQ} \xrightarrow{k_{11}} \mathrm{EQ} \qquad \mathrm{EPQ} \xrightarrow{k_{11}} \mathrm{EQ} \xrightarrow{k_{13}} \mathrm{X} \qquad (4\text{-}66)$$

The initial rate can be defined as $v = k_{13}\,\mathbf{EQ}$. The distribution equations are written as shown for the algorithm method above, and the expression for **EQ** is substituted into the equation for v. Finally, the expression for f_9 is substituted to obtain the final rate equation:

$$V = \frac{\left[\dfrac{k_9 k_{11}}{k_{10}+k_{11}}\right]\mathbf{ABE_t}}{K_\mathrm{ia}K_\mathrm{b} + K_\mathrm{b}\mathbf{A} + K_\mathrm{a}\mathbf{B} + \left[1+\dfrac{k_9(k_{11}+k_{13})}{k_{13}(k_{10}+k_{11})}\right]\mathbf{AB}} \qquad (4\text{-}67)$$

Note that the final equation has the same algebraic form as equation 4-60, and the addition of intermediates did not change the form.

Cha's method can be used in nonrandom mechanisms such as an equilibrium ordered mechanism where addition of A is at equilibrium:

$$(\mathrm{E} \underset{K_{ia}}{\rightleftharpoons} \mathrm{EA}) \underset{k_4}{\overset{k_3\mathbf{B}}{\rightleftharpoons}} \mathrm{EAB} \xrightarrow{k_5} \mathrm{E} + \text{products} \tag{4-68}$$

This can be replaced by

$$(\mathrm{X}) \underset{k_4}{\overset{k_3 f_3 \mathbf{B}}{\rightleftharpoons}} \mathrm{EAB} \xrightarrow{k_5} \mathrm{E} + \text{products} \tag{4-69}$$

where $f_3 = \mathbf{A}/(K_{ia} + \mathbf{A})$. Then the net rate constants are k_5 and $k_3 f_3 \mathbf{B} k_5/(k_4 + k_5)$ and

$$v = \frac{\mathbf{E_t}}{\dfrac{k_4 + k_5}{k_3 f_3 \mathbf{B} k_5} + \dfrac{1}{k_5}} = \frac{k_5 \mathbf{E_t AB}}{\dfrac{K_{ia}(k_4 + k_5)}{k_3} + \dfrac{(k_4 + k_5)\mathbf{A}}{k_3} + \mathbf{AB}} = \frac{V\mathbf{AB}}{K_{ia}K_b + K_b\mathbf{A} + \mathbf{AB}} \tag{4-70}$$

where $V = k_5 \mathbf{E_t}$ and $K_b = (k_4 + k_5)/k_3$.

There can be more than one rapid equilibrium segment. When substrates A and B add in obligate order, but substrate C adds randomly and independently, one has

$$\underset{\mathrm{X}}{\begin{pmatrix} \mathrm{EC} \\ K_{ic} \updownarrow \\ \mathrm{E} \end{pmatrix}} \underset{k_2}{\overset{k_1\mathbf{A}}{\rightleftharpoons}} \underset{\mathrm{Y}}{\begin{pmatrix} \mathrm{EAC} \\ K_{ic} \updownarrow \\ \mathrm{EA} \end{pmatrix}} \underset{k_4}{\overset{k_3\mathbf{B}}{\rightleftharpoons}} \underset{\mathrm{Z}}{\begin{pmatrix} \mathrm{EABC} \\ K_{ic} \updownarrow \\ \mathrm{EAB} \end{pmatrix}} \quad \mathrm{EABC} \xrightarrow{k_5 f_5} \mathrm{E} + \text{products} \tag{4-71}$$

Note that f factors are not needed for rate constants k_1–k_4, since these steps are independent of the presence of C. But since only EABC from rapid equilibrium segment Z reacts to give products, f_5, which equals $\mathbf{C}/(K_{ic} + \mathbf{C})$, is needed to modify k_5. This mechanism is then modeled:

$$\mathrm{X} \underset{k_2}{\overset{k_1\mathbf{A}}{\rightleftharpoons}} \mathrm{Y} \underset{k_4}{\overset{k_3\mathbf{B}}{\rightleftharpoons}} \mathrm{Z} \xrightarrow{k_5 f_5} \mathrm{E} + \text{products} \tag{4-72}$$

The net rate constant method then gives

$$v = \frac{k_1 k_3 k_5 f_5 \mathbf{A}\mathbf{B}\mathbf{E_t}}{k_2 k_4 + k_2 k_5 f_5 + k_3 k_5 f_5 \mathbf{B} + k_1 k_4 \mathbf{A} + k_1 k_5 f_5 \mathbf{A} + k_1 k_3 \mathbf{A}\mathbf{B}} \quad (4\text{-}73)$$

After substitution for f_5, this becomes

$$v = \frac{V\mathbf{ABC}}{K_{ia}K_{ib}K_{ic} + K_{ib}K_{ic}\mathbf{A} + K_{ia}K_{b}\mathbf{C} + K_{a}\mathbf{BC} + K_{b}\mathbf{AC} + K_{ic}\mathbf{AB} + \mathbf{ABC}} \quad (4\text{-}74)$$

where $K_{ia} = k_2/k_1$, $K_{ib} = k_4/k_3$, $K_a = k_5/k_1$, and $K_b = (k_4 + k_5)/k_3$.

This equation has the same form as that for a fully ordered mechanism with the exception that K_{ic} replaces K_c.

A further example of Cha's method is for a random mechanism where only the binary EA and EB complexes are in rapid equilibrium with free enzyme. [We derived the rate equation for the case where EAB is also in rapid equilibrium (mechanism 4-54) in equations 4-55 to 4-60.]

$$\left(\begin{array}{ccc} & \overset{K_{ia}}{\rightleftharpoons} \mathrm{EA} & \\ \mathrm{E} & & \\ & \underset{K_{ib}}{\rightleftharpoons} \mathrm{EB} & \end{array} \right)_{\mathrm{X}} \begin{array}{c} \xrightleftharpoons[k_6]{k_5 f_5 \mathbf{B}} \\ \\ \xrightleftharpoons[k_8]{k_7 f_7 \mathbf{A}} \end{array} \mathrm{EAB} \xrightarrow{k_9} \mathrm{E} + \text{products} \quad (4\text{-}75)$$

This reduces to

$$\mathrm{X} \xrightleftharpoons[k_6 + k_8]{k_5 f_5 \mathbf{B} + k_7 f_7 \mathbf{A}} \mathrm{EAB} \xrightarrow{k_9} \mathrm{E} + \text{products} \quad (4\text{-}76)$$

where $f_5 = K_{ib}\mathbf{A}/(K_{ia}K_{ib} + K_{ib}\mathbf{A} + K_{ia}\mathbf{B})$ and $f_7 = K_{ia}\mathbf{B}/(K_{ia}K_{ib} + K_{ib}\mathbf{A} + K_{ia}\mathbf{B})$.

The rate equation for this mechanism is

$$v = \frac{V\mathbf{AB}}{K_{ia}K_{b} + K_{b}\mathbf{A} + K_{a}\mathbf{B} + \mathbf{AB}} \quad (4\text{-}77)$$

where $V = k_9\mathbf{E_t}$, $\quad K_a = \dfrac{k_8(k_6 + k_8 + k_9)}{k_7(k_6 + k_8)}$, and $K_b = \dfrac{k_6(k_6 + k_8 + k_9)}{k_5(k_6 + k_8)}$

This equation is the same as that derived for mechanism 4-54 or for an ordered mechanism (the rules that lead to this identical rate equation will be discussed in Chapter 5). If $k_9 \ll (k_6 + k_8)$, mechanism 4-54 is obtained, and K_a and K_b are defined by the dissociation constants of A and B from EAB.

Derivation of Equations for Isotope Exchange

Although we will not consider isotope exchange until Chapter 8, derivations of rate equations for isotope exchange are presented here. The simplest method for deriving rate equations for isotopic exchange is to use net rate constants for isotope transfer. Starting with the step in which labeled product is released (which will be irreversible under initial exchange conditions), one works backwards to the step where labeled substrate adds and writes net rate constants. The net rate constant for the initial addition of labeled substrate, multiplied by the enzyme form with which it combines, is the initial exchange rate. For example, to derive the rate equation for A–Q exchange in mechanism 4-78:

$$\mathrm{E} \underset{k_2}{\overset{k_1\mathbf{A}^*}{\rightleftharpoons}} \mathrm{EA} \underset{k_4}{\overset{k_3\mathbf{B}}{\rightleftharpoons}} \begin{pmatrix}\mathrm{EAB}\\ \mathrm{FPQ}\end{pmatrix} \underset{k_6\mathbf{P}}{\overset{k_5}{\rightleftharpoons}} \mathrm{FQ} \underset{k_8\mathbf{Q}}{\overset{k_7}{\rightleftharpoons}} \mathrm{F} \underset{k_{10}}{\overset{k_3\mathbf{C}}{\rightleftharpoons}} \begin{pmatrix}\mathrm{FC}\\ \mathrm{ER}\end{pmatrix} \underset{k_{12}\mathbf{R}}{\overset{k_{11}}{\rightleftharpoons}} \mathrm{E} \quad (4\text{-}78)$$

the net rate constants needed are

$$k_7' = k_7 \quad k_5' = \frac{k_5k_7}{k_6\mathbf{P} + k_7} \quad k_3' = \frac{k_3k_5k_7\mathbf{B}}{k_4k_6\mathbf{P} + k_7(k_4 + k_5)}$$
$$k_1' = \frac{k_1k_3k_5k_7\mathbf{A}^*\mathbf{B}}{k_2k_4k_6\mathbf{P} + k_2(k_4 + k_5)k_7 + k_3k_5k_7\mathbf{B}} \quad (4\text{-}79)$$

where $\mathbf{A}^*$ is the concentration of labeled A. The equation for exchange is then

$$v^*_{\mathrm{A}\to\mathrm{Q}} = \frac{k_1k_3k_5k_7\mathbf{A}^*\mathbf{B}(\mathbf{E})}{k_2k_4k_6\mathbf{P} + k_2(k_4 + k_5)k_7 + k_3k_5k_7\mathbf{B}} \quad (4\text{-}80)$$

Or changing into kinetic constants by use of the following definitions for a Bi Bi Uni Uni mechanism:

$$V_1 = \frac{k_1k_3k_5k_7k_9k_{11}}{\text{coef ABC}}, \quad K_a = \frac{k_3k_5k_7k_9k_{11}}{\text{coef ABC}}, \quad K_b = \frac{k_1(k_4 + k_5)k_7k_9k_{11}}{\text{coef ABC}},$$
$$K_p = \frac{k_2(k_4 + k_5)k_8k_{10}k_{12}}{\text{coef PQR}}, \quad K_q = \frac{k_2k_4k_6k_{10}k_{12}}{\text{coef PQR}}, \quad K_{ia} = \frac{k_2}{k_1}, \quad K_{ib} = \frac{k_4}{k_3},$$
$$K_{ip} = \frac{k_5}{k_6}, \text{ and } K_{iq} = \frac{k_7}{k_8}$$

gives equation 4-81 as the rate equation:

$$v^*_{A \to Q} = \frac{V_1 \mathbf{A}^* \mathbf{B}(\mathbf{E}/\mathbf{E_t})}{K_{ia}K_b + K_a\mathbf{B} + \dfrac{K_{ia}K_bK_q\mathbf{P}}{K_pK_{iq}}} \tag{4-81}$$

It requires only that a value for $\mathbf{E}/\mathbf{E_t}$ is substituted to get the complete equation. $\mathbf{E}/\mathbf{E_t}$ can be obtained from the distribution equations derived for the chemical mechanism or, if the reaction is at equilibrium, from a simple equilibrium analysis involving dissociation constants. In the present case, when the absence of C and R is assumed so that the sequence from E to F is at equilibrium:

$$\frac{\mathbf{E}}{\mathbf{E_t}} = \frac{K_{ia}K_{ib}\mathbf{PQ}}{K_{ia}K_{ib}\mathbf{PQ} + K_{ip}K_{iq}\mathbf{AB} + K_{ip}\mathbf{ABQ} + K_{ib}\mathbf{APQ} + \mathbf{ABPQ}} \tag{4-82}$$

Substitution into equation 4-81 gives

$$v^*_{A \to Q} = \frac{V_1 K_{ia}K_{ib}\mathbf{A}^*\mathbf{BPQ}}{\Delta\left(K_{ia}K_b + K_a\mathbf{B} + \dfrac{K_{ia}K_bK_q\mathbf{P}}{K_pK_{iq}}\right)} \tag{4-83}$$

where Δ is the denominator of equation 4-82. This method will also work for isotopic exchange from P* to A in the presence of the forward chemical reaction. Thus for

$$\mathrm{E} \underset{k_2}{\overset{k_1\mathbf{A}}{\rightleftharpoons}} \begin{pmatrix}\mathrm{EA}\\ \mathrm{FPQ}\end{pmatrix} \underset{k_4\mathbf{P}^*}{\overset{k_3}{\rightleftharpoons}} \mathrm{EQ} \xrightarrow{k_5} \mathrm{E} \tag{4-84}$$

the rate equation is

$$v^*_{P \to A} = \frac{k_2k_4\mathbf{P}^*(\mathbf{EQ})}{k_2 + k_3} \tag{4-85}$$

Equation 4-85 represents the net rate constant for reaction of P* with EQ to give E and labeled A. The distribution equation for EQ in the steady state is

$$\frac{\mathbf{EQ}}{\mathbf{E_t}} = \frac{k_1k_3\mathbf{A}}{k_1k_4\mathbf{AP} + k_1(k_3 + k_5)\mathbf{A} + (k_2 + k_3)k_5 + k_2k_4\mathbf{P}} \tag{4-86}$$

If one wants to compare the P* → A exchange rate to the chemical rate in the forward direction, one divides the equation for the exchange rate, given in equation 4-87 by combination of equations 4-85 and 4-86, by equation 4-88 for the forward chemical reaction:

$$v^*_{P\rightarrow A} = \frac{k_1k_2k_3k_4\mathbf{AP}^*\mathbf{E_t}}{(k_2+k_3)\Delta} \tag{4-87}$$

$$v_{A\rightarrow P+Q} = \frac{k_1k_3k_5\mathbf{AE_t}}{\Delta} \tag{4-88}$$

where Δ is the denominator of equation 4-86. This gives equation 4-89:

$$\frac{v^*_{P\rightarrow A}}{v_{A\rightarrow P+Q}} = \frac{k_2k_4\mathbf{P}^*}{k_5(k_2+k_3)} = \frac{\mathbf{P}^*}{K_{is}} \tag{4-89}$$

where K_{is} is the slope inhibition constant for the noncompetitive product inhibition by P.

For mechanisms with branches, random mechanisms, one should use Cleland's modification of the King–Altman method for deriving isotope exchange equations (ARB *36*, 77).

Derivation of Equations for Isotope Effects

As an example we will derive the equation for a primary deuterium kinetic isotope effect, but the same procedure can be used for any kinetic isotope effect, primary or secondary, deuterium or heavier atom. Consider the unireactant mechanism depicted in equation 4-90:

$$\mathrm{E} \underset{k_2}{\overset{k_1\mathbf{A}}{\rightleftharpoons}} \mathrm{EA} \underset{k_4}{\overset{k_3}{\rightleftharpoons}} \mathrm{EP} \xrightarrow{k_5} \mathrm{E} \tag{4-90}$$

where k_3 is the chemical step, which includes cleavage of a C–H bond. Of interest for isotope effects are the independent parameters V and V/K. By the method of net rate constants, the rate expressions for V/K and V are given in equations 4-48 and 4-49. The rate constants for the chemical step will differ when deuterium (D) is substituted for protium (H) in the substrate. The equation for V/K for the reaction with deuterated substrate, equation 4-48, becomes

$$\left(\frac{V}{K_a\mathbf{E_t}}\right)_D = \frac{k_1k_{3D}k_5}{k_2k_{4D}+k_2k_5+k_{3D}k_5} \tag{4-91}$$

where the D added to the rate constant reflects the deuterium sensitivity of the step.

The isotope effect is measured as the ratio of the rate obtained with H and D, and thus

$$\frac{\left(\frac{V}{K_a\mathbf{E_t}}\right)_H}{\left(\frac{V}{K_a\mathbf{E_t}}\right)_D} = \frac{\frac{k_1 k_{3H} k_5}{k_2 k_{4H} + k_2 k_5 + k_{3H} k_5}}{\frac{k_1 k_{3D} k_5}{k_2 k_{4D} + k_2 k_5 + k_{3D} k_5}} = \frac{k_{3H}(k_2 k_{4D} + k_2 k_5 + k_{3D} k_5)}{k_{3D}(k_2 k_{4H} + k_2 k_5 + k_{3H} k_5)} \tag{4-92}$$

Using the nomenclature of Northrop and Cook and Cleland (B *14*, 2644;B *20*, 1790) where a leading superscript denotes an isotope effect on the parameter shown, dividing the numerator and denominator of equation 4-92 by $k_2 k_5$ gives:

$$^D\left(\frac{V}{K_a}\right) = \frac{k_{3H}\left(\frac{k_{4D}}{k_5} + 1 + \frac{k_{3D}}{k_2}\right)}{k_{3D}\left(\frac{k_{4H}}{k_5} + 1 + \frac{k_{3H}}{k_2}\right)} \tag{4-93}$$

Defining terms, $^Dk_3 = k_{3H}/k_{3D}$, $^Dk_4 = k_{4H}/k_{4D}$, and $^DK_{eq} = k_{3H}k_{4D}/k_{3D}k_{4H}$ (see Chapter 9). Equation 4-93 is then simplified by substituting definitions, and dropping the subscript H, giving

$$^D\left(\frac{V}{K_a}\right) = \frac{^Dk_3 + \frac{k_3}{k_2} + \left(\frac{k_4}{k_5}\right)^D k_{eq}}{1 + \frac{k_3}{k_2} + \frac{k_4}{k_5}} \tag{4-94}$$

similarly, it can be shown that from equation 4-49:

$$^DV = \frac{\frac{k_{3H} k_5}{k_{3H} + k_{4H} + k_5}}{\frac{k_{3D} k_5}{k_{3D} + k_{4D} + k_5}} = \frac{k_{3H} k_5 (k_{3D} + k_{4D} + k_5)}{k_{3D} k_5 (k_{3H} + k_{4H} + k_5)} = \frac{^Dk_3 + \frac{k_3}{k_5} + \left(\frac{k_4}{k_5}\right)^D k_{eq}}{1 + \frac{k_3}{k_5} + \frac{k_4}{k_5}} \tag{4-95}$$

Shorthand Notation for Rate Constants

In this book we will write all rate constants as k with a numeral subscript. Forward rate constants will have odd numbers and reverse rate constants even ones. But when deriving equations it is easier to use a shorthand notation, in which rate constants are denoted by their number and numbers are indicated by a following decimal point. Thus an expression such as $(k_5/k_4)(1 + k_3/k_2)$ would be written $(5/4)(1. + 3/2)$. Use of this shorthand notation can save a lot of writing (and typing of subscripts!) and is routinely used by the authors in working with kinetic equations.

Data Processing

While one should always plot one's data graphically to see what they look like (usually as double reciprocal plots), for final analysis the data should be fitted to an appropriate rate equation by the least-squares method. The graphical analysis will suggest one or more rate equations to try, and one can then compare the fits to several rate equations to decide which one is best. What one should *not* do is to make an unweighted least-squares fit to the double reciprocal form of a rate equation. Taking the reciprocal gives the lower and generally least accurate velocities more weight than is justified statistically, and thus fits to the reciprocal form require v^4 weights, if the errors in the velocities are constant, or v^2 weights, if errors are proportional to velocities.

Because rate equations are not linear, one must use nonlinear least-squares methods for the fitting (AdvEnz *29*, 1). Computer programs to make such fits written in Fortran are available (ME *63*, 103), and commercial computer programs that can make such fits are available. We will not discuss the details of these programs but rather how to use them.

Regardless of what computer program one uses, one must decide on the weighting of the velocities. If one assumes that errors are independent of the size of the velocities, one fits an equation in a form such as

$$v = \frac{V\mathbf{A}}{K + \mathbf{A}} \tag{4-96}$$

This is normally the form to use if the range of the velocities is 5 or less. If one assumes that errors are proportional to the size of the velocities, one fits the equation in the log form:

$$\log v = \log\left[\frac{V\mathbf{A}}{K + \mathbf{A}}\right] \tag{4-97}$$

This is the form to use if the range of velocities is 10 or more (and for all pH profiles) or if different levels of enzyme are used for different substrate concentrations and the velocities are adjusted to a constant enzyme level. If one is in doubt about which equation to use, the residuals should be examined (that is, the differences between fitted and experimental velocities). If the residuals are independent of the size of the velocity, or nearly so, equation 4-96 is the one to use. If the residuals are larger for the higher velocities but the differences in log v are independent of velocity, equation 4-97 should be used.

When comparing fits to two possible rate equations (competitive or noncompetitive inhibition, for example), one compares the σ values, which are the sum of squares of residuals, divided by degrees of freedom (number of data points minus number of fitted constants). A lower σ indicates a better fit. One should also see if the fitted constants have significant values. Good fits will have errors that are less than 10–15% of the values, and acceptable errors may be as high as 25%. Thus a fit to noncompetitive inhibition that gives $K_{is} = 7 \pm 1$, but $K_{ii} = 40 \pm 60$, is likely to represent competitive inhibition, and the fit to the equation for competitive inhibition should give a precise value of K_{is} and a σ that is not appreciably larger than the one from the noncompetitive fit.

As an example of this approach, consider a reaction with three substrates. If one varies the level of all three (four levels of each gives 64 data points), one can then fit the initial velocities to the general equation:

$$v = \frac{V\mathbf{ABC}}{\Delta} \tag{4-98}$$

where Δ, the denominator, contains a constant term, as well as terms in **A**, **B**, **C**, **AB**, **AC**, **BC**, and **ABC**. If any of these terms are not determined (negative value; standard error larger than value), leave them out and fit the data to the equation without them. If σ is not significantly increased, one is justified in deleting the term. For example, an ordered mechanism will lack the **B** term, and ping-pong mechanisms will lack the constant term, as well as others.

Some rate equations are difficult to fit without supplying preliminary estimates of one or more constants. The Fortran programs discussed above invert the rate equation to make it linear and generate a preliminary fit using v^4 weights (for equal size errors in v) or v^2 weights (if a log fit is to be made). The constants obtained are then used as the preliminary estimates for the nonlinear fit, which is an iterative process; 3–5 iterations are usually enough.

For an equation such as equation 4-99, however, which describes hyperbolic competitive inhibition, this approach does not work.

$$v = \frac{V\mathbf{A}}{K\left[\dfrac{1 + \dfrac{\mathbf{I}}{K_{in}}}{1 + \dfrac{\mathbf{I}}{K_{id}}}\right] + \mathbf{A}} \tag{4-99}$$

One then constructs a least-squares surface and looks for a minimum. To do this, one assumes an 11×11 grid of K_{in} and K_{id} values stepped by 1.26 (the 10th root of 10) and at each point on the grid makes a least-squares fit to equation 4-99 with the chosen values of K_{in} and K_{id} fixed. The values of σ, V, and K are then printed out in a grid pattern and a map of contours of σ is made. The value of K_{in} and/or K_{id} used as starting points for the grid may have to be adjusted to find the true minimum point. Once such a point is found, the values of K_{in} and K_{id} at the minimum are used as preliminary estimates for a least-squares fit to the full equation 4-99 or its log form.

A least-squares surface is also useful for finding out why there is no minimum if none is found. A long horizontal valley shows that one constant is well determined but the other is not. If the valley has a headwall at the left end but is open at the other, the second constant has a minimum value but may be infinite (which suggests that it is not present and one should pick a different rate equation). A long diagonal valley indicates that the two nonlinear constants do not have discrete values but can get large or small together. The shape of the least-squares surface is thus an excellent diagnostic tool.

In some cases, graphical analysis can also give preliminary estimates to use for least-squares fitting of rate equations that are not made linear by inversion. A typical example is

$$v = \frac{V(d + \mathbf{A})}{c + b\mathbf{A} + \mathbf{A}^2} \tag{4-100}$$

which in reciprocal form is a 2/1 function:

$$\frac{1}{v} = \left(\frac{1}{V}\right)\left[\frac{1 + b\left(\frac{1}{\mathbf{A}}\right) + c\left(\frac{1}{\mathbf{A}^2}\right)}{1 + d\left(\frac{1}{\mathbf{A}}\right)}\right] \tag{4-101}$$

This is the equation that applies to a steady-state random mechanism with two substrates unless the second substrate is saturating.

Preliminary estimates of the constants b, c, and d can be obtained from a plot of equation 4-101 from the slopes and intercepts of the asymptote and the tangent to the curve at $1/\mathbf{A} = 0$. Thus

$d =$ (tangent slope – asymptote slope)/(asymptote intercept – $1/v$ intercept)

$b = d +$ (tangent slope)/($1/v$ intercept)

$c = d$ (asymptote slope)/($1/v$ intercept)

5

INITIAL VELOCITY STUDIES IN THE ABSENCE OF ADDED INHIBITORS

In Chapter 2 the overall initial velocity rate equation was discussed in terms of obtaining information on the kinetic mechanism of an enzyme-catalyzed reaction. The denominator of the rate equation was identified as providing information on the kinetic mechanism, while the numerator provides information on the thermodynamic driving force. The denominator of the rate equation represents a distribution of the enzyme among all the possible forms that could exist for a given enzyme-catalyzed reaction. Thus depending on the kinetic mechanism for an enzyme-catalyzed reaction, the overall rate equation will differ, and one attempts to determine what terms exist in the denominator of the rate equation to elucidate kinetic mechanism. In this chapter the general background for interpreting initial velocity data will be established by using initial velocity patterns in the absence of added inhibitors, exploiting differences that exist in the overall initial velocity rate equation, depending on the kinetic mechanism of the enzyme-catalyzed reaction. The Uni Bi mechanism is applicable to, among others, enzyme reactions such as those catalyzed by phosphatases like alkaline phosphatase and ammonia lyases such as aspartate ammonia-lyase. To illustrate the general procedure, two different Uni Bi kinetic mechanisms will be used, the

Uni Bi steady-state ordered and Uni Bi rapid equilibrium random. The rate equation will be analyzed in the first section, and in the last section a procedure for collecting the initial velocity data will be outlined. Complete rate equations, in terms of rate constants and kinetic constants, along with distribution equations and Haldane relationships for a number of Uni-, Bi-, and Terreactant mechanisms, are provided in Appendix II.

Uni Bi Enzyme Reactions

Two possible Uni Bi mechanisms are depicted in shorthand notation in equations 5-1 and 5-2. In both cases the substrate A binds to the enzyme active site and is converted to the products. However, there are two possibilities for the release of P and Q. The first is an ordered release of P prior to Q, equation 5-1, while the other allows for release of either P or Q first. In the case of the random kinetic mechanism illustrated, the slow step along the reaction pathway will be defined as the interconversion of EA and EPQ.

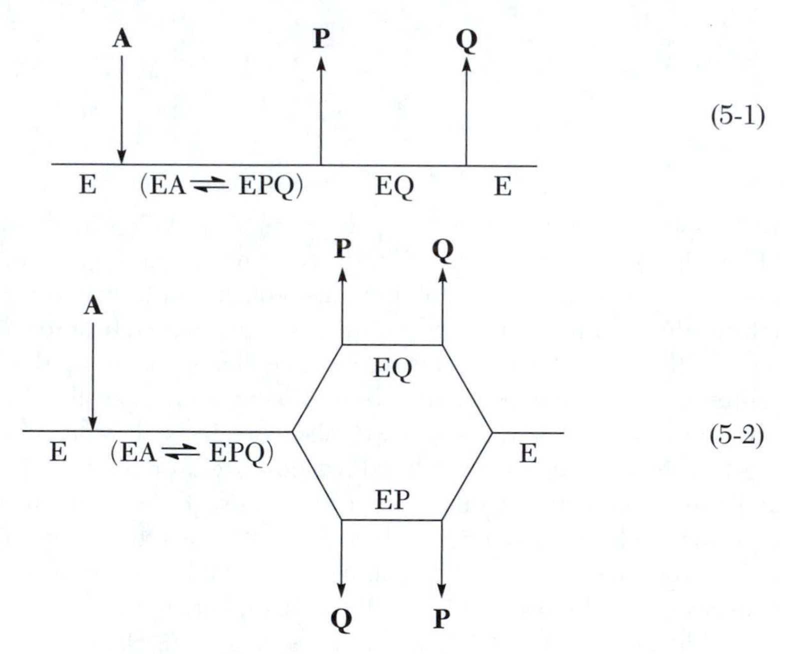

The overall rate equations for mechanisms 5-1 and 5-2 are given in equations 5-3 and 5-4. (Letters shown in boldface type throughout the chapter depict reactant concentrations.) There is an extra term in the denominator of the rate equation for mechanism 5-1. However, the initial velocity rate equations in either reaction direction are identical for both of the above kinetic mechanisms. Although the mechanisms cannot be distinguished in the absence of inhibitors, the use of inhibitors will allow a distinction in most cases. The initial velocity rate equations are obtained by setting the concentrations of products to zero, that is, to adhere to initial velocity conditions. Thus, when **P** and **Q** are zero, the Michaelis–Menten equation is obtained in the unireactant direction A to P + Q, equation 5-5. Allowing **A** to become equal to zero generates equation 5-6 in the bireactant direction P + Q to A. The initial velocity rate equations are identical for these two kinetic mechanisms. Equations 5-3 and 5-4 appear to differ in the **P** term, but the two terms are equivalent as will become clear after a discussion of Haldane relationships later in this chapter.

$$v = \frac{V_1\mathbf{A} - \dfrac{V_2\mathbf{PQ}K_a}{K_pK_{iq}}}{K_a + \mathbf{A} + \dfrac{\mathbf{AP}}{K_{ip}} + \dfrac{K_a\mathbf{Q}}{K_{iq}} + \dfrac{K_{ia}\mathbf{P}}{K_{ip}} + \dfrac{K_a\mathbf{PQ}}{K_pK_{iq}}} \tag{5-3}$$

$$v = \frac{V_1\mathbf{A} - \dfrac{V_2\mathbf{PQ}K_a}{K_pK_{iq}}}{K_a + \mathbf{A} + \dfrac{K_a\mathbf{P}}{K_{ip}} + \dfrac{K_a\mathbf{Q}}{K_{iq}} + \dfrac{K_a\mathbf{PQ}}{K_pK_{iq}}} \tag{5-4}$$

$$v = \frac{V_1\mathbf{A}}{K_a + \mathbf{A}} \tag{5-5}$$

$$v = \frac{V_2\mathbf{PQ}}{K_pK_{iq} + K_q\mathbf{P} + K_p\mathbf{Q} + \mathbf{PQ}} \tag{5-6}$$

We have assumed in mechanism 5-2 that EPQ formation was at equilibrium. We will get a rate equation of the same algebraic form as equation 5-4 if we assume that only EP and EQ formation are in rapid equilibrium, with the rest of the steps in steady state. That is, the denominator of the rate equation will contain only K_a, A, P, Q, and PQ terms. The presence of the AP term in equation 5-3 results from the existence of EQ in the steady state. In a fully steady-state Random Uni Bi mechanism where an EP complex forms, an AQ term will also be present, as well as terms in P^2 and Q^2 and a more complicated numerator.

The initial velocity rate equations will be explored first in order to understand why and how the initial rate changes with reactant concentration and to determine what can and cannot be learned from a study of the initial rate in the absence of added inhibitors alone. A consideration of how one is able to distinguish between the two mechanisms using inhibition studies will be discussed in Chapter 6.

An explanation of the saturation curve for A, equation 5-5, can be found in any general text and will be considered only briefly here. At concentrations of $\mathbf{A} < K_a$, the rate of combination of A and E limits the overall reaction. The reaction rate is second-order (first-order in both A and E) and is depicted by the second-order rate constant $V/K\mathbf{E}_t$, one of the limits of the Michaelis–Menten equation as discussed in Chapter 3. (The measured parameter at a fixed enzyme concentration, V/K with units of seconds^{-1}, is divided by $\mathbf{E}_t$ to generate a second-order rate constant with units of molar^{-1}seconds^{-1}.) At $\mathbf{A} > K_a$, the rate is limited by whatever is the slowest step, interconversion of EA and EPQ or release of one or both of the products P and Q. The reaction is first-order in whatever enzyme form(s) precede(s) the slow step(s). [The measured parameter at a fixed concentration of A, V with units of molar · seconds^{-1}, is divided by $\mathbf{E}_t$ to generate the first order rate constant with units of seconds^{-1} ($V/\mathbf{E}_t$ or k_{cat}).]

Equation 5-6 gives the initial rate equation for the reverse of the Uni Bi reaction. The reverse reaction is bireactant and thus has a rate equation for a bireactant sequential mechanism. The rate varies with **P** and **Q**. For purposes of demonstration, we will first consider application of equation 5-6 to the *rapid equilibrium random* Uni Bi kinetic mechanism, that is, the interconversion of EA and EPQ is the slowest step along the reaction pathway in mechanism 5-2 above. (As noted above, the same rate equation applies if only EP and EQ formation are in rapid equilibrium.) The reciprocal of equation 5-6 is given as equation 5-7:

$$\frac{1}{v} = \frac{K_{iq}K_p}{V_2\mathbf{PQ}} + \frac{K_p}{V_2\mathbf{P}} + \frac{K_q}{V_2\mathbf{Q}} + \frac{1}{V_2} \tag{5-7}$$

If **P** is varied at different fixed levels of **Q**, equation 5-8 is obtained:

$$\frac{1}{v} = \left(\frac{K_p}{V_2}\right)\left(1 + \frac{K_{iq}}{\mathbf{Q}}\right)\left(\frac{1}{\mathbf{P}}\right) + \left(\frac{1}{V_2}\right)\left(1 + \frac{K_q}{\mathbf{Q}}\right) \tag{5-8}$$

A double reciprocal plot of equation 5-8 with **P** varied at each of four different values of **Q**, from **Q**1 to **Q**4, is shown in Figure 5-1. The initial double reciprocal representation of the data is called a *primary plot*. A family of lines is observed that intersects to the left of the ordinate. The slope and intercept of the plot shown in Figure 5-1 depend linearly on the reciprocal of the concentration of the fixed reactant, Q. The expressions for the slope and intercept are derived from equation 5-8 and shown as equations 5-9 and 5-10.

$$\lim_{\mathbf{P} \to 0}$$

$$\text{slope} = \left(\frac{K_p}{V_2}\right)\left(1 + \frac{K_{iq}}{\mathbf{Q}}\right) = \left(\frac{K_p K_{iq}}{V_2}\right)\left(\frac{1}{\mathbf{Q}}\right) + \frac{K_p}{V_2} \tag{5-9}$$

$$\lim_{\mathbf{P} \to \infty}$$

$$\text{intercept} = \left(\frac{1}{V_2}\right)\left(1 + \frac{K_q}{\mathbf{Q}}\right) = \left(\frac{K_q}{V_2}\right)\left(\frac{1}{\mathbf{Q}}\right) + \frac{1}{V_2} \tag{5-10}$$

As discussed in Chapter 3, the slope is the reciprocal of V/K ($v = (V/K)\mathbf{A}$ as the substrate concentration approaches zero), while the intercept is the reciprocal of V ($v = V$ as the varied substrate concentration approaches infinity). Plots of equations 5-9 or 5-10 are also shown in the inset to Figure 5-1. The slope and intercept plots are called *secondary plots* or *replots*. Slopes and intercepts of the secondary plots represent the limits in which the concentration of Q approaches

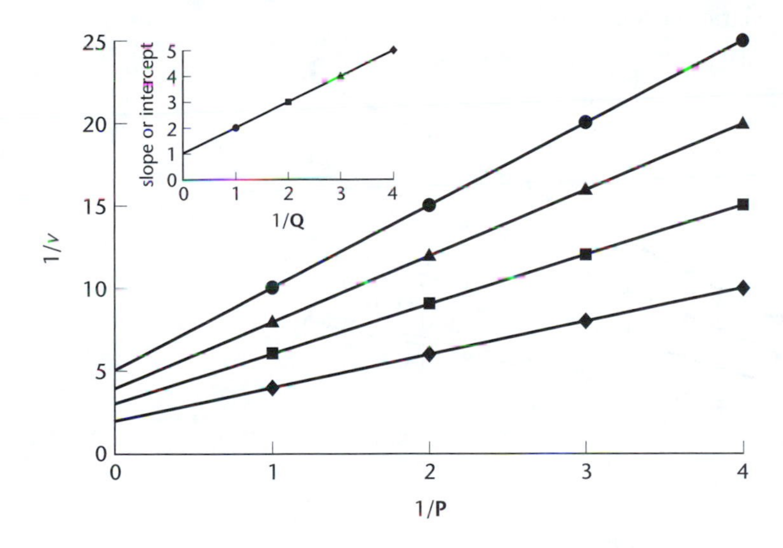

Figure 5-1. Primary double reciprocal plot for the rapid equilibrium random kinetic mechanism. The varied reactant is P, and Q is the fixed varying reactant. (Inset) Secondary plot of slopes and intercepts against the reciprocal concentration of the fixed variable reactant. The values of K_{iq}, K_p, and K_q used in generating the above plots are 1 mM, while V_2 is 1 mM/min.

zero and infinity, respectively. The slope of the slope replot is $K_{iq}K_p/V_2$ and is approximated when concentrations of both reactants approach zero, conditions in which the E form of the enzyme predominates for mechanisms 5-1 and 5-2. The intercept of the slope replot is the reciprocal of K_p/V_2 and corresponds to the condition where the concentration of P approaches zero, and the concentration of Q approaches infinity, and thus reflects the EQ form of the enzyme for mechanisms 5-1 and 5-2. The slope of the intercept replot is the reciprocal of V_2/K_q and corresponds to a concentration of P approaching infinity and a concentration of Q approaching zero, representing the EP form of the enzyme for the random mechanism but the E form of the enzyme for the ordered mechanism. The intercept of the intercept replot is the reciprocal of V_2, corresponding to both P and Q concentrations approaching infinity as depicted in the inset in Figure 5-1. The analysis just provided indicates that generally there are four regions to a double reciprocal plot representing the four limits of equation 5-8, as shown in Figure 5-2. One can thus read any double reciprocal plot in this manner, and know the enzyme form (or forms) that may predominate under any conditions. (Knowledge of kinetic mechanism is required to know the exact enzyme forms, since there is a certain amount of degeneracy in the rate equation for some kinetic mechanisms.)

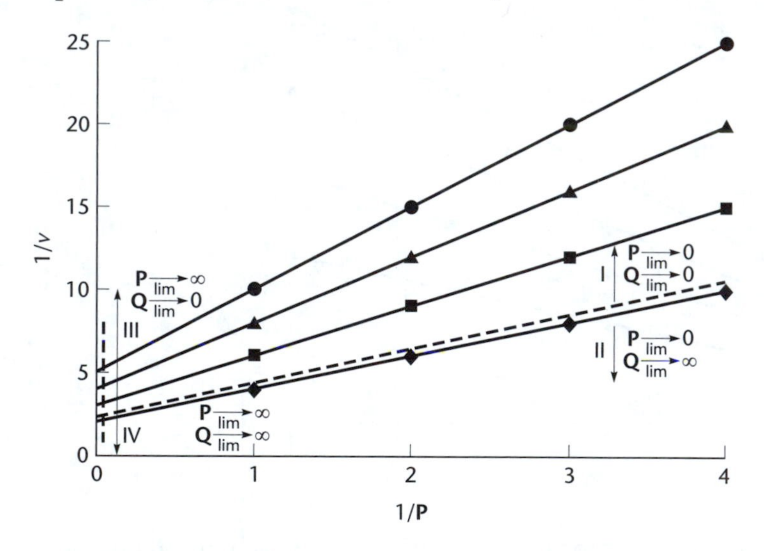

Figure 5-2. Graphic representation of reading double reciprocal plots. The dotted lines are used to separate the regions of the plot in which given enzyme forms predominate as shown. The plot is a repeat of that shown in Figure 5-1.

From the inset in Figure 5-1, one can see that the kinetic constants can be estimated graphically. The maximum velocity V and the V/K for each reactant are obtained directly from the reciprocal of the intercepts of both replots and the slope of the intercept replot. The K_m values are obtained as the ratio of K/V and $1/V$ (K_q is also obtained as the abscissa intercept of the intercept replot), while the K_{iq} value is obtained as the abscissa intercept of the slope replot. To obtain confidence limits in the form of standard errors on the kinetic parameters, one should fit the initial velocity data to the proper rate equation, in this case equation 5-6. Computer programs are available to perform the data fitting, as discussed in Chapter 4. However, one should always plot the data as in Figure 5-1 to obtain a visual idea of the goodness of fit and an assessment of the proper rate equation to be used for data fitting.

Equation 5-7 is symmetric and can also be rearranged to reflect Q as the variable reactant, with P as the varying fixed reactant, equation 5-11. It would appear that the term obtained as the abscissa intercept of the slope replot is of little use, since it is still a ratio of kinetic constants. However, the ratio $K_{iq}K_p/K_q$ is also equal to K_{ip} in a rapid equilibrium mechanism. It does not matter which path is taken to go from E to EPQ. The equilibrium constant must be the same and will be the product of the equilibrium constants for addition of P and Q by the top path and Q and P by the bottom path, so that $K_{ip}K_q = K_{iq}K_p$ and $K_{ip} = K_{iq}K_p/K_q$.

$$\frac{1}{v} = \left(\frac{K_q}{V_2}\right)\left(1 + \frac{K_{ip}}{\mathbf{P}}\right)\left(\frac{1}{\mathbf{Q}}\right) + \left(\frac{1}{V_2}\right)\left(1 + \frac{K_p}{\mathbf{P}}\right) \tag{5-11}$$

The slope and intercept are now a function of **P**.

$$\mathbf{Q} \xrightarrow[\text{lim}]{} 0$$

$$\text{slope} = \left(\frac{K_q}{V_2}\right)\left(1 + \frac{K_{ip}}{\mathbf{P}}\right) = \left(\frac{K_{ip}K_q}{V_2}\right)\left(\frac{1}{\mathbf{P}}\right) + \frac{K_q}{V_2} \tag{5-12}$$

$$\mathbf{Q} \xrightarrow[\text{lim}]{} \infty$$

$$\text{intercept} = \left(\frac{1}{V_2}\right)\left(1 + \frac{K_p}{\mathbf{P}}\right) = \left(\frac{K_p}{V_2}\right)\left(\frac{1}{\mathbf{P}}\right) + \frac{1}{V_2} \tag{5-13}$$

Thus, the same data set can and should be plotted as shown in equations 5-8 to 5-10 and 5-11 to 5-13. The plots shown in Figure 5-3 will be qualitatively identical to those in Figure 5-1.

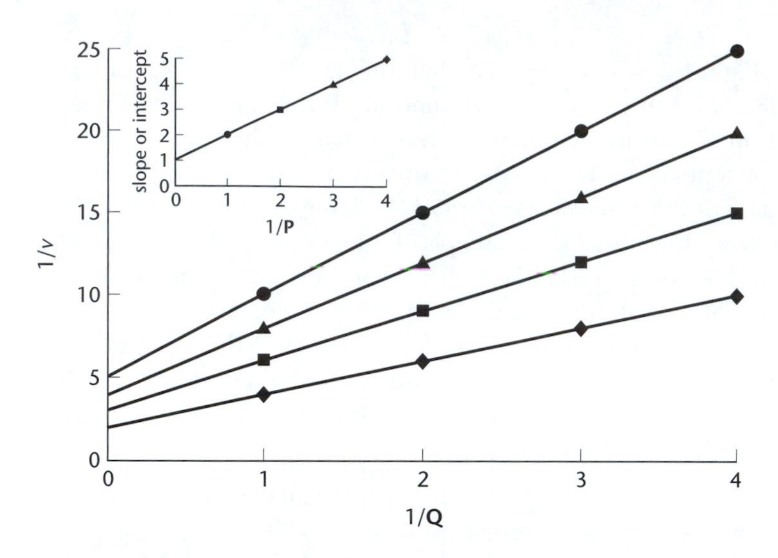

Figure 5-3. (A) Primary double reciprocal plot for the rapid equilibrium random kinetic mechanism. The varied reactant is Q, and P is the fixed variable reactant. (Inset) Slope or intercept replot. The values of the kinetic constants are identical to those given in the caption to Figure 5-1.

Note that the slope of the slope replot and the intercept of the intercept replot terms remain the same, since both reactants are extrapolated to zero and infinity, whatever the varied reactant, and it is the K/V terms that are switched. As a result, the abscissa intercepts are different, with the negative reciprocal of K_{ip} and K_p obtained from the slope and intercept replots, respectively. Thus, the rate equation is truly symmetric, giving estimates of the dissociation constant for EQ when **P** is extrapolated to zero and **Q** is varied and of the dissociation constant for EQ when **P** is extrapolated to zero and **Q** is varied.

It should be clear at this point that one can read primary double reciprocal plots in terms of the enzyme form that predominates in different regions of the plot. A demonstration of this is given in Figure 5-2. For the rapid equilibrium random mechanism in the upper right region (I), both **P** and **Q** are limiting and E predominates, while in the lower right region (II), **P** is saturating and EP predominates. Likewise, near the ordinate, EQ and EPQ predominate depending on whether **P** is limiting [upper region (III)] or saturating [lower region (IV)]. The ability to determine the enzyme form that predominates under any set of conditions is useful in qualitatively diagnosing anomalous effects such as curvature in double reciprocal plots, as will be seen later.

To this point, discussion has centered on dissection of the rate equation of the Bi Uni rapid equilibrium random mechanism. How data are plotted, the enzyme forms that predominate under given conditions of reactant concentrations, and an isolation and quantitation of the kinetic parameters graphically have all been considered. However, we have not undertaken a consideration of why the rate increases as the reactant concentration increases. In the case of a rapid equilibrium random mechanism, the explanation is straightforward as shown below:

$$E + Q \longrightarrow EQ + P \longrightarrow EPQ \tag{5-14}$$

At limiting concentrations of both P and Q, the E form of the enzyme predominates, while the catalytic form of the enzyme (Michaelis complex) is EPQ, which is present at very low concentrations, giving a very low rate. At a fixed, limiting concentration of P, increasing the concentration of Q leads to the formation of the EQ complex and at the same time more EPQ, and an increase in the rate. The opposite is also true; that is, increasing the concentration of **P** at limiting **Q** gives the EP complex and also gives more EPQ, only by the other pathway in which P adds before Q.

$$E + P \longrightarrow EP + Q \longrightarrow EPQ \tag{5-15}$$

Once either EQ or EP is formed, an increase in the concentration of either P or Q will produce an increase in EPQ and thus the rate, until all of the enzyme is present as the EPQ complex.

The initial velocity rate equation of a Bi Uni *steady-state ordered* mechanism is identical to that of the Bi Uni rapid equilibrium random mechanism. Thus, all of the primary initial velocity patterns and secondary replots discussed above will be identical for both mechanisms. However, the enzyme forms represented by the kinetic terms in the denominator of the rate equation differ, as does the explanation of why the rate increases with increasing reactant concentration. Since there is only one pathway for the formation of EPQ, which requires Q bound prior to P according to equation 5-1, there will be no EP complex. Instead, both the $K_{iq}K_p$ term and the $K_q\mathbf{P}$ term will represent E. The reason the **P** term represents free enzyme is that the small amount of EQ formed is continually converted to EPQ, which forms EA, resulting in E. Thus, as long as **Q** is limiting, the predominant enzyme form will be E whatever the concentration of P.

In a Bi Uni steady-state ordered mechanism, the rate changes from a low value to a higher one as **P** increases because whatever amount of EQ is present is converted to EPQ and the product A. As **Q** increases, the initial rate increases at all values of **P** since more EQ is available for P to bind to. There is also an increase in the initial rate with **P** at a saturating concentration and **Q** is increased. Under these conditions, the rate of conversion of EPQ to A plus E is relatively fast compared to combination of P with the small amount of the EQ complex that exists at nonsaturating **Q**. In other words, P cannot force all of the enzyme into the EPQ complex, since it continually turns over to regenerate E, and the rate is dependent on **Q** and EQ.

Bireactant Enzyme Reactions

To this point we have considered Uni Bi enzyme reactions in both reaction directions. As the abbreviation suggests, the reaction is unireactant in one reaction direction and bireactant in the other under initial velocity conditions in the absence of added product. As a result, a consideration of the initial velocity patterns for bireactant enzymes has already been initiated. The previous section served to show the origin of the initial velocity rate equation, how data appear when plotted, the interpretation of the initial velocity patterns in terms of predominant enzyme forms, and a graphical estimation of kinetic constants. The reaction in only one reaction direction will be considered, and the procedure developed in the previous section will be applied to the remaining kinetic mechanisms to be considered. The common kinetic mechanisms will be considered, but one should bear in mind that because an enzyme has an ordered mechanism in one reaction direction, it need not be ordered in the opposite direction. In the following discussion based on previously published theory (BBA 67, 188), an initial velocity pattern will be illustrated only if it contains something unusual, and therefore diagnostic.

Ordered Sequential Mechanisms

The ordered bireactant mechanism has three common variants that represent a continuous spectrum from rate-limiting interconversion of the central complexes, *rapid equilibrium ordered*, to rate-limiting release of the last product, *Theorell–Chance*. The *steady-state ordered* mechanism considered in the previous section lies between the two extremes. A common feature of all the ordered mechanisms is the implication that no site for B exists initially (or the affinity for B is too low in the absence of A). Thus, the binding of A to enzyme must induce a conformational change in E that allows B to bind or A, once bound to enzyme, actually forms a portion of the binding site for B. A random mechanism will look ordered if there is a high degree of synergism of binding of the two substrates and one binds more tightly than the other. The classic example of this is the glycerol kinase reaction, which appears ordered with glycerol adding to enzyme first, but appears equilibrium ordered, with MgATP adding to enzyme first, when the substrate is changed from glycerol to an aminopropanediol (B 28, 5728).

Equilibrium Ordered Mechanism. This mechanism will be considered first, followed by a consideration of the other two. The mechanism is pictured schematically in a manner identical to the steady-state ordered mechanism, equation 5-16, but with the requirement that the off rate constant

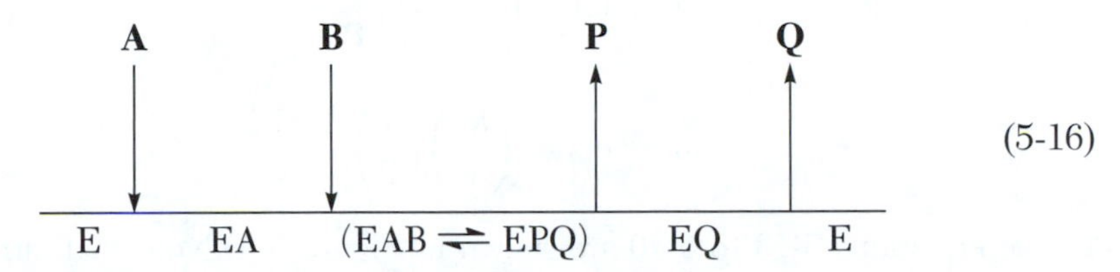

(5-16)

of A be much greater than $V/\mathbf{E_t}$. Thus, the addition of A is at equilibrium but the addition of B need not be, and $V/\mathbf{E_t}$ can be limited by release of P or Q or by the interconversion of EAB and EPQ. Because of the equilibrium addition of A, the kinetic mechanism has a rate equation that differs from that of the steady-state ordered mechanism. The rate equation is asymmetric, that is, gives a different algebraic form dependent on which reactant concentration is varied. We will thus consider it in detail. The initial velocity rate equation is given in equation 5-17:

$$v = \frac{V\mathbf{AB}}{K_{ia}K_b + K_b\mathbf{A} + \mathbf{AB}} \tag{5-17}$$

Note that there is no $K_a\mathbf{B}$ term present as in the steady-state mechanism. As long as $\mathbf{A} \geq \mathbf{E_t}$, extrapolation of **B** to infinity forces all of the enzyme into the EAB complex even if $\mathbf{A} \ll K_{ia}$. (Although the latter is correct theoretically, it would not be possible to measure an initial rate with $\mathbf{A} = \mathbf{E_t}$.) Enzyme forms represented in the denominator of the rate equation are thus as follows:

Enzyme Form	Denominator Term
E	$K_{ia}K_b$
EA	$K_b\mathbf{A}$
EAB	$\mathbf{AB}$

The asymmetric rate equation will not give qualitatively identical primary plots depending on the reactant varied. To illustrate, consider the reciprocal form of the rate equation arranged so that A is the varied reactant:

$$\frac{1}{v} = \left(\frac{K_{ia}K_b}{V\mathbf{B}}\right)\left(\frac{1}{\mathbf{A}}\right) + \left(\frac{1}{V}\right)\left(1 + \frac{K_b}{\mathbf{B}}\right) \tag{5-18}$$

Expressions for slope and intercept are

$$\text{slope} = \left(\frac{K_{ia}K_b}{V}\right)\left(\frac{1}{\mathbf{B}}\right) \tag{5-19}$$

$$\text{intercept} = \left(\frac{K_b}{V}\right)\left(\frac{1}{\mathbf{B}}\right) + \frac{1}{V} \tag{5-20}$$

Plots of equations 5-18 to 5-20 are shown in Figure 5-4. Note that although the primary plot appears identical to one that would be obtained for a steady-state ordered mechanism, the slope replot passes through the origin. The slope replot can readily be explained by the following scheme:

$$\left[\mathrm{E} + \mathrm{A} \underset{K_{ia}}{\rightleftharpoons} \mathrm{EA}\right] \underset{k_4}{\overset{k_3\mathbf{B}}{\rightleftharpoons}} \mathrm{EAB} \xrightarrow{k_{cat}} \text{products} \tag{5-21}$$

At infinite concentrations of B, no free enzyme exists, as indicated by the value of zero for the slope in equation 5-19, since B traps A on enzyme to form the EAB complex. The slope replot is diagnostic for a rapid equilibrium ordered mechanism. Kinetic parameters are estimated from the slope and intercept replots as discussed in the previous section. At infinite **B** the double reciprocal plot, Figure 5-4, will give a horizontal line, that is, K_a is zero.

If B is varied at fixed concentrations of A, equations 5-18 to 5-20 become equations 5-22 to 5-24:

$$\frac{1}{v} = \left(\frac{K_b}{V}\right)\left(1 + \frac{K_{ia}}{\mathbf{A}}\right)\left(\frac{1}{\mathbf{B}}\right) + \frac{1}{V} \tag{5-22}$$

$$\text{slope} = \left(\frac{K_{ia}K_b}{V}\right)\left(\frac{1}{\mathbf{A}}\right) + \frac{K_b}{V} \tag{5-23}$$

$$\text{intercept} = \left(\frac{1}{V}\right) \tag{5-24}$$

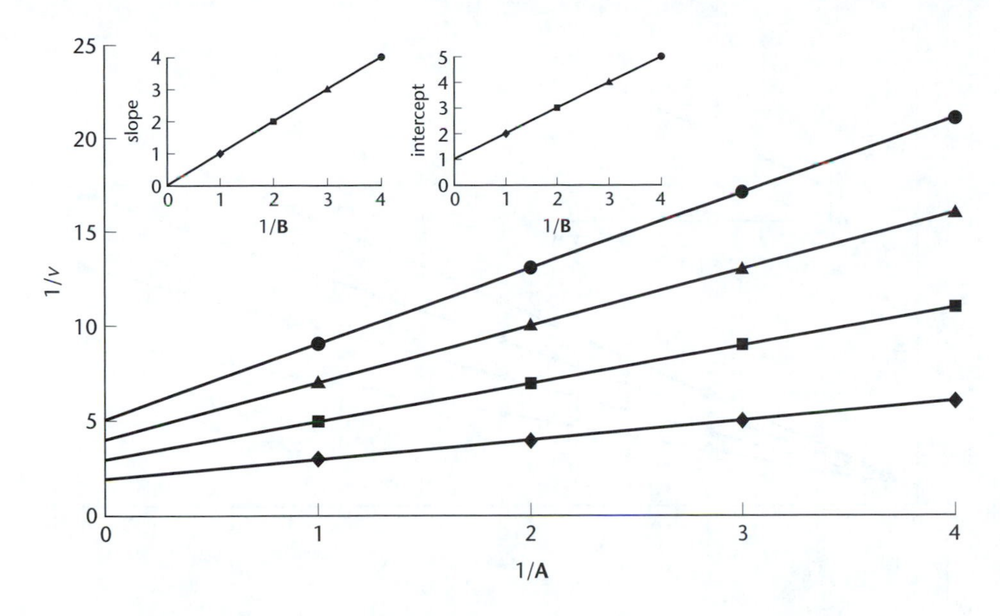

Figure 5-4. Primary and secondary plots for a rapid equilibrium ordered mechanism with A as the varied reactant. The slope replot is a diagnostic. Values of K_{ia} and K_b used in generating the above plots are 1 mM, while V_1 is 1 mM/min.

When plotted in this manner, as shown in Figure 5-5, the primary plot intersects on the ordinate, also diagnostic for a rapid equilibrium ordered mechanism. The explanation is the same as that provided above; that is, regardless of the concentration of A, extrapolating **B** to infinity causes all of the enzyme to become EAB, and thus the maximum rate is attained. When the diagnostic plots are observed, one knows not only that the mechanism is rapid equilibrium ordered but also which of the two reactants adds to enzyme first. The second reactant to add is the one that, when varied, gives a primary plot that intersects on the ordinate. The first reactant to add is the one that, when varied, gives a slope replot that goes through the origin.

Creatine kinase exhibits an equilibrium ordered kinetic mechanism at pH 7 with MgATP adding prior to creatine, but the back reaction shows random addition of MgADP and phosphocreatine (JBC *248*, 8418). At pH 8, the kinetic mechanism is random in both reaction directions (B *20*, 1204). The change in mechanism as the pH is decreased from 8 to 7 is due to a loss of affinity for creatine as the active-site general base becomes protonated.

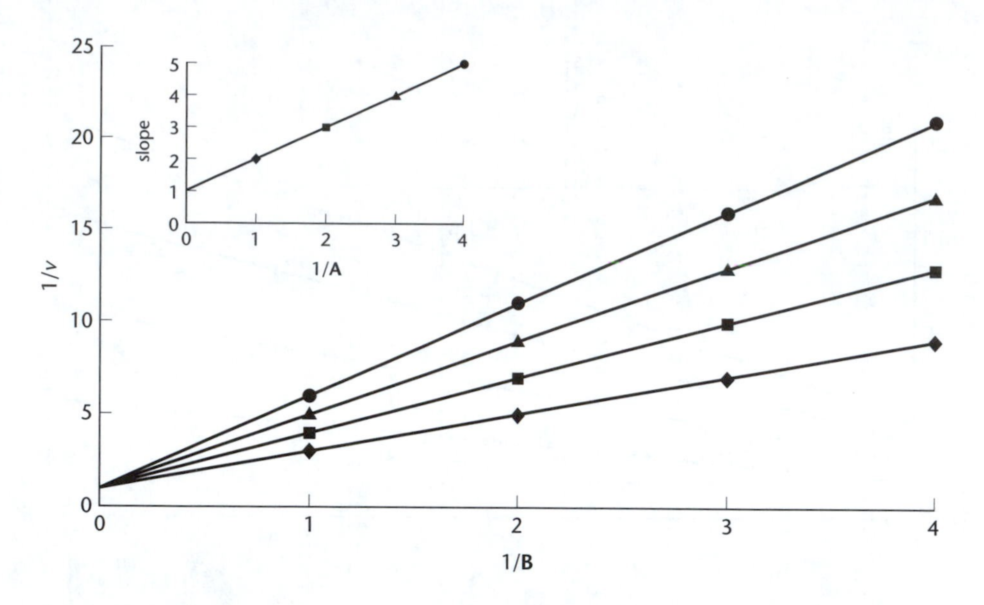

Figure 5-5. Primary plot for a rapid equilibrium ordered mechanism with B as the varied reactant. (Inset) Replot of the slope of the lines from the main plot against the reciprocal of the concentration of A. Kinetic constants are as indicated in the caption to Figure 5-4.

Steady-State Ordered Mechanism. This mechanism can be pictured schematically by use of equation 5-16. However, since the products are present at zero concentration under initial velocity conditions, the mechanism has been considered in detail as the reverse reaction of the Uni Bi steady-state ordered mechanism (simply substitute A and B for Q and P). The steady-state ordered case, as stated above, is intermediate between the rapid equilibrium and Theorell–Chance mechanisms. In terms of enzyme forms, the steady-state case is reiterated below.

Enzyme Form	Denominator Term
E	$K_{ia}K_b$, $K_a\mathbf{B}$
EA	$K_b\mathbf{A}$
EAB, EPQ, EQ	**AB**

Thus two terms represent E, and the **AB** term represents all of the enzyme forms from EAB to release of the last product (including EAB, EPQ, and EQ), since several steps can contribute to limitation of the rate in this mechanism.

In addition, the rate equation is symmetric and the primary plots obtained for either varied reactant are similar.

An example of the above behavior is shikimate dehydrogenase, which catalyzes the oxidation of shikimate to dehydroshikimate (B *10*, 1947). Initial velocity patterns are observed that intersect to the left of the ordinate in both reaction directions, consistent with a sequential kinetic mechanism. The ordered nature of the mechanism was confirmed by product inhibition and isotope exchange at equilibrium: $NADP^+$ and NADPH compete with one another, while all other product inhibition patterns are noncompetitive (see Chapter 6), and total substrate inhibition of the $NADP^+$–NADPH isotope exchange reaction is obtained by raising the concentrations of shikimate and dehydroshikimate together (see Chapter 8).

Theorell–Chance Mechanism. The last of the ordered mechanisms to be considered is the Theorell–Chance mechanism, named after the investigators who developed the theory to explain the kinetics of alcohol dehydrogenase from equine liver. The mechanism is depicted schematically as shown in equation 5-25. There are no ternary complexes, and reaction of EA and B gives EQ and P directly.

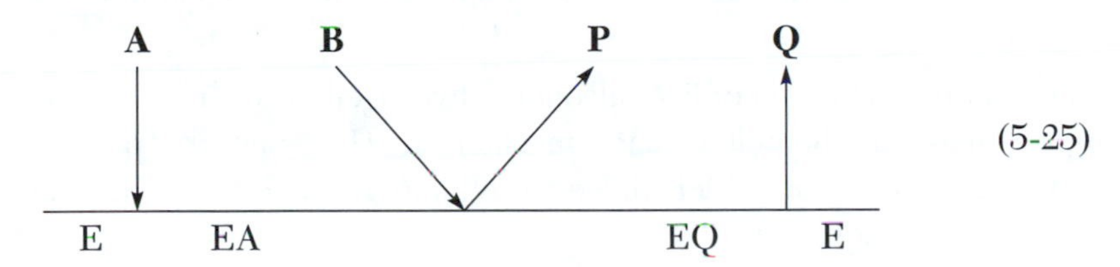

(5-25)

The initial velocity rate equation for the Theorell–Chance mechanism is identical to that for the steady-state mechanism (equation 5-6) but with **A** and **B** substituted for **Q** and **P** [$v = V\mathbf{AB}/(K_{ia}K_b + K_a\mathbf{B} + K_b\mathbf{A} + \mathbf{AB})$]. Thus, the

Enzyme Form	Denominator Term
E	$K_{ia}K_b$, $K_a\mathbf{B}$
EA	$K_b\mathbf{A}$
EQ	$\mathbf{AB}$

graphical treatment and estimation of kinetic constants will be identical to that for the steady-state ordered mechanism. The enzyme forms represented by the terms in the denominator of the rate equation are also identical to those in the steady-state ordered case, except for the absence of central complexes, and V is limited by release of Q from EQ.

Initial velocity patterns in the absence of products alone are not sufficient to distinguish between the steady-state ordered and Theorell–Chance mechanisms, and other information, including a crossover point analysis of the initial velocity patterns (see below) or inhibition studies, for example, is needed. Note that the difference in enzyme forms between the steady-state and Theorell–Chance mechanisms is in the AB term, which comprises several enzyme forms for the former but represents just the EQ form for the latter.

The Theorell–Chance mechanism is a limiting case of the ordered mechanism where no central complexes, EAB and EPQ, exist. In reality, the concentration of the central complexes is very low, and their interconversion is very fast with respect to release of the last reactant from the enzyme; at saturating concentrations of A and B, the enzyme is almost all in the EQ form. With the concentration of B maintained below its K_m, the initial rate increases as the concentration of A increases as a result of an increase in the concentration of EA to which B binds. At low concentrations of A, the rate increases with increasing concentrations of B, which traps more of the small amount of EA that is available.

The kinetic mechanism of liver alcohol dehydrogenase with ethanol as a substrate approximates a Theorell–Chance mechanism. However, the product inhibition (see Chapter 6) by acetaldehyde versus ethanol is noncompetitive, indicating that central complexes exist.

Random Sequential Reactions. Just as there is a continuum for ordered mechanisms, there are also extremes for a random mechanism, with *rapid equilibrium random* and *steady-state random* being the common ones. Both mechanisms can be depicted schematically as shown in equation 5-26. The differences between the two again are dependent on the location of the slow step(s). In a rapid equilibrium random mechanism the formation of EA and EB complexes is at equilibrium (off rate constants for A and B are both much greater than k_{cat}). But the off rate constants of A and B from the EAB complex do not have to be greater than k_{cat} in order to observe the simple rate equation that corresponds to the rapid equilibrium mechanism. This is because the last reactant to add to the enzyme to form a central complex can add in steady-state or rapid equilibrium fashion without a change in the form of the rate equation, while it does matter for any reactant that cannot add last.

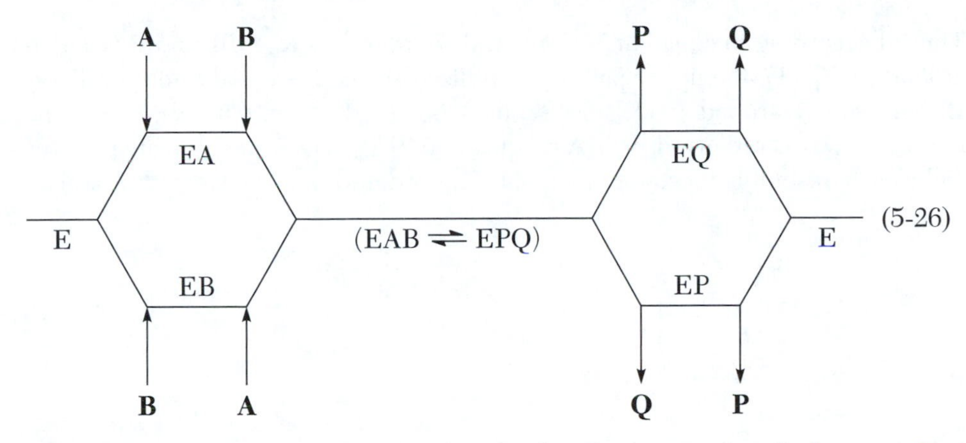

The rapid equilibrium random mechanism has been considered above under Uni-Bi reactions. The enzyme forms that are represented by the terms in the denominator of the initial velocity rate equation are given below:

Enzyme Form	Denominator Term
E	$K_{ia}K_b$
EB	$K_a\mathbf{B}$
EA	$K_b\mathbf{A}$
EAB, EPQ, EP, EQ	$\mathbf{AB}$

A number of examples of this kinetic mechanism can be found in the literature. Included among the reported examples are a phosphofructokinase from *Propionibacterium freudenreichii* (B *23*, 4101) that utilizes $MgPP_i$ instead of MgATP and 6-phosphogluconate dehydrogenases from *Candida utilis* (B *32*, 2036) and sheep liver (ABB *336*, 215). In all three cases, the mechanism was confirmed by product and dead-end inhibition studies (see Chapter 6).

Caveat. The primary and secondary plots obtained for a rapid equilibrium random mechanism can under certain conditions look like those discussed above as being diagnostic for a rapid equilibrium ordered kinetic mechanism. This occurs when only one of the two pathways in the random mechanism is observed. If $K_{ia} > K_a$, A binds much tighter when B is bound, termed synergistic binding. If marked synergistic binding is observed, and depending on the concentration range used for the varied reactant, a rapid equilibrium random mechanism can degenerate into a rapid equilibrium ordered mechanism.

Thus, if according to equation 5-27 **A** is varied around its K_{ia} (10), and **B** is varied around its K_b (1), the upper pathway is utilized since **B** is equal to only $0.01K_{ib}$. If **A** is varied around its K_a (0.1) and **B** is varied around its K_{ib} (100), the lower pathway is preferred since **A** is equal to $0.01K_{ia}$. The initial velocity patterns will closely resemble those for an equilibrium ordered mechanism with A adding first.

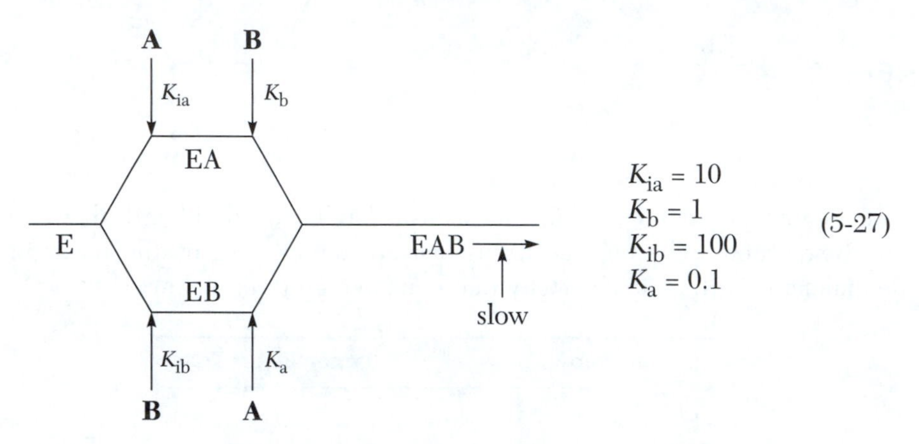

(5-27)

The last sequential bireactant mechanism to be discussed is the steady-state random mechanism. It differs from the rapid equilibrium random case in that the off rate constants of A from EA and B from EB do not greatly exceed k_{cat}. The rates of the two pathways from E to EAB often differ significantly in rate. The general form of the steady-state random rate equation is given in equation 5-28:

$$v = \frac{a\mathbf{AB} + b\mathbf{A}^2\mathbf{B} + c\mathbf{AB}^2}{d + e\mathbf{A} + f\mathbf{B} + g\mathbf{A}^2 + h\mathbf{B}^2 + j\mathbf{AB} + k\mathbf{A}^2\mathbf{B} + m\mathbf{AB}^2} \tag{5-28}$$

The coefficients a, b, c, etc., are all ratios of rate constants. Rearranging the above gives equation 5-29:

$$v = \frac{(a\mathbf{B} + c\mathbf{B}^2)\mathbf{A} + b\mathbf{BA}^2}{d + f\mathbf{B} + h\mathbf{B}^2 + (e + j\mathbf{B} + m\mathbf{B}^2)\mathbf{A} + (g + k\mathbf{B})\mathbf{A}^2} \tag{5-29}$$

At constant **B**

$$v = \frac{V\mathbf{A}^2 + p\mathbf{A}}{\mathbf{A}^2 + q\mathbf{A} + r} \tag{5-30}$$

where V, p, q, and r are combinations of the terms shown in the equation above. The reciprocal form of the equation is a 2/1 function as shown in equation 5-31:

$$\frac{1}{v} = \frac{1 + \dfrac{q}{\mathbf{A}} + \dfrac{r}{\mathbf{A}^2}}{V\left(1 + \dfrac{p}{\mathbf{A}}\right)} \tag{5-31}$$

Thus, at high concentrations of A, curvature either upward, suggestive of *partial substrate inhibition*, or downward, suggestive of *substrate activation*, may be observed, as in Figure 5-6. The substrate inhibition is partial, as indicated by intersection at the ordinate of the concave upward curve. (Complete substrate inhibition is indicated by a concave upward curve that is asymptotic to the ordinate, as will be seen in Chapter 6.) If partial substrate inhibition is observed, the pathway in which A adds to enzyme prior to B is slower than that in which B adds first, while if substrate activation is observed, the opposite is true. In most cases in the literature where a steady-state random mechanism is indicated, the reactant concentrations were not increased to a level high enough to observe the above behavior. As a result, the data most closely approximate a rapid equilibrium random mechanism.

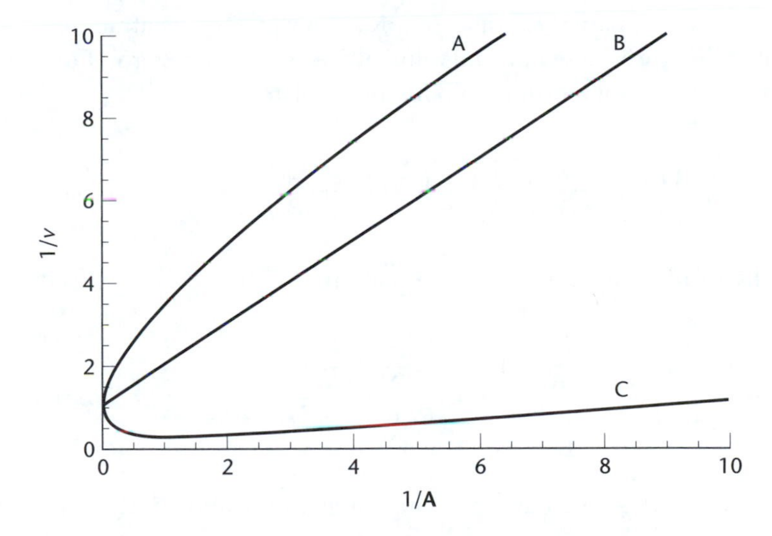

Figure 5-6. Partial substrate inhibition or substrate activation in a steady-state random mechanism: A plot of equation 5-31 with (A) $p = r = 1$, $q = 5$; (B) $p = q = 11$, $r = 10$; (C) $q = r = 1$, $p = 10$. $V = 1$ in all cases.

If the numerator and denominator of equation 5-29 are divided by $\mathbf{B}^2$, equation 5-32 is obtained, and as $\mathbf{B}$ approaches ∞ it reduces to equation 5-33, which has the same form as the Michaelis–Menten equation:

$$v = \frac{\left(\frac{a}{\mathbf{B}} + c\right)\mathbf{A} + \frac{\mathbf{A}^2 b}{\mathbf{B}}}{\left(\frac{d}{\mathbf{B}^2} + \frac{f}{\mathbf{B}} + h\right) + \left(\frac{e}{\mathbf{B}^2} + \frac{j}{\mathbf{B}} + m\right)\mathbf{A} + \left(\frac{g}{\mathbf{B}^2} + \frac{k}{\mathbf{B}}\right)\mathbf{A}^2} \tag{5-32}$$

$$v = \frac{c\mathbf{A}}{h + m\mathbf{A}} = \frac{\left(\frac{c}{m}\right)\mathbf{A}}{\left(\frac{h}{m}\right) + \mathbf{A}} \tag{5-33}$$

Equation 5-32 reduces to the Michaelis–Menten form since, at saturating $\mathbf{B}$, all of the enzyme is funneled through the pathway E → EB → EBA → EPQ → products. A similar analysis shows that if $\mathbf{A}$ is saturating, the rate as a function of $\mathbf{B}$ also becomes a hyperbola similar to that described by equation 5-33.

Dependence of K_m on Reactant Concentration

At any concentration of B, the apparent K_m value (appK_m) for varied concentrations of A is the reciprocal of the x-axis intercept (the value of $1/\mathbf{A}$ when $1/v$ is zero). For a steady-state ordered or rapid equilibrium random mechanism:

$$\frac{1}{v} = \left(\frac{K_a}{V}\right)\left(1 + \frac{K_{ia}K_b}{K_a\mathbf{B}}\right)\left(\frac{1}{\mathbf{A}}\right) + \left(\frac{1}{V}\right)\left(1 + \frac{K_b}{\mathbf{B}}\right) \tag{5-34}$$

When $1/v = 0$, the abscissa intercept is given by equation 5-35:

$$\frac{1}{\mathbf{A}} = \frac{1}{\text{app}K_a} = \frac{\frac{K_b}{\mathbf{B}} + 1}{\frac{K_{ia}K_b}{\mathbf{B}} + K_a} = \frac{K_b + \mathbf{B}}{K_{ia}K_b + K_a\mathbf{B}} \tag{5-35}$$

Taking the reciprocal of equation 5-35 and rearranging gives equation 5-36:

$$\text{app}K_a = K_a\left[\frac{1 + (K_{ia}K_b/K_a\mathbf{B})}{1 + (K_b/\mathbf{B})}\right] = K_{ia}\left(\frac{1 + (K_a\mathbf{B}/K_{ia}K_b)}{1 + (\mathbf{B}/K_b)}\right) \tag{5-36}$$

Equation 5-36 defines a rectangular hyperbola that does not pass through the origin. Graphically this is shown in Figure 5-7, for the limiting conditions where K_{ia} is equal to, less than, or greater than K_a. When **B** is near zero, the appK_a is equal to K_{ia} (effectively the dissociation constant for EA) according to equation 5-36, while extrapolation of **B** to infinite concentration gives K_a. If one is attempting to determine K_m at a fixed concentration of the second reactant, it is thus important to be certain that the second reactant is at a sufficiently high concentration that the true K_m is determined. Conversely, if one wishes to determine K_{ia}, the level of **B** should be kept well below its K_m value.

Ping Pong Reactions

The second type of bireactant mechanism is termed Ping Pong or double-displacement as discussed in Chapter 1. In a ping pong reaction, the first substrate binds to enzyme, transfers a piece of itself to the enzyme, and dissociates as the first product before the second reactant binds, picks up the transferred piece, and dissociates as the second product. There are two kinds of Ping Pong mechanisms, and each will be considered in turn. The first is called a *one-site* or *classical Ping Pong* mechanism, while the second is a *two-site* or *nonclassical Ping Pong* mechanism.

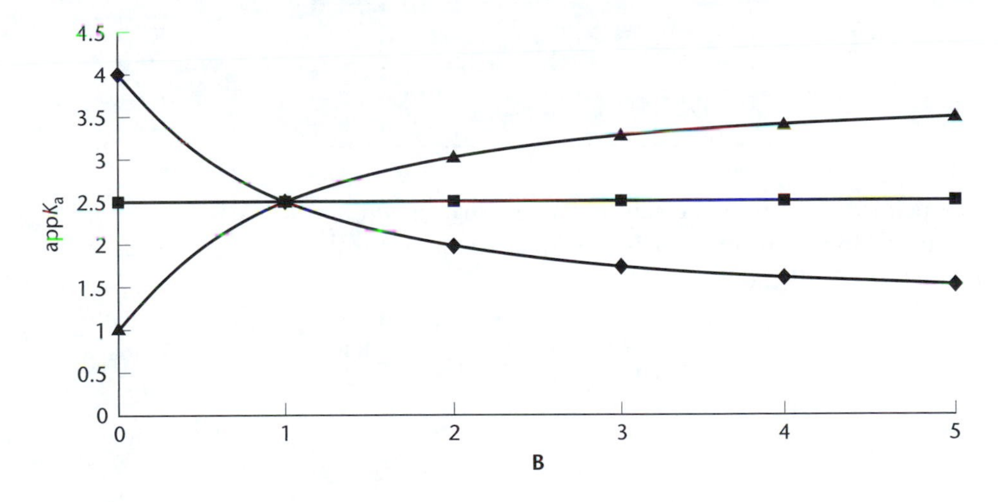

Figure 5-7. Dependence of appK_a on the concentration of B. Three limiting conditions are depicted: $K_{ia} < K_a$ (▲), $K_{ia} = K_a$ (■), and $K_{ia} > K_a$ (◆). Values used to calculate the above three curves are $K_{ia} = 1$ mM and $K_a = 4$ mM; $K_{ia} = K_a = 2.5$ mM; and $K_{ia} = 4$ mM and $K_a = 1$ mM, respectively. In all cases $K_b = 1$ mM.

An enzyme with a classical Ping Pong mechanism has a single site in which each reactant binds and thus can only bind one at a time. The classical mechanism is depicted schematically in equation 5-37:

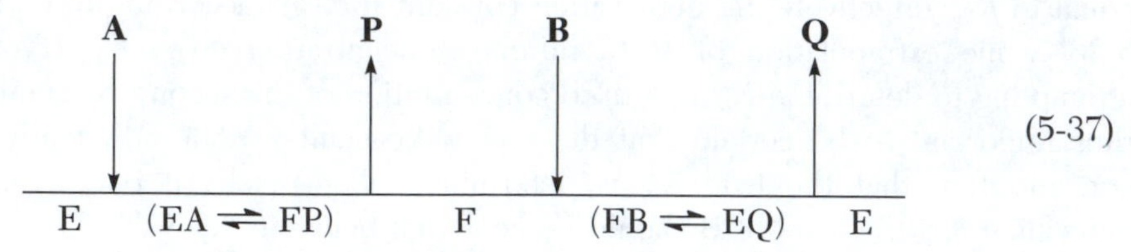

(5-37)

The rate equation for a Ping Pong kinetic mechanism differs from that of a sequential mechanism in the absence of the constant ($K_{ia}K_b$) term. The rate equation, as shown in equation 5-38, is still symmetric.

$$v = \frac{V\mathbf{AB}}{K_a\mathbf{B} + K_b\mathbf{A} + \mathbf{AB}} \tag{5-38}$$

There are two stable enzyme forms, E and F. The terms in the denominator of the rate equation are as follows:

Enzyme Form	Denominator Term
E	$K_a\mathbf{B}$
F	$K_b\mathbf{A}$
EA, FP, FB, EQ	**AB**

The reciprocal form of the rate equation, rearranged so that A is the variable reactant, is shown in equation 5-39, with expressions for slope and intercept given in equations 5-40 and 5-41:

$$\frac{1}{v} = \left(\frac{K_a}{V}\right)\left(\frac{1}{\mathbf{A}}\right) + \left(\frac{1}{V}\right)\left(1 + \frac{K_b}{\mathbf{B}}\right) \tag{5-39}$$

$$\text{slope} = \frac{K_a}{V} \tag{5-40}$$

$$\text{intercept} = \left(\frac{1}{V}\right)\left(1 + \frac{K_b}{\mathbf{B}}\right) \tag{5-41}$$

Primary and secondary plots are shown in Figure 5-8. A series of parallel lines is observed in the primary double reciprocal plot, so that the slope replot, indicated by equation 5-40, will not be a function of **B** and gives the reciprocal of the *V/K* for A. The predominant enzyme form under conditions where **A** is limiting is thus E at any concentration of B, and the *V/K* reflects only the first half-reaction. As **A** is increased toward saturation, however, the rate does increase with the concentration of B, as shown by equation 5-41 and the intercept replot in Figure 5-8. At low concentrations of B, the enzyme is predominantly in the other stable form, F, and the *V/K* for B reflects only the second half-reaction. Saturation with **A** and **B** gives the central complexes for both half-reactions as the only enzyme forms. This is not to say that enzyme will be equally distributed in both central complexes at saturating reactants. Depending on which of the two half-reactions is slower, the enzyme may be predominantly in one or the other central complex form or, if both half-reactions contribute to rate limitation, in both.

The rate equation is symmetric as shown by equations 5-42 to 5-44. A plot of $1/v$ against $1/\mathbf{B}$ will also give a set of parallel lines with the slope equal to the reciprocal of the *V/K* for B and reflecting the F form of the enzyme. The intercept replot now gives the *V/K* value for A, and E is the predominant enzyme form. The explanation is identical to that provided above.

$$\frac{1}{v} = \left(\frac{K_b}{V}\right)\left(\frac{1}{\mathbf{B}}\right) + \left(\frac{1}{V}\right)\left(1 + \frac{K_a}{\mathbf{A}}\right) \tag{5-42}$$

$$\text{slope} = \frac{K_b}{V} \tag{5-43}$$

$$\text{intercept} = \left(\frac{1}{V}\right)\left(1 + \frac{K_a}{\mathbf{A}}\right) \tag{5-44}$$

Examples. Aspartate aminotransferase catalyzes reversible transfer of the α-amino group of L-aspartate to α-ketoglutarate to give L-glutamate and oxaloacetate. The enzyme has a covalently bound pyridoxal 5′-phosphate, which serves as a cofactor to accept the amino group from aspartate as it is converted to oxaloacetate. The pyridoxamine 5′-phosphate thus formed transfers the amino group onto α-ketoglutarate (JBC *237*, 2109). Another enzyme that exhibits a Ping Pong kinetic mechanism is nucleoside-diphosphate kinase, which catalyzes the transfer of a phosphoryl group from an XTP to a YDP to give YTP and XDP, where X and Y can be any of the common nucleosides (B *8*, 633).

The nonclassical or two-site Ping Pong mechanism has an initial velocity rate equation that is identical to that of the classical Ping Pong mechanism, but now

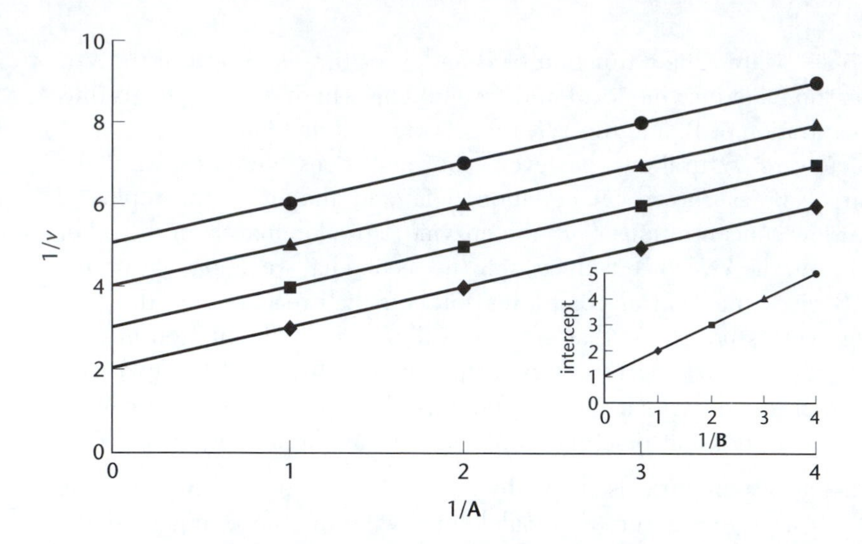

Figure 5-8. Primary and secondary plots for a Ping Pong mechanism with A as the varied reactant. Values of K_a, and K_b are 1 mM. The value of V is 1 mM/min. The primary plot is a diagnostic.

there are two sites, with A and P being adsorbed at one and B and Q at the other. The physical connection between the two sites is usually made by a carrier such as biotin or lipoic acid, but in redox enzymes, electrons may transfer between sites through the protein and/or cofactor(s). The differences between the two kinds of mechanism will become evident when product and dead-end inhibition are considered.

The two-site Ping Pong mechanism can be depicted schematically as shown in equation 5-45:

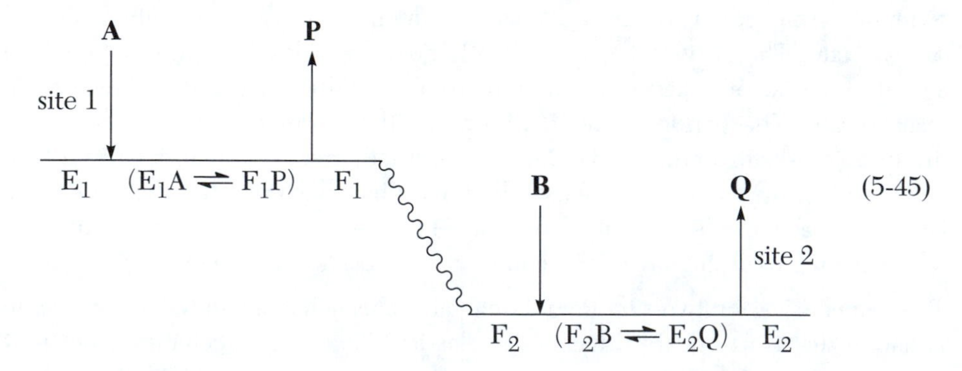

(5-45)

Although the rate equations are identical, the denominator terms represent different enzyme forms, as shown below.

Enzyme Form	Denominator Term
E_1, E_2	$K_a\mathbf{B}$
F_1, F_2	$K_b\mathbf{A}$
E_1A, F_1P, F_2B, E_2Q	**AB**

At limiting concentrations of A, all of the enzyme is in the E form at sites 1 and 2. At limiting **B**, the enzyme is predominantly in the F forms. The transferred piece of A can bind in both site 1 and site 2. In addition to the interconversion of central complexes, as listed for the classical Ping Pong mechanism, it is possible that transfer between sites may limit the overall reaction.

The first example of a multisite Ping Pong kinetic mechanism was that described by Northrop for transcarboxylase (JBC *244*, 5808; 5820). The epitome of known multisite Ping Pong mechanisms is that of eukaryotic fatty acid synthetase, which has a seven-site Ping Pong mechanism (JBC *250*, 2709). Biotin-containing carboxylases have two-site Ping Pong mechanisms. Glutathione and lipoate dehydrogenases have this type of mechanism, with the pyridine nucleotide reducing the flavin from one side and the flavin reducing the other substrate on the opposite face of the flavin.

Table 5-1 provides a summary of the Bi Bi mechanisms considered, with respect to requirements for specific rate processes, differences on rate equations, and reciprocal plots.

Determination of V_{max}

V_{max} can be estimated by measuring the rate at saturating substrate concentration, but one must be certain that saturation is reached. As an example, consider a measurement at $10K_m$ which is close to saturation (91%). If **B** is set at $10K_b$ and **A** is set equal to $10K_a$, and these values are substituted into the equation for a sequential mechanism $[v = V\mathbf{AB}/(K_{ia}K_b + K_a\mathbf{B} + K_b\mathbf{A} + \mathbf{AB})]$, v will be equal to $100V/(K_{ia}/K_a + 120)$. Under all conditions, that is, $K_a > K_{ia}$, $K_a = K_{ia}$, and $K_a < K_{ia}$, v will at best be equal to 83–85% of V_{max}.

In many cases, initial velocity data are collected in two reaction directions, but on different days, and making use of different enzyme stock solutions or preparations. Thus, one cannot easily compare the two reaction directions or

Table 5-1. Summary of Bireactant Kinetic Mechanisms

Mechanism	Conditions	Rate equation	Reciprocal plots
Steady-state random (SSR)	no restrictions on any rate constants	contains square terms in numerator and denominator	curvature near the y-axis; uncommonly observed
rapid equilibrium random (RER)	dissociation of binary complexes fast; central complex interconversion may be slow or fast	$v = \frac{V\mathbf{AB}}{K_{ia}K_b + K_b\mathbf{A} + K_a\mathbf{B} + \mathbf{AB}}$	both plots intersect to the left of the y-axis[a]
steady-state ordered (SSO)	compulsory order of addition of A and B	$v = \frac{V\mathbf{AB}}{K_{ia}K_b + K_b\mathbf{A} + K_a\mathbf{B} + \mathbf{AB}}$	same as RER[a]
equilibrium ordered (EO)	off rate constant for A much greater than k_{cat}	$v = \frac{V\mathbf{AB}}{K_{ia}K_b + K_b\mathbf{A} + \mathbf{AB}}$	A varied, slope replot passes through origin; B varied, intersection on Ordinate[b]
Theorell–Chance (TC)	second product release very slow; appears that no central complex exists	$v = \frac{V\mathbf{AB}}{K_{ia}K_b + K_b\mathbf{A} + K_a\mathbf{B} + \mathbf{AB}}$	same as RER and SSO[a]
Ping Pong (PP)	product released between addition of substrates	$v = \frac{V\mathbf{AB}}{K_b\mathbf{A} + K_a\mathbf{B} + \mathbf{AB}}$	parallel lines[c]

[a]If $K_a > K_{ia}$, the double reciprocal plot could appear parallel.
[b]If strong synergism of binding occurs between A and B, then RER degenerates to EO.
[c]RER, SSO, and TC could give this pattern if $K_a > K_{ia}$.

determine the K_{eq} using the Haldane relationship, which equates K_{eq} to a ratio of kinetic constants in both reaction directions (to be discussed later). If the kinetic constants are known, one may need only the ratio V_1/V_2 to obtain an estimate of K_{eq} via the Haldane relationship. Measuring V in both reaction directions would normally require that many assays be performed since an initial velocity pattern must be done in both directions. A shortcut to obtaining the ratio V_1/V_2 is to vary substrates in constant ratio as demonstrated below. Let x be equal to a constant ratio of **B**/**A**, and thus **B** is equal to x**A**.

In the case of a sequential mechanism, $1/v$ is given by equation 5-34. Substituting x**A** for **B** and collecting terms gives equation 5-46:

$$\frac{1}{v} = \left(\frac{K_{ia}K_b}{xV}\right)\left(\frac{1}{\mathbf{A}^2}\right) + \left(\frac{K_a}{V} + \frac{K_b}{xV}\right)\left(\frac{1}{\mathbf{A}}\right) + \frac{1}{V} \tag{5-46}$$

which has the form of a parabola, as shown in Figure 5-9. The data are fitted to equation 5-47 to determine V. To minimize the curvature and thus obtain better estimates of V, the ratio x should be chosen so that one substrate is at a higher ratio to its K_m than the other one.

$$v = \frac{V\mathbf{A}^2}{a + b\mathbf{A} + \mathbf{A}^2} \tag{5-47}$$

Experimentally, the above experiment would be carried out in both reaction directions, with the same enzyme concentration. If different enzyme levels have to be used because one V is much higher than the other, the values have to be corrected to the same enzyme level.

Similarly, for a Ping Pong mechanism, equation 5-48 is obtained, when both substrates are varied in constant ratio, depicting a straight line as shown in Figure 5-9.

$$\frac{1}{v} = \left(\frac{K_a}{V} + \frac{K_b}{xV}\right)\left(\frac{1}{\mathbf{A}}\right) + \frac{1}{V} \tag{5-48}$$

Thus, potentially the Ping Pong and sequential mechanisms, even with $\mathbf{A} \gg K_{ia}$, can be differentiated if the reactant concentrations are varied over a wide enough range.

Can Sequential and Ping Pong Mechanisms Resemble One Another?

A parallel initial velocity pattern is not necessarily indicative of a Ping Pong mechanism. In the case of an ordered or random sequential mechanism rate

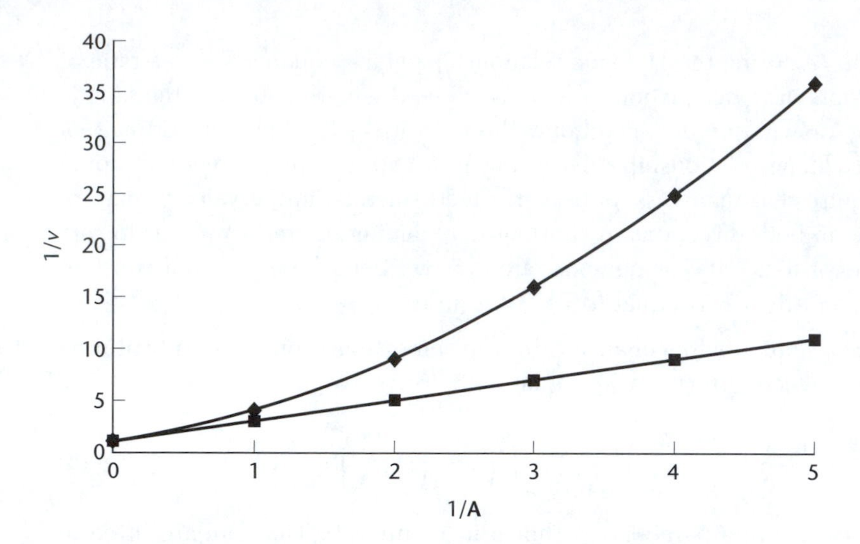

Figure 5-9. Plot of reciprocal of initial velocity against the reciprocal of the concentration of A varied in constant ratio with **B**: sequential mechanism (◆) or Ping Pong mechanism (■). Values of K_{ia}, K_a, and K_b are 1 mM. The value of V is 1 mM/min.

equation 5-34 applies. If $K_{ia} \ll K_a$ and the concentration of A is varied around K_a, the constant term $K_{ia}K_b/V\mathbf{AB}$ will approach zero, since $\mathbf{A} \gg K_{ia}$. As a result, the pattern will approximate a series of parallel lines and appear ping pong. The pattern is really intersecting, but the intersection point is so far below the abscissa that the convergence is not observable. This is the case only in the faster reaction direction. If the initial velocity pattern is determined in the reverse reaction direction, a highly intersecting initial velocity pattern will be observed. An example of this behavior is observed with acetate kinase (JBC *251*, 6775).

In the case of a Ping Pong kinetic mechanism, if the intermediate F is unstable and decomposes on the time scale of the steady-state assay, as depicted in mechanism 5-49, the initial velocity pattern in the absence of added inhibitors will intersect to the left of the ordinate. This will be true when the appearance of Q is monitored; the kinetics differ if formation of P is monitored.

$$\mathrm{E} \underset{k_2}{\overset{k_1\mathbf{A}}{\rightleftharpoons}} \mathrm{EA} \xrightarrow[\mathrm{P}]{k_3} \mathrm{F} \underset{k_8}{\overset{k_7\mathbf{B}}{\rightleftharpoons}} \mathrm{FB} \xrightarrow{k_9} \mathrm{E} + \mathrm{Q}; \quad \mathrm{F} \xrightarrow{k_5} \mathrm{E} \tag{5-49}$$

The rate equation for mechanism 5-49 is qualitatively identical to that of equation 5-6 with **A** and **B** replacing **P** and **Q**. The mechanism in terms of rate constants is given in equation 5-50:

$$v = \frac{d\mathbf{Q}}{dt} = \frac{k_1k_3k_7k_9\mathbf{ABE_t}}{\left\{\begin{array}{c} k_5(k_8+k_9)(k_2+k_3)+k_1\mathbf{A}(k_8+k_9)(k_3+k_5) \\ +k_7k_9\mathbf{B}(k_2+k_3)+k_1k_7\mathbf{AB}(k_3+k_9) \end{array}\right\}} \quad (5\text{-}50)$$

Note that, unlike the rate equation for a Ping Pong mechanism in which F is stable, there is a constant term in the denominator of the rate equation. The expressions for V, V/K_a, and V/K_b are given in equations 5-51 to 5-53, with the asterisk indicating parameters derived for mechanism 5-49, while the other

$$\left(\frac{V}{\mathbf{E_t}}\right)* = \frac{k_3k_9}{k_3+k_9} \qquad \left(\frac{V}{\mathbf{E_t}}\right) = \frac{k_3k_9}{k_3+k_9} \quad (5\text{-}51)$$

$$\left(\frac{V}{K_a\mathbf{E_t}}\right)* = \frac{k_1k_3}{k_2+k_3} \qquad \left(\frac{V}{K_a\mathbf{E_t}}\right) = \frac{k_1k_3}{k_2+k_3} \quad (5\text{-}52)$$

$$\left(\frac{V}{K_b\mathbf{E_t}}\right)* = \frac{k_3k_7k_9}{(k_8+k_9)(k_3+k_5)} \qquad \left(\frac{V}{K_b\mathbf{E_t}}\right) = \frac{k_7k_9}{k_8+k_9} \quad (5\text{-}53)$$

expressions are derived for a Ping Pong mechanism in which F is stable over the steady-state time course. There is no difference in the expressions for V and V/K_a as a result of the instability of F. The rate at limiting **A** but saturating **B**, (V/K_a), reflects only the first half-reaction, as stated above when the normal ping pong mechanism was considered. Once the first product is released, the infinite level of B converts F to FB as fast as it is formed. At limiting **B**, however, the decomposition of F becomes significant, and thus the $V/K_b\mathbf{E_t}$ value for mechanism 5-49 is lower than that for a mechanism where F is stable. An example of this behavior is seen with *O*-acetylserine sulfhydrylase reconstituted with PLP analogues (ABB *324*, 71).

The rate equation for the appearance of P in mechanism 5-49 differs from that derived for appearance of Q in that the numerator is $k_1k_3[k_5(k_8+k_9)+k_7k_9\mathbf{B}]\mathbf{AE_t}$. The double reciprocal plot with **A** varied is linear, while that with varied **B** is a hyperbola with a horizontal asymptote. The asymptote lies above the vertical intercept if $k_5 > k_9$ and below it if $k_5 < k_9$. The rate at low **B** is dominated by the rate constants k_3 and k_5, while at high **B** the rate is dominated by the rate constants k_3 and k_9. A finite rate is obtained at zero **B** as a result of decomposition of F via k_5.

Crossover Point Analysis

A sequential mechanism with two substrates and products normally has a rate equation in the forward direction given by equation 5-54:

$$v = \frac{V\mathbf{AB}}{K_{ia}K_b + K_a\mathbf{B} + K_b\mathbf{A} + \mathbf{AB}} \tag{5-54}$$

When $1/v$ is plotted versus $1/\mathbf{A}$ at different levels of **B**, the result is a pattern of lines intersecting with a vertical coordinate given by equation 5-55:

$$\frac{1}{v_{\text{Xover}}} = \left(\frac{1}{V_1}\right)\left(1 - \frac{K_a}{K_{ia}}\right) \tag{5-55}$$

The same vertical coordinate is seen if $1/v$ is plotted versus $1/\mathbf{B}$ at different levels of **A**. In this equation, K_a is the Michaelis constant for A and K_{ia} is the dissociation constant of A from EA for the first substrate to add in an ordered mechanism or either substrate in a rapid equilibrium random mechanism.

In the back reaction, the vertical coordinate of the crossover point is given by equation 5-56:

$$\frac{1}{v_{\text{Xover}}} = \left(\frac{1}{V_2}\right)\left(1 - \frac{K_q}{K_{iq}}\right) \tag{5-56}$$

where K_q is the Michaelis constant and K_{iq} is the dissociation constant of Q from EQ for the last product released in an ordered mechanism or either substrate in a random one.

If V_1 and V_2 are measured at equal enzyme levels or the rates are corrected to equal enzyme levels, the dimensionless parameter R is given by equation 5-57:

$$R = \frac{\sum(\text{vertical coordinates of crossover points})}{\left(\frac{1}{V_1} + \frac{1}{V_2}\right)} \tag{5-57}$$

R tells where the rate-limiting step is in the direction with the slower maximum velocity. In an ordered mechanism, R varies from 0 for a Theorell–Chance mechanism, where second product release is rate-limiting, to 1.0 for an equilibrium ordered mechanism, where the part of the reaction that includes catalysis and release of the first product is rate-limiting. For a Theorell–Chance mechanism, in fact, the crossover point is $1/V_1 - 1/V_2$ in the forward direction and $1/V_2 - 1/V_1$ in the reverse direction. The crossover point is thus well below the horizontal axis in the direction with the faster maximum velocity and above it in the slower direction. Failure to match this prediction can be used to distinguish a random mechanism with two dead-end complexes from a Theorell–Chance mechanism (they give identical product inhibition patterns).

In the general case of an ordered mechanism, R is the proportion of enzyme in central complexes in the slower direction when substrates are saturating, and $R/(1 - R)$ is the ratio of central complex concentration to that of the EQ complex. R is also the ratio of the rate constant for second product release and the rate constant for catalysis and first product release. With glycerokinase, R was 5/6, and thus in the slow reaction of glycerol-P with MgADP, the release of glycerol was 5 times faster than catalysis and release of MgATP (JBC *249*, 2562).

The justification for our statements about the interpretation of R can be explained by using the following derivation based on the rate equation for an ordered mechanism, equation 5-58:

$$\mathrm{E} \underset{k_2}{\overset{k_1\mathbf{A}}{\rightleftharpoons}} \mathrm{EA} \underset{k_4}{\overset{k_3\mathbf{B}}{\rightleftharpoons}} \mathrm{EAB} \xrightarrow{k_5} \mathrm{EQ} \xrightarrow{k_7} \mathrm{E} \tag{5-58}$$

The expression for R is

$$R = \frac{\dfrac{1}{V_1} - \dfrac{K_a}{K_{ia}V_1} + \dfrac{1}{V_2} - \dfrac{K_q}{K_{iq}V_2}}{\dfrac{1}{V_1} + \dfrac{1}{V_2}} = \frac{\dfrac{1}{V_1} - \dfrac{1}{k_2} + \dfrac{1}{V_2} - \dfrac{1}{k_7}}{\dfrac{1}{V_1} + \dfrac{1}{V_2}} \tag{5-59}$$

This follows from the definitions of k_2 and k_7 (see Appendix II). Then, since $1/V_1 = (1/k_5 + 1/k_7)$ and $1/V_2 = (1/k_2 + 1/k_4)$

$$R = \frac{\dfrac{1}{k_4} + \dfrac{1}{k_5}}{\dfrac{1}{k_2} + \dfrac{1}{k_4} + \dfrac{1}{k_5} + \dfrac{1}{k_7}} \tag{5-60}$$

The equation shows that if $k_4, k_5 \gg k_2, k_7$, R will be zero, while if the reverse is true, R will be 1.0. Further, the smallest rate constant (corresponding to the reaction in the slowest direction) will have the greatest effect on the expression.

Terreactant Enzyme Mechanisms

Thus far, the chapter has been devoted to uni- and bireactant enzyme mechanisms. A systematic study of a terreactant enzyme-catalyzed reaction requires the collection of more data, but more information is also gained. One method of obtaining the initial velocity data for a terreactant mechanism is to break the reaction down into bireactant pieces. Thus, initial velocity data are collected at varied concentrations of A and B but a fixed concentration of C; at varied concentrations of A and C but a fixed concentration of B; and at varied concentrations of B and C but a fixed level of A. Data are then interpreted similarly to those obtained for a bireactant mechanism.

In mechanisms with random segments, the full steady-state rate equation will contain squared or cubed terms, but in practice, most reciprocal plots will look linear and we will discuss terreactant mechanisms in terms of equation 5-61. Most terreactant mechanisms will give rate equations lacking one or more of the denominator terms in equation 5-61. We will discuss the most common of these, and the reader should consult Viola and Cleland (ME *87*, 353) where all possible terreactant initial velocity patterns and mechanisms are discussed.

$$v = \frac{V\mathbf{ABC}}{\left\{\begin{matrix} \text{const} + (\text{coef A})\mathbf{A} + (\text{coef B})\mathbf{B} + (\text{coef C})\mathbf{C} + K_a\mathbf{BC} \\ + K_b\mathbf{AC} + K_c\mathbf{AB} + \mathbf{ABC} \end{matrix}\right\}} \tag{5-61}$$

No Constant Term—Ping Pong Mechanisms

This indicates a Ping Pong mechanism. The Hexa-Uni Ping Pong mechanism lacks denominator A, B, and C terms, and thus all initial velocity patterns are parallel.

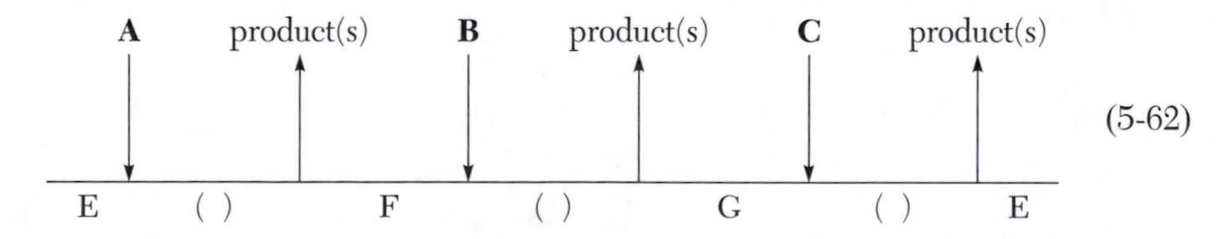

(5-62)

The pyruvate dehydrogenase complex shows these patterns with pyruvate, CoA, and NAD as the three substrates (JBC *248*, 8348).

A number of synthetases and carboxylases have Ping Pong mechanisms with two half-reactions:

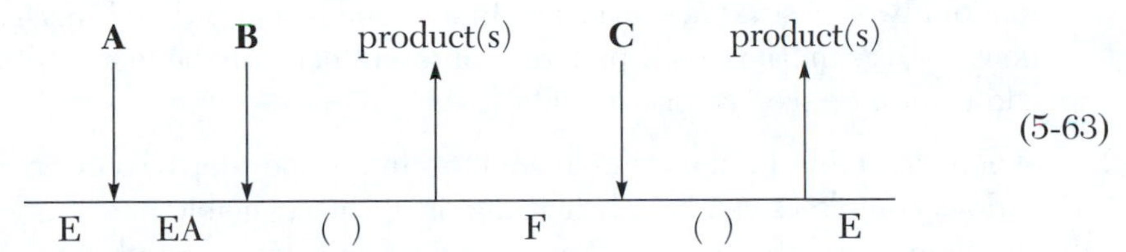

(5-63)

In these mechanisms the constant, **A**, and **B** terms are missing from the denominator of the rate equation but the **C** term is present. If A and B add in steady-state or random fashion, the **A** and **B** terms are the only missing ones, but if addition of A and B is ordered and at rapid equilibrium, the **BC** term is also missing. The A-C and B-C initial velocity patterns are parallel, while the A-B one is intersecting, or equilibrium ordered if the **BC** term is missing. Examples of enzymes with this mechanism are pyrophosphorylytic synthetases and biotin-containing carboxylases.

If a bireactant Ping Pong mechanism has a dissociable activator and the concentrations of all three reactants are varied, one has a pseudo-terreactant mechanism. If the activator C can dissociate from both E and F, one sees a parallel A-B initial velocity pattern but intersecting A-C and B-C patterns. In this case the constant and **C** terms are missing in the denominator of equation 5-61. Tyrosine aminotransferase shows these patterns (B *7*, 2072).

Constant Term Present—Sequential Mechanisms

These mechanisms are sequential and proceed through an EABC complex. The **ABC** term is present except in a Theorell–Chance-like mechanism where the third substrate adds and reacts so rapidly that the maximum velocity and K_ms for B and C appear infinite [i.e., too large to measure; reverse reaction of malic enzyme at high pH (B *16*, 571)]. We will thus consider the meaning of the presence or absence of the **A**, **B**, **C**, **AB**, **BC**, and **AC** terms in the denominator of equation 5-61.

There are several rules that apply to bireactant, terreactant, or higher-order initial velocity patterns (ME *87*, 353):

(1) It matters whether the first reactant(s) to add (which cannot add last) does so in steady-state fashion (k_{off} less than, or does not greatly exceed, $V/\mathbf{E_t}$) or in rapid equilibrium ($k_{off} \gg V/\mathbf{E_t}$), but it does not matter whether the last substrate to add to the enzyme does so in steady-state or rapid equilibrium

fashion. Since both substrates can add last, this explains why the initial velocity rate equation looks the same for rapid equilibrium and steady-state random two-substrate mechanisms. In an equilibrium ordered mechanism, however, the initial velocity pattern is different because the first substrate to add cannot be the last one to add.

(2) When reactants bind in rapid equilibrium, some steps connecting the various complexes can be missing (that is, go at negligible rates) as long as all complexes present can be interconverted by a rapid equilibrium pathway.

(3) Where a substrate binds in a steady-state fashion, it always has a finite Michaelis constant (that is, for A the $K_a\mathbf{BC}$ term is present). Conversely, the lack of a $K_a\mathbf{BC}$, $K_b\mathbf{AC}$, or $K_c\mathbf{AB}$ term indicates rapid equilibrium addition of substrate A, B, or C and the fact that this substrate cannot add last. Further, the lack of a $K_a\mathbf{BC}$ term shows that there is no EBC complex in the mechanism. When a Michaelis constant is finite, however, binding of a substrate may be either steady-state or rapid equilibrium, and the ternary complex may be absent if the addition of this substrate is in steady state.

(4) The lack of an **A**, **B**, or **C** term indicates the absence of the corresponding binary complex and thus an obligatory order of addition for at least one substrate (that is, it cannot add until one of the others does and then prevents the other one from leaving the enzyme). The presence of an **A**, **B**, or **C** term does not require that the corresponding binary complex form, however, if the mechanism is a steady-state one.

(5) The last substrate to add to the enzyme can do so in ordered or random fashion without a change in the rate equation. Thus the initial velocity pattern is the same for a steady-state ordered and a random bireactant mechanism. In fact, in a rapid equilibrium random mechanism the substrates can be quite sticky and dissociate slowly from the ternary complex as long as they dissociate rapidly from their binary complexes.

With these rules in mind, we will summarize the possible terreactant initial velocity patterns.

(1) All denominator terms present. One possibility is a completely random mechanism, equation 5-64, where each of the eight terms in the

denominator of the rate equation represents an enzyme form (thus the constant term represents E, the **A** term represents EA, and so forth). A second possible mechanism involves random addition of A and B in steady-state fashion followed by ordered addition of C (rule 5).

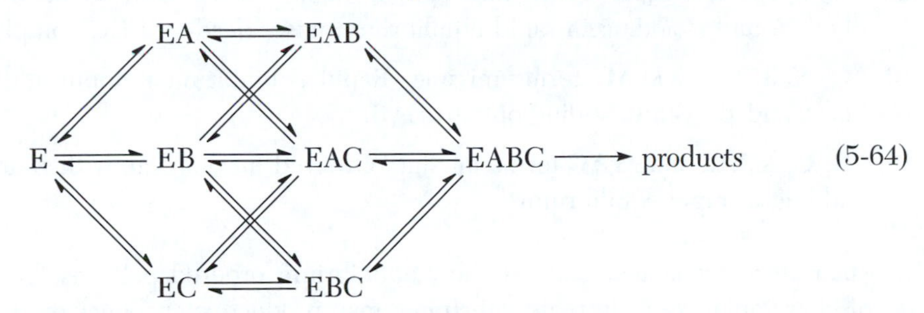

(5-64)

(2) **B** term missing. The two possible mechanisms are a fully ordered steady-state mechanism [glutamate dehydrogenase, with NADH, α-ketoglutarate, and ammonia adding in that order (B *19*, 2321)] or one in which substrates A and B add in obligate order but C can add randomly (again, rule 5). β-Hydroxy-β-methylglutaryl-CoA reductase shows this mechanism, with CoA adding before mevaldate and NADP binding randomly (B *15*, 4191).

(3) **B** and **C** terms missing. This mechanism involves addition of A, followed by random addition of B and C. Citrate cleavage enzyme shows this pattern, with MgATP adding first and CoA and citrate as B and C (JBC *242*, 4239).

(4) K_a**BC**, or K_a**BC** and K_b**AC** terms missing. These are unlikely mechanisms with rapid equilibrium addition of A, or A and B, with these substrates unable to be the third substrate to add and C always forming a binary complex (see ME *87*, 353).

(5) **B** and K_a**BC** terms missing. Rapid equilibrium addition of A prior to B in an ordered mechanism or in a mechanism where C adds randomly. With A as a metal activator that has to add first (Mn^{2+}), galactosyltransferase has an ordered mechanism with B and C being UDP-galactose and *N*-acetylglucosamine (JBC *246*, 3977).

(6) **B** and K_b**AC** terms missing. An unlikely mechanism involving rapid equilibrium addition of B after A and an EC, but not an EAC complex.

(7) **B**, **C**, and K_a**BC** terms missing. Rapid equilibrium addition of A, followed by random addition of B and C.

(8) **B**, **C**, and K_b**AC** terms missing. An ordered mechanism with A adding in steady state but B in rapid equilibrium. Glutamate dehydrogenase with α-ketovalerate or α-ketobutyrate as substrate shows this pattern.

(9) **B**, K_a**BC**, and K_b**AC** terms missing. An unlikely ordered mechanism with both A and B adding in rapid equilibrium plus a dead-end EC complex.

(10) **C**, K_a**BC**, and K_b**AC** terms missing. Rapid equilibrium random addition of A and B, with C adding only to EAB.

(11) **B**, **C**, K_a**BC**, and K_b**AC** terms missing. Ordered mechanism, with A and B adding in rapid equilibrium.

Not many terreactant mechanisms show equilibrium ordered patterns, but the use of slow mutants or alternate substrates may produce such behavior. Rapid equilibrium mechanisms may also be produced by altering the pH to eliminate stickiness.

Methods for Telling Which Denominator Terms Are Missing

If a **B** term is missing, the A-C pattern becomes parallel when B is saturating, and likewise saturation with C causes the A-B pattern to become parallel if a **C** term is not present. Equilibrium ordered patterns show the absence of **AB**, **BC**, or **AC** terms. A systematic graphical approach for determining the form of the rate equation requires that all reactant concentrations be varied.

Thus, as for a bireactant mechanism, the initial rate is obtained at varied concentrations of A and different fixed concentrations of B and a single fixed concentrations of C, as shown in Figure 5-10. The initial velocity pattern is then repeated at different fixed concentrations of C. Thus several initial velocity patterns are generated as in Figure 5-10. Taking the reciprocal of equation 5-61 for a terreactant mechanism and arranging it so that **A** is varied gives the following:

$$\frac{1}{v} = \left(\frac{\text{const}}{\mathbf{BC}} + \frac{\text{coef B}}{\mathbf{C}} + \frac{\text{coef C}}{\mathbf{B}} + K_a\right)\left(\frac{1}{V\mathbf{A}}\right) + \left(1 + \frac{\text{coef A}}{\mathbf{BC}} + \frac{K_b}{\mathbf{B}} + \frac{K_c}{\mathbf{C}}\right)\left(\frac{1}{V}\right) \tag{5-65}$$

Linear reciprocal plots are obtained, and both slopes and intercepts are a function of **B** and **C**. To determine which terms in the denominator of the rate equation are present, each of the terms must be isolated. Expressions for the slope and intercept of equation 5-65, with 1/**B** factored out, are given in equations 5-66

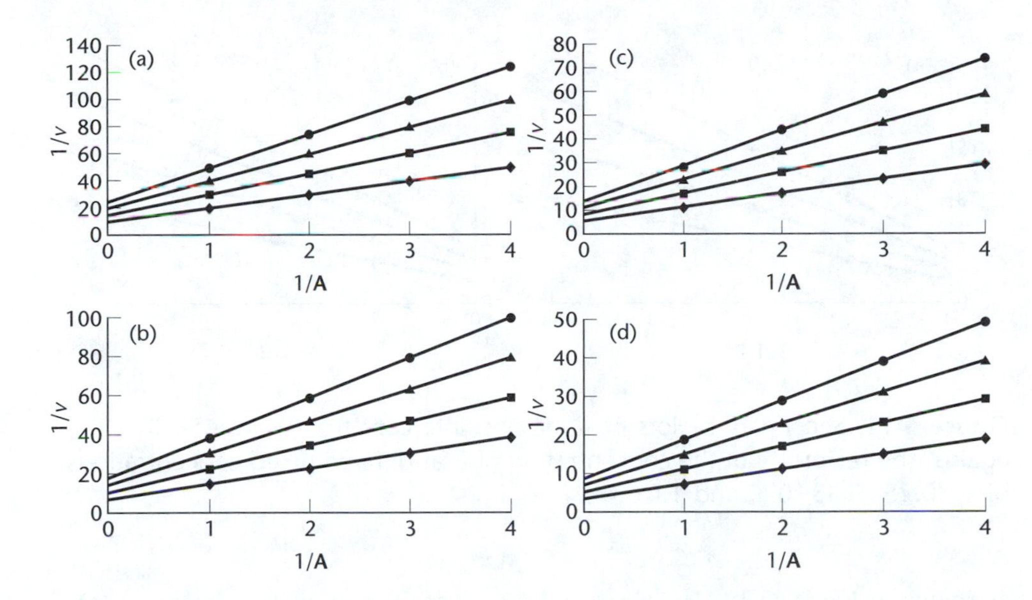

Figure 5-10. Double reciprocal plots obtained for a terreactant enzyme reaction. The rate is measured at the concentrations of A indicated, at several concentrations (0.25 mM, •; 0.33 mM, ▲; 0.5 mM, ■; and 1 mM, ◆) of B, and a single fixed concentration of C (0.25 mM, panel A). The pattern is then repeated at three different additional concentrations of C (0.33 mM, panel B; 0.5 mM, panel C; and 1 mM, panel D). To simplify the calculations, all kinetic constants were fixed at a value of 1 and *V* was fixed at 1 mM/min. Note the change in the values on the ordinate.

and 5-67. Each of the replots gives a family of straight lines like an initial velocity pattern, as shown in Figure 5-11.

$$\text{slope} = \left(\frac{\text{const}}{\mathbf{C}} + \text{coef C}\right)\left(\frac{1}{V\mathbf{B}}\right) + \left(\frac{\text{coef B}}{\mathbf{C}} + K_a\right)\left(\frac{1}{V}\right) \tag{5-66}$$

$$\text{intercept} = \left(\frac{\text{coef A}}{\mathbf{C}} + K_b\right)\left(\frac{1}{V\mathbf{B}}\right) + \left(1 + \frac{K_c}{\mathbf{C}}\right)\left(\frac{1}{V}\right) \tag{5-67}$$

Expressions for the slopes and intercepts from equations 5-66 and 5-67 are dependent on the concentration of C as shown in equations 5-68 to 5-71. As can be seen by inspection, the slopes and intercepts are linear functions of the reciprocal of the concentration of C, and thus four tertiary replots are obtained

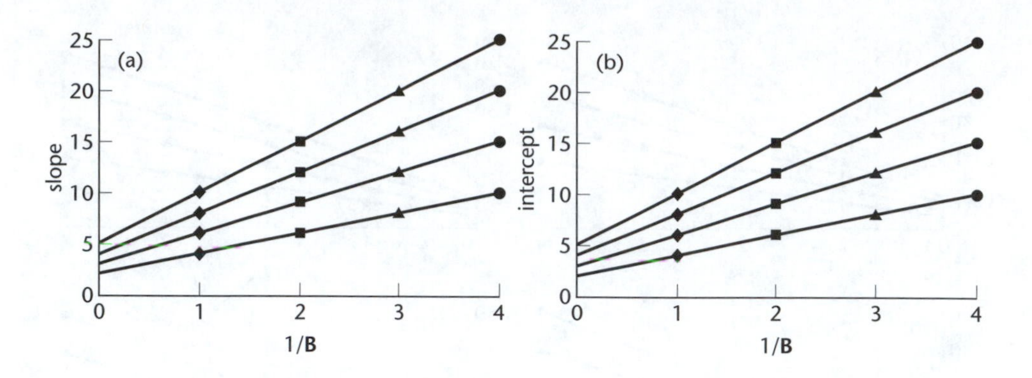

Figure 5-11. Secondary replots of slope and intercept from Figure 5-10 against the reciprocal of the concentration of B at different fixed concentrations of C (0.25, 0.33, 0.5, and 1.0 mM).

as shown in Figure 5-12. The slope and intercept of each of the tertiary replots gives an isolated kinetic parameter reflecting the individual terms in the denominator of the rate equation, equation 5-61.

$$\text{slope of slope replot} = \left(\frac{\text{const}}{V}\right)\left(\frac{1}{\mathbf{C}}\right) + \frac{\text{coef C}}{V} \tag{5-68}$$

$$\text{intercept of slope replot} = \left(\frac{\text{coef B}}{V}\right)\left(\frac{1}{\mathbf{C}}\right) + \frac{K_a}{V} \tag{5-69}$$

$$\text{slope of intercept replot} = \left(\frac{\text{coef A}}{V}\right)\left(\frac{1}{\mathbf{C}}\right) + \frac{K_b}{V} \tag{5-70}$$

$$\text{intercept of intercept replot} = \left(\frac{K_c}{V}\right)\left(\frac{1}{\mathbf{C}}\right) + \frac{1}{V} \tag{5-71}$$

Missing terms will show as a slope of zero or a line that passes through the origin. The terms present are indicators of the kinetic mechanism. For example, an ordered mechanism in which A, B, and C add in that order does not have a **B** term, while the fully random mechanism would require the presence of all denominator terms.

For statistical analysis of a terreactant mechanism, one should fit all of the data to equation 5-61 and then make fits of the data to equations that are missing

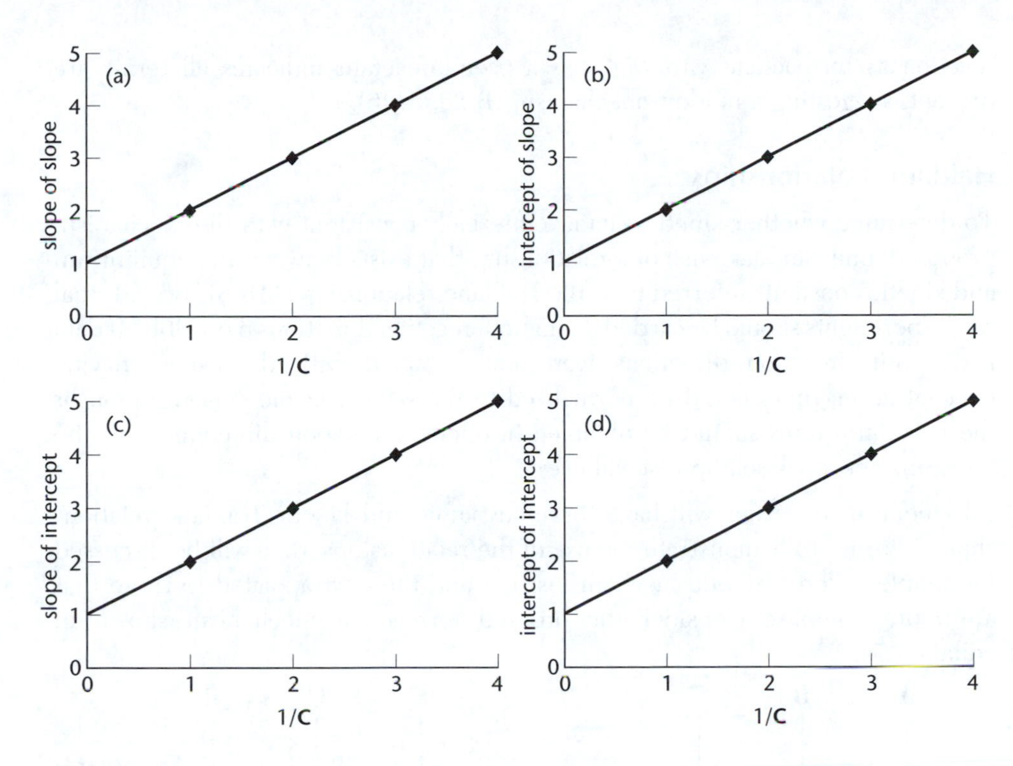

Figure 5-12. Tertiary replots obtained from the secondary replots of Figure 5-11.

denominator terms that have coefficients not significantly different from zero. Statistically, σ provides an indicator of goodness of fit and is equal to the sum of squares of the residuals divided by degrees of freedom, where the degrees of freedom are equal to the number of data points minus the number of parameters in the equation being fitted. If σ is not increased by deleting a term and refitting the data, one is justified in assuming this term is absent from the denominator of the rate equation.

Examples. Glutamate dehydrogenase has a terreactant reaction in the direction of reductive amination of α-ketoglutarate with ammonia and NADPH. In this case the α-ketoglutarate term is absent, consistent with case 2 above (B *19*, 2321). The *Ascaris suum* NAD-malic enzyme catalyzes the metal ion-dependent oxidative decarboxylation of L-malate with NAD^+ as the oxidant. Treatment of the reaction as terreactant with Mg^{2+} as a pseudoreactant indicates all terms are present, suggesting a random mechanism (B 23, 5446).

Haldane Relationships

To determine whether kinetic data are internally consistent with the mechanism proposed, one can make use of a relationship that exists between the equilibrium and kinetic constants referred to as the Haldane relationship (ME 87, 366). Initial rate experiments should be carried out in both reaction directions if possible. If data in opposite reaction directions were not obtained with the same enzyme concentration, they should be normalized to the same enzyme concentration or the maximum rates should be obtained in opposite reaction directions with the same enzyme stock solution (see above).

All kinetic mechanisms will have thermodynamic and kinetic Haldane relationships. In order to demonstrate the use of the relationships, they will be discussed for simple ordered kinetic mechanisms first and then extrapolated to those that are more complex. Consider the ordered terreactant mechanism shown in equation 5-72:

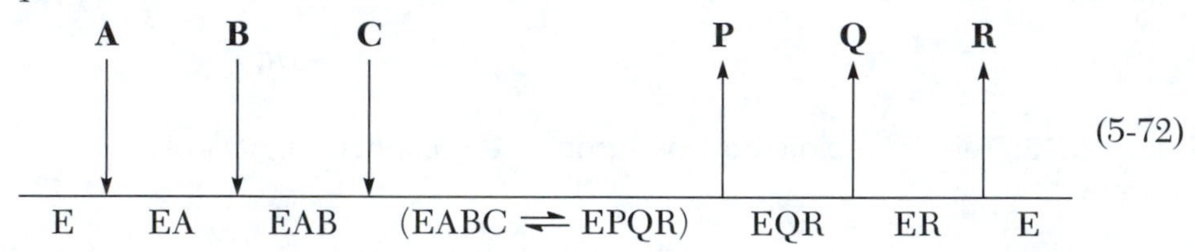

The thermodynamic Haldane relationship is the product of the equilibrium constants for each of the individual steps along the kinetic pathway. In the nomenclature of Cleland (Chapter 2), these are the K_i values (dissociation constants) for each of the reactants, with two exceptions as stated below. Thus, for mechanism 5-72,

$$K_{eq} = \frac{K_{ip}K_{iq}K_{ir}}{K_{ia}K_{ib}K_{ic}} = \frac{\left(\frac{\mathbf{(E)(R)}}{\mathbf{(ER)}}\right)\left(\frac{\mathbf{(ER)(Q)}}{\mathbf{(EQR)}}\right)\left(\frac{\mathbf{(EQR)(P)}}{\mathbf{(EABC+EPQR)}}\right)}{\left(\frac{\mathbf{(E)(A)}}{\mathbf{(EA)}}\right)\left(\frac{\mathbf{(EA)(B)}}{\mathbf{(EAB)}}\right)\left(\frac{\mathbf{(EAB)(C)}}{\mathbf{(EABC+EPQR)}}\right)} = \frac{\mathbf{PQR}}{\mathbf{ABC}} \qquad (5\text{-}73)$$

There are two exceptions, as stated above, to the definition of K_i as a dissociation constant, K_{ib} and K_{ip} in Ordered Bi Bi and Ordered Uni Bi reactions. In these cases the K_ds for B and P are defined as $V_2K_{ib}K_a/V_1K_{ia}$ and $V_1K_{ip}K_q/V_2K_{iq}$, respectively. Based on these definitions, the thermodynamic Haldanes for the Ordered Bi Bi and Ordered Uni Bi mechanisms are given in equations 5-74 and 5-75, respectively:

$$K_{eq} = \left(\frac{K_{iq}}{K_{ia}}\right)\left(\frac{V_1K_{ia}}{V_2K_{ib}K_a}\right)\left(\frac{V_1K_{ip}K_q}{V_2K_{iq}}\right) = \left(\frac{V_1}{V_2}\right)^2\left(\frac{K_{ip}K_q}{K_aK_{ib}}\right) \quad (5\text{-}74)$$

$$K_{eq} = \left(\frac{K_{iq}}{K_{ia}}\right)\left(\frac{V_1K_{ip}K_q}{V_2K_{iq}}\right) = \left(\frac{V_1}{V_2}\right)\left(\frac{K_{ip}K_q}{K_{ia}}\right) \quad (5\text{-}75)$$

The kinetic Haldane relationship is the ratio of the apparent rate constants in the forward and reverse reaction directions with all substrates at low concentration. For a unireactant kinetic mechanism, the rate constant at the limit where reactants approach zero concentration is the *V/K* for the reactant; for example, in the A to P direction, $v = (V/K_a)\mathbf{A}$. Thus

$$K_{eq} = \frac{\mathbf{P}_{eq}}{\mathbf{A}_{eq}} = \frac{V_1K_p}{V_2K_a} \quad (5\text{-}76)$$

For a bireactant mechanism, the Haldane is determined as the ratio of the *V/K*s for the last reactant to bind and the first product to be released times the ratio of the reciprocals of the dissociation constants for all other reactants to bind. Thus, for an ordered bireactant mechanism the Haldane is given in equation 5-77:

$$K_{eq} = \frac{\left(\frac{V_1}{K_b}\right)\left(\frac{1}{K_{ia}}\right)}{\left(\frac{V_2}{K_p}\right)\left(\frac{1}{K_{iq}}\right)} = \frac{V_1K_pK_{iq}}{V_2K_bK_{ia}} \quad (5\text{-}77)$$

The logic is that the *V/K* is the apparent first-order rate constant for reaction of the final reactant to add, and the dissociation constant indicates the level of the enzyme form with which it reacts. Under conditions where **B** is limiting at a fixed **A**, $v = (V_1/K_b)\mathbf{B}(\mathbf{EA})$, and $\mathbf{A}/K_{ia} = (\mathbf{EA})/(\mathbf{E})$. It follows that extending the above treatment to an ordered terreactant mechanism gives equation 5-78:

$$K_{eq} = \frac{\left(\frac{V_1}{K_c}\right)\left(\frac{1}{K_{ib}}\right)\left(\frac{1}{K_{ia}}\right)}{\left(\frac{V_2}{K_p}\right)\left(\frac{1}{K_{iq}}\right)\left(\frac{1}{K_{ir}}\right)} = \left(\frac{V_1}{V_2}\right)\left(\frac{K_pK_{iq}K_{ir}}{K_cK_{ib}K_{ia}}\right) \quad (5\text{-}78)$$

For a random kinetic mechanism such as that shown in equation 5-26, there are four possible Haldanes. One is for the top pathway where the order of addition of reactants and release of products is A-B-P-Q, given by equation 5-77 above. The other three are for the bottom pathway (B-A-Q-P) and the other two combinations (A-B-Q-P and B-A-P-Q). Equations for the latter three Haldanes are determined according to the above general rules, so for the B-A-P-Q pathway, $K_{eq} = (V/K_a)(1/K_{ib})/(V/K_p)(1/K_{iq})$.

A Ping Pong kinetic mechanism, such as that shown in equation 5-40, will have a Haldane relationship for each half-reaction. The kinetic Haldane relationship for each of the half reactions is the ratio of the V/K values for A and P for the first half-reaction and B and Q for the second. The overall Haldane is then the product of those for the individual half-reactions, as shown in equation 5-79:

$$K_{eq} = \frac{\left(\frac{V_1}{K_a}\right)\left(\frac{V_1}{K_b}\right)}{\left(\frac{V_2}{K_p}\right)\left(\frac{V_2}{K_q}\right)} = \frac{V_1^2 K_p K_q}{V_2^2 K_a K_b} \quad (5\text{-}79)$$

Each half-reaction will also have a thermodynamic Haldane relationship, which will be the ratio of the dissociation constants for P and A (first half-reaction) to those for Q and B (second half-reaction). The dissociation constants are readily determined by patterns of isotope exchange for the half-reactions; see Chapter 8. A combination of thermodynamic and kinetic Haldanes can be used, for example, a kinetic Haldane for the first half-reaction and a thermodynamic Haldane for the second half-reaction or vice versa.

For more complex reactions, the same process discussed above can be used. For a Bi Uni Uni Bi ping pong mechanism, the first half-reaction will have a Haldane equal to the ratio of the V/K values for B and P, corrected by multiplying by $1/K_{ia}$, etc. (ME *87*, 366).

It is important to point out that the Haldane relationship puts limits on the values of $V/\mathbf{E_t}$. The ratio of the V/K values in both reaction directions for the inner reactants in an ordered mechanism or a pair of reactants in a random mechanism, along with binding constants for other reactants, is equal to the solution equilibrium constant. To optimize $V/\mathbf{E_t}$ in the forward direction of a reaction, one can raise the $V/K\mathbf{E_t}$ values in the two reaction directions together in constant ratio until one hits the diffusion limit. The K_m in the forward direction cannot be greater than the physiologic level of the substrate in the cell, or the catalytic power of the enzyme is wasted. The maximum values of $V/K\mathbf{E_t}$ and K thus set the upper limit for $V/\mathbf{E_t}$. Enzymes designed to carry high flux in cells are always optimized within the Haldane limit.

Alternative Substrate Studies

Useful information can often be obtained by the use of alternative substrates. Substrates showing less than 10% the rate of the normal ones will usually not be sticky and their binding will be in rapid equilibrium. This simplifies the equations for pH profiles and isotope effects. The mechanism may even change. Thus fructose 6-sulfate is a slow alternative substrate for 6-phosphofructokinase, which binds it only after MgATP so that the normal random mechanism of this enzyme is converted into an equilibrium ordered one (JBC *251*, 3604).

Alternative substrates with fixed geometry have proven very useful in determining the anomeric specificity of enzymes for sugar substrates. The only crystalline form of D-fructose is the β-pyranose form as shown in Figure 5-13, and the two furanose anomers equilibrate so fast that it is difficult to tell which might be a substrate.

Fructokinase phosphorylates the 1-position of D-fructose and its anomeric specificity was determined by using anhydrosugar alcohols that could not mutarotate. 2,5-Anhydro-D-mannitol (an analogue of the β-furanose form) was an excellent substrate; 2,5-anhydro-D-glucitol (an analogue of the α-furanose form) was phosphorylated at the 6-position and is an analogue of α-L-sorbose, which is also a substrate of fructokinase (JBC *248*, 8174). [The difference between L-sorbose and D-fructose is the orientation of the 6-hydroxymethyl group.] Neither 2,6-anhydro-D-mannitol nor 2,6-anhydro-D-glucitol, which are analogues of the pyranoses, was a substrate. Similar studies with 2,5-anhydro-D-mannitol 6-P and 2,5-anhydro-D-glucitol 6-P established that 6-phosphofructokinase also uses only the β-furanose anomer of D-fructose 6-P as a substrate (JBC *249*, 1265).

Because MgATP exists as a rapidly equilibrating mixture of isomers (Figure 5-14), it is not easy to determine which of these is the substrate for a given enzyme. This has been approached two ways. First, inert coordination complexes of ATP with Cr(III), Co(III), or Rh(III) have been separated into their screw-sense isomers and used to determine substrate specificity for enzymes where turnover can be observed (ME *87*, 159). Thus hexokinase uses the Λ isomer of β,γ-bidentate CrATP as a substrate, while adenylate kinase uses the Δ isomer (B *19*, 1496). When reaction is fast enough for multiple turnovers to be observed, the change in the CD spectrum of a mixture of isomers can be used to tell which is the active one, since the Λ and Δ isomers have opposite CD spectra.

The other method for determining the screw-sense specificity of MgATP is to use chiral sulfur-substituted nucleotides with either Mg^{2+} (which prefers to coordinate to oxygen versus sulfur by a factor of 31,000; B *23*, 5262) or Cd^{2+}

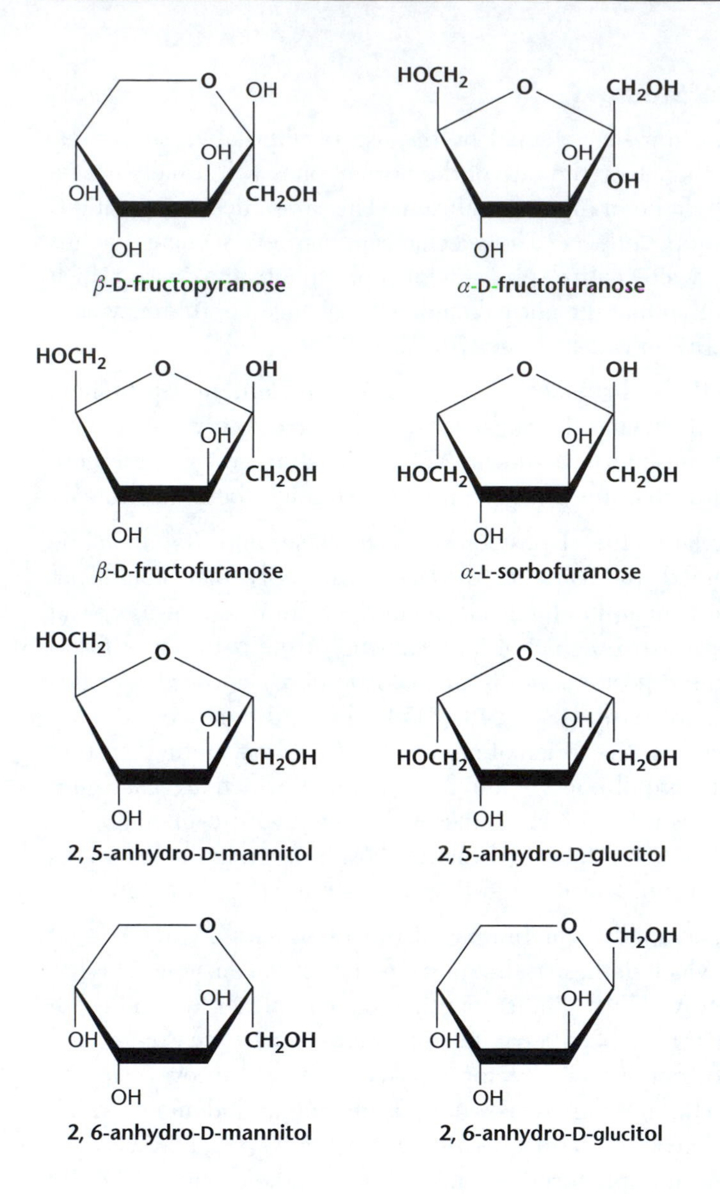

Figure 5-13. Anhydro analogues of α- and β-fructofuranoses and α- and β-fructopyranoses.

(which prefers sulfur over oxygen by a factor of 60) (Figure 5-15). The preference of one isomer of ATPßS over the other as the metal ion is switched defines the screw-sense specificity, since each metal ion causes the complex to have one predominant screw-sense (ARB 54, 367). To use this method, the ATPβS must be free of ATP, which often reacts faster than the sulfur analogue, and both *V/K* and *V* values must be compared.

Use of α-S nucleotides allows one to tell whether the α-phosphate is coordinated to the metal ion during the reaction. With hexokinase, the α-S isomer specificity was the same with Mg^{2+} or Cd^{2+} and thus the substrate is bidentate ATP (JBC *254*, 10839). With creatine kinase, however, a reversal in the α-S isomer specificity was observed upon changing from Mg^{2+} to Cd^{2+}, and thus the substrate is tridentate MgATP (JBC *255*, 8229).

A kinetic method for determining the kinetic mechanism by use of alternative substrates has been developed by Radika and Northrop (AB *141*, 413). The method is based on potential changes in the second-order rate constants in a bireactant mechanism as one of the substrates is changed. The method requires measurement of the initial rate under conditions where one substrate is saturating

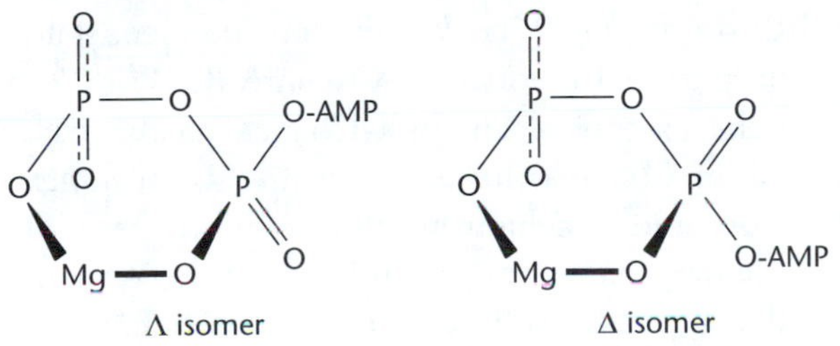

Λ isomer

Δ isomer

Figure 5-14. Screw-sense isomers of bidentate MgATP. MgATP also forms tridentate complexes, and some kinases have been shown to use these.

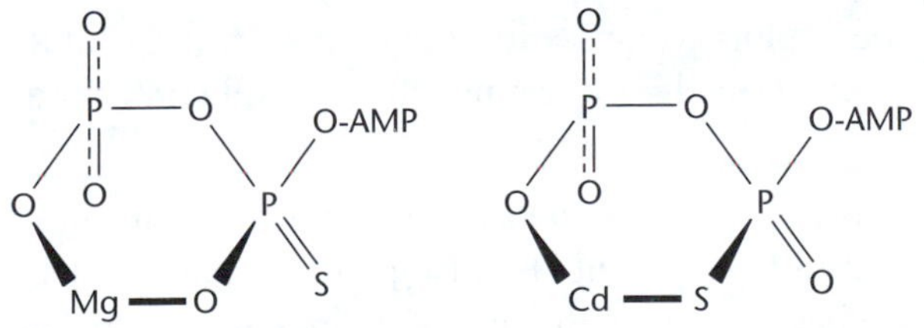

Figure 5-15. Isomers of ATPβS that are substrates for yeast hexokinase when coordinated to Mg^{2+} or Cd^{2+} to give Λ screw-sense complexes.

and invariant (e.g., A), while the concentration of the second (B) is varied. This experiment is then repeated for several alternative reactants (B′, B″, etc.), and the process is repeated with the other substrate (B) saturating and A, A′, and A″ varied. The patterns obtained are specific for a given mechanism. There are several caveats that must be kept in mind when this method is used. First, substrates that exhibit substrate inhibition should be avoided since these can produce apparent changes in slope depending on the ratio of substrate concentration to its substrate inhibition constant. Second, one should use alternative substrates that have moderate structural differences to avoid causing a change in mechanism. Third, one should use alternative substrates that exhibit differences in V_{max} to maximize the probability of observing changes in the slope of the double reciprocal plot.

In the case of a Ping Pong mechanism, the second-order rate constant for a reactant reflects one of the two halves of the reaction. Thus, the patterns obtained with **A** varied and **B**, **B′**, and **B″** saturating or with **B** varied and **A**, **A′**, and **A″** saturating will be parallel. Of course alternative reactants must be chosen that have different second-order rate constants. In addition, significant breakdown of F with any of the substrates will generate intersection.

The V/K_a in the ordered mechanism is equal to k_1, the on rate constant for substrate. Thus, the slope of the plot of $1/v$ versus 1/**A** (where **B**, **B′**, or **B″** is saturating) will be invariant. In the case of a Theorell–Chance mechanism, a single line (same K/V and $1/V$) will be obtained for all alternative B reactants. The opposite experiment for an ordered mechanism, however, will exhibit a change in slope (K/V) with **A**, **A′**, and **A″**, since the second-order rate constant in this case includes the catalytic steps as well as the rate constants for release of B and P from the central complexes. Thus, the reciprocal plot of 1/**B** with saturating **A**, **A′**, or **A″** will show intersection to the left of the vertical axis, although not at the same point. Finally, the V/Ks for both reactants in a rapid equilibrium random mechanism will include the catalytic steps, and thus reciprocal plots versus either substrate will exhibit intersection (again not at the same point) as the saturating alternate substrates are changed.

This method is very useful for supporting a kinetic mechanism for a given enzyme. However, one loses information as compared to using the conventional approach outlined earlier in the chapter, since the method does not give the same amount of quantitative data. Furthermore, mechanisms can change with different substrates since the rate constants differ.

Kinetics of Metal Ions

Metal ion cofactors are not used up during enzymatic reactions, and their kinetics are thus somewhat different from those of the organic substrates. A metal ion can either combine separately with the enzyme or form a complex with a substrate, which then serves as the actual substrate (MgATP is an example of the latter). In some enzymatic reactions two metal ions are required, one to complex ATP and the other to form a complex with the enzyme [pyruvate kinase and biotin carboxylase are examples (B 27, 4317)]. Patterns that are observed when the complex concentration is varied at different levels of free metal ion or when the level of free substrate is varied at different levels of free metal ion are summarized below. The concentration of the metal–substrate complex, **C**, is $\mathbf{MA}/K_d$, where K_d is the dissociation constant of the complex, while **M** and **A** are metal ion and substrate concentrations. The value of K_d must be known in order to carry out these experiments. Fortunately, the dissociation constants of MgATP, MgADP, and other complexes are tabulated in the literature (Martell, 1979). The K_d is a function of ionic strength, pH, and competing monovalent ion (especially Na^+) concentration, so it is best to determine the value under the experimental conditions used. In the equations below, K_m is the Michaelis constant of the metal ion, K_{ia} is the dissociation constant of A from EA, K_{im} is the dissociation constant of M from EM, and K_c is the Michaelis constant of the metal ion–substrate complex.

Case 1. The denominator of the rate equation when other reactant concentrations are held constant contains only a term in **AM** and a constant term. This corresponds to a case where only the complex combines with enzyme or where combination of A and M is highly synergistic. If one varies the complex concentration at different levels of free metal ion, the reciprocal plots will be coincident (equation 5-80):

$$\frac{1}{v} = \left(\frac{K_c}{V}\right)\left(\frac{1}{\mathbf{C}}\right) + \frac{1}{V} \tag{5-80}$$

If one varies the concentration of free A at different levels of free metal ion, the slopes of reciprocal plots vary with $1/\mathbf{M}$, and the slope replot goes through the origin. The intercepts do not vary with metal ion concentration:

$$\frac{1}{v} = \left(\frac{K_m}{K_d V}\right)\left(\frac{1}{\mathbf{M}}\right)\left(\frac{1}{\mathbf{A}}\right) + \frac{1}{V} \tag{5-81}$$

This case will apply when MgATP is the substrate and neither free Mg^{2+} nor free ATP binds to the enzyme.

Case 2. The denominator of the rate equation contains **AM** and **A** terms. Either there is equilibrium ordered addition of A and M or, more likely, the complex is the substrate and there is dead-end combination of A to form EA. If the concentration of the complex is varied at different levels of free metal ion, the slopes of reciprocal plots are the same. The intercepts are linear with 1/**M**, but the replot does not go through the origin (equation 5-82):

$$\frac{1}{v} = \left(\frac{K_m}{V}\right)\left(\frac{1}{\mathbf{C}}\right) + \left[1 + \left(\frac{K_m K_d}{K_{ia}}\right)\left(\frac{1}{\mathbf{M}}\right)\left(\frac{1}{V}\right)\right] \tag{5-82}$$

If the concentration of free substrate is varied at different levels of free metal ion, the slopes are linear with 1/**M** and the slope replot goes through the origin. The intercepts are linear with 1/**M**, but the replot does not go through the origin (equation 5-83):

$$\frac{1}{v} = \left(\frac{K_m K_d}{V}\right)\left(\frac{1}{\mathbf{M}}\right)\left(\frac{1}{\mathbf{A}}\right) + \left[1 + \left(\frac{K_m K_d}{K_{ia}}\right)\left(\frac{1}{\mathbf{M}}\right)\left(\frac{1}{V}\right)\right] \tag{5-83}$$

The kinetics of pyruvate kinase when MgADP and free ADP are C and A and PEP is saturating shows this pattern, because the high level of PEP lowers the K_d of the second metal ion required by the reaction to the point that it does not affect the kinetics (ABB *217*, 491; B *26*, 2243).

Case 3. The denominator of the rate equation contains **AM** and **M** terms. This corresponds to equilibrium ordered combination of M followed by A, or the complex is the substrate and there is a dead-end EM complex. If the complex concentration is varied at different levels of free metal ion, the slope is linear with **M**, but the replot does not go through the origin. The intercepts are the same regardless of **M** (equation 5-84):

$$\frac{1}{v} = \left(\frac{K_m}{V}\right)\left(1 + \frac{\mathbf{M}}{K_{im}}\right)\left(\frac{1}{\mathbf{C}}\right) + \frac{1}{V} \tag{5-84}$$

If the concentration of free A is varied at different levels of free metal ion, the slope is linear with 1/**M**, but the replot does not go through the origin. The intercepts are independent of **M** (equation 5-85):

$$\frac{1}{v} = \left(\frac{K_m K_d}{V K_{im}}\right)\left(1 + \frac{K_{im}}{\mathbf{M}}\right)\left(\frac{1}{\mathbf{A}}\right) + \frac{1}{V} \tag{5-85}$$

The case where the substrate combines with the EM complex and not free enzyme is fairly common. Phosphoenolpyruvate carboxylase shows equilibrium ordered addition of Mg^{2+} followed by PEP (bicarbonate then adds third) (B *31*, 6421). These patterns will be seen, however, only when the substrate prevents release of the metal ion until another substrate molecule is ready to add after catalysis occurs. If the metal ion can escape from the enzyme after the substrate adds, one has case 4. Note that the equilibration of E and EM does not have to be fast, since M is not consumed. This step will come to equilibrium in the steady state and give the equilibrium ordered pattern even if the step is slow.

Case 4. The denominator of the rate equation contains terms in **AM**, **A**, and **M**. This corresponds to random addition of A and M or ordered addition of A followed by M (note the reverse order gives case 3). If the complex concentration is varied at different levels of free metal ion, the slope is linear with **M**, but the replot does not go through the origin. The intercept is linear with 1/**M** and the replot does not go through the origin (equation 5-86):

$$\frac{1}{v} = \left(\frac{K_m}{V}\right)\left(1 + \frac{\mathbf{M}}{K_{im}}\right)\left(\frac{1}{\mathbf{C}}\right) + \left[1 + \left(\frac{K_m K_d}{K_{ia}}\right)\left(\frac{1}{\mathbf{M}}\right)\right]\left(\frac{1}{V}\right) \qquad (5\text{-}86)$$

If the concentration of free A is varied at different levels of free metal ion, both slope and intercept are linear with 1/**M**, and neither replot goes through the origin (equation 5-87):

$$\frac{1}{v} = \left(\frac{K_m K_d}{V K_{im}}\right)\left(1 + \frac{K_{im}}{\mathbf{M}}\right)\left(\frac{1}{\mathbf{A}}\right) + \left[1 + \left(\frac{K_m K_d}{K_{ia}}\right)\left(\frac{1}{\mathbf{M}}\right)\right]\left(\frac{1}{V}\right) \qquad (5\text{-}87)$$

This is the pattern shown by enolase (B *31*, 7166), where 2-P-glycerate or P-enolpyruvate adds first, followed by Mg^{2+}. In this case, Mg^{2+} has to dissociate before the organic product, so high metal ion gives substrate inhibition by preventing this. If the mechanism is random, high metal ion concentration is unlikely to cause substrate inhibition.

When two metal ions are required for an enzymatic reaction, kinetic analysis depends on how tightly they are bound. If the dissociation constant of the second metal ion from the enzyme is higher than that of MgATP, the requirement for the second metal ion can be detected by varying MgATP concentration at different levels of free Mg^{2+}. If it is lower, one must use rapid reaction methods with high levels of enzyme and similar levels of ATP and demonstrate that full activity is only obtained when 2 equiv of metal ion are added. Alternatively, if CrATP will act as a substrate, one can see if free Mg^{2+} is required as well. Enolase requires two metal ions, but the first is bound so tightly in the presence of the organic substrate that only the kinetics of the second metal ion, which adds after the substrate, are observed.

Cooperativity and Allosterism

Positive Cooperativity

Not all enzymes show classical Michaelis kinetics where v is equal to $V\mathbf{A}/(K_a + \mathbf{A})$. Some show sigmoid plots of v versus **A** (Figure 5-16) and concave upwards reciprocal plots (Figure 5-17).

Figures 5-16 and 5-17 describe a positive cooperativity and suggest that the affinity of substrate for the enzyme increases as the substrate concentration increases. These patterns usually result from oligomeric proteins that undergo a concerted transition in each subunit from a low- (or no) affinity form to a high-affinity form. One of the best-studied examples is aspartate transcarbamoylase, which catalyzes the transfer of a carbamoyl group from carbamoyl-P to aspartate, the first step in pyrimidine biosynthesis.

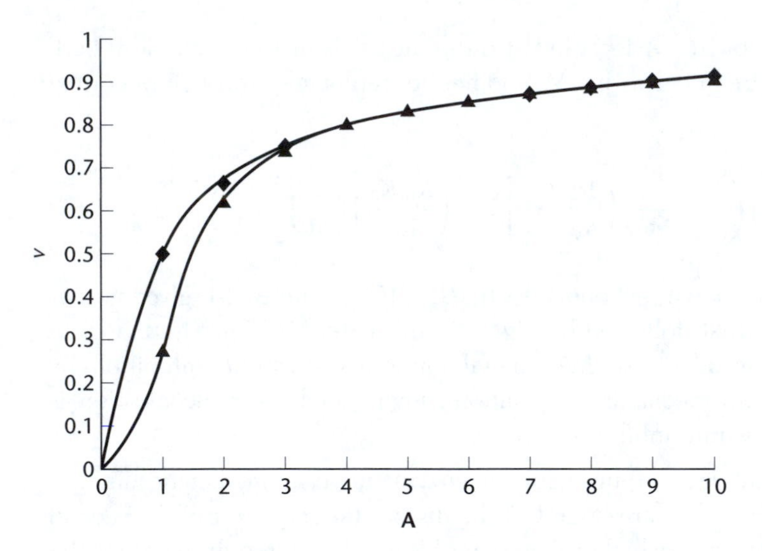

Figure 5-16. Michaelis–Menten kinetics (◆) with K_m and V both unity, and positively cooperative kinetics for an enzyme with six active sites and a ratio of inactive T form to active R form (L) in the absence of substrate of 50 (▲). V and the dissociation constant of substrate from an active site of the R form are both unity. It is assumed that substrate binds only to the R form. The value of V was fixed at 1 mM/min.

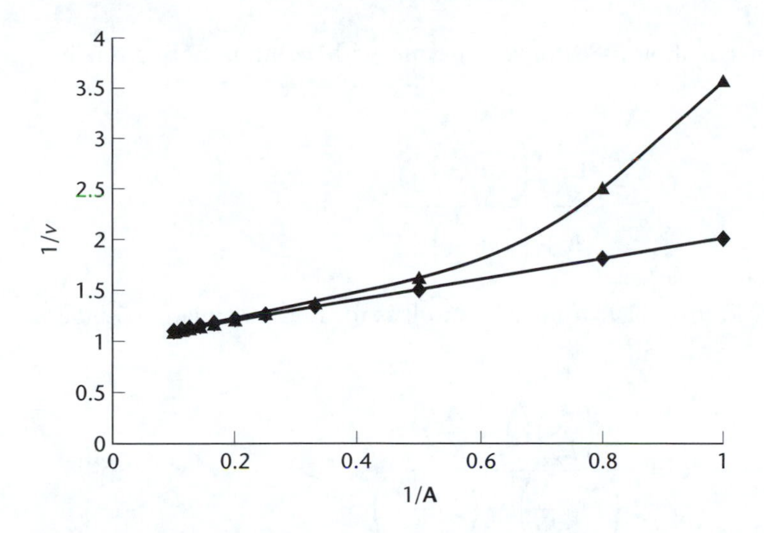

Figure 5-17. Reciprocal plot of the cooperative data from Figure 5-16.

The enzyme consists of six catalytic sites, arranged as two catalytic trimers connected by three dimeric regulatory subunits ($2C_33R_2$). In the absence of ligands other than carbamoyl-P, most of the enzyme is in an inactive form, called T, but a small amount is in the R form in which all six catalytic sites are active. Combination of aspartate with the R form displaces the R–T equilibrium, generating five more active catalytic sites. The last aspartates to bind thus find the enzyme largely in the active R form, and this behavior leads to the observed cooperativity (JBC *263*, 18583).

If one assumes that (1) only the R form is active and (2) the dissociation constant of A from the R form and the turnover number are not affected by other substrate molecules bound to the R form, the rate equation, assuming rapid equilibrium binding, is given by equation 5-88:

$$v = \frac{V\left(\frac{\mathbf{A}}{K}\right)\left(1+\frac{\mathbf{A}}{K}\right)^5}{L+\left(1+\frac{\mathbf{A}}{K}\right)^6} \tag{5-88}$$

In equation 5-88, L is the equilibrium ratio of T/R forms in the absence of A and K is the dissociation constant of A. A plot of equation 5-88 with $V=1$ is shown in Figure 5-16 for an L value of 50. Shown for comparison is a plot of $v=V\mathbf{A}/(K+\mathbf{A})$, also with $V=1$.

The general form of equation 5-88 for an enzyme with n subunits is given in equation 5-89:

$$v = \frac{V\left(\frac{\mathbf{A}}{K}\right)\left(1+\frac{\mathbf{A}}{K}\right)^{n-1}}{L+\left(1+\frac{\mathbf{A}}{K}\right)^{n}} \tag{5-89}$$

If A has some affinity for the inactive T form as well as for R, then equation 5-90 is obtained:

$$v = \frac{V\left(\frac{\mathbf{A}}{K_r}\right)\left(1+\frac{\mathbf{A}}{K_r}\right)^{n-1}}{L\left(1+\frac{\mathbf{A}}{K_t}\right)^{n}+\left(1+\frac{\mathbf{A}}{K_r}\right)^{n}} \tag{5-90}$$

In equation 5-90, K_r is the dissociation constant of A from the R form and K_t is that from the T form. Equation 5-90 reduces to equation 5-85 if $K_t \gg K_r$. For most enzymes studied to date, K_r/K_t is 0.01–0.04, and thus equation 5-85 suffices to fit the data. L values are reported from 60 to 10^4.

The model described by equation 5-89 or 5-90 is called the Monod model (JMB *12*, 88), and the data for aspartate transcarbamoylase fit it very well. If the dissociation constant of the substrate is not independent of the occupancy of the other active sites and/or if the turnover number differs, this model will not fit. In that case it is necessary to invoke the Koshland model, in which the restrictions of the Monod model do not apply. The resulting rate equation is not as simple as equation 5-89 or 5-90, and requires many more kinetic constants to express it (B *5*, 365).

Behavior corresponding to the Koshland model is shown by several enzymes in binding their substrates. Thus the sequential binding of NAD^+ to rabbit muscle glyceraldehyde-3-phosphate dehydrogenase occurs with K_d values of $< 10^{-11}$, $< 10^{-9}$, 3×10^{-7}, and 2.6×10^{-5} M (B *7*, 4011). In this case the cooperativity is negative, and the binding of one NAD^+ hinders the binding of the next by inducing a conformation change in the remaining subunits.

Negative Cooperativity

As noted above, the binding of ligands to enzymes may show negative as well as positive cooperativity. The rate equations may also show the phenomenon, in which case reciprocal plots are concave downwards rather than upwards.

Antagonistic effects on binding or catalysis in oligomeric proteins may cause this effect, and in the extreme may cause "half of the sites reactivity," in which only a portion of the active sites are capable of reacting at a given time. Then the reaction may cycle from one site to the next.

Apparent negative cooperativity can also result from a mixture of enzymes with different K_m values. For example, a mixture of isozymes, or a mixture of active plus damaged enzyme, or a mutant enzyme contaminated by wild type could result in such a mixture. The apparent negative cooperativity is readily shown by equation 5-91:

$$v = \frac{V_1\mathbf{A}}{K_1 + \mathbf{A}} + \frac{V_2\mathbf{A}}{K_2 + \mathbf{A}} \tag{5-91}$$

In equation 5-91, V_1 and K_1 represent the kinetic parameters for one enzyme in the mix, while V_2 and K_2 represent those for the other enzyme. If $K_1 = K_2$, reciprocal plots are still linear, but if $K_1 \neq K_2$, equation 5-92 is obtained:

$$v = \frac{(V_1K_2 + V_2K_1) + (V_1 + V_2)\mathbf{A}^2}{K_1K_2 + (K_1 + K_2)\mathbf{A} + \mathbf{A}^2} \tag{5-92}$$

In reciprocal form equation 5-92 is a 2/1 function (a hyperbola with a nonhorizontal asymptote), as given in equation 5-93:

$$\frac{1}{v} = \frac{1 + \dfrac{(K_1 + K_2)}{\mathbf{A}} + \dfrac{K_1K_2}{\mathbf{A}^2}}{(V_1 + V_2) + \dfrac{(V_1K_2 + V_2K_1)}{\mathbf{A}}} \tag{5-93}$$

The resulting double reciprocal plot is concave downwards and has an initial slope (at low $1/\mathbf{A}$) given by equation 5-94 and an asymptote (at high $1/\mathbf{A}$) given by equation 5-95. The vertical intercept is $1/(V_1 + V_2)$.

$$\text{initial slope} = \frac{V_1K_1 + V_2K_2}{(V_1 + V_2)^2} \tag{5-94}$$

$$\text{asymptote} = \left(\frac{K_1K_2}{V_1K_2 + V_2K_1}\right)\left(\frac{1}{\mathbf{A}}\right) + \frac{V_1K_2^2 + V_2K_1^2}{(V_1K_2 + V_2K_1)^2} \tag{5-95}$$

It can be readily shown that as long as K_1 and K_2 are different, the initial slope will be greater than the asymptote slope, and thus the curve is concave downwards.

The Hill Plot

An empirical plot that is often used to characterize cooperativity is based on the following simple model:

$$v = \frac{V_1\mathbf{A}^n}{K_1 + \mathbf{A}^n} \tag{5-96}$$

If one solves for $\mathbf{A}^n/K_1$ and takes the log of both sides of the equation:

$$\log\left(\frac{v}{V_1 - v}\right) = n[\log\ \mathbf{A}] - \log\ K_1 \tag{5-97}$$

However, more terms will be included in the true rate equation than $\mathbf{A}^n$, unless n is equal to 1. Thus, for $n = 4$, one expects $\mathbf{A}^3$, $\mathbf{A}^2$, and $\mathbf{A}$ terms in addition to $\mathbf{A}^4$. As a result, in practice a Hill plot of log $[v/(V - v)]$ against log **A** is not a straight line. For positive cooperativity, there will be a maximum slope in the vicinity of zero on the vertical axis. The slope at high log **A** will approach unity, although this part of the plot is very inaccurate since one is dividing v by the very small difference between V and v, and it is difficult to estimate V accurately. The slope at low values of log **A** approaches a unit slope (1, 2, etc.) corresponding to the number of substrate molecules that have to be present to initiate reaction. Thus for adenylate kinase, where two molecules of ADP react to form ATP and AMP, the Hill plot slope at low log **ADP** will be 2.

In practice, tetrameric enzymes usually give maximum slopes in a Hill plot of $\sim$2.8, even with very large L factors in equation 5-88, and slopes approaching 4 are never seen. Nonetheless, a slope above 1 in the Hill plot does indicate positive cooperativity, and if the number of subunits is known from structural studies, one can compare the slope with the number of subunits to get some idea of the degree of cooperativity.

Just as positive cooperativity gives slopes above unity in a Hill plot, negative cooperativity leads to slopes less than 1. Thus with the model in equation 5-91, the Hill plot will have unit slopes at high log **A** and at low log **A** and a slope less than unity when the velocity is about half of V.

Allosterism

An allosteric site is one other than the active site where a modifier adds and causes an effect. With aspartate transcarbamoylase, for example, the modifiers ATP and CTP bind competitively to the regulatory subunits that are between the catalytic subunits. ATP has a strong affinity for the R form of the enzyme, and in the presence of 5 mM ATP almost all of the enzyme is in the R form and the kinetics are of the normal Michaelis–Menten type. CTP, on the other hand, has a strong preference for the T form of the enzyme and thus increases the L factor in equation 5-88. The increased value of L results in an increase in the degree of cooperativity and inhibition at nonsaturating levels of aspartate. Thus, the balance between ATP and CTP concentrations, rather than changes in the concentration of aspartate, regulates the rate of the enzyme and ensures that the level of pyrimidine biosynthesis matches the needs of the cell.

The rate equation in the presence of CTP will have L in the denominator of equation 5-88 multiplied by $(1+\mathbf{I}/K_I)^6$, where **I** is CTP concentration and K_I is its dissociation constant from the T form. [If the number of regulatory subunits were different, the exponent would correspond to that number.] In the presence of ATP, L will be divided by $(1+\mathbf{B}/K_B)^6$, where **B** is the ATP concentration and K_B is its dissociation constant from the R form. The L term in the denominator of equation 5-89 is similarly modified by terms for inhibitors or activators, with exponents equal to the number of allosteric sites for the modifiers.

The Monod model for aspartate transcarbamoylase is strongly supported by the ^{13}C isotope effects in carbamoyl-P (B *31*, 6570). At low aspartate, the ^{13}C isotope effect is 1.022, but as aspartate levels are increased, the value decreases hyperbolically to unity at infinite aspartate concentration, with ~4 mM aspartate causing half of the change in the isotope effect. These data show that the mechanism is ordered with aspartate adding second (see Chapter 9). When 5 mM ATP or 5 mM CTP is present, exactly the same curve is seen, showing that (1) only the R form is active and (2) the R form shows the same kinetic properties whether ATP is present or absent. Presumably CTP binds only to the T form.

With aspartate transcarbamoylase, CTP and ATP are called K-type allosteric modifiers, since the maximum velocity is not appreciably affected but only the apparent affinity for aspartate. In other cases the presence of the modifier at the allosteric site may affect V (a V-type modifier) or both V and K by induced conformation changes in the active site. The Monod model does not predict such behavior, and the Koshland model has to be used to explain these cases.

An example of this type of allosteric behavior is shown by prephenate dehydrogenase, which is a dimeric enzyme inhibited by tyrosine. Combination of prephenate at one site affects the binding at the other site at low levels of NAD, so that positive cooperativity is observed. At saturating levels of NAD^+, however, cooperativity disappears. The binding of tyrosine at the allosteric sites (it does not bind at the active site) is antagonistic, with the K_i value for the second combination 40–70-fold higher than that for the first combination. The enzyme with two tyrosines bound is inactive, while enzyme with one prephenate and one tyrosine bound has 6% of the activity of enzyme with only a single prephenate present (B *30*, 7783).

Kinetic Mechanism of Regulation

As suggested above, not all regulated enzymes exhibit behavior that differs from Michaelis–Menten kinetics. Even when they do exhibit positive or negative cooperativity, the limiting rate constants V and V/K can be estimated. In order to determine the kinetic mechanism of regulation, one can treat the allosteric effector as an additional reactant. This is a reasonable treatment since the enzyme activity will vary with the concentration of the effector. In this section we will briefly consider the kinetics of regulated enzymes that exhibit Michaelis–Menten kinetics but are activated or inhibited by an effector (B *21*, 113). Effectors can activate, inhibit, or do both depending on reactant concentration, which determines the enzyme form that predominates under given reactant concentrations.

In the case of a bireactant mechanism, with an effector treated as a reactant, one obtains a terreactant reaction mechanism. Initial velocity patterns are then obtained by measuring the initial rate at different concentrations of A and different fixed concentrations of B and zero concentration of effector. The experiment is then repeated as a function of the effector concentration. Replots of slopes and intercepts versus the reciprocal of the fixed varying reactant are then constructed to give a secondary initial velocity pattern from which tertiary replots of slope of slope, intercept of slope, slope of intercept, and intercept of intercept versus effector concentration are constructed. In this way the kinetic constants $K_{ia}K_b/V$, K_a/V, K_b/V, and $1/V$ (or K_a/V, K_b/V, and $1/V$ in the case of a ping pong mechanism) affected by the allosteric modulator, and thus the predominant enzyme form to which the effector binds are determined, as is the dissociation constant for the allosteric effector binding to each of the enzyme forms (as the concentration of the effector that gives half the change in each of the rate constants).

Additional information is obtained concerning the rate processes affected by the allosteric modulator depending on the rate constants increased or decreased. To illustrate this, consider the ordered pathway below:

$$\mathrm{EA} \underset{k_4}{\overset{k_3\mathbf{B}}{\rightleftharpoons}} \mathrm{EAB} \underset{k_6}{\overset{k_5}{\rightleftharpoons}} \mathrm{E^*AB} \underset{k_8}{\overset{k_7}{\rightleftharpoons}} \mathrm{E^*PQ} \underset{k_{10}}{\overset{k_9}{\rightleftharpoons}} \mathrm{EPQ} \xrightarrow{k_{11}} \mathrm{EQ} \xrightarrow{k_{13}} \quad (5\text{-}98)$$

In mechanism 5-98, k_3 and k_4 represent addition and release of B; k_5, k_6, k_9, and k_{10} represent conformation changes in the central complexes; k_7 and k_8 represent the forward and reverse catalytic steps; while k_{11} and k_{13} represent off rate constants for products P and Q, respectively. The rate expressions for V and V/K, respectively, are given in equations 5-99 and 5-100:

$$\frac{V}{E_t} = \frac{\dfrac{k_7}{\left(1+\dfrac{k_6}{k_5}\right)}}{1+\dfrac{k_7\left[\dfrac{1}{k_5}+\left(\dfrac{1}{k_9}\right)\left(1+\dfrac{k_{10}}{k_{11}}\right)+\dfrac{1}{k_{11}}+\dfrac{1}{k_{13}}\right]}{1+\dfrac{k_6}{k_5}}+\left(\dfrac{k_8}{k_9}\right)\left(1+\dfrac{k_{10}}{k_{11}}\right)} \tag{5-99}$$

$$\frac{V}{KE_t} = \frac{\dfrac{k_3k_5k_7}{k_4k_6}}{1+\left(\dfrac{k_7}{k_6}\right)\left(1+\dfrac{k_5}{k_4}\right)+\left(\dfrac{k_8}{k_9}\right)\left(1+\dfrac{k_{10}}{k_{11}}\right)} \tag{5-100}$$

On the basis of changes in the magnitude of V and V/K, the following can be stated generally regarding the effect of the allosteric modulator on specific rate constants.

(1) An increase or decrease in V and V/K in the presence of an allosteric effector likely results from an increase in the forward rate constants for the catalytic pathway, k_5, k_7, and k_9, and/or the rate constant for release of P, k_{11}, or a decrease in the reverse rate constants for the catalytic pathway, k_6, k_8, and k_{10}, and/or release of the reactant off rate constant, k_4, or some combination of these. The most common effect is on the rate constants for the conformation changes preceding and following the chemical steps (see below). The least common effect is on the rate constants for the chemical steps.

(2) When V is increased or decreased with no effect on V/K, the most likely effect is on the off rate constant for the last product, k_{13}.

(3) When V/K is increased or decreased with no effect on V, the most likely effect is on the off rate constant for the reactant, k_4.

Glutamate dehydrogenase is allosterically activated by ADP and allosterically inhibited by GTP (B *21*, 113). ADP increases V by about 3-fold but decreases the $V/K_{glutamate}$ and V/K_{NADP} by 1.5-fold and 3-fold, respectively. GTP, on the other hand, decreases V but increases the V/K for glutamate. These effectors thus likely exert their effect on the net rate constants for product and reactant release.

ADP thus increases the net off rates for glutamate and NADP, giving a decrease in the amount of ternary complex and thus a decrease in the rate under *V/K* conditions. When reactants are saturating, however, the rate is increased as a result of an increase in the rate of release of products (the biggest effect here is likely on NADPH release). GTP has the opposite effect, decreasing the net off rate for glutamate and NADP and increasing the amount of ternary complex under *V/K* conditions but decreasing the off rate for products at saturating reactant concentrations. Thus, ADP is a *V*-type activator and *K*-type inhibitor, while GTP is a *V*-type inhibitor and *K*-type activator.

Isotope effects can be used in conjunction with the rate studies to better pinpoint steps affected by the allosteric modulators (see Chapter 9).

Transmission of Allosteric Effects—Coupling of Active and Allosteric Sites

A more useful quantitative way of determining the interaction between active and allosteric sites has been developed by Reinhart (ABB *224*, 389; ME *380*, 187). Experiments allow an estimate of the thermodynamic linkage between two sites, that is, an estimate of the free energy of coupling (Q) between the sites.

Consider the following mechanism:

$$\begin{array}{ccccc} \mathrm{E} & \underset{k_2}{\overset{k_1\mathbf{A}}{\rightleftharpoons}} & \mathrm{EA} & \xrightarrow{k_3} & \mathrm{E}+\mathbf{P} \\ k_5\mathbf{X}\downarrow\uparrow k_6 & & k_{12}\uparrow\downarrow k_{11}\mathbf{X} & & \\ \mathrm{XE} & \underset{k_8}{\overset{k_7\mathbf{A}}{\rightleftharpoons}} & \mathrm{XEA} & \xrightarrow{k_9} & \mathrm{XE}+\mathbf{P} \end{array} \tag{5-101}$$

The rate equation derived for mechanism 5-101, assuming that binding of the allosteric ligand comes to equilibrium, is given in equation 5-102:

$$v = \frac{\left(k_3 + \dfrac{k_9 X}{K_{\mathrm{ix}}^0}\right)A}{\dfrac{\left(\dfrac{K_{\mathrm{a}}^0}{k_7}\right)\left(1+\dfrac{X}{K_{\mathrm{ix}}^0}\right)+\left(\dfrac{K_{\mathrm{a}}^\infty X}{k_1 K_{\mathrm{ix}}^\infty}\right)\left(1+\dfrac{X}{K_{\mathrm{ix}}^0}\right)}{\dfrac{1}{k_7}+\dfrac{X}{k_1 K_{\mathrm{ix}}^\infty}} + A\left(1+\dfrac{X}{K_{\mathrm{ix}}^\infty}\right)} \tag{5-102}$$

In equation 5-102, K_{ix}^0 and K_{a}^0 represent the dissociation constant for X at zero **A** and the K_{m} for A at zero **X**, respectively, and V^0 is the maximum rate at zero

X (k_3, a net rate constant, in equation 5-101). The term Q is the coupling free energy between allosteric and activator sites and is equal to $K_{\rm a}^0/K_{\rm a}^\infty$, $K_{\rm ix}^0/K_{\rm ix}^\infty$, or $K_{\rm ia}^0/K_{\rm ia}^\infty$, based on the limits of equation 5-102 and the thermodynamic cycle shown in equation 5-101. Finally, W is the ratio of V^∞/V^0, that is, the ratio of k_9/k_3 in mechanism 5-101. The value of W tends to zero for a complete allosteric inhibitor and can have any value > 1 for an allosteric activator. The above theory has been applied to the allosterically regulated phosphofructokinase reaction. This mechanism-independent general theoretical approach has the advantage that one obtains a quantitative estimate of the allosteric effect, the value of Q, from a simple steady-state analysis. The method has also been extended to systems with multiple allosteric ligands.

Practical Considerations

When working with a new enzyme, there is a general plan that one should follow in order to collect initial velocity data in the absence of added inhibitors, as outlined below for a bireactant enzyme-catalyzed reaction:

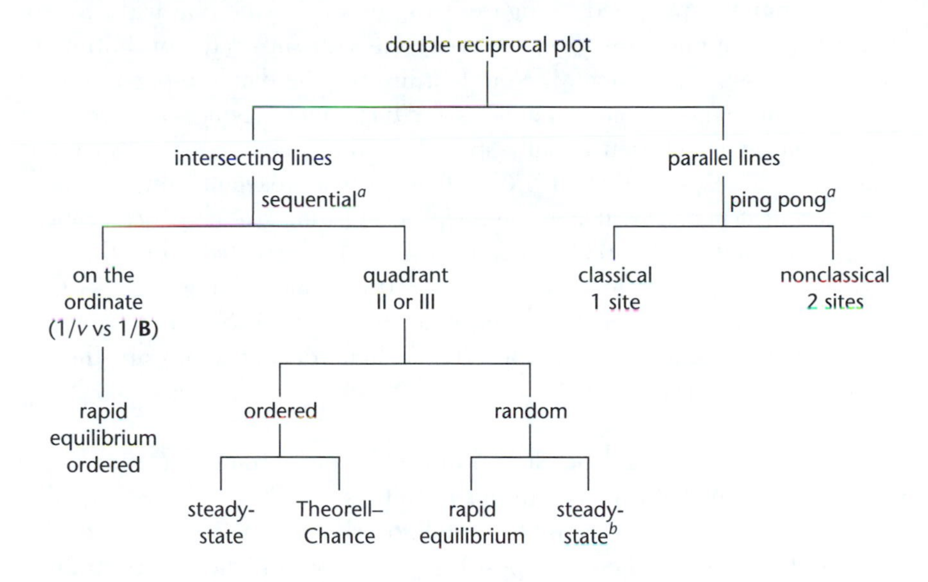

[a]Exceptions for parallel lines indicating a ping pong mechanism and intersecting lines indicating a sequential mechanism have been discussed above.

[b]Steady-state random mechanisms may additionally exhibit curvature near the ordinate in double reciprocal plots.

(1) Estimate values for the reactant K_ms by measuring the initial rate at a fixed concentration of one reactant (relatively high, in the millimolar range if possible) at different levels of the other reactant. Cover as wide a range as possible to show anomalies such as substrate inhibition (observed as a decrease in the rate with increasing reactant concentration) or positive cooperativity (observed as a pronounced decrease in the rate at low reactant concentrations).

(2) Plot the data in double reciprocal form, so that the K_m can be estimated as the reciprocal of the abscissa intercept (see Chapter 2).

(3) When initial velocity studies are carried out, it is best to use a substrate concentration range from $0.5K_m$ to $5–10K_m$. For multisubstrate reactions, each of the substrates should be varied over this range, with one substrate varied at different fixed levels of the other. The protocol can be expanded to include as many reactants as necessary. The actual substrate concentrations used should be equally spaced on the double reciprocal plot. This will make it easier to see the quality of the data prior to fitting. An illustration is provided in Figure 5-18, for an enzyme that has a K_m for A of 4 mM. In one case, data are collected with substrate concentrations chosen as 1, 4, 7, and 10 mM. Note that most of the data are piled up near the ordinate. The second example is for the same enzyme, but with four 4 concentrations chosen equally spaced on the reciprocal scale spanning the range $0.5K_m$ (2 mM) to $5K_m$ (20 mM). The reactant concentrations are easily calculated by taking the reciprocal of the extreme concentrations ($1/2 = 0.5$ and $1/20 = 0.05$), taking the difference between the two ($0.5–0.05 = 0.45$), dividing by 3 ($0.45/3 = 0.15$), and adding 0.15 sequentially to the low reciprocal concentration (0.05, 0.2, 0.35, 0.5) to generate the four reciprocal concentrations. The actual concentrations are then 2, 2.86, 5, and 20 mM.

Inhibition experiments should be carried out with concentrations of the varied substrate as determined above. The fixed substrate is usually maintained equal to its K_m, while the concentration of the inhibitor should be fixed at 0, appK_i, 2(appK_i), and 4(appK_i) whenever possible. The use of the concentrations indicated will generate a doubling of the slope (competitive or noncompetitive inhibition) or intercept (noncompetitive or uncompetitive inhibition) at each of the fixed inhibitor concentrations (see Chapter 6). The apparent K_i is estimated by fixing both reactants at their K_m values and measuring the initial velocity as a function of the inhibitor concentration. A plot of $1/v$ versus **I** will give the abscissa intercept as the appK_i.

(4) Obtain the initial velocity pattern by measuring the initial rate with one reactant (arbitrarily termed B) fixed at $0.5K_m$ as discussed above (step 3). [If anomalies such as curvature in the reciprocal plots are observed, be sure to adjust the concentration range to include the anomalies.] Repeat the process at additional fixed concentrations of the fixed reactant up to a value of 5–10 times its K_m. If no anomalies were observed in steps 1–4, it is satisfactory to use four concentrations of each reactant, so that 16 points are measured.

(5) Plot the data in double reciprocal form as $1/v$ versus 1/**A** at different fixed levels of B and as $1/v$ versus 1/**B** at different fixed levels of A. Plot slopes and intercepts versus the reciprocal of the fixed reactant to obtain the secondary plots from both primary plots. It is important to plot the data and not simply fit an equation to the data by using the computer. The visualization of the data helps to determine whether they are well conditioned; that is, the primary and secondary plots are linear. On the basis of the plots, the proper equation can be selected to fit to the data. The secondary replots also allow a graphical estimate of the kinetic constants. As a final step, the data must be fitted by use of the appropriate rate equation, so that standard errors are

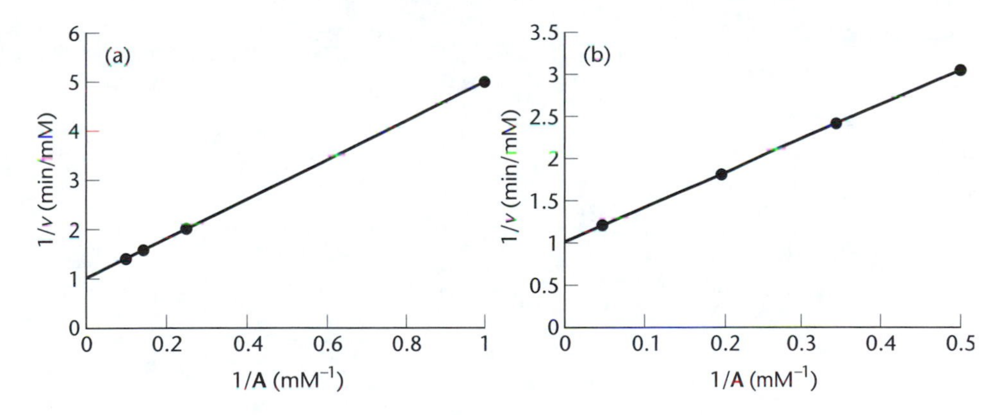

Figure 5-18. Illustration of spacing of substrate concentrations to determine kinetic parameters. An arbitrary K_m of 4 mM was chosen for A, and V is 1 mM/min. (A) Substrate concentrations are 1, 4, 7, and 10 mM. (B) Concentrations, equally spaced on the reciprocal scale, are 2, 2.86, 5, and 20 mM.

obtained. If there is a question concerning the appropriate equation, fit with all possible rate equations.

(6) The above, steps 1–5 should be repeated in the opposite reaction direction if possible, measuring the initial rate as a function of P and Q concentration.

From Data to Interpretation

The consideration of initial velocity data in this chapter has utilized a theoretical approach, discussing the initial velocity patterns that are predicted for a given kinetic mechanism. However, this approach is opposite to the way data are collected and interpreted in the laboratory; that is, data collection precedes interpretation. The flow chart above provides some indication of the process one mentally uses to provide an indication of kinetic mechanism from the data.

6

INITIAL VELOCITY STUDIES: PRESENCE OF ADDED INHIBITORS

In Chapter 5, initial velocity patterns obtained in the absence of added inhibitors were discussed for the most commonly encountered kinetic mechanisms. Evident is the fact that, in most cases, kinetic mechanism cannot be assigned solely on the basis of initial velocity patterns in the absence of added inhibitors. This inability to assign kinetic mechanism is partially a result of the degenerative nature of the initial velocity rate equation for a number of mechanisms, for example, rapid equilibrium random, steady-state ordered, and Theorell–Chance mechanisms. There are instances when an apparently diagnostic initial velocity pattern is obtained, as in the case of a rapid equilibrium ordered kinetic mechanism where intersection is on the ordinate when B is the varied reactant and A is the fixed varied reactant. However, even in these cases the interpretation remains ambiguous. In the example utilized, strong synergism in a rapid equilibrium random mechanism can also yield an initial velocity pattern that intersects on the ordinate. Other experimental approaches are required to further distinguish between the possible kinetic mechanisms. One such additional approach, which still makes use of initial velocity studies, is the use of inhibitors. In this chapter, inhibition studies will be considered in detail, first in general with respect to the kinds of inhibition and then specifically with respect to the kinds of inhibition patterns (double reciprocal plots) expected for a given kinetic mechanism for product and dead-end inhibition. Finally, other kinds of inhibition experiments, such as alternate substrate and product inhibition, will be discussed.

Irreversible Step

Before the kinds of inhibitors are considered, it is important to introduce the concepts of *irreversible step* and *reversible connection*. In the context of initial velocity, one must consider practical irreversibility. There are three conditions that constitute irreversibility. For a unimolecular step that does not result in release of a product from enzyme, the higher the equilibrium constant, the more it tends to be practically irreversible. (A step with an equilibrium constant of about 100 can be considered irreversible.) In addition, since initial velocities are measured, and the product concentration is zero under these conditions, a step in which a product is released is practically irreversible. Finally, a step in which a reactant adds to the enzyme is irreversible when that reactant is saturating. A reactant concentration 50 times its K_m (98% saturating) can be considered saturating. (It is generally accepted that initial velocity studies have an associated error of ~5%, and thus one cannot distinguish between 95% and 100% saturation.) All steps that do not meet the above criteria are reversible, and if there are two or more such steps adjacent to one another in a reaction pathway, they are considered reversibly connected.

Types of Inhibition

There are three kinds of inhibitors, called *competitive*, *noncompetitive*, and *uncompetitive*. Some investigators refer to a mixed or mixed noncompetitive inhibition, but as will be shown, these are types of noncompetitive inhibition. The kinds of inhibitor are distinguished qualitatively on the basis of the initial velocity patterns observed in the presence of the inhibitor, and following directly from the qualitative pattern, the enzyme form or forms to which the inhibitor binds are determined.

Competitive Inhibition

A competitive inhibitor is one that competes with the varied substrate for enzyme, generally suggesting that the inhibitor and the varied substrate combine with the same form of the enzyme. There are, however, exceptions to the generality as will be pointed out below. At high concentrations of the inhibitor, it takes much higher reactant concentrations to obtain a rate, while at high substrate concentrations, it requires much higher inhibitor concentrations to observe inhibition. It follows that extrapolation of the substrate concentration to infinity eliminates the effect of the inhibitor.

A scheme describing competitive inhibition by I versus A in a unireactant mechanism is provided in equation 6-1:

$$\begin{array}{ccc} E & \rightleftharpoons EA \longrightarrow & E + P \\ \updownarrow & & \\ EI & & \end{array} \tag{6-1}$$

The rate equation describing equation 6-1 is given in equation 6-2, and its reciprocal form is given in equation 6-3:

$$v = \frac{V\mathbf{A}}{K_a(1 + \mathbf{I}/K_{is}) + \mathbf{A}} \tag{6-2}$$

$$\frac{1}{v} = \left(\frac{K_a}{V}\right)\left(1 + \frac{\mathbf{I}}{K_{is}}\right)\left(\frac{1}{\mathbf{A}}\right) + \frac{1}{V} \tag{6-3}$$

As can be seen, the **A** term in the denominator of equation 6-2, representing the EA form of the enzyme, is unaffected by the presence of inhibitor. The expression for the slope, given in equation 6-4, representing the E form of the enzyme, increases linearly with the concentration of inhibitor, consistent with the competition between A and I for E. The term that modifies the slope is $(1 + \mathbf{I}/K_{is})$. Thus, when **I** is zero, no inhibition is observed, while as **I** increases compared to its K_{is}, inhibition is observed. Since K_{is} is the dissociation constant for the EI complex given by the equilibrium between free E and I and the EI complex, the ratio $\mathbf{I}/K_{is}$ represents **EI/E**, and thus inhibition is observed as EI is formed. The apparent K_m obtained as a function of **I** increases linearly, reflecting the increased concentration of substrate required to attain the maximum rate in the presence of I.

$$\text{slope} = \left(\frac{K_a}{VK_{is}}\right)\mathbf{I} + \frac{K_a}{V} = \left(\frac{K_a}{V}\right)\left(1 + \frac{\mathbf{I}}{K_{is}}\right) \tag{6-4}$$

A graphic representation of competitive inhibition is provided in Figure 6-1, along with a secondary slope replot.

An estimate of the competitive inhibition constant is obtained as the abscissa intercept of the slope replot.

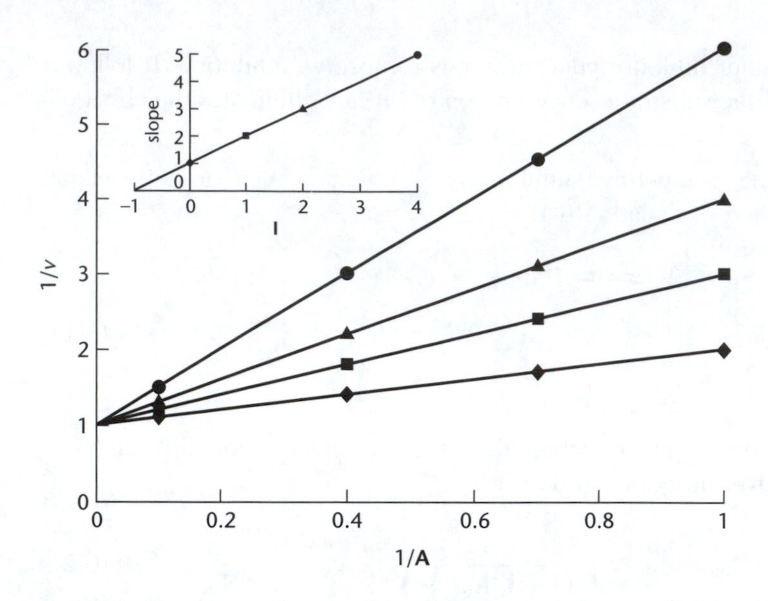

Figure 6-1. Primary double reciprocal plot for competitive inhibition. Units on the ordinate and abscissa are minutes/millimolar and millimolar^{-1}, respectively. The varied reactant is A, and the concentrations of I are 0, K_{is}, $2K_{is}$, and $4K_{is}$. The values of V, K_a, and K_{is} are 1 mM/min, 1 mM, and 1 mM, respectively. The inset shows the secondary plot of the slope of the primary plot versus **I**. The abscissa intercept is $-K_{is}$.

Noncompetitive Inhibition

A noncompetitive inhibitor is one which binds to an enzyme form different than that to which the varied substrate binds, and there is a reversible connection between the addition of varied substrate and that of the inhibitor. Alternatively, a noncompetitive inhibitor can combine both with the same enzyme form as does the varied substrate and with another enzyme form. Thus, increasing the reactant to an infinite concentration will not eliminate all of the inhibition by the inhibitor.

A scheme describing noncompetitive inhibition by I versus B in an Ordered Bi Bi reaction with **A** fixed and **B** varied is provided in equation 6-5:

$$\begin{array}{ccccccc} E & \rightleftharpoons & EA & \rightleftharpoons & EAB & \longrightarrow & E + P \\ \updownarrow & & & & & & \\ EI & & & & & & \end{array} \qquad (6\text{-}5)$$

The rate equation describing equation 6-5 is given in equation 6-6, and its reciprocal form is given in equation 6-7:

$$v = \frac{\mathrm{app}V\mathbf{B}}{\mathrm{app}K(1 + \mathbf{I}/K_{is}) + \mathbf{B}(1 + \mathbf{I}/K_{ii})} \quad (6\text{-}6)$$

$$\frac{1}{v} = \left(\frac{\mathrm{app}K}{\mathrm{app}V}\right)\left(1 + \frac{\mathbf{I}}{K_{is}}\right)\left(\frac{1}{\mathbf{B}}\right) + \left(\frac{1}{\mathrm{app}V}\right)\left(1 + \frac{\mathbf{I}}{K_{ii}}\right) \quad (6\text{-}7)$$

where $\mathrm{app}V = V/(1 + K_a/\mathbf{A})$, $\mathrm{app}K = K_b[(1 + K_{ia}/\mathbf{A})/(1 + K_a/\mathbf{A})]$, $K_{is} = K_i(1 + \mathbf{A}/K_{ia})$, and $K_{ii} = K_i(1 + \mathbf{A}/K_a)$. As can be seen, both the **B** term, representing EAB and part of E, and the app*K* term, representing EA and part of E, are affected by the presence of inhibitor. The expressions for the slope and intercept are given in equation 6-8. Both the slope and intercept increase linearly with the concentration of inhibitor. The $(1 + \mathbf{I}/K_{is})$ and $(1 + \mathbf{I}/K_{ii})$ terms have meanings similar to those discussed above for the competitive case.

$$\text{intercept} = \left(\frac{1}{\mathrm{app}V}\right)\left(1 + \frac{\mathbf{I}}{K_{ii}}\right); \quad \text{slope} = \left(\frac{\mathrm{app}K}{\mathrm{app}V}\right)\left(1 + \frac{\mathbf{I}}{K_{is}}\right) \quad (6\text{-}8)$$

A graphic representation of noncompetitive inhibition is provided in Figure 6-2, along with secondary slope and intercept replots.

Estimates of the noncompetitive inhibition constants, K_{is} and K_{ii}, are obtained as the abscissa intercepts of the slope and intercept replots, respectively. The lines in the primary plot illustrated in Figure 6-2 intersect below the abscissa, indicating that $K_{is} > K_{ii}$. The plot intersects on the abscissa if $K_{is} = K_{ii}$, while if $K_{ii} > K_{is}$, the plot will intersect above the abscissa. Cases where $K_{is} \neq K_{ii}$ are sometimes referred to as mixed inhibition by some investigators. Since there is no *a priori* reason K_{is} and K_{ii} should be equal, these cases should also simply be termed noncompetitive.

Uncompetitive Inhibition

An uncompetitive inhibitor combines with an enzyme form that results from combination of the varied substrate with enzyme; that is, either the enzyme–substrate complex or some other form downstream from it. Thus, inhibition is observed when the varied substrate is saturating, conditions that favor production of the enzyme form to which inhibitor binds. As the concentration of the varied substrate is extrapolated to zero, the enzyme form to which inhibitor binds is not present, and no inhibition is observed.

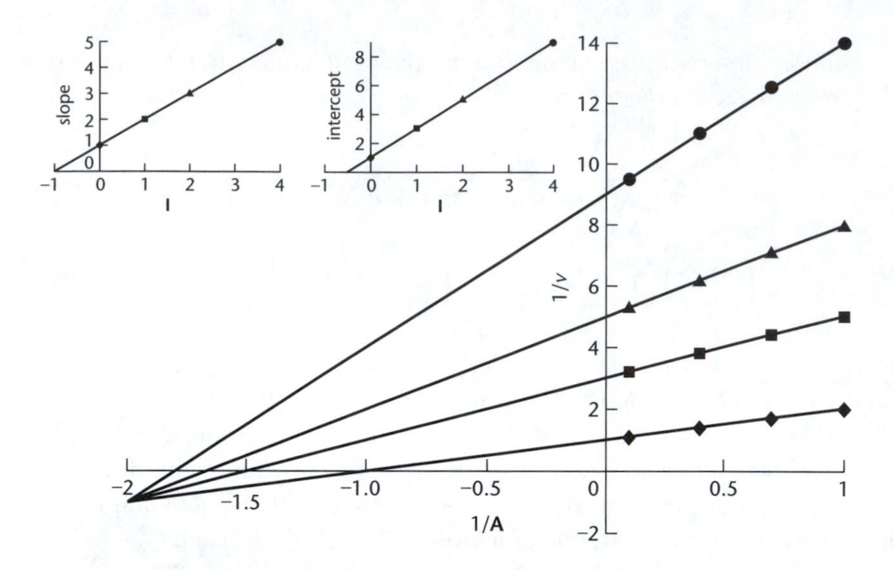

Figure 6-2. Primary double reciprocal plot for noncompetitive inhibition. Units on the ordinate and abscissa are minutes/millimolar and millimolar^{-1}, respectively. The varied reactant is A, and the concentrations of I are 0, K_{is}, $2K_{is}$, and $4K_{is}$; the concentration of B is fixed equal to K_b. The values of V_{app}, $K_{a\ app}$, K_{is}, and K_{ii} are 1 mM/min, 1 mM, 1 mM, and 0.5 mM, respectively. Insets show the secondary plots of the slopes and intercepts of the primary plots versus **I**. The abscissa intercepts for the slope and intercept replots are –appK_{is} and –appK_{ii}, respectively.

A scheme describing uncompetitive inhibition by I versus A in an Ordered Bi Bi kinetic mechanism with **B** maintained constant is provided in equation 6-9:

$$\begin{array}{ccccccc} E & \rightleftharpoons & EA & \rightleftharpoons & EAB & \longrightarrow & E + \text{products} \\ & & \updownarrow & & & & \\ & & EAI & & & & \end{array} \tag{6-9}$$

The rate equation describing mechanism 6-9 is given in equation 6-10, and its reciprocal form is given in equation 6-11:

$$v = \frac{\text{app}V\mathbf{A}}{\text{app}K + \mathbf{A}(1 + \mathbf{I}/K_{ii})} \tag{6-10}$$

$$\frac{1}{v} = \left(\frac{\text{app}K}{\text{app}V}\right)\left(\frac{1}{\mathbf{A}}\right) + \left(\frac{1}{\text{app}V}\right)\left(1 + \frac{\mathbf{I}}{K_{ii}}\right) \tag{6-11}$$

where $\text{app}\,V = V/(1+K_a/\mathbf{A})$, $\text{app}K = K_a[(1 + (K_{ia}K_b/K_a\mathbf{B}))/(1 + (K_b/\mathbf{B}))]$, and $K_{ii} = K_i(1 + (\mathbf{B}/K_b))$; K_i is the dissociation constant for I from EAI. As can be seen, only the **A** term representing the EA form of the enzyme is affected by the presence of inhibitor. The intercept increases linearly with the concentration of inhibitor, as shown in equation 6-11.

A graphical representation of uncompetitive inhibition is provided in Figure 6-3, along with secondary intercept replot.

A set of parallel lines is obtained in the double reciprocal plot when the initial rate is measured at different fixed levels of the varied reactant. An estimate of the

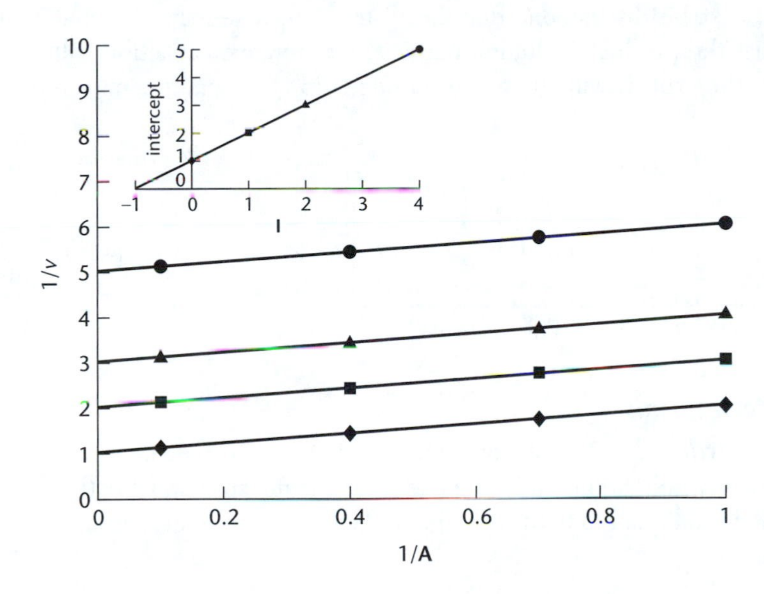

Figure 6-3. Primary double reciprocal plot for uncompetitive inhibition. Units on the ordinate and abscissa are minutes/millimolar and millimolar^{-1}, respectively. The varied reactant is A, and the concentrations of I are 0, K_{ii}, $2K_{is}$, and $4K_{is}$; the concentration of B is fixed equal to K_b. The values of V_{app}, K_{app}, and K_{ii} are 1 mM/min, 1 mM, and 1 mM, respectively. The inset shows the secondary plot of the intercepts of the primary plots versus **I**. The abscissa intercept of the replot is appK_{ii}.

uncompetitive inhibition constant, K_{ii}, is obtained as the abscissa intercept of the intercept replot.

Product Inhibition

As suggested in Chapter 5, inhibition of the reaction upon addition of a product can be used to rule out kinetic mechanisms that could not be ruled out on the basis of consideration of initial velocity studies in the absence of added inhibitors alone. The ability of product inhibition to define the kinetic mechanism derives directly from the overall rate equation for a given enzyme-catalyzed reaction. Specifically, terms in the denominator of the rate equation (distribution of enzyme into different species) that are invisible when product is not present upon initiation of the reaction are visualized when the product is added prior to initiation of the reaction. Thus, the basis for a specific product inhibition pattern being predicted for a given product in a specific Uni Bi reaction will first be discussed. After the discussion, a set of rules that can be used to predict the product inhibition patterns for any given reaction will be presented, and the rules will then be applied to bireactant enzyme mechanisms.

$$\begin{array}{ccccccc} & \mathbf{A} & & \mathbf{P} & & \mathbf{Q} & \\ & \downarrow k_1 \mid k_2 & & \uparrow k_3 \mid k_4 & & \uparrow k_5 \mid k_6 & \\ \text{E} & & (\text{EA} \rightleftharpoons \text{EPQ}) & & \text{EQ} & & \text{E} \end{array} \qquad (6\text{-}12)$$

Uni Bi Reaction Mechanisms

Uni Bi Steady-State Ordered. The steady-state ordered Uni Bi mechanism is presented schematically in equation 6-12, and the overall rate equation for this mechanism is presented in equation 6-13 (see Appendix 2 for the rate equation in terms of rate constants):

$$v = \frac{V_1\mathbf{A} - V_2\mathbf{PQ}K_a/K_pK_{iq}}{K_a + \mathbf{A} + \mathbf{AP}/K_{ip} + K_a\mathbf{P}K_q/K_pK_{iq} + K_a\mathbf{PQ}/K_pK_{iq} + K_a\mathbf{Q}/K_{iq}} \qquad (6\text{-}13)$$

(The $\mathbf{P}$ term can also be written $K_{ia}\mathbf{P}/K_{ip}$.) There are two products, P and Q, which can be used to obtain inhibition patterns versus A. Experiments are carried out by measuring the initial rate as a function of $\mathbf{A}$ at each of several different concentrations of one inhibitor including zero, with the other product maintained at zero concentration. Setting **P** equal to zero in equation 6-13 will thus generate the equation for product inhibition by Q, as shown in equation 6-14, with its reciprocal form given in equation 6-15:

$$v = \frac{V_1\mathbf{A}}{K_a(1+\mathbf{Q}/K_{iq}) + \mathbf{A}} \tag{6-14}$$

$$\frac{1}{v} = \left(\frac{K_a}{V_1}\right)\left(1+\frac{\mathbf{Q}}{K_{iq}}\right)\left(\frac{1}{\mathbf{A}}\right) + \frac{1}{V_1} \tag{6-15}$$

Note that in equation 6-14 it is the K_a term that is modified by the presence of Q, and thus the inhibition by Q is competitive versus A with both combining to free E. A replot of the slope versus **Q** gives the value of K_{iq} as the abscissa intercept.

If **Q** is set equal to zero, equation 6-13 reduces to equation 6-16 and its reciprocal, equation 6-17, the equations for product inhibition by P. In this case both the **A** and K_a terms are modified by the presence of P, and thus the inhibition is noncompetitive. The K_pK_{iq}/K_q term in equations 6-13 to 6-16 can be expressed as $K_{ip}K_a/K_{ia}$ by use of the Haldane relationships; see Chapter 5.

$$v = \frac{V_1\mathbf{A}}{K_a(1+K_q\mathbf{P}/K_pK_{iq}) + \mathbf{A}(1+\mathbf{P}/K_{ip})} \tag{6-16}$$

$$\frac{1}{v} = \left(\frac{K_a}{V_1}\right)\left(1+\frac{K_q\mathbf{P}}{K_pK_{iq}}\right)\left(\frac{1}{\mathbf{A}}\right) + \left(\frac{1}{V_1}\right)\left(1+\frac{\mathbf{P}}{K_{ip}}\right) \tag{6-17}$$

An estimate of K_pK_{iq}/K_q or K_aK_{ip}/K_{ia} is obtained as the abscissa intercept of the slope replot against **P**, while an estimate of K_{ip} is obtained as the abscissa intercept of the intercept replot against **P**. The enzyme form with which P combines is the EQ complex, and thus in order to observe inhibition by P, EQ must be present under conditions of steady-state turnover of the enzyme. The latter requires that one of the steps that at least contributes to rate limitation for the overall reaction at saturating **A** must be dissociation of the EQ complex (see Chapter 2 for rate processes included in the expressions for V and V/K). The noncompetitive pattern is a result of two phenomena in the present case. At limiting **A**, combination of P with the EQ complex results in at least a partial prevention of the release of Q and reversal of the reaction to produce A, decreasing the amount of product produced (equation 6-18):

$$\overset{k_5}{\longleftarrow} \mathrm{EQ} \underset{k_3}{\overset{k_4\mathbf{P}}{\rightleftharpoons}} EPQ \rightleftharpoons EA \overset{k_2}{\longrightarrow} \mathrm{E} \qquad (6\text{-}18)$$

The observation of inhibition at limiting **A** requires that **P** be high enough to trap the relatively low level of EQ produced. The expression for K_{ip} is $(k_3+k_5)/k_4$. Thus, if $k_5 < k_3$, that is, the off rate constant for Q is slower than the off rate constant for P, K_{ip} will equal k_3/k_4, the true dissociation constant of EPQ to EQ and P. From the intercept of equation 6-17, it can be seen that K_{ii} is equal to K_{ip}, while K_{is} is equal to K_aK_{ip}/K_{ia}, which in terms of rate constants is equal to $(k_2+k_3)k_5/k_2k_4$. The value of K_{is} relative to K_{ip} will be determined by the ratio of K_{ia} and K_a. If k_2 and k_5 are equal to one another, $K_{ia}=K_a$ and $K_{is}=K_{ip}$. This also occurs if $k_3 < k_2$ and k_5, where $K_{ip}=k_5/k_4$, much higher than the true dissociation constant of P. The inhibition by P at low **A** can be thought of as a pressure against the flow to produce products, much as a small dam in a brook. The slope component of the noncompetitive inhibition results from the pressure against reaction flow. Increasing **A** overcomes the latter pressure but cannot overcome the inhibition by P, because the EQ complex is still produced, and binding of P to EQ decreases the effective concentration of enzyme producing P and Q. According to the analogy, increasing **A** increases flow in the direction of P and Q but over the dam, which is still present and still impedes flow. The intercept component of the noncompetitive inhibition results from the ability of P to combine with EQ, even at infinite **A**; that is, infinite **A** does not eliminate the binding site for P as an inhibitor.

Product inhibition patterns will thus provide additional information concerning the kinetic mechanism for a Uni Bi Ordered kinetic mechanism, with competitive inhibition predicted for Q versus A and noncompetitive predicted for P versus A. In Chapter 5, initial velocity patterns in the absence of added inhibitors were unable to distinguish between Uni Bi steady-state ordered and rapid equilibrium random kinetic mechanisms. Can product inhibition be used to distinguish between these kinetic mechanisms?

Uni Bi Rapid Equilibrium Random. The rapid equilibrium random Uni Bi mechanism is presented schematically in equation 6-19. In this mechanism,

E, EP, and EQ must be in equilibrium.

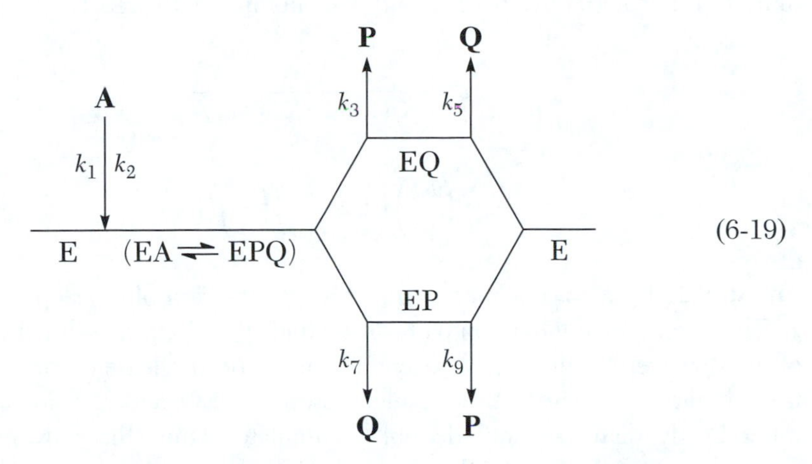

(6-19)

The overall rate for this mechanism is given in equation 6-20 (see Appendix 2 for the rate equation in terms of rate constants).

$$v = \frac{V_1\mathbf{A} - V_2\mathbf{PQ}K_a/K_pK_{iq}}{K_a + \mathbf{A} + K_a\mathbf{P}/K_{ip} + K_a\mathbf{Q}/K_{iq} + K_a\mathbf{PQ}/K_pK_{iq}} \qquad (6\text{-}20)$$

Setting **P** equal to zero in equation 6-20 generates the equation for product inhibition by Q, as shown in equation 6-21, and its reciprocal form is given in equation 6-22:

$$v = \frac{V_1\mathbf{A}}{K_a(1 + \mathbf{Q}/K_{iq}) + \mathbf{A}} \qquad (6\text{-}21)$$

$$\frac{1}{v} = \left(\frac{K_a}{V_1}\right)\left(1 + \frac{\mathbf{Q}}{K_{iq}}\right)\left(\frac{1}{\mathbf{A}}\right) + \frac{1}{V_1} \qquad (6\text{-}22)$$

The K_a term of equation 6-22 is modified by the presence of Q, and thus the inhibition by Q is competitive versus A, with both combining with free E. A replot of the slope versus **Q** gives the value of K_{iq} as the abscissa intercept. This pattern is the same as that seen in the ordered case (equations 6-14 and 6-15).

If the concentration of Q is set equal to zero, equation 6-20 reduces to equation 6-23 and its reciprocal, equation 6-24, the equations for product inhibition by P.

Again it is the K_a term that is modified by the presence of P, and thus the inhibition is competitive, with P and A combining with free E.

$$v = \frac{V_1\mathbf{A}}{K_a(1+\mathbf{P}/K_{ip}) + \mathbf{A}} \tag{6-23}$$

$$\frac{1}{v} = \left(\frac{K_a}{V_1}\right)\left(1+\frac{\mathbf{P}}{K_{ip}}\right)\left(\frac{1}{\mathbf{A}}\right)+\frac{1}{V_1} \tag{6-24}$$

An estimate for K_{ip} is obtained as the abscissa intercept of the slope replot against **P**. The competitive inhibition patterns by both P and Q are a direct consequence of the slow step for the overall reaction being prior to release of the products from their binary complexes. As a result, as soon as EQ and EP are formed, they immediately dissociate and do not accumulate. Thus, the only enzyme form available for combination with P and Q is E, the same enzyme form with which A must combine for reaction.

Product inhibition patterns can thus provide a means to distinguish between different kinetic mechanisms. As will be seen below, however, the distinction cannot be made in every case by product inhibition, and still other techniques must be utilized.

Rules for Predicting Product Inhibition Patterns

While the above inhibition patterns for P and Q versus A were explained, several conclusions were reached concerning the reasons slope and intercept inhibitions were observed. In the case of inhibition by Q in the Ordered Uni Bi mechanism, A binds to the same enzyme form and provides a slope effect. In the case of inhibition by P, the substrate combines with a different enzyme form, but the two are reversibly connected to one another. In addition, increasing **A** to infinity will not overcome the inhibition by P. These and other concepts were utilized by Cleland (BBA 67, 188) to devise a set of simple rules one can use to predict the product inhbition patterns for any given kinetic mechanism. The rules are as follows.

(1) A product inhibitor affects the slope of a double reciprocal plot when (a) the product inhibitor and the varied substrate combine reversibly with the same enzyme form or (b) the product inhibitor and the varied substrate combine reversibly with different enzyme forms and the two forms are connected by a series of reversible steps along which reaction can occur.

(2) A product inhibitor affects the intercept of a double reciprocal plot when it combines reversibly with an enzyme form other than that with which the varied substrate combines and saturation with the varied substrate does not overcome the inhibition.

As a result, prediction of product inhibition patterns for any mechanism becomes a simple and systematic process. Several questions are asked of each inhibitor at each of its combinations, and the pattern is determined as a result of the presence of slope and intercept effects.

The rules can be applied to the two Uni Bi reaction mechanisms already considered. By reference to mechanism 6-12 above, the following table can be completed. The steady-state random case can be very complex and will not be considered here. However, in many cases the data for this mechanism will closely fit the equations for the rapid equilibrium mechanism discussed below.

In Table 6-1, the substrate/product column refers to the varied substrate and the product added to determine the inhibition pattern. The second column refers to the enzyme form with which each combines, while the third column refers to whether a reversible connection exists between the combination of substrate and product. The fourth column refers to whether increasing the concentration of the varied substrate to infinity eliminates the inhibition, and the last column gives the predicted inhibition pattern. Thus, combination with the same enzyme form for E and Q gives only a slope effect and competitive inhibition (abbreviated C). Combination of A and P with different enzyme forms and the inability of A to overcome its inhibition at infinite concentrations indicates an intercept effect, while the reversible connection between the two points of combination indicate a slope effect, giving an overall noncompetitive pattern (abbreviated NC).

The case of mechanism 6-19 is considered in Table 6-2.

Table 6-1. Prediction of Product Inhibition Patterns in a Uni Bi Ordered Mechanism

Substrate/ product	Enzyme form bound	Reversible connection	Overcome inhibition	Inhibition pattern
A	E			
Q	E	–	yes	C
A	E			
P	EQ	yes	no	NC

Table 6-2. Prediction of Product Inhibition Patterns in a Uni Bi Rapid Equilibrium Random Mechanism

Substrate/ product	Enzyme form bound	Reversible connection	Overcome inhibition	Inhibition pattern
A	E			
Q	E, EP	yes	yes	C
A	E			
P	E, EQ	yes	yes	C

The products combine with E and the binary complex containing the other product. However, E, EP, and EQ are in equilibrium so inhibition can be overcome as a result of saturation with A, which will eliminate all three enzyme forms. The interpretation in this instance is quite simple since the substrate and both products combine with E, giving competitive inhibition patterns. A steady-state random mechanism could give NC for one or both.

Bireactant Enzyme Mechanisms

By the approach developed for the Uni Bi mechanisms above, the Bi Bi reaction mechanisms will be considered. The presence of a second substrate in the case of the bireactant mechanisms complicates matters slightly. One of the two reactant concentrations will remain fixed at a constant level, while that of the other is varied at different fixed levels of the inhibitor. However, since saturation with the fixed reactant establishes an irreversible step in the mechanism, the inhibition pattern must be obtained under conditions in which the fixed reactant is saturating and repeated under conditions where it is not. The number of possible inhibition patterns is thus increased for the Bi Bi reactions.

Ordered Kinetic Mechanisms

Steady-State Ordered. The schematic representation of a steady-state ordered mechanism is given in equation 6-25:

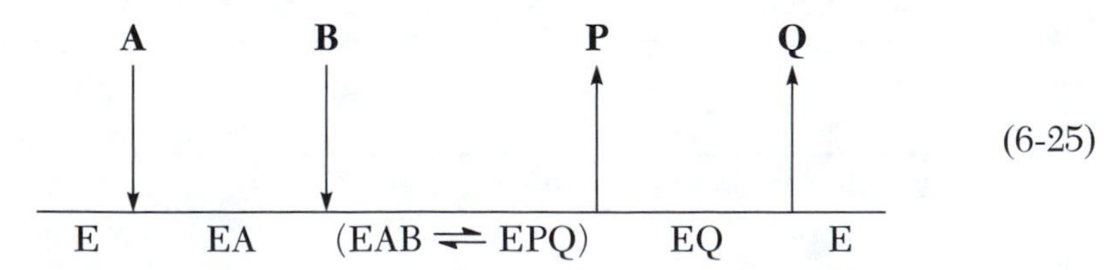

(6-25)

The inhibition patterns predicted for the steady-state ordered mechanism are provided in Table 6-3. There are eight possible product inhibition patterns that can be carried out for a bireactant mechanism, with each substrate fixed at a nonsaturating and saturating concentration and the other substrate varied for each of the two products. Note that seven of the eight possible product inhibition patterns can be measured for the steady-state ordered mechanism. Since Q competes with A, the pattern with A saturating and Q as a product inhibitor versus B gives no inhibition. (In practice, a very high level of A causes a proportional increase in the inhibition constant of Q, so weak inhibition may still be observed.) As a general rule, any time a competitive pattern is observed, one of the possible inhibition patterns will not show inhibition.

The competitive inhibition by Q versus A will be qualitatively and quantitatively independent of the concentration of B. An increase in the concentration of B will change the amount of E available, and thus there are two terms in the denominator of the rate equation for E (see Chapter 5). However, E is continually produced by the reaction, since the conversion of EAB to EPQ cannot be the sole rate-determining step for this mechanism (if it is the sole rate-determining step,

Table 6-3. Inhibition Patterns Predicted for Bireactant Enzyme-Catalyzed Reactions[a]

Substrate/ product	Enzyme form bound	Reversible connection	Overcome inhibition	Inhibition pattern
		Steady-State Ordered		
B = K_b or $50K_b$				
A	E			
Q	E (E′)	(yes)	yes (no)	C (NC)[a]
A = K_a (at saturating **A** there is no inhibition by Q)				
B	EA			
Q	E	yes	no	NC[a]
B = K_b				
A	E			
P	EQ	yes	no	NC
B = $50K_b$				
A	E			
P	EQ	no	no	UC

(Continued)

Table 6-3. Continued

Substrate/ product	Enzyme form bound	Reversible connection	Overcome inhibition	Inhibition Pattern
$\mathbf{A} = K_a$ or $50K_a$				
B	EA			
P	EQ	yes	no	NC
	Theorell–Chance Mechanism			
$\mathbf{B} = K_b$ or $50K_b$				
A	E			
Q	E (E′)	(yes)	yes (no)	C (NC)[a]
$\mathbf{A} = K_a$ (at saturating **A** there is no inhibition by Q)				
B	EA			
Q	E	yes	no	NC[a]
$\mathbf{B} = K_b$				
A	E			
P	EQ	yes	no	NC
$\mathbf{A} = K_a$ or $50K_a$				
B	EA			
P	EQ	yes	yes	C
	Equilibrium Ordered[b]			
$\mathbf{B} = K_b$ (at saturating **B**, V_{max} is observed at all concentrations of A)				
A	E			
Q	E	-	yes	C
$\mathbf{A} = K_a$ (at saturating **A** there is no inhibition by Q)				
B	EA			
Q	E	yes	yes	C
$\mathbf{B} = K_b$ or $50K_b$				
A	E			
P	EQ	-	N/A	no inhib.
$\mathbf{A} = K_a$ or $50K_a$				
B	EA			
P	EQ	-	N/A	no inhib.
	Rapid Equilibrium Random with Dead-End EBQ Complex[c]			
$\mathbf{B} = K_b$ or $50K_b$				
A	E, EB			
Q	E, EB	-	yes	C

Table 6-3. Continued

Substrate/ product	Enzyme form bound	Reversible connection	Overcome inhibition	Inhibition pattern
A = K_a (at saturating **A** there is no inhibition by Q)				
B	E, EA			
Q	E	-	yes	C
	EB	no	no	UC
			net	NC
B = K_b (at saturating **B** there is no inhibition by P)				
A	E, EB			
P	E	-	yes	C
A = K_a (at saturating **A** there is no inhibition by P)				
B	E, EA			
P	E	-	yes	C
Rapid Equilibrium Random with Dead-End EAP and EBQ Complexes[d]				
B = K_b or 50K_b				
A	E, EB			
Q	E, EB	-	yes	C
A = K_a (at saturating **A** there is no inhibition by Q)				
B	E, EA			
Q	E	-	yes	C
	EB	no	no	UC
			net	NC
B = K_b or 50K_b				
A	E, EB			
P	E	-	yes	C
P	EA	no	no	UC
			net	NC
A = K_a or 50K_a				
B	E, EA			
P	E, EA	-	yes	C
Classical (One-Site) Ping Pong				
B = K_b or 50K_b				
A	E			
Q	E (E′)	(yes)	yes (no)	C (NC)[a]
A = K_a (at saturating **A** there is no inhibition by Q)				
B	F			
Q	E	yes	no	NC[a]

(*Continued*)

Table 6-3. Continued

Substrate/ product	Enzyme form bound	Reversible connection	Overcome inhibition	Inhibition pattern
B = K_b (at saturating **B** there is no inhibition by P)				
A	E			
P	F	yes	no	NC
A = K_a or 50K_a				
B	F			
P	F	-	yes	C
		Nonclassical (Two-Site) Ping Pong		
B = K_b (at saturating **B** there is no inhibition by Q)				
A	site 1			
Q	site 2	yes	no	NC
A = K_a or 50K_a				
B	site 2			
Q	site 2	-	yes	C
B = K_b or 50K_b				
A	site 1			
P	site 1	-	yes	C
A = K_a (at saturating **A** there is no inhibition by P)				
B	site 2			
P	site 1	yes	no	NC

[a]The terms in parentheses are those that would differ in the case of an Iso mechanism, that is, isomerization of free enzyme. The other three cases, for example, Q versus A at saturating **B** in an ordered mechanism, can also be carried out and give the same NC pattern since Q and A no longer bind to the same enzyme form.

[b]It is assumed that E and EQ are in equilibrium.

[c]The steady-state random case will be considered later in the chapter.

[d]One of the dead-end complexes, the one with the substrate/product pair that is missing the piece to be transferred, is expected to be observed. The second complex may be observed depending on the size of the piece transferred. If the piece is small, for example, transfer of a hydride from NADH to FAD, the second complex should form. As the piece becomes larger, for example, phosphoryl or acetyl transfer, the complex may form but with reduced affinity. Ultimately, with a large piece transferred, for example, succinyl or glycosyl transfer, the product will be excluded from the active site by the substrate with which it overlaps.

the mechanism will be equilibrium ordered). In addition, both A and Q can compete equally for any E available. The independence of the inhibition by Q of the concentration of B is shown in equation 6-26, the reciprocal form of the steady-state initial velocity rate equation in the presence of Q. [Both of the E terms in the denominator of the rate equation are modified by $(1 + \mathbf{Q}/K_{iq})$. The rate equation is derived from the overall rate equation for a steady-state ordered mechanism; see Appendix 2.]

$$\frac{1}{v} = \left(\frac{K_a}{V}\right)\left(1 + \frac{K_{ia}K_b}{K_a\mathbf{B}}\right)\left(1 + \frac{\mathbf{Q}}{K_{iq}}\right)\left(\frac{1}{\mathbf{A}}\right) + \left(\frac{1}{V}\right)\left(1 + \frac{K_b}{\mathbf{B}}\right) \tag{6-26}$$

The slope term is thus equal to $(K_a/V)(1 + K_{ia}K_b/K_a\mathbf{B})(1 + \mathbf{Q}/K_{iq})$, and the absolute value of the abscissa intercept of the slope replot will be equal to K_{iq}.

The noncompetitive inhibition by Q versus B results from two phenomena: the first is the combination of Q with E and B with EA, and the second is the inability of B to overcome the inhibition by Q, since E is continually produced. Thus, inhibition is still obtained at infinite **B**, giving an effect on the intercept. The effect on the slope results from the fact that B can affect the amount of E present for combination with Q. A rearrangement of equation 6-26 above so that B is the varied reactant is given as equation 6-27, with slope and intercept expressions as equations 6-28 and 6-29:

$$\frac{1}{v} = \left(\frac{K_b}{V\mathbf{B}}\right)\left[1 + \left(\frac{K_{ia}}{\mathbf{A}}\right)\left(1 + \frac{\mathbf{Q}}{K_{iq}}\right)\right] + \left(\frac{1}{V}\right)\left[1 + \left(\frac{K_a}{\mathbf{A}}\right)\left(1 + \frac{\mathbf{Q}}{K_{iq}}\right)\right] \tag{6-27}$$

$$\text{slope} = \left(\frac{K_{ia}K_b}{V\mathbf{A}K_{iq}}\right)\mathbf{Q} + \left(\frac{K_b}{V}\right)\left(1 + \frac{K_{ia}}{\mathbf{A}}\right) = \left(\frac{K_b}{V}\right)\left(1 + \frac{K_{ia}}{\mathbf{A}}\right)\left(1 + \frac{\mathbf{Q}}{K_{iq}(1 + \mathbf{A}/K_{ia})}\right) \tag{6-28}$$

$$\text{intercept} = \left(\frac{K_a}{V\mathbf{A}K_{iq}}\right)\mathbf{Q} + \frac{K_a}{V\mathbf{A}} + \frac{1}{V} = \left(\frac{1}{V}\right)\left(1 + \frac{K_a}{\mathbf{A}}\right)\left(1 + \frac{\mathbf{Q}}{K_{iq}(1 + \mathbf{A}/K_a)}\right) \tag{6-29}$$

The apparent inhibition constants of Q in equations 6-28 and 6-29 are $K_{iq}(1 + \mathbf{A}/K_{ia})$ and $K_{iq}(1 + \mathbf{A}/K_a)$, respectively. The apparent value will be greater than the true K_{iq} depending on **A** compared to K_{ia} (the dissociation constant for EA) at low **B** and on **A** compared to K_a (the steady-state dissociation constant for EA) at infinite **B**. The values $(1+\mathbf{A}/K_{ia})$ and $(1+\mathbf{A}/K_a)$ can be considered difficulty factors for the binding of Q (just as $1+\mathbf{Q}/K_{iq}$ is a difficulty factor for the binding of A), with higher and higher **A** depleting **E** available for Q. It should be noted at this point that the product inhibition patterns obtained with Q versus A give a true K_{iq} value and that the values of the apparent K_{iq} versus **B** can be corrected to true K_{iq} values if the **A** used and values for K_a and K_{ia} are known. Identity of the three values (within error) is an important test for the assignment of a kinetic mechanism, just as is adherence of kinetic data to the Haldane relationship (see Chapter 5).

Inhibition by P, as suggested for the Uni Bi Ordered mechanism, requires that EQ exist in the steady-state, so that release of Q must be at least one of the slow steps along the reaction pathway. All four patterns are observed with P as a product inhibitor, since none are competitive (Table 6-3). With B as the varied substrate, at any **A**, P inhibits at low **B** by its pressure against forward reaction flux. Increasing **B** overcomes this effect partially (slope effect), but infinite **B** cannot eliminate the enzyme form (EQ) to which P binds, and thus an intercept effect is observed (see unireactant mechanisms above). Increasing **A** increases the amount of EA to which B binds and thus facilitates the reversal of the P effect by B. The rate equation for product inhibition by P with **B** varied is given in equation 6-30, with slope and intercept expressions given in equations 6-31 and 6-32:

$$\frac{1}{v}=\left(\frac{K_b}{V\mathbf{B}}\right)\left(1+\frac{K_{ia}}{\mathbf{A}}\right)\left(1+\frac{\mathbf{P}}{(K_pK_{iq}/K_q)}\right)+\left(\frac{1}{V}\right)\left(1+\frac{K_a}{\mathbf{A}}\right)\left(1+\frac{\mathbf{P}}{K_{ip}(1+K_a/\mathbf{A})}\right) \tag{6-30}$$

$$\begin{aligned}\text{slope}&=\left(\frac{K_{ia}K_b}{V\mathbf{A}}+\frac{K_b}{V}\right)\left(\frac{K_q\mathbf{P}}{K_pK_{iq}}\right)+\frac{K_{ia}K_b}{V\mathbf{A}}+\frac{K_b}{V}\\&=\left(\frac{K_b}{V}\right)\left(1+\frac{K_{ia}}{\mathbf{A}}\right)\left(1+\frac{\mathbf{P}}{K_pK_{iq}/K_q}\right)\end{aligned} \tag{6-31}$$

$$\text{intercept}=\left(\frac{1}{VK_{ip}}\right)\mathbf{P}+\frac{K_a}{V\mathbf{A}}+\frac{1}{V}=\left(\frac{1}{V}\right)\left(1+\frac{K_a}{\mathbf{A}}\right)\left(1+\frac{\mathbf{P}}{K_{ip}(1+K_a/\mathbf{A})}\right) \tag{6-32}$$

The apparent inhibition constants of P in equations 6-31 and 6-32 are K_pK_{iq}/K_q and $K_{ip}(1+K_a/\mathbf{A})$, respectively. Thus the apparent inhibition constant obtained from the slope term is independent of **A**, while the apparent value obtained from the intercept term will be dependent on **A**. When **A** is lower than K_a

(the steady-state dissociation constant for EA), the apparent value will be greater than K_{ip}, while at infinite **A**, a condition that favors the maximum production of EQ, the value will be equal to K_{ip}. It should be noted that K_{ip} is not a true dissociation constant for P dissociating from EPQ and will depend on the level of EQ that builds up in the steady state, determined by the rate constant for release of Q.

With A as the varied substrate, at low **A** and **B**, the predominant enzyme form is E, but as suggested above, the **P** used is sufficient to slow down the release of Q so that it becomes slow enough to be seen. Increasing **A** provides more EA to which B binds and thus facilitates overcoming the negative pressure by P (slope effect). At infinite **A**, the EQ form of the enzyme still exists, and there is still inhibition by P (intercept effect). The end result is the noncompetitive pattern by P versus A.

The rate equation for product inhibition by P with **A** varied is given in equation 6-33, with slope and intercept expressions given in equations 6-34 and 6-35:

$$\frac{1}{v} = \left(\frac{K_a}{V\mathbf{A}}\right)\left(1+\frac{K_{ia}K_b}{K_a\mathbf{B}}\right)\left[1+\frac{\mathbf{P}}{(K_pK_{iq}/K_q)(1+K_a\mathbf{B}/K_{ia}K_b)}\right] + \left(\frac{1}{V}\right)\left(1+\frac{K_b}{\mathbf{B}}\right)\left[1+\frac{\mathbf{P}}{\left(1+\frac{K_b}{\mathbf{B}}\right)\Big/\left(\frac{1}{K_{ip}}+\frac{K_qK_b}{K_pK_{iq}\mathbf{B}}\right)}\right] \tag{6-33}$$

$$\text{slope} = \left(\frac{K_{ia}K_bK_q}{K_pK_{iq}V\mathbf{B}}\right)\mathbf{P}+\frac{K_{ia}K_b}{V\mathbf{B}}+\frac{K_a}{V} = \left(\frac{K_a}{V}\right)\left(1+\frac{K_{ia}K_b}{K_a\mathbf{B}}\right)\left[1+\frac{\mathbf{P}}{\frac{K_pK_{iq}}{K_q}\left(1+\frac{K_a\mathbf{B}}{K_{ia}K_b}\right)}\right] \tag{6-34}$$

$$\text{intercept} = \left(\frac{K_bK_q}{K_pK_{iq}V\mathbf{B}}+\frac{1}{VK_{ip}}\right)\mathbf{P}+\frac{K_b}{V\mathbf{B}}+\frac{1}{V} = \left(\frac{1}{V}\right)\left(1+\frac{K_b}{\mathbf{B}}\right)\left[1+\frac{\mathbf{P}}{\left(1+\frac{K_b}{\mathbf{B}}\right)\Big/\left(\frac{1}{K_{ip}}+\frac{K_qK_b}{K_pK_{iq}\mathbf{B}}\right)}\right] \tag{6-35}$$

This inhibition is noncompetitive if **B** is not saturating. The apparent inhibition constants of P in equations 6-34 and 6-35 are $(K_pK_{iq}/K_q)[1 + \mathbf{B}/(K_{ia}K_b/K_a)]$ and $(1 + K_b/\mathbf{B})/[1/K_{ip} + K_qK_b/K_pK_{iq}\mathbf{B}]$, respectively. Thus the apparent inhibition constant obtained from the slope term is equal to K_pK_{iq}/K_q only when **B** is less than $K_{ia}K_b/K_a$, that is, when both **A** and **B** are low, favoring E and EQ. The apparent value obtained from the intercept term varies from K_pK_{iq}/K_q at low **B** to K_{ip} at high **B**. The amount of EQ formed will increase as a hyperbolic function of **B**.

At saturating **B**, however, with A as the varied substrate, a distinctly different inhibition pattern is obtained by P. At low **A**, the predominant enzyme form is E, and as soon as a small amount of EA is formed it is converted back to EAB (the rate of production of products under these conditions is k_3[**B**][**EA**], and **B** is very high), and as a result no inhibition by P is observed. At infinite **A**, however, EQ is present in the steady state, and P inhibits. The inhibition pattern thus changes from noncompetitive at low **B** to uncompetitive at saturating **B**; saturating **B** interrupts the reversible connection from P to A (see rules above). The change in inhibition pattern is diagnostic for a steady-state ordered mechanism.

Theorell–Chance. A schematic representation of a Theorell–Chance mechanism is given in equation 6-36 (see Appendix 2 for the rate equation in terms of rate constants).

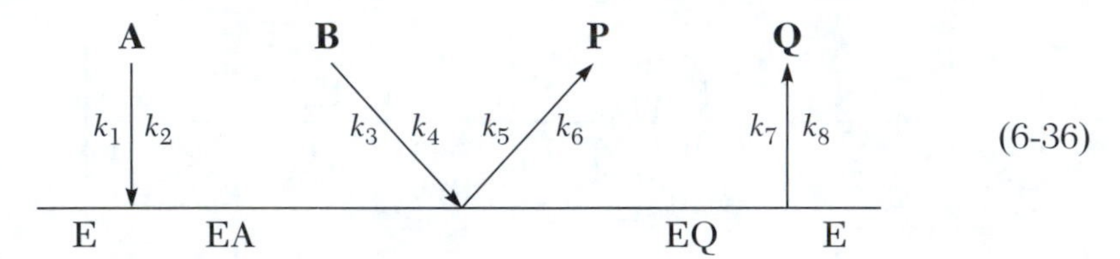

(6-36)

The inhibition patterns predicted for this mechanism are provided in Table 6-3. Six of the eight possible product inhibition patterns can be measured for the Theorell–Chance mechanism. Since Q competes with A and P competes with B, the patterns with **A** saturating and Q as a product inhibitor versus B, and with **B** saturating and P as a product inhibitor of A, give no inhibition. The competitive inhibition by Q versus A at any **B** is interpreted in a manner qualitatively and quantitatively (see equations 6-26 to 6-29 above) identical to the inhibition obtained for the steady-state ordered mechanism.

As suggested for the steady-state ordered mechanism above, inhibition by P requires that EQ exist in the steady state and that release of Q must be at least one of the slow steps along the reaction pathway. In the case of a Theorell–Chance mechanism, release of Q is the only slow step along the reaction pathway for the overall reaction with reactants saturating. With B as the varied substrate at any **A**, P inhibits at low **B** by its pressure against forward reaction flux. Since there is no central complex, increasing **B** will completely overcome the effect of P, resulting in a competitive inhibition. Increasing **A** increases the **EA** to which B binds and thus facilitates the reversal of the P effect by B. The rate equation for product inhibition by P with B varied is given in equation 6-37, with the slope expression given in equation 6-38:

$$\frac{1}{v}=\left(\frac{K_b}{V}\right)\left(1+\frac{K_{ia}}{\mathbf{A}}\right)\left[1+\frac{\mathbf{P}}{K_pK_{iq}/K_q}\right]\left(\frac{1}{\mathbf{B}}\right)+\left(\frac{1}{V}\right)\left(1+\frac{K_a}{\mathbf{A}}\right) \tag{6-37}$$

$$\text{slope}=\left(\frac{K_b}{V}\right)\left(1+\frac{K_{ia}}{\mathbf{A}}\right)\left[1+\frac{\mathbf{P}}{K_pK_{iq}/K_q}\right] \tag{6-38}$$

Note that, as discussed above, the inhibition is competitive. Setting the slope equal to zero and solving for the absolute value of **P** in equation 6-38 gives an apparent K_{ip} value equal to K_pK_{iq}/K_q. The inhibition is independent of **A**.

With A as the varied substrate, at low **A** and **B**, the qualitative and quantitative interpretation of the inhibition by P is identical to that for the steady-state ordered mechanism (see equation 6-34 above). The inhibition by P will also depend on **B** when **A** is saturating as shown in equation 6-37 with the $K_{ia}/\mathbf{A}$ and $K_a/\mathbf{A}$ terms equal to zero. An increase in **B** causes more EQ to form, while an increase in **P** causes more EA to form, and inhibition will depend on the relative concentrations of P and Q compared to their dissociation constants.

Equilibrium Ordered. The schematic representation for the equilibrium ordered kinetic mechanism is identical to that provided above for the steady-state ordered mechanism, equation 6-25, except that the binding of A and Q (but not necessarily B and P) is assumed to be at equilibrium. The predicted product inhibition patterns are provided in Table 6-3. In the case of the rapid equilibrium ordered mechanism, only the two inhibition patterns with Q as the product will provide inhibition. As discussed in Chapter 5, this mechanism requires rapid equilibration of E with A and Q, that is, rapid release of Q from EQ once it is formed from EPQ. As a result, no EQ will build up in the steady state. Since Q competes with both A and B, the patterns with **A** or **B** saturating and Q as a product inhibitor versus B or A give no inhibition. The initial velocity rate equation is asymmetric with respect to A or B used as the varied reactant (see Chapter 5). The reciprocal form of the initial velocity rate equation for inhibition by Q with A as the varied reactant is given by equation 6-39:

$$\frac{1}{v}=\left(\frac{K_{ia}K_b}{V\mathbf{B}}\right)\left(1+\frac{\mathbf{Q}}{K_{iq}}\right)\left(\frac{1}{\mathbf{A}}\right)+\left(\frac{1}{V}\right)\left(1+\frac{K_b}{\mathbf{B}}\right) \tag{6-39}$$

The slope term $(K_{ia}K_b/V\mathbf{B})(1+\mathbf{Q}/K_{iq})$ is affected by **Q**, and the absolute value of the abscissa intercept of the slope replot is K_{iq}. The inhibition is interpreted as for the other ordered mechanisms in terms of Q affecting the **E** available for combination with A. Note also that increasing **B** to infinity will cause the slope term to tend to zero, eliminating the inhibition by Q. Rearrangement of equation 6-39 so that B is the varied reactant, however, also gives a competitive inhibition pattern, as shown in equation 6-40:

$$\frac{1}{v}=\left(\frac{K_b}{V}\right)\left(1+\frac{K_{ia}}{\mathbf{A}}\right)\left[1+\frac{\mathbf{Q}}{K_{iq}(1+\mathbf{A}/K_{ia})}\right]\left(\frac{1}{\mathbf{B}}\right)+\frac{1}{V} \qquad (6\text{-}40)$$

The slope term, $(K_b/V)(1+K_{ia}/\mathbf{A})(1+\mathbf{Q}/[K_{iq}(1+\mathbf{A}/K_{ia})])$, is more complex in this case, and the absolute value of the abscissa intercept will be equal to $K_{iq}(1+\mathbf{A}/K_{ia})$; that is, the apparent inhibition constant for Q will be a linear function of the ratio of $\mathbf{A}/K_{ia}$, which is equal to the ratio of **EA/E**. Increasing **B** to infinity at any **Q** will result, according to the following coupled equilibria, in EAB or EAB and EPQ and leave no E for Q to combine with:

$$\mathbf{E}+\mathbf{A} \rightleftharpoons \mathbf{EA}+\mathbf{B} \rightleftharpoons \mathbf{EAB} \qquad (6\text{-}41)$$

Random Kinetic Mechanisms

Rapid Equilibrium Random with EBQ Dead-End Complex. A schematic representation of the rapid equilibrium random mechanism is given in equation 6-42 (see Appendix II for the rate equation in terms of rate constants).

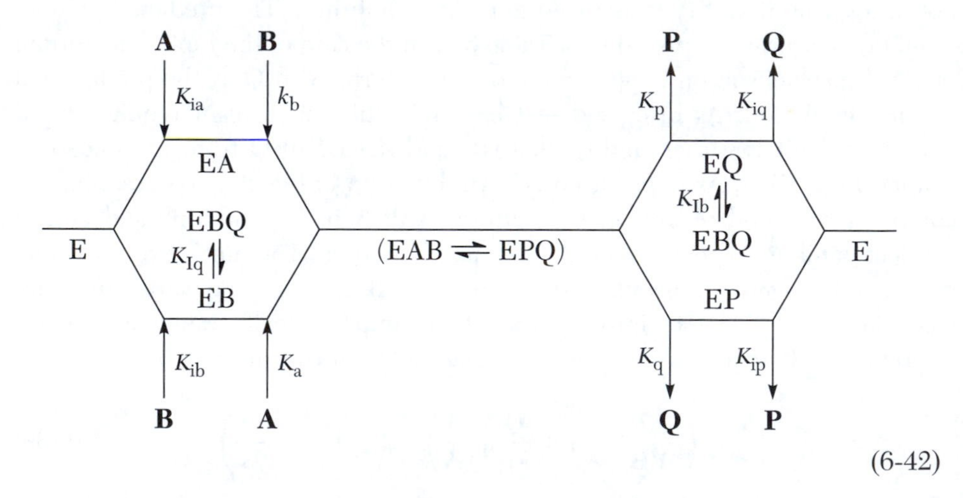

(6-42)

The predicted product inhibition patterns are provided in Table 6-3. Since Q will compete with A and P will compete with B, the inhibition patterns with the fixed reactant maintained at a saturating concentration give no inhibition. The interpretation of the competitive inhibition patterns by either P or Q versus either A or B is the same, with product competing with reactant for free enzyme. Neither the EQ nor the EP forms of the enzyme build up in the steady state as a result of the interconversion of EAB and EPQ and/or release of P and Q from EPQ being the rate-limiting step. A dead-end complex is predicted, however, in which the reactant and product that do not have the piece to be transferred (B and Q) can both bind to enzyme, as shown in Figure 6-4.

With **A** maintained at a nonsaturating level, increasing **B** favors the pathway in which B adds before A, and P will compete with B for E. Under these conditions Q will also compete with B for E, but it will additionally bind to the EB complex, and thus inhibition is observed at low **B** and at infinite **B**, a net noncompetitive inhibition. With **B** constant, increasing **A** favors the pathway

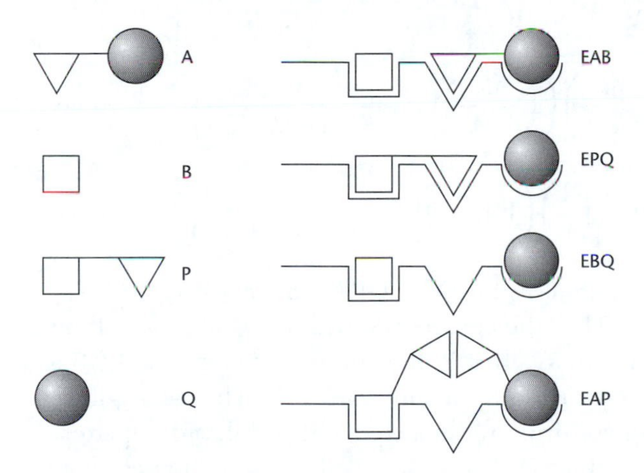

Figure 6-4. Graphical illustration of dead-end complexes in a rapid equilibrium random kinetic mechanism. The reactants, A, B, P, and Q, are shown on the left. The substrate and product ternary complexes are shown at the top right. The EBQ complex can form because the subsites for reactants without the group to be transferred are readily accommodated. The EAP complex, however, can form only with difficulty, since both reactants contain the group to be transferred.

in which A adds prior to B, and P and Q compete with A for E. The reciprocal form of the initial velocity rate equations for inhibition by Q with A or B as the varied reactants are given by equations 6-43 and 6-44 (note that $K_{ia}K_b = K_{ib}K_a$).

$$\frac{1}{v} = \left(\frac{K_a}{V\mathbf{A}}\right)\left(1+\frac{K_{ib}}{\mathbf{B}}\right)\left[1+\frac{\mathbf{Q}(K_{ib}/K_{iq}\mathbf{B}+1/K_{Iq})}{(1+K_{ib}/\mathbf{B})}\right]+\left(\frac{1}{V}\right)\left(1+\frac{K_b}{\mathbf{B}}\right) \tag{6-43}$$

$$\frac{1}{v} = \left(\frac{K_b}{V\mathbf{B}}\right)\left(1+\frac{K_{ia}}{\mathbf{A}}\right)\left[1+\frac{\mathbf{Q}}{K_{iq}(1+\mathbf{A}/K_{ia})}\right]+\left(\frac{1}{V}\right)\left(1+\frac{K_a}{\mathbf{A}}\right)\left[1+\frac{\mathbf{Q}}{K_{Iq}(1+\mathbf{A}/K_a)}\right] \tag{6-44}$$

Expessions for the slope of equation 6-43 and the slope and intercepts of equation 6-44 are given in equations 6-45 to 6-47 and include the expressions for apparent K_{iq}s.

$$\text{slope of equation 6-43} = \left(\frac{K_a}{V}\right)\left(1+\frac{K_{ib}}{\mathbf{B}}\right)\left[1+\frac{\mathbf{Q}(K_{ib}/\mathbf{B}K_{iq}+1/K_{Iq})}{1+K_{ib}/\mathbf{B}}\right] \tag{6-45}$$

$$\text{slope of equation 6-44} = \left(\frac{K_b}{V}\right)\left(1+\frac{K_{ia}}{\mathbf{A}}\right)\left[1+\frac{\mathbf{Q}}{K_{iq}(1+\mathbf{A}/K_{ia})}\right] \tag{6-46}$$

$$\text{intercept of equation 6-44} = \left(\frac{1}{V}\right)\left(1+\frac{K_a}{\mathbf{A}}\right)\left[1+\frac{\mathbf{Q}}{K_{Iq}(1+\mathbf{A}/K_a)}\right] \tag{6-47}$$

Equation 6-45 shows that the apparent K_i for Q is $(1+K_{ib}/\mathbf{B})/(K_{ib}/\mathbf{B}K_{iq} + 1/K_{Iq})$ or $(1+\mathbf{B}/K_{ib})/(1/K_{iq}+\mathbf{B}/K_{ib}K_{Iq})$. Thus the apparent K_i will be equal to K_{iq} at low **B** but K_{Iq} at high **B**. If K_{iq} is equal to K_{Iq}, that is, the affinity of Q for E and EB is the same, the apparent K_i will be equal to K_{iq} at any level of **B**. The values of K_{iq} and K_{Iq} are more easily determined from the noncompetitive inhibition pattern for Q versus B (equations 6-46 and 6-47). The apparent slope inhibition constant is equal to $K_{iq}(1+\mathbf{A}/K_{ia})$, where $\mathbf{A}/K_{ia}$ (**EA/E** from the dissociation constant for the EA complex) is a difficulty factor for binding of Q, based on the competition between A and Q. The apparent intercept inhibition constant is equal to $K_{Iq}(1+\mathbf{A}/K_a)$, where $\mathbf{A}/K_a$ has a meaning similar to A/K_{ia} (K_a is the steady-state dissociation constant; see Chapter 3).

Inhibition by P in mechanism 6-42 versus A and B is competitive, and binding is only to E. The rate equation varying **A** is equation 6-48, with the slope expression given in equation 6-49:

$$\frac{1}{v} = \left(\frac{K_a}{V}\right)\left(1+\frac{K_{ib}}{\mathbf{B}}\right)\left[1+\frac{\mathbf{P}}{K_{ip}(1+\mathbf{B}/K_{ib})}\right]\left(\frac{1}{\mathbf{A}}\right) + \left(\frac{1}{V}\right)\left(1+\frac{K_b}{\mathbf{B}}\right) \quad (6\text{-}48)$$

$$\text{slope} = \left(\frac{K_a}{V}\right)\left(1+\frac{K_{ib}}{\mathbf{B}}\right)\left[1+\frac{\mathbf{P}}{K_{ip}(1+\mathbf{B}/K_{ib})}\right] \quad (6\text{-}49)$$

The apparent K_i is equal to $K_{ip}(1+\mathbf{B}/K_{ib})$. With **B** varied the equation is similar, and the apparent K_i is equal to $K_{ip}(1+\mathbf{A}/K_{ia})$. There is also a possibility of P combining with EA to give a dead-end EAP complex (Figure 6-4, Table 6-3). The rate equations are very similar to those discussed above for the combination of Q with E and EB.

Steady-State Random. The steady-state random mechanism adheres schematically to equation 6-42, but unlike the rapid equilibrium mechanism, there are no restrictions on any of the rate constants. Patterns can be quite complex, but in most cases the data for this mechanism will closely fit the equations for the rapid equilibrium random mechanism.

Ping Pong Mechanisms

Classical (One-Site) Ping Pong Mechanism. The schematic representation of a one-site Ping Pong mechanism is given in equation 6-50 (see Appendix 2 for the rate equation in terms of rate constants).

$$\begin{array}{ccccccc} & \mathbf{A} & & \mathbf{P} \quad \mathbf{B} & & \mathbf{Q} & \\ & k_1 \downarrow k_2 & & k_3 \uparrow k_4 \quad k_5 \downarrow k_6 & & k_7 \uparrow k_8 & \\ \mathrm{E} & & (\mathrm{EA} \rightleftharpoons \mathrm{FP}) & \mathrm{F} & (\mathrm{FB} \rightleftharpoons \mathrm{EQ}) & & \mathrm{E} \end{array} \quad (6\text{-}50)$$

The inhibition patterns predicted for the classical Ping Pong mechanism are provided in Table 6-3. Six of the eight possible product inhibition patterns give inhibition, since Q competes with A and P competes with B. The product inhibition patterns obtained are qualitatively identical to those obtained for the Theorell–Chance mechanism (see above).

The competitive inhibition by Q versus A will depend on **B**. An increase in **B** will increase the **E** available, as the rate of conversion of F to E increases.

Both A and Q can compete equally for any E available. The dependence of the inhibition by Q on **B** is shown in equation 6-51, the reciprocal form of the steady-state initial velocity rate equation with A as the varied substrate in the presence of Q. Two terms containing Q are present in the denominator of the rate equation, both of which represent free E. The K_a/V term reflects **E** under conditions in which **B** is high compared to K_b and **A** is limiting, a condition that favors E as a result of the rapid formation of E from F and B. The second term, $K_aK_{ib}/V\mathbf{B}$, reflects free E at limiting concentrations of A and B. (The rate equation is derived from the overall rate equation for a Ping Pong mechanism; see Appendix 2.)

$$\frac{1}{v} = \left(\frac{K_a}{V}\right)\left[1 + \frac{\mathbf{Q}(1 + K_{ib}/\mathbf{B})}{K_{iq}}\right]\left(\frac{1}{\mathbf{A}}\right) + \left(\frac{1}{V}\right)\left(1 + \frac{K_b}{\mathbf{B}}\right) \qquad (6\text{-}51)$$

The slope of equation 6-51 is affected by Q and is shown in equation 6-52:

$$\text{slope} = \left(\frac{K_a}{V}\right)\left[1 + \frac{\mathbf{Q}(1 + K_{ib}/\mathbf{B})}{K_{iq}}\right] \qquad (6\text{-}52)$$

The apparent inhibition constant of Q in equation 6-52 is $K_{iq}/(1 + K_{ib}/\mathbf{B})$. Thus the apparent inhibition constant from the slope term will be equal to K_{iq} only when **B** is much greater than K_{ib}; that is, when **B** is high, favoring E alone. When **A** and **B** are both low, Q converts E to F and its apparent K_i value approaches zero.

In mechanism 6-50, the noncompetitive inhibition by Q versus B results from two phenomena. First is the combination of Q to E and B to F and the inability of increasing **B** to overcome the inhibition by Q, since E is continually produced from F and B. Thus, inhibition is still obtained at infinite **B** giving an effect on the intercept. Second is the effect on the slope resulting from the fact that **B** can affect the **E** present for combination with Q. A rearrangement of equation 6-51 above, so that B is the varied reactant, is shown in equation 6-53, with slope and intercept expressions in equations 6-54 and 6-55:

$$\frac{1}{v} = \left(\frac{K_b}{V}\right)\left(1 + \frac{K_aK_{ib}\mathbf{Q}}{\mathbf{A}K_bK_{iq}}\right)\left(\frac{1}{\mathbf{B}}\right) + \left(\frac{1}{V}\right)\left(1 + \frac{K_a}{\mathbf{A}}\right)\left[1 + \frac{\mathbf{Q}}{K_{iq}(1 + \mathbf{A}/K_a)}\right] \qquad (6\text{-}53)$$

$$\text{slope} = \left(\frac{K_b}{V}\right)\left[1 + \frac{\mathbf{Q}}{(K_{iq}K_b\mathbf{A}/K_aK_{ib})}\right] \qquad (6\text{-}54)$$

$$\text{intercept} = \left(\frac{1}{V}\right)\left(1 + \frac{K_a}{\mathbf{A}}\right)\left[1 + \frac{\mathbf{Q}}{K_{iq}(1 + \mathbf{A}/K_a)}\right] \qquad (6\text{-}55)$$

The apparent inhibition constants for Q in equations 6-54 and 6-55 are $(K_{iq}K_b\mathbf{A}/K_aK_{ib})$ (or $K_qK_{ip}\mathbf{A}/K_{ia}K_p$) and $K_{iq}(1 + \mathbf{A}/K_a)$, respectively. The slope inhibition constant is directly dependent on the ratio of **A** to K_aK_{ib}/K_b at low **B**. At low levels of **A** and **B**, any Q converts E back to F, so the apparent inhibition constant approaches zero. At saturating **B**, K_{iq} is dependent on **A** compared to its K_a (the steady-state dissociation constant for EA). Decreasing **A** to zero and **B** to infinity again ensures that all of the enzyme is present as E, the form to which Q binds. But at high **A**, there is no E in the steady state and inhibition by Q is eliminated.

The Ping Pong mechanism is symmetric with respect to the inhibition by products. Thus, P competes with B for the F form of the enzyme. An increase in A will increase the amount of F available for B, as the rate of conversion of E to F increases, and B and P can compete equally for the available F. Inhibition by P with **B** varied is shown in equation 6-56, the reciprocal form of the steady-state initial velocity rate equation in the presence of P. As for Q inhibition, two terms in the denominator of the rate equation for mechanism 6-50 contain **P**.

$$\frac{1}{v} = \left(\frac{K_b}{V}\right)\left[1 + \frac{\mathbf{P}}{[K_{ip}/(1 + K_{ia}/\mathbf{A})]}\right]\left(\frac{1}{\mathbf{B}}\right) + \left(\frac{1}{V}\right)\left(1 + \frac{K_a}{\mathbf{A}}\right) \tag{6-56}$$

The slope of equation 6-56 is a function of **P** and is shown in equation 6-57:

$$\text{slope} = \left(\frac{K_b}{V}\right)\left[1 + \frac{\mathbf{P}(1 + K_{ia}/\mathbf{A})}{K_{ip}}\right] \tag{6-57}$$

The apparent inhibition constant of P in equation 6-57 is $K_{ip}/(1 + K_{ia}/\mathbf{A})$. Thus the apparent inhibition constant from the slope term will be equal to K_{ip} only when **A** is much greater than K_{ia}, favoring formation of F to which P binds. As **A** approaches zero, the inhibition constant approaches zero.

With A as the varied substrate, P inhibits at low **B** by binding to the F form of the enzyme. The rate equation for inhibition by P with **A** varied is in equation 6-58, with slope and intercept expressions in equations 6-59 and 6-60:

$$\frac{1}{v} = \left(\frac{K_a}{V}\right)\left(1 + \frac{K_{ia}K_b\mathbf{P}}{K_a\mathbf{B}K_{ip}}\right)\left(\frac{1}{\mathbf{A}}\right) + \left(\frac{1}{V}\right)\left(1 + \frac{K_b}{\mathbf{B}}\right)\left[1 + \frac{\mathbf{P}}{K_{ip}(1 + \mathbf{B}/K_b)}\right] \tag{6-58}$$

$$\text{slope} = \left(\frac{K_a}{V}\right)\left[1 + \frac{\mathbf{P}}{(K_a\mathbf{B}K_{ip})/(K_{ia}K_b)}\right] \tag{6-59}$$

$$\text{intercept} = \left(\frac{1}{V}\right)\left(1+\frac{K_b}{\mathbf{B}}\right)\left[1+\frac{\mathbf{P}}{K_{ip}(1+\mathbf{B}/K_b)}\right] \qquad (6\text{-}60)$$

The apparent inhibition constants of P in equations 6-58 and 6-59, are $K_{ip}(K_a\mathbf{B}/K_{ia}K_b)$ and $K_{ip}(1+\mathbf{B}/K_b)$, respectively. The apparent inhibition constant is directly dependent on the ratio of **B** to $K_a/K_{ia}K_b$ at low **A**. As both **A** and **B** decrease to zero, P converts F back to E, so its apparent inhibition constant approaches zero. At saturating **A**, K_{ip} depends on the ratio of **B** and K_b (the steady-state dissociation constant for FB). Decreasing **B** to zero and **A** to infinity again ensures that all of the enzyme is present as F, the form to which P binds.

Nonclassical (Two-Site) Ping Pong Mechanism. The schematic representation of a two-site Ping Pong mechanism is given in equation 6-61 (see BBA 67, 173).

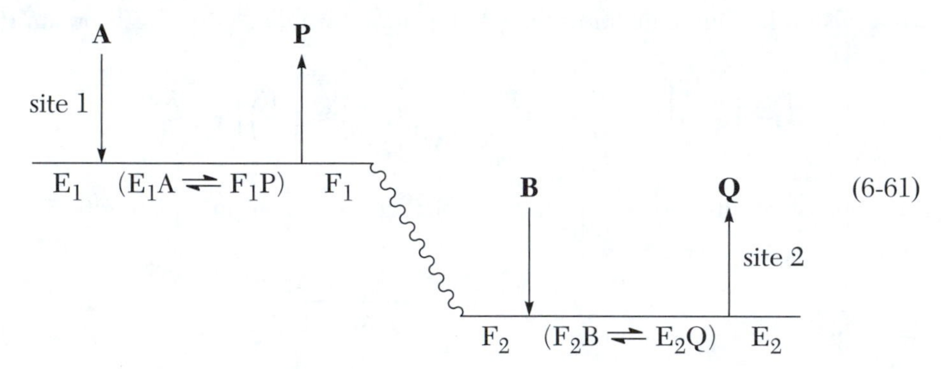

(6-61)

where the subscripts 1 and 2 reflect the two distinct sites. The inhibition patterns predicted for the nonclassical Ping Pong mechanism are provided in Table 6-3. Six of the eight possible product inhibition patterns can be measured, since P competes with A and Q competes with B. Note that the predicted product inhibition patterns are opposite those predicted for the classical Ping Pong mechanism (see above). The competitive reactant–product pairs are those that bind to the same site. As shown in mechanism 6-61, A produces P at site 1, while B picks up the piece of A that is transferred from site 1 to site 2 and produces Q.

Alternate Product Inhibition

An alternate product is one that would be produced from an alternate substrate. If product release is ordered (P followed by Q) and the alternate substrate produces products S and Q in that order, S as an alternate product produces interesting kinetic patterns. If production of Q is measured, S will act as a noncompetitive inhibitor. This is true whether there is only one substrate, as in a phosphatase, or whether there are two, as in an alcohol dehydrogenase. In this case, an alternate second substrate will produce an alternate first product and the same second product (NADH).

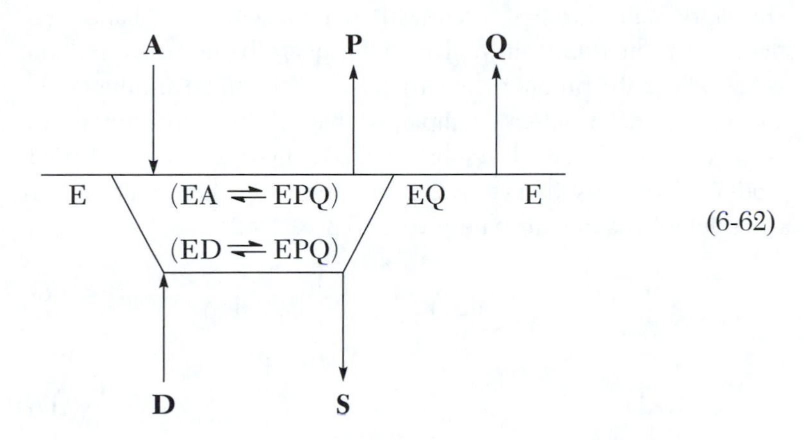

(6-62)

If production of P is measured, however, S produces hyperbolic uncompetitive inhibition or activation, depending on whether breakdown of EQ is slower than reaction of S with EQ to reverse the reaction (activation) or faster (inhibition). For the single substrate case, the rate equation for production of P is given as equation 6-63:

$$\frac{d\mathbf{t}}{d\mathbf{P}} = \frac{1}{v} = \left(\frac{K_a}{V_1}\right)\left(\frac{1}{\mathbf{A}}\right) + \frac{(1/V_1)(1 + \mathbf{S}/K_{iss})}{(1 + K_{id}\mathbf{S}/K_d K_{is})} \tag{6-63}$$

In equation 6-63, K_a and K_d are the Michaelis constants for substrate A (which produces P and Q) and the alternate substrate D (which produces S and Q), K_{id} is the dissociation constant of D, V_1 is the maximum velocity with substrate A, and K_{is} and K_{iss} are inhibition constants for S. The lack of a slope effect is explained by the fact that at low substrate concentration the collision of the substrate with the enzyme limits the rate, and the path by which EQ returns to be available for the next collision is unimportant. The intercept effect is hyperbolic because increasing levels of S divert the reaction of EQ from dissociation to E and Q to reaction with S and regeneration of free enzyme, or the EA complex in the two-substrate case.

Observation of the hyperbolic uncompetitive pattern shows that S does not combine with free enzyme (or with the EA complex in the two-substrate case). Any slope effect observed in the noncompetitive inhibition of S versus D thus comes solely from product inhibition and not from dead-end combination with E or EA. If S does combine with E or EA, it will give a linear slope inhibition in the alternate product inhibition study.

This approach was first used with potato acid phosphatase (B 5, 799). With glycerol production from β-glycerophosphate being measured and *p*-nitrophenol as the alternate product, hyperbolic uncompetitive inhibition by *p*-nitrophenol was observed. Another example is that of liver alcohol dehydrogenase with *p*-hydroxybenzyl alcohol as substrate and *p*-hydroxybenzaldehyde as the measured product (from absorbance at 282 nm, the isosbestic point of NAD–NADH) and acetaldehyde as an alternate product (B *4*, 2442).

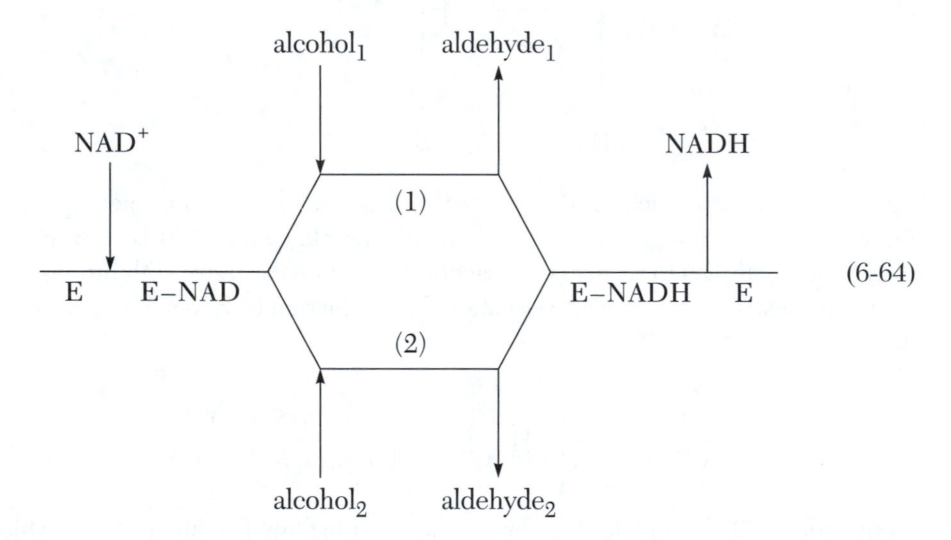

(6-64)

The pattern observed was S-linear inhibition I-hyperbolic activation. Thus acetaldehyde does bind to E–NAD, although weakly, and cause a linear slope effect. This experiment rules out a completely Theorell–Chance mechanism with these reactants. Such a mechanism as diagrammed in equation 6-65 involves only bimolecular steps for the conversion of EA to EQ and back. Thus the alternate product inhibition pattern should show an infinite maximum velocity for formation of P when both **B** and **S** are saturating. The fact that the intercept replot had a finite asymptote greater than zero shows that there *are* central complexes present in the mechanism.

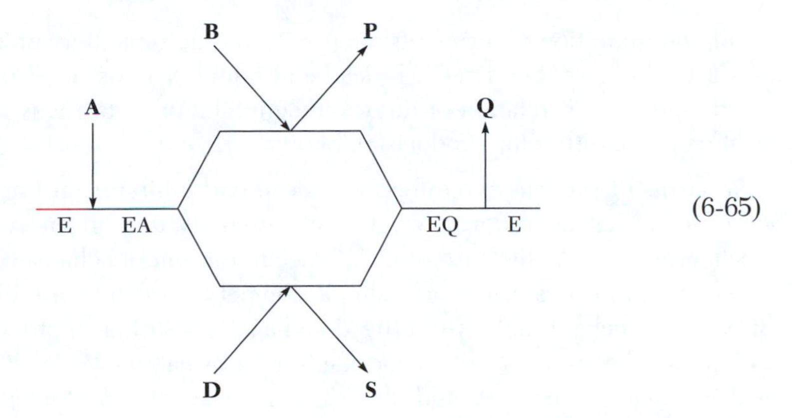

(6-65)

Dead-End Inhibition

Although product inhibition aids significantly in ruling out possible kinetic mechanisms, it is not sufficient, as shown by Table 6-3, to rule out all but a single possibility. Another initial velocity tool is the use of dead-end inhibition. A dead-end inhibitor is an analogue of a substrate that normally resembles a reactant in structure and competes with a reactant for its binding site(s) on the enzyme. However, once bound, the analogue does not undergo the chemical transformation undergone by the substrates, and thus the label dead-end. Binding of the inhibitor to enzyme effectively and reversibly ties up a portion of the enzyme population. The amount of the enzyme that is not available for reaction will depend on the concentration of the inhibitor compared to its K_i value and the amount of the substrate compared to its apparent K_m.

The prototypical example of a dead-end inhibitor is malonate, a dead-end analogue of succinate, which competes with succinate for its binding site on succinate dehydrogenase. The structure of malonate resembles that of succinate, but malonate cannot be oxidized once bound to succinate dehydrogenase.

FAD + succinate $\xrightleftharpoons{\text{succinate dehydrogenase}}$ fumarate + $FADH_2$

succinate

fumarate

malonate

Dead-end inhibition experiments are carried out in a manner similar to those for product inhibitors, but since the dead-end inhibitor does not participate in the reaction, the interpretation of the resulting inhibition patterns is less complex in most cases than that for product inhibitors.

As a means of showing the effects of a dead-end inhibitor on the rate equation, let us consider an ordered mechanism, such as that given schematically in mechanism 6-25. In the case of a Unireactant enzyme mechanism, the dead-end inhibitor will in most cases be a simple competitor of the varied substrate. In a Bireactant mechanism, however, dead-end analogues of each of the reactants can and should be used to obtain mechanistic information. We will call dead-end analogues of reactants A and B I_a and I_b, respectively. Since the dead-end analogue competes with the substrate that it resembles, it will bind to the same enzyme forms to which the substrate binds. I_a competes with A and thus must bind to the E form of the enzyme. Mechanism 6-25 is modified as shown in mechanism 6-66:

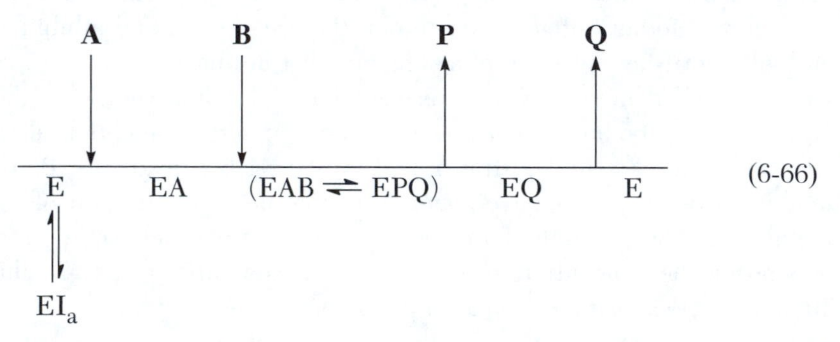

(6-66)

There are two terms in the denominator of the rate equation for a steady-state ordered mechanism that represent E, $K_{ia}K_b$ and $K_a\mathbf{B}$ (see Chapter 5). Inhibition by I_a changes the rate equation by modifying the E terms by the factor $(1 + \mathbf{I_a}/K_i)$, where K_i is equal to $(\mathbf{E})(\mathbf{I_a})/(\mathbf{EI_a})$, and $\mathbf{I_a}/K_i$ is thus equal to $(\mathbf{EI_a})/(\mathbf{E})$. At low $\mathbf{I_a}$ compared to K_i, no inhibition will be observed, but as $\mathbf{I_a}$ increases compared to K_i, the amount of E available for the reactant A to bind to is decreased, and inhibition is observed. The reciprocal of the rate equation for inhibition by I_a with A as the varied substrate is given in equation 6-67:

$$\frac{1}{v} = \left(\frac{K_a}{V}\right)\left(1 + \frac{K_{ia}K_b}{K_a\mathbf{B}}\right)\left(1 + \frac{\mathbf{I_a}}{K_i}\right)\left(\frac{1}{\mathbf{A}}\right) + \left(\frac{1}{V}\right)\left(1 + \frac{K_b}{\mathbf{B}}\right) \qquad (6\text{-}67)$$

As expected, only the slope is affected, since I_a competes with A. A secondary replot is constructed of slope versus $\mathbf{I_a}$, and the apparent K_{is} (the slope inhibition constant) is given as the value of the abscissa intercept. In the case of inhibition by I_a versus A, inhibition is obtained under conditions where **A** is limiting, and thus binding of I_a to E comes to equilibrium, and the apparent K_{is} is equal to the true K_i. The latter is independent of **B**, since B does not affect the **E** available to I_a and A.

With B as the varied reactant, inhibition can only be observed when **A** is not saturating. (In practice **A** is normally maintained close to K_a.) Rearranging equation 6-67 gives equation 6-68:

$$\frac{1}{v} = \left(\frac{K_b}{V}\right)\left(1+\frac{K_{ia}}{\mathbf{A}}\right)\left[1+\frac{\mathbf{I_a}}{K_i(1+\mathbf{A}/K_{ia})}\right]\left(\frac{1}{\mathbf{B}}\right) + \left(\frac{1}{V}\right)\left(1+\frac{K_a}{\mathbf{A}}\right)\left[1+\frac{\mathbf{I_a}}{K_i(1+\mathbf{A}/K_a)}\right] \tag{6-68}$$

In equation 6-68, the slope and intercept expressions are affected by the presence of I_a, and thus inhibition is noncompetitive. With **B** limiting, the following equilibria are established:

$$EI_a \rightleftharpoons E \rightleftharpoons EA \tag{6-69}$$

and I_a ties up some of the existing enzyme. At infinite **B**, E is still available to bind I_a, but the amount of inhibition by I_a will depend on the rate of production of E from the EAB complex. For example, if the rate of formation of E is very slow, no inhibition by I_a will be observed at infinite **B**, and one has an equilibrium ordered kinetic mechanism as will be discussed below. The slope and intercept expressions in equation 6-68 as a function of $\mathbf{I_a}$ are given in equations 6-70 and 6-71:

$$\text{slope} = \left(\frac{K_b}{V}\right)\left(1+\frac{K_{ia}}{\mathbf{A}}\right)\left[1+\frac{\mathbf{I_a}}{K_i(1+\mathbf{A}/K_{ia})}\right] \tag{6-70}$$

$$\text{intercept} = \left(\frac{1}{V}\right)\left(1+\frac{K_a}{\mathbf{A}}\right)\left[1+\frac{\mathbf{I_a}}{K_i(1+\mathbf{A}/K_a)}\right] \tag{6-71}$$

The apparent K_{is} and K_{ii} (the slope and intercept inhibition constants), are $K_i(1+\mathbf{A}/K_{ia})$ and $K_i(1+\mathbf{A}/K_a)$, respectively, where $(1+\mathbf{A}/K_{ia})$ and $(1+\mathbf{A}/K_a)$ are difficulty factors for I_a binding. The apparent K_i values depend on **A** compared to its K_{ia} (at limiting **B**) or K_a (at saturating **B**). At near zero **A**, all of the enzyme is present as E and the true K_i for I_a is measured at any level of **B**. As **A** is increased, however, more E is converted to EA and other forms, shown by an increase in the ratios of $\mathbf{A}/K_{ia}$ and $\mathbf{A}/K_a$, which reflect the ratio of **EA/E** at low and high **B**. The values of true K_i calculated from slope and intercept inhibition constants, after correction for the **A** present, should be identical within error and represent a mechanism-dependent test of the internal consistency of the data.

A dead-end analogue of the second substrate, I_b, in mechanism 6-66 will compete with B and thus will bind to the EA complex. There is a single term in the denominator of the rate equation for a steady-state ordered mechanism that represents EA, $K_b\mathbf{A}$ (see Chapter 5). Inhibition by I_b changes the rate equation by modifying the EA term by the factor $(1+\mathbf{I_b}/K_i)$, where K_i, by analogy to above, is equal to $[\mathbf{EA}][\mathbf{I_b}]/[\mathbf{EAI_b}]$, and $\mathbf{I_b}/K_i$ is thus equal to $[\mathbf{EAI_b}]/[\mathbf{EA}]$. The reciprocal of the rate equation for inhibition by I_b with A as the varied substrate is given in equation 6-72, and the expression for the intercept is provided in equation 6-73:

$$\frac{1}{v}=\left(\frac{K_a}{V}\right)\left(1+\frac{K_{ia}K_b}{K_a\mathbf{B}}\right)\left(\frac{1}{\mathbf{A}}\right)+\left(\frac{1}{V}\right)\left(1+\frac{K_b}{\mathbf{B}}\right)\left[1+\frac{\mathbf{I_b}}{K_i(1+\mathbf{B}/K_b)}\right] \quad (6\text{-}72)$$

$$\text{intercept}=\left(\frac{1}{V}\right)\left(1+\frac{K_b}{\mathbf{B}}\right)\left[1+\frac{\mathbf{I_b}}{K_i(1+\mathbf{B}/K_b)}\right] \quad (6\text{-}73)$$

Note that only the intercept is affected and the inhibition is uncompetitive. This is expected, since I_b competes with B for EA, which is present only at finite **A**. The apparent K_{ii} is obtained as the value of the abscissa intercept of the secondary replot of intercept versus $\mathbf{I_b}$. The apparent K_{ii} is $K_i(1+\mathbf{B}/K_b)$. As is the case for inhibition by I_a versus B, the apparent K_{ii} for I_b must be corrected for the competing substrate B. The uncompetitive inhibition by I_b versus A is diagnostic for an ordered kinetic mechanism.

With B as the varied reactant, equation 6-73 is rearranged to give equation 6-74:

$$\frac{1}{v}=\left(\frac{K_b}{V}\right)\left(1+\frac{K_{ia}}{\mathbf{A}}\right)\left[1+\frac{\mathbf{I_b}}{K_i(1+K_{ia}/\mathbf{A})}\right]\left(\frac{1}{\mathbf{B}}\right)+\left(\frac{1}{V}\right)\left(1+\frac{K_a}{\mathbf{A}}\right) \quad (6\text{-}74)$$

$$\text{slope}=\left(\frac{K_b}{V}\right)\left(1+\frac{K_{ia}}{\mathbf{A}}\right)\left[1+\frac{\mathbf{I_b}}{K_i(1+K_{ia}/\mathbf{A})}\right] \quad (6\text{-}75)$$

As expected, only the slope is affected by the presence of I_b (equation 6-75) and thus inhibition is competitive. With **A** limiting, the predominant term in equation 6-75 is $K_{ia}/\mathbf{A}$, and no inhibition is observed. At infinite **A**, however, the apparent K_{is} is equal to the true K_i. The apparent K_{is} is dependent on **A**, and the difficulty factor in this case is $(1+K_{ia}/\mathbf{A})$, where $K_{ia}/\mathbf{A}$ reflects **E/EA**.

Rules for Predicting Dead-End Inhibition Patterns

As for product inhibition, an analysis of dead-end inhibition such as that for the ordered mechanism above allowed the development of a set of simple rules one can use to predict the dead-end inhibition patterns for any given kinetic mechanism. The rules are as follows.

(1) A dead-end inhibitor affects the slope of a double reciprocal plot when (a) both I and varied substrate combine with the same enzyme form or (b) when there is a reversible connection between the points of addition of I and the varied substrate, and the varied substrate adds after I.

(2) A dead-end inhibitor affects the intercept of a double reciprocal plot when I and varied substrate add to different enzyme forms unless I adds prior to the varied substrate and both additive steps are at thermodynamic equilibrium.

Prediction of dead-end inhibition patterns for any mechanism is, as stated above, a very straightforward process. As for a product inhibitor, several questions are asked of each inhibitor at each of its combinations, and the pattern is determined as a result of the presence of slope and intercept effects. The results obtained above for a steady-state ordered kinetic mechanism are shown in Table 6-4. The major difference in the inhibition patterns of product and dead-end inhibitors is that the latter cannot cause slope effects by partial reversal of the reaction. Dead-end inhibitors thus give more UC patterns.

In Table 6-4, the first column refers to the varied substrate and the inhibitory analogue, the second column refers to the enzyme form with which each combines, the third column is self-explanatory, the fourth column is applicable to the case where the answer in column 3 is yes, and the last column gives the predicted inhibition pattern. Thus, combination with the same enzyme form for A and I_a, and for B and I_b, gives only a slope effect and competitive inhibition. Combination of B and I_a with different enzyme forms with I_a combining before B, and for the steady-state ordered case where combination of the two does not come to thermodynamic equilibrium, gives both an intercept and slope effect and a noncompetitive pattern. Combination of I_b after A leads to only an intercept effect and an uncompetitive pattern. The additional combination of I_b with EQ gives an additional intercept effect, which causes a change in the inhibition versus B from competitive to noncompetitive; that is, inhibition resulting from combination with the two different enzyme forms gives slope and intercept effects and a net noncompetitive inhibition.

Table 6-4. Dead-End Inhibition Patterns Predicted for Bireactant Enzyme-Catalyzed Reactions

Substrate/ inhibitor	Enzyme form bound	I combines before substrate	Thermodynamic equilibrium	Inhibition pattern
		Steady-State Ordered		
A	E			
I_a	E	N/A	N/A	C
B	EA			
I_a	E	yes	no	NC
A	E			
I_b	EA	no	N/A	UC
I_b	EQ	no	N/A	UC
	if I_b combines with both EA and EQ			UC
B	EA			
I_b	EA	N/A	N/A	C
I_b	EQ	no	N/A	UC
	if I_b combines with both EA and EQ			NC
		Theorell–Chance		
$\mathbf{B} = K_b$ or $> 50K_b$				
A	E			
I_a	E	N/A	N/A	C
$\mathbf{A} = K_a$ (at saturating **A** there is no inhibition by I_a)				
B	EA			
I_a	E	yes	no	NC
$\mathbf{B} = K_b$ (at saturating **B** there is no inhibition by I_a)				
A	E			
I_b	EA	N/A	N/A	UC
$\mathbf{A} = K_a$ or $> 50K_a$				
B	EA			
I_b	EA	N/A	N/A	C
		Equilibrium Ordered		
$\mathbf{B} = K_b$ (at saturating **B**, V_{max} is observed at all concentrations of A)				
A	E			
I_a	E	N/A	N/A	C
$\mathbf{A} = K_a$ (at saturating **A** there is no inhibition by I_a)				
B	EA			
I_a	E	yes	yes	C

(*Continued*)

Table 6-4. Continued

Substrate/ inhibitor	Enzyme form bound	I combines before substrate	Thermodynamic equilibrium	Inhibition pattern
B = K_b (at saturating **B** there is no inhibition by I_b)				
A	E			
I_b	EA	N/A	N/A	UC
A = finite (at limiting **A** there is no inhibition by I_b)				
B	EA			
I_b	EA	N/A	N/A	C
		Rapid Equilibrium or Steady-State Random		
B = K_b				
A	E, EB			
I_a	E, EB	N/A	N/A	C
B = $>50K_b$				
A	EB			
I_a	EB	N/A	N/A	C
A = K_a (at saturating **A** there is no inhibition by I_a)				
B	E, EA			
I_a	E	N/A	N/A	C
	EB	N/A	N/A	UC
			net	NC[a]
A = K_a				
B	E, EA			
I_b	E, EA	N/A	N/A	C
A = $>50K_a$				
B	EA			
I_b	EA	N/A	N/A	C
B = K_b (at saturating **B** there is no inhibition by I_b)				
A	E, EB			
I_b	E	N/A	N/A	C
	EA	N/A	N/A	UC
			net	NC[b]
		Classical (One-Site) Ping Pong		
B = any concentration				
A	E			
I_a	E	N/A	N/A	C

(*Continued*)

Table 6-4. Continued

Substrate/ inhibitor	Enzyme form bound	I combines before substrate	Thermodynamic equilibrium	Inhibition pattern
A = K_a (at saturating **A** there is no inhibition by I_a)				
B	F			
I_a	E	N/A	N/A	UC
B = K_b (at saturating **B** there is no inhibition by I_b)				
A	E			
I_b	F	N/A	N/A	UC
A = any concentration				
B	F			
I_b	F	N/A	N/A	C
		Nonclassical (Two-Site) Ping Pong		
B = any concentration				
A	site 1			
I_a	site 1	N/A	N/A	C
A = K_a (at saturating **A** there is no inhibition by I_a)				
B	site 2			
I_a	site 1	N/A	N/A	UC
B = K_b (at saturating **B** there is no inhibition by I_b)				
A	site 1			
I_b	site 2	N/A	N/A	UC
A = any concentration				
B	site 2			
I_b	site 2	N/A	N/A	C

[a]Unless overlap between B and I_a prevents formation of EBI_a; then C.
[b]Unless overlap between A and I_b prevents formation of EAI_b; then C.

Using the analysis shown above, one finds the same inhibition patterns for the Theorell–Chance mechanism. The equilibrium ordered mechanism, however, differs in one respect; that is, inhibition by I_a is competitive versus B because it combines prior to the reactant and addition of inhibitor and reactant comes to equilibrium. A summary and analysis of patterns obtained for the commonly observed kinetic mechanisms are shown in Table 6-4.

In the case of random mechanisms, the inhibitory analogue, for example, I_a binds to all forms of the enzyme to which the reactant, A, binds and thus gives competitive inhibition. With the other substrate, B, varied at finite **A**, I_a cannot be prevented by B from combining in the A site and thus gives noncompetitive inhibition. The sole exception occurs when there is too much overlap between B and I_a to allow both to bind at the same time. I_a is then competitive versus B. Since the mechanism is symmetric, the same explanation holds for I_b.

The classical and nonclassical Ping Pong mechanisms also give qualitatively identical results. I_a and A compete for E in the one-site reaction and for site 1 in the two-site reaction, while I_b and B compete for F and site 2, respectively. Since the other reactant and the inhibitory analogue, I_a and B or I_b and A, combine with different enzyme forms (sites), the inhibition patterns will be uncompetitive. It is at a saturating concentration of the other substrate (for example B) that the form (E), with which the inhibitory analogue (I_a) and its like substrate (A) combine, predominates.

Combination of Inhibitor with More Than a Single Enzyme Form

It is not uncommon for an inhibitor to combine with two different enzyme forms along the reaction pathway. To determine the kind of inhibition, each combination of the inhibitor is considered separately. The following rule is thus added to those listed above for prediction of product and dead-end inhibition patterns.

When an inhibitor combines with one enzyme form as a product inhibitor and a different one as a dead-end inhibitor, each inhibition is considered separately. If both combinations predict a slope or intercept effect and a reversible connection exists between the two points of product addition with the combination of product as a product following that as a dead-end inhibitor, then the slopes or intercepts will vary as a parabolic function of the product concentration.

When a dead-end inhibitor combines with two different enzyme forms along the reaction pathway, each inhibition is considered separately, and the effects on slope and intercept are additive, but there are no parabolic slope effects. The exception is formation of an EI_2 complex, which gives a parabolic slope or intercept effect.

Mixed Product and Dead-End Inhibition

Combination as both a product and a dead-end inhibitor is also possible in the ordered mechanism, where P combines with the EA complex, as shown in equation 6-76:

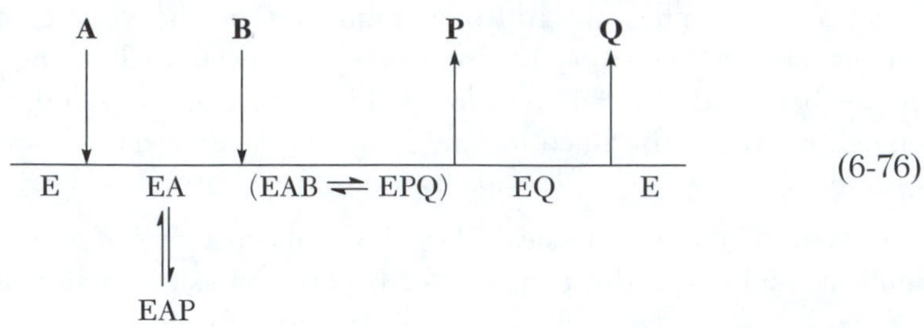

(6-76)

Consider inhibition by P with A as the varied substrate and **B** fixed equal to K_b. In mechanism 6-76, P combines as a product with EQ, while A combines with E. There is a reversible connection between E and EQ, giving both slope and intercept effects. Combination of P with EA is a dead-end inhibition, and the rules for dead-end inhibition apply. The varied substrate, A, combines with E, while P combines with EA after the varied substrate, producing an intercept effect. There is a reversible connection between the two combinations of P, with EA and EQ, and addition of P to EQ will produce more EA to which P also binds. Thus, the intercept effects are multiplicative and the inhibition pattern is S-linear, I-parabolic noncompetitive. In tabular form, the example is given in Table 6-5.

Table 6-5. Prediction of Mixed Product and Dead-End Inhibition Patterns

Substrate/ product	Enzyme form bound	Reversible connect.	Infinite S overcomes inhib. by P?	I combines before S?	Therm. equil. S and P?	Inhibition pattern
			Steady-State Ordered with EAP			
B = K_b						
A	E					
P	EQ	yes	no	no	N/A	NC
P	EA	no	no	no	N/A	UC
reversible connection between the two combinations of P				net	S-lin, I-par	NC
A = any concentration						
B	EA					
P	EQ	yes	no	no	N/A	NC
P	EA	N/A	N/A	N/A	N/A	C
reversible connection between the two combinations of P				net	S-par, I-lin	NC

With B as the varied substrate and **A** fixed, B combines with EA and P combines with EQ, giving both slope and intercept effects. In addition, both B and P combine with EA, giving a slope effect. The two points of addition are again reversibly connected, and binding of P to the EQ that exists in the steady state increases the amount of EA present at the fixed concentration of A. Thus, the slope will be a parabolic function of P. Saturation with B eliminates the EA form of the enzyme, giving a linear intercept effect. The overall inhibition pattern is termed S-parabolic, I-linear noncompetitive, Table 6-5.

The rate expression for inhibition by P as a product inhibitor is given as equation 6-33 above. The equation can be modified to add the dead-end inhibition resulting from P binding to the EA complex by multiplying the EA term (K_b/V) by $1 + \mathbf{P}/K_{Ip}$. The modified equation is given below as equation 6-77, while equations 6-78 and 6-79 give the slope and intercept expressions.

$$\frac{1}{v} = \left(\frac{K_a}{V}\right)\left(1 + \frac{K_{ia}K_b}{K_a\mathbf{B}}\right)\left[1 + \frac{\mathbf{P}}{(K_pK_{iq}/K_q)(1 + K_a\mathbf{B}/K_{ia}K_b)}\right]\left(\frac{1}{\mathbf{A}}\right) + \left(\frac{1}{V}\right)\left(1 + \frac{K_b}{\mathbf{B}}\right)\left[1 + \frac{\mathbf{P}(1/K_{Ip} + K_q/K_pK_{iq} + \mathbf{B}/K_bK_{ip})}{1 + \mathbf{B}/K_b} + \frac{\mathbf{P}^2}{(K_pK_{iq}/K_q)(K_{Ip}(1 + \mathbf{B}/K_b))}\right] \tag{6-77}$$

$$\text{slope} = \left(\frac{K_a}{V}\right)\left(1 + \frac{K_{ia}K_b}{K_a\mathbf{B}}\right)\left[1 + \frac{\mathbf{P}}{(K_pK_{iq}/K_q)(1 + K_a\mathbf{B}/K_{ia}K_b)}\right] \tag{6-78}$$

$$\text{intercept} = \left(\frac{1}{V}\right)\left(1 + \frac{K_b}{\mathbf{B}}\right)\left[1 + \frac{\mathbf{P}(1/K_{Ip} + K_q/K_pK_{iq} + \mathbf{B}/K_bK_{ip})}{1 + \mathbf{B}/K_b} + \frac{\mathbf{P}^2}{(K_pK_{iq}/K_q)(K_{Ip}(1 + \mathbf{B}/K_b))}\right] \tag{6-79}$$

The apparent K_{is} value derived from equation 6-78 is $(K_pK_{iq}/K_q)(1 + K_a\mathbf{B}/K_{ia}K_b)$, where K_pK_{iq}/K_q and $K_{ia}K_b/K_a$ are effective steady-state dissociation constants for P as a product inhibitor and B as a reactant at limiting **A**. Increasing **B** compared to $K_{ia}K_b/K_a$ counteracts the ability of P to reverse the reaction, causing inhibition. The intercept expression is, as stated above, a parabolic function of **P**, with the $\mathbf{P}^2$ term providing for the dependence of inhibition by combination with EA on the amount of reversal of the reaction generated by combination of P with EQ.

Note, however, that saturation with B eliminates all inhibition by P with the exception of its combination to EQ. Quantitative analysis is best obtained by computer fitting of the data to equation 6-77.

Varying **B** at different fixed **A** gives opposite effects on slope and intercept, as shown in equations 6-80 to 6-82:

$$\frac{1}{v} = \left(\frac{K_b}{V}\right)\left(1+\frac{K_{ia}}{\mathbf{A}}\right)\left[1+\mathbf{P}\left(\frac{K_q}{K_pK_{iq}}+\frac{1}{K_{Ip}(1+K_{ia}/\mathbf{A})}\right) + \frac{\mathbf{P}^2}{(K_pK_{iq}/K_q)(K_{Ip}(1+K_{ia}/\mathbf{A}))}\right]\left(\frac{1}{\mathbf{B}}\right) + \left(\frac{1}{V}\right)\left(1+\frac{K_a}{\mathbf{A}}\right)\left[1+\frac{\mathbf{P}}{K_{ip}(1+K_a/\mathbf{A})}\right] \tag{6-80}$$

$$\text{slope} = \left(\frac{K_b}{V}\right)\left(1+\frac{K_{ia}}{\mathbf{A}}\right)\left[1+\mathbf{P}\left(\frac{K_q}{K_pK_{iq}}+\frac{1}{K_{Ip}(1+K_{ia}/\mathbf{A})}\right) + \frac{\mathbf{P}^2}{(K_pK_{iq}/K_q)(K_{Ip}(1+K_{ia}/\mathbf{A}))}\right] \tag{6-81}$$

$$\text{intercept} = \left(\frac{1}{V}\right)\left(1+\frac{K_a}{\mathbf{A}}\right)\left[1+\frac{\mathbf{P}}{K_{ip}(1+K_a/\mathbf{A})}\right] \tag{6-82}$$

Increasing **B** eliminates the EA form of the enzyme, and the intercept reflects only the combination of P with EQ. The apparent K_{ii} derived from equation 6-82 is $K_{ip}(1+K_a/\mathbf{A})$, which indicates that increasing **A** increases **EQ** (actually **EA**, but **B** is saturating and so the increase in **A** translates into an increase in **EQ**) to which P binds to inhibit. The slope effect is now a parabolic function of **P**, since combination of P with EQ will only facilitate inhibition by P combining with EA when **B** is limiting, since B competes with P for EA. Note that in this case increasing the fixed reactant **A** facilitates the combined effect of P combining at EQ and EA, as a result of increasing **EA**.

An example is the inhibition by ethanol versus actetaldehyde in the alcohol dehydrogenase reaction (B 2, 935). The inhibition by ethanol is noncompetitive but with a parabolic slope effect that results from ethanol increasing the concentration of E–NADH, to which ethanol also binds.

Mixed product and dead-end inhibition is common in the physiologic reaction direction of a one-site Ping Pong kinetic mechanism as a result of dead-end combination of P with E or Q with F. Since these are similar in form, we consider just the case of P with E:

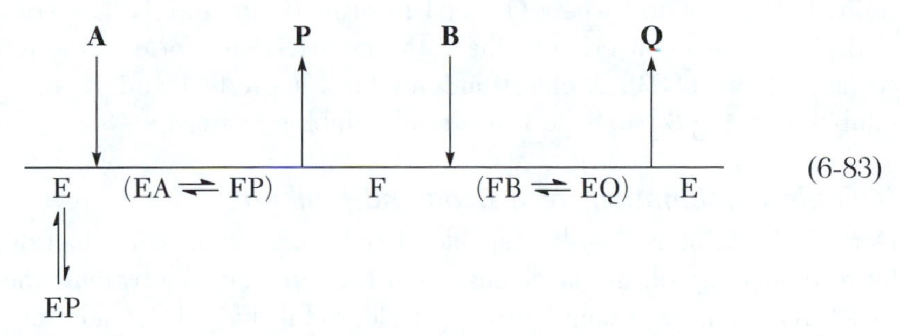

(6-83)

The product inhibition that results from combination of P with F is competitive inhibition versus B and noncompetitive inhibition versus A. The dead-end combination of P with A causes competitive inhibition versus A. A dead-end inhibitor combining with E is normally uncompetitive versus B, but since the presence of P establishes a reversible link between E and F, the dead-end combination of P with E gives noncompetitive inhibition.

Thus, regardless of which substrate is the varied one, the two combinations of P both produce slope effects, while only one intercept effect is predicted. Because the combination of P with F increases the steady-state level of E, the two slope effects result in parabolic inhibition. Thus P gives S-parabolic I-linear inhibition versus both substrates. In the reverse direction where P is a substrate, the dead-end combination with E will cause substrate inhibition that is competitive versus Q (see page 178).

The rate equations for the mixed product and dead-end inhibition are as follows. With A varied:

$$\frac{1}{v} = \left(\frac{K_a}{V}\right)\left[1 + \mathbf{P}\left(\frac{1}{K_{Ip}} + \frac{K_{ia}K_b}{K_aK_{ip}\mathbf{B}}\right) + \mathbf{P}^2\left(\frac{K_{ia}K_b}{K_aK_{ip}K_{Ip}\mathbf{B}}\right)\right]\left(\frac{1}{\mathbf{A}}\right) + \left(\frac{1}{V}\right)\left(1 + \frac{K_b}{\mathbf{B}}\right)\left[1 + \frac{\mathbf{P}}{K_{ip}(1 + \mathbf{B}/K_b)}\right] \tag{6-84}$$

With B varied:

$$\frac{1}{v} = \left(\frac{K_b}{V}\right)\left[1 + \left(1 + \frac{K_{ia}}{\mathbf{A}}\right)\left(\frac{\mathbf{P}}{K_{ip}}\right) + \left(\frac{K_{ia}}{\mathbf{A}}\right)\left(\frac{\mathbf{P}^2}{K_{ip}K_{Ip}}\right)\right]\left(\frac{1}{\mathbf{B}}\right) + \left(\frac{1}{V}\right)\left(1 + \frac{K_a}{\mathbf{A}}\right)\left[1 + \frac{\mathbf{P}}{K_{Ip}(1 + \mathbf{A}/K_a)}\right] \tag{6-85}$$

Note that, in the absence of the dead-end inhibitor ($K_{Ip} = \infty$), P is noncompetitive versus A and competitive versus B. An example is the O-acetylserine sulfhydrylase reaction, where O-acetyl-L-serine (OAS) and the E (pyridoxal) form of the enzyme is converted to the F (α-aminoacrylate) form of the enzyme and acetate (B 32, 6833). Acetate binds to F as a product and E as a dead-end inhibition giving S-parabolic I-linear NC inhibition versus OAS.

Multiple Combinations of a Dead-End Inhibitor

Dead-end inhibitors are also capable of combining with more than one enzyme form, depending on the mechanism and the number of substrates the inhibitor structurally mimics. Examples to be considered include the steady-state ordered, rapid equilibrium random, and Ping Pong kinetic mechanisms. In predicting the inhibition pattern expected for I_b as a dead-end inhibitor, the resulting inhibition from each combination of I_b is considered separately, with inhibition assessed by the rules for predicting dead-end inhibition.

Consider mechanism 6-86, in which I_b combines with both EA and EQ complexes:

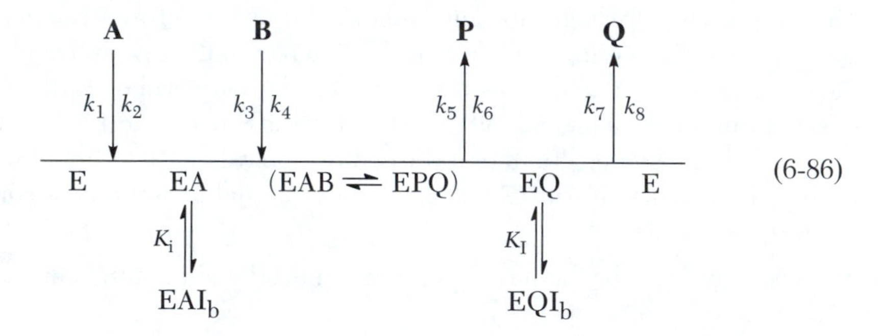

(6-86)

With A as the varied substrate, inhibition by I_b is obtained with **B** nonsaturating, since saturation would eliminate combination to EA. When **A** is limiting, the EA and EQ complexes do not, for all intents and purposes, exist. Inhibition is only observed at finite **A** and is thus uncompetitive as outlined in Table 6-6. When B is varied, however, EQ will exist in the steady-state at finite **B**, and I_b will inhibit by combination with EA and EQ. Combination with EA gives a slope effect, while combination with EQ gives an intercept effect and a net noncompetitive inhibition, as shown in Table 6-6. As a test for the combination of I_b with EQ, saturation with both **A** and **B** should still produce uncompetitive inhibition (Table 6-6).

The rate expression describing dead-end inhibition by I_b combining with EA and EQ in a steady-state ordered mechanism when A is varied is shown in equation

Table 6-6. Predicted Inhibition Patterns for Combination of a Dead-End Inhibitor with Two Enzyme Forms in Bireactant Enzyme-Catalyzed Reactions

Substrate/ inhibitor	Enzyme form bound	$S \rightarrow \infty$ overcome inhibition?	I combines before S?	Therm. equil. S and I?	Inhibition pattern
		Steady-State Ordered with EQI_b *and* EAI_b			
B $= K_b$					
A	E				
I_b	EA	no	no	-	UC
	EQ	no	no		UC
				net	UC
B = saturating					
A	E				
I_b	EQ	no	no		UC
A $= K_a$ or saturating					
B	EA				
I_b	EA	yes	no	-	C
I_b	EQ	no	no	-	UC
				net	NC
		Rapid Equilibrium Random with EI_b *and* EAI_b			
B $= K_b$ (no inhibition if **B** is saturating)					
A	E, EB				
I_b	E	yes	N/A	-	C
I_b	EA	no	no	-	UC
				net	NC
A $= K_a$					
B	E, EA				
I_b	E, EA	yes	-	-	C
				net	C
A = saturating					
B	EA				
I_b	EA	yes	-	-	C
				net	C

(*Continued*)

Table 6-6. Continued

Substrate/ inhibitor	Enzyme form bound	$S \to \infty$ overcome inhibition?	I combines before S?	Therm. equil. S and I?	Inhibition pattern
		Ping Pong with EI_b, FI_b			
$\mathbf{B} = K_b$					
A	E				
I_b	F	no	-	-	UC
I_b	E	yes	-	-	C
				net	NC
$\mathbf{B}$ = saturating					
A	E		-	-	C
I_b	E	yes		net	C
$\mathbf{A} = K_a$					
B	F				
I_b	F	yes	-	-	C
I_b	E	no	-	-	UC
				net	NC
$\mathbf{A}$ = saturating					
B	F				
I_b	F	yes	-	-	C
				net	C

6-87, where appK_i is the apparent K_{ii} for I_b. The inhibition, as predicted in Table 6-6, is uncompetitive versus A, and the full intercept expression is given as equation 6-88, where K_i and K_I are the dissociation constants of I_b from EAI_b and EQI_b, respectively.

$$\frac{1}{v} = \left(\frac{K_a}{V}\right)\left(1 + \frac{K_{ia}K_b}{K_a\mathbf{B}}\right)\left(\frac{1}{\mathbf{A}}\right) + \left(\frac{1}{V}\right)\left(1 + \frac{K_b}{\mathbf{B}}\right)\left(1 + \frac{\mathbf{I_b}}{\text{app}K_i}\right) \tag{6-87}$$

$$\text{intercept} = \left(\frac{1}{V}\right)\left(1 + \frac{K_b}{\mathbf{B}}\right)\left(1 + \mathbf{I_b}\left[\frac{K_b/\mathbf{B}K_i + K_qV_1/K_{iq}V_2K_I}{(1 + K_b/\mathbf{B})}\right]\right) \tag{6-88}$$

The apparent K_{ii} is thus equal to $(1 + K_b/\mathbf{B})/(K_qV_1/K_{iq}V_2K_I + K_b/\mathbf{B}K_i)$. The expression $K_qV_1/K_{iq}V_2 = k_5/[k_5 + k_7]$ determines how much EQ is present in the steady state. At limiting **B**, the EA form of the enzyme will predominate, and the above expression reduces to K_i, while at saturating **B**, EA is no longer available and EQ is the only enzyme form with which I_b can combine. The apparent K_{ii} is then $K_{iq}V_2K_I/K_qV_1$ or $K_I([k_5+k_7]/k_5)$.

With B as the varied reactant, both slope and intercept are affected as predicted (equations 6-89 to 6-91 and Table 6-6):

$$\frac{1}{v}=\left(\frac{K_b}{V}\right)\left(1+\frac{K_{ia}}{\mathbf{A}}\right)\left[1+\frac{\mathbf{I_b}}{K_i(1+K_{ia}/\mathbf{A})}\right]\left(\frac{1}{\mathbf{B}}\right)+\left(\frac{1}{V}\right)\left(1+\frac{K_a}{\mathbf{A}}\right)\left[1+\frac{K_qV_1\mathbf{I_b}}{K_{iq}V_2K_I(1+K_a/\mathbf{A})}\right] \tag{6-89}$$

$$\text{slope}=\left(\frac{K_b}{V}\right)\left(1+\frac{K_{ia}}{A}\right)\left[1+\frac{\mathbf{I_b}}{K_i(1+K_{ia}/\mathbf{A})}\right] \tag{6-90}$$

$$\text{intercept}=\left(\frac{1}{V}\right)\left(1+\frac{K_a}{\mathbf{A}}\right)\left[1+\frac{K_qV_1\mathbf{I_b}}{K_{iq}V_2K_I(1+K_a/\mathbf{A})}\right] \tag{6-91}$$

The slope effect, evaluated at limiting **B**, reflects combination with EA, and the apparent K_{is} is $K_i(1+K_{ia}/\mathbf{A})$. Thus, as **A** approaches zero, **EA** in the steady state approaches zero, and the apparent inhibition constant approaches infinity. As **A** approaches infinity, the apparent inhibition constant approaches the true dissociation constant for EA–I_b, K_i. The intercept effect, corresponding to infinite **B**, at near zero **A** generates a very large value for appK_{ii}, because very little EA or EQ exists in the steady state. Infinite **A** and **B** produce the maximum level of EQ expected in the steady state and the minimum value, $K_I([k_5+k_7]/k_5)$, predicted for appK_{ii}. The present situation can be contrasted to combination of I_b with EA alone, where I_b is competitive versus B and uncompetitive versus A, as in Table 6-4.

The above analysis shows that when an inhibitor combines with EQ in an ordered mechanism, its apparent K_{ii} as an uncompetitive inhibitor is the true dissociation constant, K_I, multiplied by $(1+k_7/k_5)$. Thus if $k_7 \ll k_5$, one sees the true value of K_I, while if $k_7 > k_5$, the apparent K_{ii} is much higher. This provides a simple method for determining k_7/k_5 ratios and thus which step is rate-limiting (catalysis and first product release, or second product release). One simply measures the value of K_I directly from the competitive inhibition versus P in the reverse reaction with **Q** saturating, and then the value of the apparent K_{ii} value in the forward direction, and the ratio of these values gives $(1+k_7/k_5)$.

If the inhibitor combines with EA as well, it will give noncompetitive inhibition versus either B or P, and comparison of the apparent K_{ii} value as a noncompetitive inhibitor versus P (**Q** saturating) and the slope inhibition constant as a competitive inhibitor versus B (**A** saturating) gives a value of $(1 + k_2/k_4)$, which allows determination of the rate-limiting step in the reverse reaction. Thus one set of inhibition experiments determines the rate-limiting steps in both directions of the reaction.

A rapid equilibrium random mechanism is expected to allow multiple combinations of at least one of the dead-end inhibitors (see above for two combinations of a product inhibitor), a direct result of the two pathways available for generating the Michaelis complex. A rapid equilibrium random mechanism with EI_b and EAI_b dead-end complexes is shown schematically in equation 6-92:

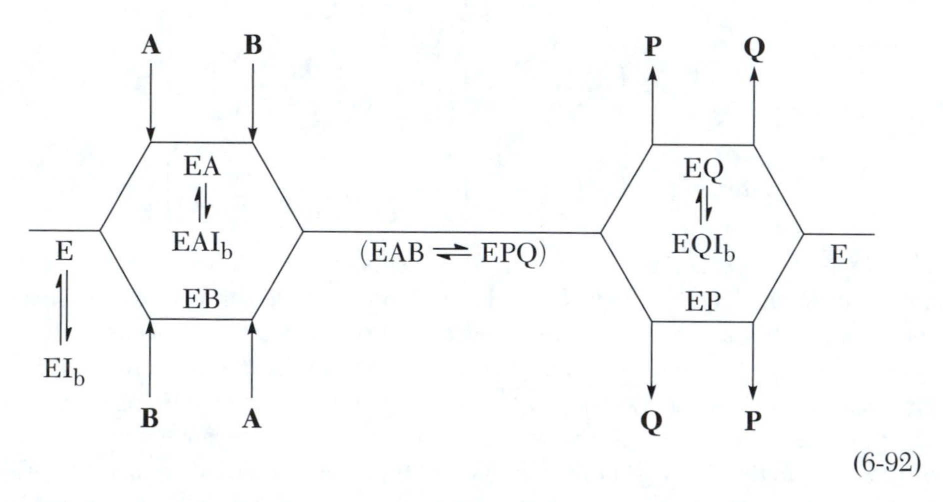

(6-92)

In mechanism 6-92, I_b will bind to E and to the EA complex, giving an overall noncompetitive inhibition versus A (Table 6-6). (If there is overlap between A and I_b in the site, the inhibitor will be competitive and not noncompetitive versus A.) However, at any **A**, I_b will compete with B. At low **A**, competition is predominantly for E, while at saturating **A**, competition is solely for EA (Table 6-5). Observation of competitive inhibition by I_a and I_b versus A and B, at saturating **B** and **A**, respectively, is diagnostic for a rapid equilibrium random kinetic mechanism with dead-end EAI_b and EBI_a complexes. Quantitative analysis of the rapid equilibrium random case is straightforward and is identical to the previous consideration of combination of P with E and EA in a rapid equilibrium random mechanism, Table 6-3, since P acts as a dead-end inhibitor in that mechanism. If EQ is present in the steady state the EQI_b complex may also form, but increasing **A** or **B** will eliminate it since E, EP, and EQ are in equilibrium.

The above behavior is exhibited by the pyrophosphate-dependent phosphofructokinase from *Propionibacterium freudenreichii* (B 23, 4101). A dead-end analogue of $MgPP_i$, MgPCP (the anhydride oxygen is replaced by carbon) is competitive versus $MgPP_i$ but noncompetitive versus F6P, while tagatose 6-phosphate, a dead-end analogue of F6P, is competitive versus F6P, but noncompetitive versus $MgPP_i$. In both cases the dead-end analogue binds to E and a binary complex, competing with the substrate it mimics. There is no EQ in the steady-state so no inhibition on the product side of the reaction is observed.

Ping Pong mechanisms are also prone to multiple combinations by dead-end inhibitors. Reactants in Ping Pong mechanisms often are structurally very similar, and since dead-end inhibitors mimic reactants, it is not unexpected to find that a dead-end analogue of B competes with both A and B. Consider the Ping Pong mechanism with dead-end $F{-}I_b$ and $E{-}I_b$ complexes shown schematically in equation 6-93:

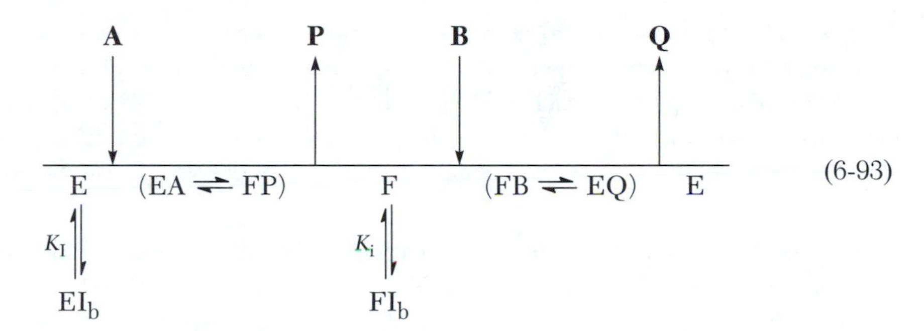

(6-93)

I_b will compete with B for F, at any **A**, generating a slope effect when **B** is varied. Since I_b also binds to E and B cannot prevent it, this generates an intercept effect for a net noncompetitive inhibition (Table 6-6). Likewise, I_b competes with A for E and also binds to the F produced at finite **A**, giving net noncompetitive inhibition when **A** is varied (Table 6-6). Saturation with either reactant, however, does not prevent inhibition by I_b, since the other stable enzyme form is still available, and under these conditions I_b will show competitive inhibition with the varied reactant (Table 6-6). The switch between noncompetitive inhibition patterns at nonsaturating concentrations of each of the fixed reactants to competitive inhibition at saturating concentrations is diagnostic for a Ping Pong kinetic mechanism with dead-end $E{-}I_b$ and $F{-}I_b$ complexes. As for the rapid equilibrium random kinetic mechanism, the Ping Pong mechanism is symmetric, and it is not unusual to observe qualitatively identical behavior for both I_a and I_b.

The rate expressions for the Ping Pong case are very straightforward, because of the competition of I_b versus A for E and versus B for F. The overall rate equation when **A** is varied is given in equation 6-94, with slope and intercept expressions as equations 6-95 and 6-96. In equations 6-94 to 6-96, K_I is the dissociation constant of I_b from EI_b and K_i is the dissociation constant of I_b from FI_b.

$$\frac{1}{v} = \left(\frac{K_a}{V}\right)\left(1 + \frac{\mathbf{I_b}}{K_I}\right)\left(\frac{1}{\mathbf{A}}\right) + \left(\frac{1}{V}\right)\left(1 + \frac{K_b}{\mathbf{B}}\right)\left[1 + \frac{\mathbf{I_b}}{K_i(1 + \mathbf{B}/K_b)}\right] \tag{6-94}$$

$$\text{slope} = \left(\frac{K_a}{V}\right)\left(1 + \frac{\mathbf{I_b}}{K_I}\right) \tag{6-95}$$

$$\text{intercept} = \left(\frac{1}{V}\right)\left(1 + \frac{K_b}{\mathbf{B}}\right)\left[1 + \frac{\mathbf{I_b}}{K_i(1 + \mathbf{B}/K_b)}\right] \tag{6-96}$$

The slope effect gives an appK_{is} equal to the true dissociation constant for E–I_b, K_I, consistent with the competition of A and I_b for E. The intercept effect, however, gives an appK_{ii} of $K_i(1 + \mathbf{B}/K_b)$, and thus inhibition is obtained only at nonsaturating **B**. Increasing **B** to infinity eliminates combination with F and converts the inhibition from noncompetitive to competitive, as predicted in Table 6-6.

With B as the varied substrate, the situation is for all intents and purposes identical, equations 6-97 and 6-98. The slope inhibition constant gives the true dissociation constant for F–I_b, K_i, while the appK_{ii} is dependent on **A** in a manner identical to that observed for A as the varied reactant, $K_I(1 + \mathbf{A}/K_a)$.

$$\text{slope} = \left(\frac{K_b}{V}\right)\left(1 + \frac{\mathbf{I_b}}{K_i}\right) \tag{6-97}$$

$$\text{intercept} = \left(\frac{1}{V}\right)\left(1 + \frac{K_a}{\mathbf{A}}\right)\left[1 + \frac{\mathbf{I_b}}{K_I(1 + \mathbf{A}/K_a)}\right] \tag{6-98}$$

Aspartate aminotransferase is a pyridoxal 5′-phosphate-dependent enzyme that catalyzes the reversible transfer of an amino group from aspartate to α-ketoglutarate. The dead-end inhibitor maleate is noncompetitive versus both aspartate and α-ketoglutarate, indicative of its combination to E (the pyridoxal form of the enzyme) and F (the pyridoxamine form of the enzyme) (B 25, 227).

Substrate Inhibition

A substrate can also act as a dead-end inhibitor when its concentration is raised to high enough levels, as a result of combining with an incorrect enzyme form. This is particularly true when the substrate structurally mimics either the other substrate or one of the products. It is more common in the nonphysiologic direction of a reaction. Although substrate inhibition complicates the graphical presentation of initial velocity data, its observation is an excellent diagnostic tool for the elucidation of kinetic mechanism.

Complete Substrate Inhibition

Substrate inhibition is very rare in a mechanism with only one substrate and one product such as shown in equation 6-99 and occurs only when an EA_2 complex forms. More commonly substrate inhibition is seen in a multireactant mechanism and is studied by varying the level of a noninhibitory substrate at different levels of the one showing substrate inhibition.

$$\begin{array}{ccccc} E + A & \rightleftharpoons & EA & \rightleftharpoons & EP \longrightarrow E + P \\ & & \upharpoonleft\downharpoonright & & \\ & & EA_2 & & \end{array} \quad (6\text{-}99)$$

The general equation for substrate inhibition is

$$v = \frac{V\mathbf{A}}{K_a + \mathbf{A}(1 + \mathbf{A}/K_I)} \quad (6\text{-}100)$$

Note that as the concentration of A approaches infinity, v becomes equal to $VK_I/\mathbf{A}$ and thus decreases as **A** increases, ultimately becoming equal to zero. In double reciprocal form, equation 6-100 is

$$\frac{1}{v} = \frac{K_a}{V\mathbf{A}} + \left(\frac{1}{V}\right)\left(1 + \frac{\mathbf{A}}{K_I}\right) \quad (6\text{-}101)$$

The equation is plotted in Figure 6-5. At high values of $1/\mathbf{A}$, the normal double reciprocal plot in the absence of inhibition is obtained, although not as a simple extrapolation of the apparent linear portion of the curve. Because of the influence of the substrate inhibition, the value of $1/V$ is overestimated and the value of $K_a/V\mathbf{A}$ is underestimated. One asymptote of equation 6-101 is given by

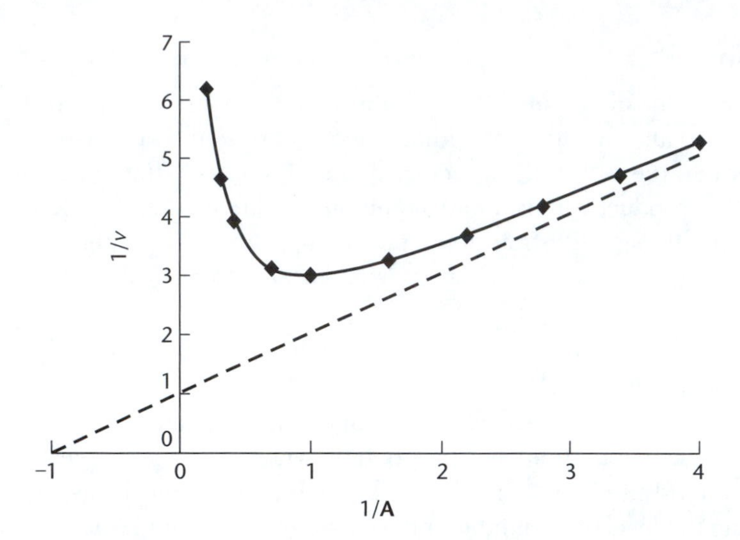

Figure 6-5. Double reciprocal plot of substrate inhibition by A. Units on the ordinate and abscissa are minutes/millimolar and millimolar^{-1}, respectively. A plot of equation 6-101 with V, K_a, and K_I equal to 1 mM/min, 1 mM, and 1 mM, respectively. The dotted line represents a plot in the absence of substrate inhibition by A.

equation 6-101 without the $\mathbf{A}/K_I$ term, and the other is on the vertical axis. To estimate the value of $1/V$ graphically, the graph paper is folded so that the two portions of the curve are superimposed, and the paper is creased. Where the crease crosses the vertical axis gives an estimate of $1/V$. The slope of the double reciprocal plot can then be drawn from this point, approaching the actual curve at high values of $1/\mathbf{A}$. Data should be plotted in this way to visualize the results qualitatively and estimate the kinetic constants. But for final analysis, the data must be fitted to equation 6-100 directly (ME *63*, 103).

There are two bireactant mechanisms that commonly exhibit substrate inhibition, Ping Pong and steady-state ordered. However, substrate inhibition can be observed in a random mechanism when one substrate has affinity for the binding site of the other (JBC *245*, 4163). Each of these will be considered separately below.

Substrate inhibition in an ordered mechanism typically results from B, which normally resembles P structurally, combining with the EQ product complex. In the opposite reaction direction, B will give mixed product and dead-end inhibition (see page 161). Consider the mechanism shown schematically in

$$\begin{array}{c} \text{A} \quad\quad \text{B} \quad\quad \text{P} \quad\quad \text{Q} \\ k_1 \downarrow k_2 \quad k_3 \downarrow k_4 \quad k_5 \uparrow k_6 \quad k_7 \uparrow k_8 \\ \text{E} \quad \text{EA} \quad (\text{EAB} \rightleftharpoons \text{EPQ}) \quad \text{EQ} \quad \text{E} \\ \quad\quad\quad\quad\quad\quad\quad\quad K_{\text{Ib}} \updownarrow \\ \quad\quad\quad\quad\quad\quad\quad\quad \text{EBQ} \end{array} \tag{6-102}$$

The rate expression reflecting mechanism 6-102 when **A** is varied is shown in equation 6-103:

$$\frac{1}{v} = \left(\frac{K_a}{V}\right)\left(1 + \frac{K_{ia}K_b}{K_a\mathbf{B}}\right)\left(\frac{1}{\mathbf{A}}\right) + \left(\frac{1}{V}\right)\left(1 + \frac{K_b}{\mathbf{B}}\right)\left[1 + \frac{\mathbf{B}}{K_{Ib}(1 + K_b/\mathbf{B})}\right] \tag{6-103}$$

where K_{Ib} is an apparent substrate inhibition constant for B combining with the EQ complex and is equal to the true dissociation constant of B from EQB multiplied by $(1 + k_7/k_5)$. Expressions for slope and intercept are given in equations 6-104 and 6-105. Note that the expression for the intercept has the same form as equation 6-101 above, the general expression for substrate inhibition. Since only the intercepts show substrate inhibition, this is called uncompetitive substrate inhibition.

$$\text{slope} = \left(\frac{K_a}{V}\right)\left(1 + \frac{K_{ia}K_b}{K_a\mathbf{B}}\right) \tag{6-104}$$

$$\text{intercept} = \left(\frac{1}{V}\right)\left(1 + \frac{K_b}{\mathbf{B}}\right)\left[1 + \frac{\mathbf{B}}{K_{Ib}(1 + K_b/\mathbf{B})}\right] = \left(\frac{1}{V}\right)\left(1 + \frac{K_b}{\mathbf{B}}\right) + \frac{\mathbf{B}}{VK_{Ib}} \tag{6-105}$$

K_{Ib} is dependent on the steady-state concentration of the EQ complex. If release of Q completely limits the overall reaction, K_{Ib} equals the true dissociation constant from EQB. A double reciprocal plot of equation 6-103 with A as the varied substrate is shown in Figure 6-6, with slope and intercept replots provided as insets. At concentrations of $\mathbf{B} < K_{Ib}$ the reciprocal plot is identical to that expected in the absence of substrate inhibition. As **B** increases above K_{Ib}, the lines

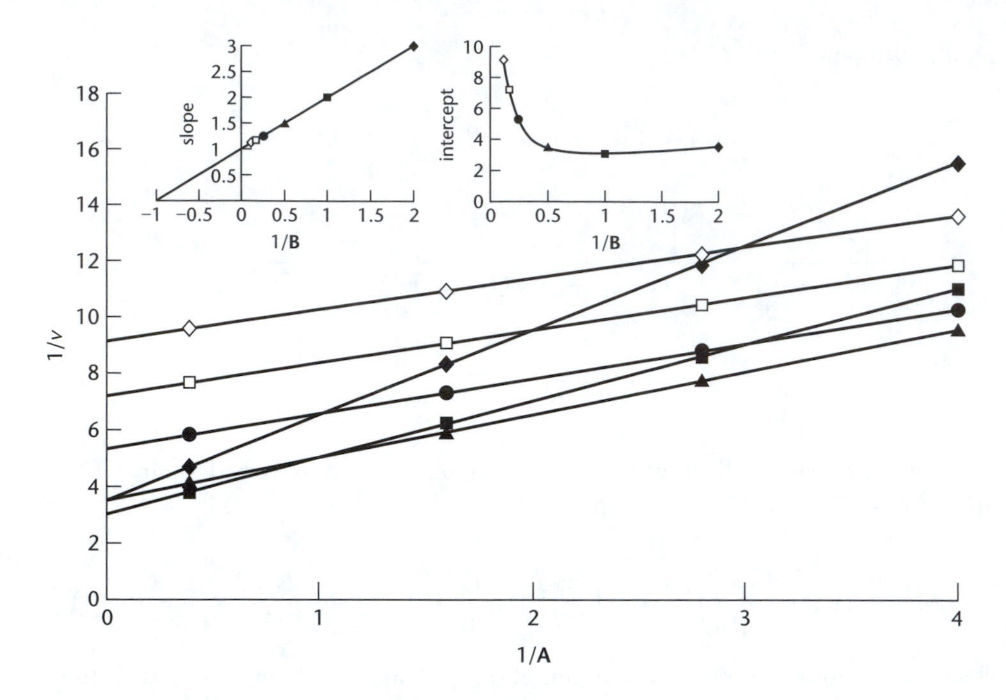

Figure 6-6. Double reciprocal plot exhibiting substrate inhibition by B in a steady-state ordered mechanism with A as the varied substrate. Units on the ordinate and abscissa are minutes/millimolar and millimolar^{-1}, respectively. The concentrations of B used are as follows: 0.5 mM, ◆; 1 mM, ■; 2 mM, ▲; 4 mM, ●; 6 mM, □; and 8 mM, ◇. Note the intersecting pattern obtained at low reactant concentration, which then becomes uncompetitive inhibition versus A as **B** approaches K_{Ib}. Insets show the linear slope replot, with the inhibition exhibited in the intercept replot. Parameters used in the plot are as follows: $V = 1$ mM/min and $K_{ia} = K_a = K_b = K_{Ib} = 1$ mM.

intersect to the right of the $1/v$ axis and are parallel to one another with increasing intercept. A plot of slope versus the reciprocal of **B** is linear. A plot of intercepts versus 1/**B**, however, exhibits substrate inhibition, and the replot is treated as discussed above for substrate inhibition. Uncompetitive substrate inhibition is common in the nonphysiological direction of a steady-state ordered mechanism and further indicates that release of Q is one of the steps that limits the overall reaction.

With B as the varied reactant the following rate expression is obtained, and the double reciprocal plot is as shown in Figure 6-7.

$$\frac{1}{v} = \left(\frac{K_b}{V}\right)\left(1+\frac{K_{ia}}{\mathbf{A}}\right)\left(\frac{1}{\mathbf{B}}\right) + \left(\frac{1}{V}\right)\left(1+\frac{K_a}{\mathbf{A}}\right)\left[1+\frac{\mathbf{B}}{K_{Ib}(1+K_a/\mathbf{A})}\right] \quad (6\text{-}106)$$

The intercept of equation 6-106 can be rearranged to give $[(1/V)(1+K_a/\mathbf{A}) + \mathbf{B}/VK_{Ib}]$. As opposed to the linear primary plots observed when A is the varied substrate, all of the reciprocal plots exhibit substrate inhibition. The inhibition results from combination of B with the EQ complex that exists in the steady state at any **A**. As **A** becomes higher compared to its K_a, the **EQ** in the steady state increases, and the **B** required to inhibit becomes lower.

The Krebs cycle enzyme malate dehydrogenase (MDH) catalyzes the reversible oxidation of L-malate to oxalacetate using NAD^+ as the oxidant. The kinetic mechanism of MDH is steady-state ordered with the dinucleotide substrates

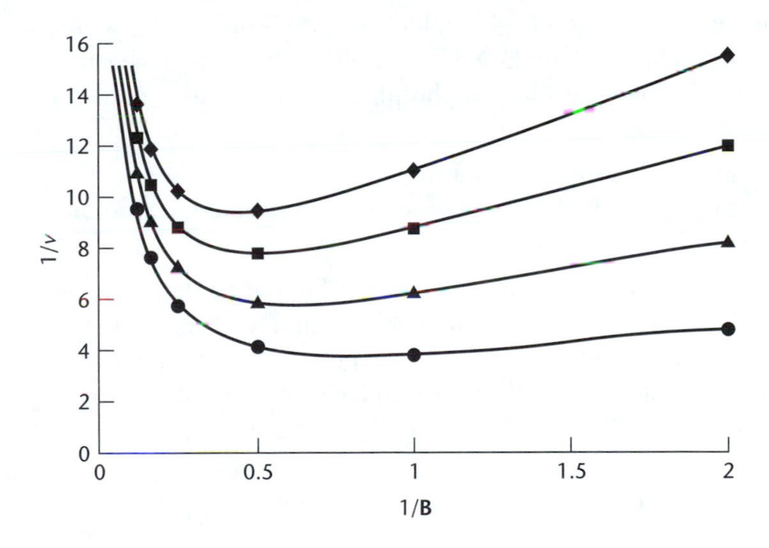

Figure 6-7. Double reciprocal plot exhibiting substrate inhibition by B in a steady-state ordered mechanism with B as the varied substrate. Units on the ordinate and abscissa are minutes/millimolar and millimolar^{-1}, respectively. The concentrations of A used are as follows: 0.25 mM, ◆; 0.357 mM, ■; 0.625 mM, ▲; and 2.5 mM, ●. Parameters used in the plot are the same as those used in Figure 6-6.

adding first and being released last. Oxalacetate exhibits uncompetitive substrate inhibition versus NADH as a result of binding to E–NAD, consistent with the ordered mechanism and suggesting slow release of NAD^+ (B *2*, 220).

The order of addition of substrates in the glyceraldehydes-3-phosphate dehydrogenase reaction is NAD^+, glyceraldehyde 3-phosphate, and inorganic phosphate. Uncompetitive substrate inhibition is observed by glyceraldehyde 3-phosphate versus NAD^+ (B *11*, 102). The substrate inhibition ($K_i = 2.7\,\text{mM}$) is observed with either arsenate or phosphate as the third substrate. 2-Deoxyglyceraldehyde 3-phosphate, however, only shows the substrate inhibition with arsenate as the third substrate. The reason for the lack of substrate inhibition with phosphate is that V_{max} with 2-deoxyglyceraldehyde 3-phosphate and phosphate is only 7.3% of that measured with glyceraldehyde 3-phosphate, while it is 47% with arsenate as the third substrate. Only when V_{max} is fast enough will there be enough E–NADH in the steady state for the substrate to combine with and give substrate inhibition. Analogues of the sugar that are neutral at C1 (glyceraldehyde 3-phosphate, glycerol 3-phosphate) bind to E–NAD and to E–NADH, while ones with a negative charge at C1 (1,3-bisphosphoglycerate, 3-phosphoglycerate) bind only to E–NADH.

In order to test whether the uncompetitive substrate inhibition by B in a steady-state ordered mechanism results from formation of an EBQ complex, a double inhibition experiment can be carried out (see below).

In the case of a one-site Ping Pong mechanism, reactants bind in a single binding pocket, and as a result they resemble one another structurally. There are two stable enzyme forms, E and F, and reactants can compete with one another for binding to the incorrect stable enzyme form. Consider a mechanism where B combines in dead-end fashion with E:

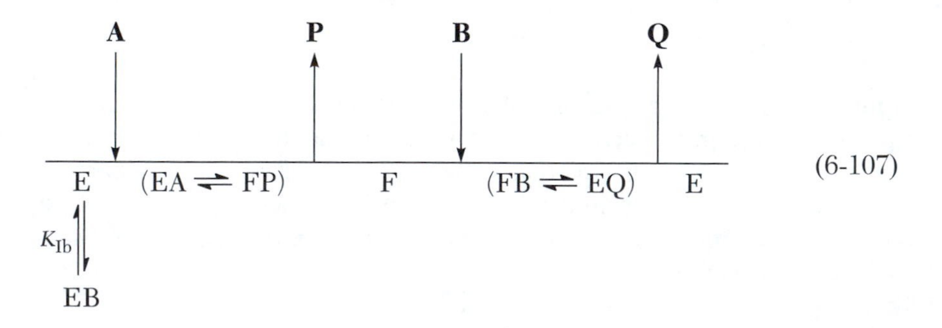

The rate expression for this case when **A** is varied is shown in equation 6-108:

$$\frac{1}{v} = \left(\frac{K_a}{V}\right)\left(1 + \frac{\mathbf{B}}{K_{Ib}}\right)\left(\frac{1}{\mathbf{A}}\right) + \left(\frac{1}{V}\right)\left(1 + \frac{K_b}{\mathbf{B}}\right) \tag{6-108}$$

A plot of equation 6-108 is shown in Figure 6-8 and clearly shows the increase in slope as **B** becomes inhibiting. Competitive substrate inhibition is typical of Ping Pong mechanisms and is most often seen in the nonphysiologic direction of the reaction.

Both substrates can also give competitive inhibition in a Ping Pong mechanism; for example, consider mechanism 6-109:

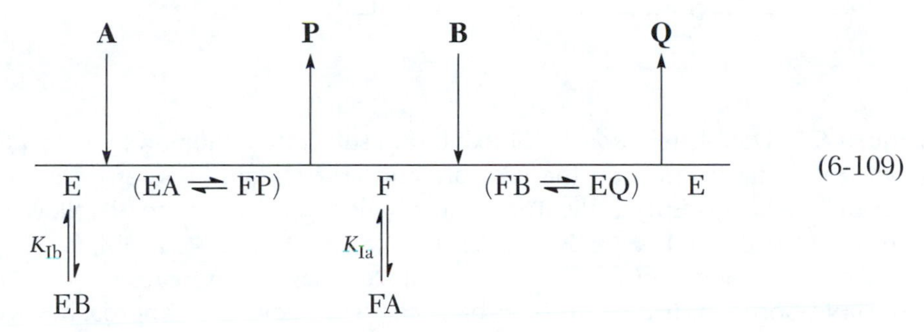

(6-109)

The rate expression for mechanism 6-109 is given as equation 6-110:

$$\frac{1}{v} = \left(\frac{K_a}{V\mathbf{A}}\right)\left(1 + \frac{\mathbf{B}}{K_{Ib}}\right) + \left(\frac{K_b}{V\mathbf{B}}\right)\left(1 + \frac{\mathbf{A}}{K_{Ia}}\right) + \frac{1}{V} \tag{6-110}$$

where K_{Ia} and K_{Ib} reflect the substrate inhibition constants for A and B combining with the F and E forms of the enzyme, respectively. A plot of the above equation with A as the varied substrate gives a complex pattern (Figure 6-9), in which lines cross to the left of the $1/v$ axis, as a result of an increase in the slope as **B** increases, and the plots at lower **B** turn up (exhibit substrate inhibition) as the $1/v$ axis is approached (**A** increases to infinity).

As both reactants increase to infinity, however, the lines at all fixed reactant concentration extrapolate to a single point on the ordinate, equal to $1/V$. Thus, inhibition by both reactants is competitive. Rough estimates of the inhibition constants can be obtained graphically by varying the concentration of the substrate inhibitor with the other reactant fixed at its K_m and plotting in Dixon format ($1/v$ versus **I**). The abscissa intercept is equal to $K_I(1 + \mathbf{A}/K_a)$. For accurate values, however, the data should be fitted directly to the rate equation that is the

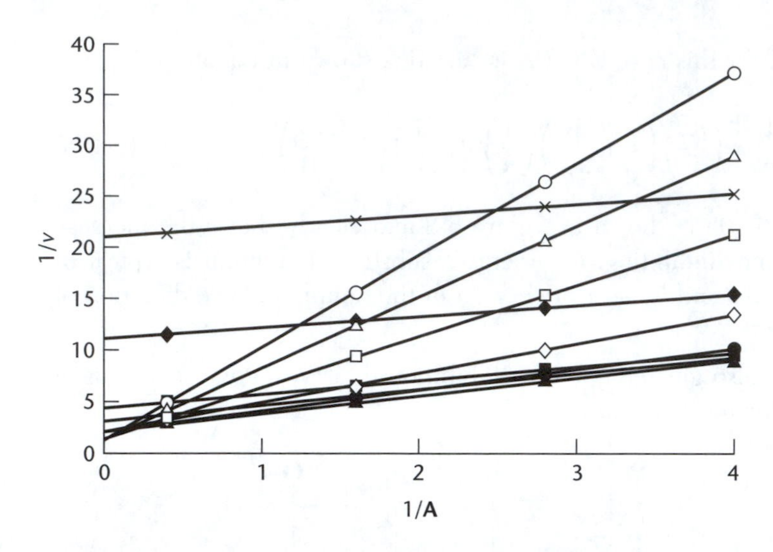

Figure 6-8. Double reciprocal plot exhibiting substrate inhibition by B in a Ping Pong mechanism. Units on the ordinate and abscissa are minutes/millimolar and millimolar^{-1}, respectively. The concentrations of B used are as follows: 0.05 mM, ×; 0.1 mM, ◆; 0.3 mM, ■; 0.5 mM, ▲; 1 mM, ●; 2 mM, ◇; 4 mM, □; 6 mM, △; and 8 mM, ○. Note the parallel lines at low reactant concentration, which then cross at high concentrations as B approaches K_{Ib}. The remainder of the lines are indicative of competitive inhibition by B versus A. Lines intersect at the true value of $1/V$. Parameters used in the plot are as follows: $V = 1$ mM/min and $K_a = K_b = K_{Ib} = 1$ mM.

reciprocal of equation 6-110. Double competitive substrate inhibition is characteristic of both one- and two-site Ping Pong mechanisms and in fact is more common in the latter.

A prototypical example of the above behavior is observed in the aspartate aminotransferase reaction (B 25, 227). In this case, both aspartate and α-ketoglutarate exhibit competitive substrate inhibition as a result of binding of reactants to the incorrect enzyme form, aspartate to the F or pyridoxamine form of the enzyme and α-ketoglutarate to the E or pyridoxal form of the enzyme, competing with the other substrate. This is not surprising, given the structural similarity of the reactants, which are all 4- or 5-carbon dicarboxylic acids.

Several other enzymes with Ping Pong mechanisms have been shown to give double competitive inhibition. In addition to aspartate aminotransferase, enzymes include thioketolase, which catalyzes the reversible condensation of two

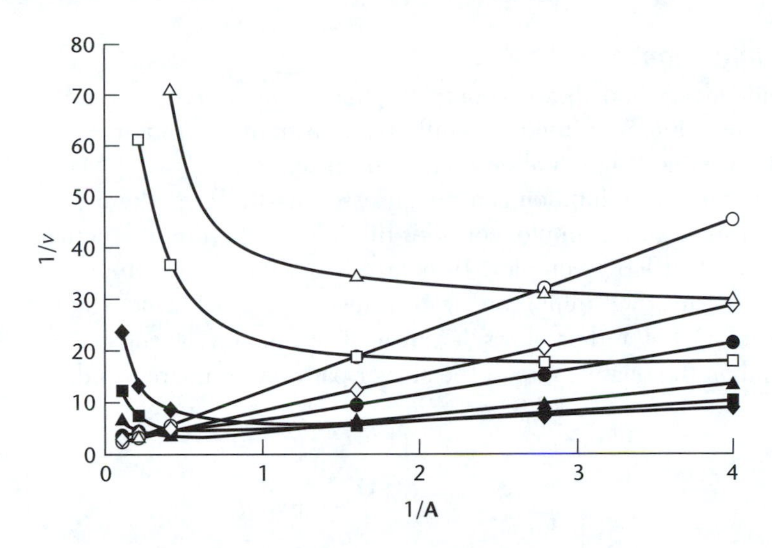

Figure 6-9. Double reciprocal plot exhibiting double competitive substrate inhibition in a Ping Pong mechanism. Units on the ordinate and abscissa are minutes/millimolar and millimolar^{-1}, respectively. The concentrations of B used are as follows: 0.05 mM, △; 0.1 mM, □; 0.5 mM, ◆; 1 mM, ■; 2 mM, ▲; 4 mM, ●; 6 mM, ◇; and 10 mM, ○. The lines, even at low reactant concentration, exhibit substrate inhibition as A increases. Overall, the plot is similar to that shown in Figure 6-8, but with inhibition by A added. Parameters used in the plot are as follows: $V = 1$ mM/min and $K_a = K_b = K_{Ib} = K_{Ia} = 1$ mM.

acetyl-CoA molecules to acetoacetyl-CoA and CoA-SH (JBC *241*, 1222); nucleoside diphosphate kinase, which catalyzes the reversible phosphorylation of any nucleoside diphosphate, for example, ADP, GDP, CDP, etc., by any nucleoside triphosphate, for example, CTP, ATP, etc. (B *8*, 633); and *O*-acetylserine sulfhydrylase, which catalyzes the formation of L-cysteine from *O*-acetyl-L-serine and sulfide (JBC *251*, 2023). It is interesting to note that all of the above Ping Pong enzymes have substrates and products that resemble one another structurally with the exception of the sulfhydrylase.

Competitive substrate inhibition can also be observed in random sequential mechanisms, for example, adenylate kinase, where AMP has affinity for the MgATP site that is strongest at lower pHs (JBC *245*, 4163). In that case the slope expression in equation 6-108 has a $K_{ib}/\mathbf{B}$ term present, so that the slopes are not parallel at low **B** but increase as **B** is decreased. (In addition, free ADP gives competitive substrate inhibition versus MgADP in the opposite reaction direction.)

Partial Substrate Inhibition

The substrate inhibition discussed above is complete, that is, the initial rate tends to zero as the substrate inhibitor tends to infinity. Substrate inhibition can, however, be partial, resulting from an alteration in the reaction pathway at high substrate concentration. Partial inhibition is normally associated with randomness in the kinetic mechanism. As an example, consider the following ordered kinetic mechanism with a dead-end EBQ complex. In contrast to the case of complete substrate inhibition, the substrate inhibition by B in mechanism 6-111 does not prevent the release of Q but rather slows it down. The amount of substrate inhibition will depend on the relative rate of the slower pathway compared to the uninhibited pathway.

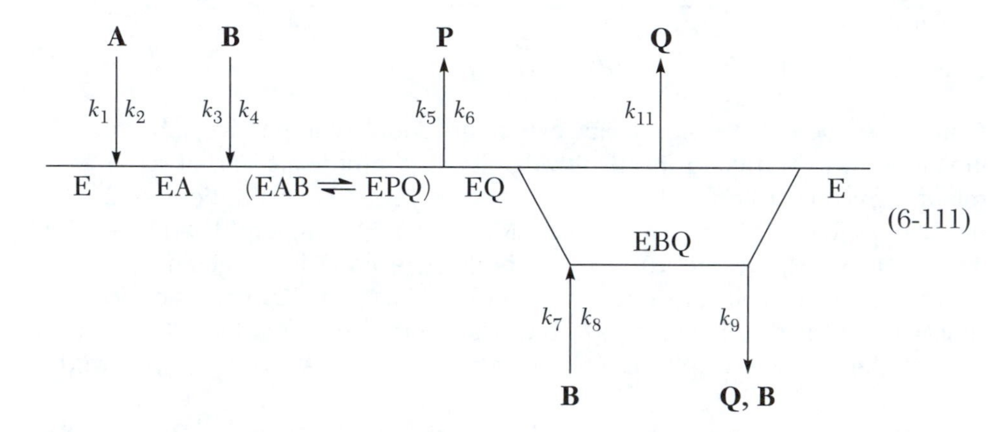

(6-111)

The substrate inhibition by B will be uncompetitive versus A and partial, since B cannot trap Q on the enzyme. The rate expression for mechanism 6-111 is given below. If release of Q from EBQ gives an EB complex and A adds to it to give EAB, there will also be a slope effect in the inhibition pattern.

$$\frac{1}{v} = \left(\frac{K_a}{V}\right)\left(1 + \frac{K_{ia}K_b}{K_a\mathbf{B}}\right)\left(\frac{1}{\mathbf{A}}\right) + \frac{K_b}{V\mathbf{B}} + \left(\frac{1}{V}\right)\left[\frac{1 + \mathbf{B}/K_{in}}{1 + \mathbf{B}/K_{id}}\right] \tag{6-112}$$

where K_{in} and K_{id} are apparent inhibition constants for B with the following expressions:

$$K_{in} = \frac{(k_5 + k_{11})(k_8 + k_9)}{k_7(k_5 + k_9)}; \quad K_{id} = \frac{k_{11}(k_8 + k_9)}{k_7 k_9} \tag{6-113}$$

As **A** and **B** approach infinity, the apparent V is

$$\text{app}V = V\frac{K_{\text{in}}}{K_{\text{id}}} \tag{6-114}$$

Thus, the apparent V is determined by the ratio $K_{\text{in}}/K_{\text{id}}$. A primary plot of equation 6-112 is shown in Figure 6-10. The intercept replot exhibits a value of 0.33 for V, consistent with the ratio of K_{in} to K_{id} (V is 1 in Figure 6-10).

Yeast alcohol dehydrogenase has a predominantly ordered kinetic mechanism with dinucleotide substrates adding to enzyme first and being released last. In the direction of alcohol oxidation, the alcohol substrate can combine with the E–NADH complex to give substrate inhibition that is partial (BJ *100*, 34). Thus, NADH can be released from the E–NADH–alcohol complex, and $K_{\text{id}} > K_{\text{in}}$. Note that if addition of B *increases* the rate of release of Q,

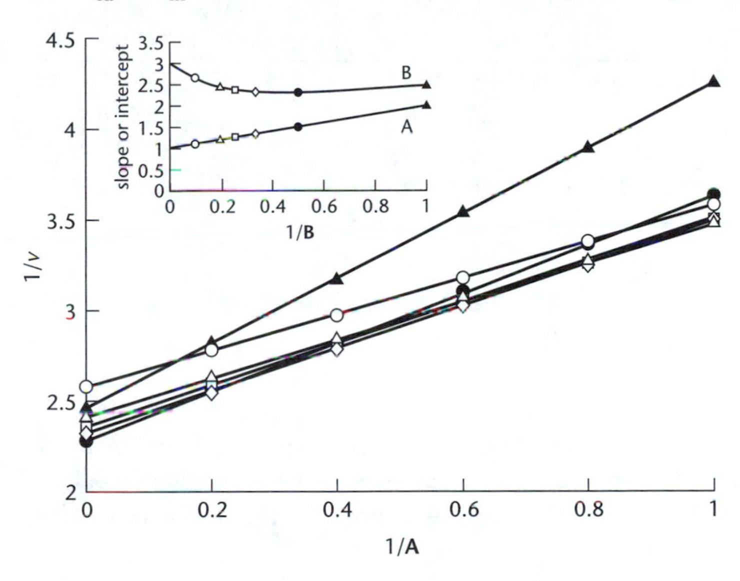

Figure 6-10. Partial substrate inhibition by B in an ordered mechanism. Units on the ordinate and abscissa are minutes/millimolar and millimolar^{-1}, respectively. The concentrations of B used are as follows: 1 mM, ▲; 2 mM, ●; 3 mM, ◇; 4 mM, □; 5 mM, △; and 10 mM, ○. The inset shows the replot of slope (A) and intercept (B) versus the reciprocal of B concentration. The substrate inhibition is exhibited on the intercept, while the slope replot is linear. Parameters used in the plot are as follows: $V = 1$ mM/min; $K_{ia} = K_a = K_b = 1$ mM; $K_{\text{in}} = 1$ mM; and $K_{\text{id}} = 3$ mM.

substrate activation will be observed, as shown for the equine liver alcohol dehydrogenase reaction with cyclohexanol as the alcohol substrate (B *20*, 1790). In the case of substrate activation, the ratio K_{in}/K_{id} will be greater than unity.

Induced Substrate Inhibition

The inhibition patterns produced by dead-end inhibitors have been discussed above. If an inhibitor resembling the first substrate in an ordered mechanism permits addition of the second substrate in a fashion that traps the inhibitor on the enzyme, this inhibitor induces substrate inhibition by the second substrate. This can be diagrammed as follows:

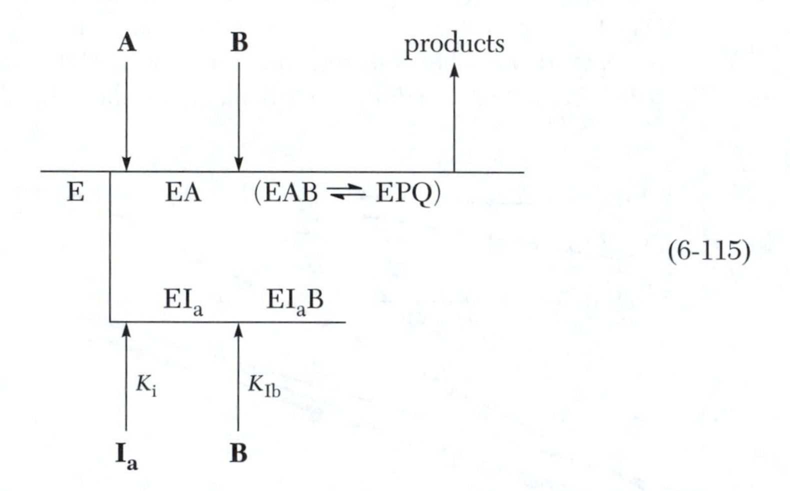

(6-115)

The addition of the first substrate cannot be in rapid equilibrium, and it is essential that the second substrate must prevent the release of the inhibitor from the enzyme, that is, reduce the inhibitor off rate constant to a negligible value. This is thus a very sensitive test for the degree of order in the mechanism. The rate equation for this case is

$$v = \frac{V\mathbf{AB}}{(K_{ia}K_b + K_a\mathbf{B})\left[1 + \left(\dfrac{\mathbf{I}}{K_i}\right)\left(1 + \dfrac{\mathbf{B}}{K_{Ib}}\right)\right] + K_b\mathbf{A} + \mathbf{AB}} \tag{6-116}$$

where K_i and K_{Ib} are dissociation constants of I from EI and of B from EIB, and the other kinetic constants are the usual ones for an ordered mechanism. The substrate inhibition occurs only in the presence of I and is competitive when **A** is varied at a fixed level of I.

Few cases of induced substrate inhibition that adhere fully to this model are known, since most "ordered" mechanisms result from high synergism in the binding of the substrates and in many cases the inhibitor may fail to induce the conformation change that permits tight binding of the second substrate. Induced substrate inhibition by methylene tetrahydrofolate is seen in the reaction catalyzed by thymidylate synthase when bromodeoxyuridine 5′-monophosphate is present (B *17*, 4018). This inhibitor forms a covalent complex with methylene tetrahydrofolate and enzyme (EIB complex in the model above) that cannot complete the reaction and from which the inhibitor cannot dissociate until methylene tetrahydrofolate does.

Induced substrate inhibition could also occur in a three-substrate ordered mechanism. An inhibitor combining with E will induce substrate inhibition by B if an EIB complex forms from which I cannot dissociate. The inhibition will be competitive versus A and uncompetitive versus C. If C could add to the EIB complex and prevent release of B and I, substrate inhibition by C would also be induced that would be competitive versus A and uncompetitive versus B. The competitive inhibitions versus A result from the ability of A to prevent I from binding to free enzyme, while the uncompetitive patterns result from all steps prior to combination of the varied substrate coming to equilibrium. Induced substrate inhibition occurs only under steady-state conditions and not under equilibrium conditions.

An inhibitor combining with the EA complex in a three-substrate ordered mechanism will induce substrate inhibition by C if an EAIC complex forms from which I cannot dissociate. The substrate inhibition will be competitive versus B and uncompetitive versus A. Oxalylglycine, which is an excellent mimic of α-ketoglutarate, does not induce substrate inhibition by ammonia in the reaction catalyzed by α-ketoglutarate dehydrogenase, although the reaction is ordered [NADH, α-ketoglutarate, ammonia (B *19*, 2321)]. Presumably ammonia combines only when it can react chemically with a keto group, and it does not react with the amide carbonyl group of oxalylglycine.

Most cases of induced substrate inhibition give partial, as opposed to total, substrate inhibition. This results from the escape of the inhibitor from the EIB complex at a reduced but still finite rate. These mechanisms are thus partly random, although with a strongly preferred order of addition of substrates. Three well-established examples are known. With yeast hexokinase, lyxose induces substrate inhibition by MgATP that is competitive, but partial, versus glucose (B *14*, 28). The binding of sugars and MgATP is highly synergistic, with the dissociation constant of MgATP being 5 mM in the absence and 0.1 mM in the presence of glucose. MgATP slows the off rate constant of lyxose but not to zero.

The second example is PEP carboxylase, where phosphoglycolate induced substrate inhibition by bicarbonate that was competitive versus PEP (B *31*, 6421). Although the effect was partial, the rate of dissociation of phosphoglycolate was decreased by a factor of at least 170 by the presence of bicarbonate. The third example is L-ribulokinase from *Escherichia coli*, where L-erythrulose induced substrate inhibition by MgATP that was competitive versus L-ribulose (ABB *396*, 219). Again, the inhibition was partial, showing that MgATP slowed down but did not totally prevent the release of L-erythrulose from the enzyme. It is likely that most induced substrate inhibitions will be of the partial type, unless covalent complexes are formed, as with thymidylate synthase. The method is a very sensitive test for the actual degree of order in a kinetic mechanism.

Alternate Substrate Inhibition

If the reaction rate is measured by the formation of products from the varied substrate, an alternate substrate giving a different (and nonmeasured) first product will cause inhibition by tying up some of the enzyme. Some of these inhibition patterns are unique enough to be diagnostic of mechanism and thus could be quite useful. This method has not been widely used, which is a pity.

Ping Pong Mechanism. If A and B are substrates that produce P and Q as products, and the formation of P is measured, a substrate C that is an alternative one to A (but still produces F and later Q) will be competitive versus A and noncompetitive versus B:

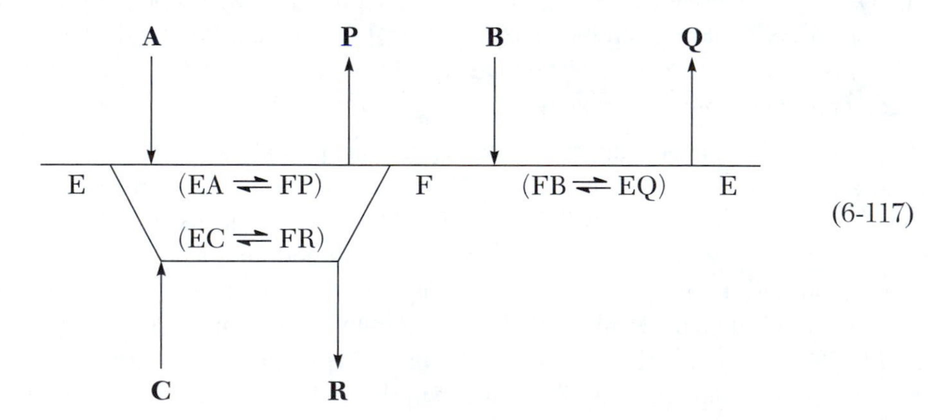

(6-117)

The competitive inhibition constant versus A is

$$K_{is} = \frac{K_c}{1 + \frac{K_{b(c)}}{\mathbf{B}}} \tag{6-118}$$

where $K_{b(c)}$ is the Michaelis constant for B with C as a substrate. The value of K_{is} thus varies from K_c (the Michaelis constant for C) at high **B** to zero as **B** approaches zero. The value at high **B** simply reflects the affinity of C for the enzyme in the steady state, as a Michaelis constant always does. The zero value at low **B** results from the fact that at low **B** most of the enzyme is in the F form. The level of C required to keep it there, and not in the E form that reacts with A, gets lower and lower as **B** is decreased.

When **B** is varied, the noncompetitive inhibition constants are

$$K_{is} = \frac{K_c K_{b(a)} \mathbf{A}}{K_{b(c)} K_a} \tag{6-119}$$

$$K_{ii} = K_c\left(1 + \frac{\mathbf{A}}{K_a}\right) \tag{6-120}$$

where $K_{b(a)}$ is the Michaelis constant for B with A as substrate, and K_a is the Michaelis constant of A. Both inhibition constants increase linearly with **A** because of the competition between A and C for free enzyme, and at saturating **B** the K_{ii} value at low **A** is the Michaelis constant of C. K_{is}, however, approaches zero at very low **A**. At low **B** where most of the enzyme is in the F form, it takes only a very low level of C to keep it there so that there is no E for A to react with. This unique variation of K_{is} values in both the competitive and noncompetitive patterns is diagnostic for a Ping Pong mechanism.

Sequential Mechanisms. In a random sequential mechanism, an alternate substrate is competitive versus the substrate it replaces and noncompetitive versus the other one:

$$\mathbf{A} + \mathbf{B} \rightleftharpoons \mathbf{P} + \mathbf{Q} \qquad \mathbf{C} + \mathbf{B} \rightleftharpoons \mathbf{P} + \mathbf{R} \tag{6-121}$$

If C is an alternative substrate for A, but the reaction rate is determined by formation of Q, the competitive inhibition constant is

$$K_{is} = \frac{K_c\left(1 + \frac{K_{ib}}{\mathbf{B}}\right)}{1 + \frac{K_{b(c)}}{\mathbf{B}}} \tag{6-122}$$

where K_{ib} and $K_{b(c)}$ are the dissociation constant of B from EB and the Michaelis constant for B with C as the other substrate. For the noncompetitive inhibition of C versus B:

$$K_{is} = K_{ic}\left(1 + \frac{\mathbf{A}}{K_{ia}}\right) \qquad (6\text{-}123)$$

$$K_{ii} = K_{c}\left(1 + \frac{\mathbf{A}}{K_{a}}\right) \qquad (6\text{-}124)$$

where K_{ic} and K_{ia} are dissociation constants of C from EC and A from EA. Note that the K_{is} values are finite at low levels of the other substrate, unlike in the Ping Pong case. This is one way to distinguish a random sequential mechanism with a nearly parallel initial velocity pattern from a Ping Pong one.

In an ordered sequential mechanism where C is an alternate substrate for the second substrate B, one observes competitive inhibition versus B and noncompetitive inhibition versus A if the reaction rate is determined from appearance of P:

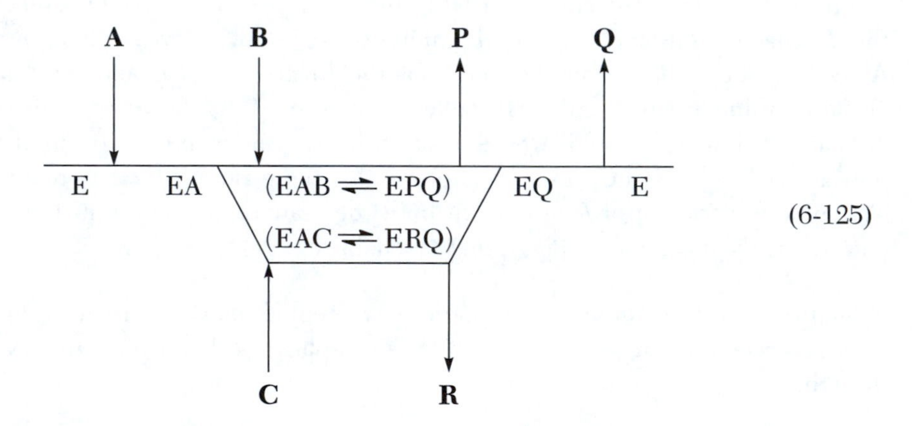

(6-125)

For the competitive inhibition:

$$K_{is} = \frac{K_{c}\left(1 + \frac{K_{ia}}{\mathbf{A}}\right)}{1 + \frac{K_{a(c)}}{\mathbf{A}}} \qquad (6\text{-}126)$$

and for the noncompetitive inhibition:

$$K_{is} = \left(\frac{K_{ia}K_c}{K_{a(c)}}\right)\left(1 + \frac{K_{a(b)}\mathbf{B}}{K_{ia}K_b}\right) \quad (6\text{-}127)$$

$$K_{ii} = K_c\left(1 + \frac{\mathbf{B}}{K_b}\right) \quad (6\text{-}128)$$

These patterns are similar to those for the random mechanism.

When C is an alternate substrate for A in an ordered mechanism, more complex patterns are observed when formation of Q (the specific product from A) is measured as the reaction rate.

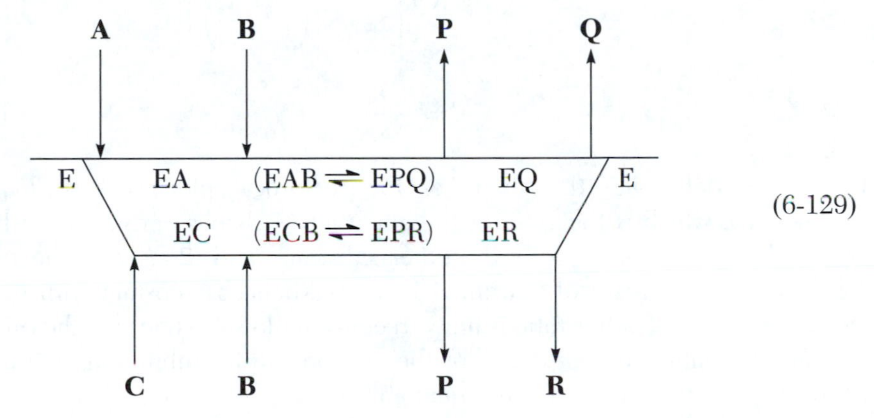

(6-129)

In this mechanism, C is competitive versus A and the K_i is given by equation 6-130:

$$K_{is} = \frac{K_c\left(1 + \dfrac{K_{ic}K_{b(c)}}{K_c\mathbf{B}}\right)}{1 + \dfrac{K_{b(c)}}{\mathbf{B}}} \quad (6\text{-}130)$$

When B is varied, however, reciprocal plots are 2/1 functions as shown in equation 6-131:

$$\frac{1}{v} = \frac{a + b\left(\dfrac{1}{\mathbf{B}}\right) + c\left(\dfrac{1}{\mathbf{B}^2}\right)}{V\left[1 + d\left(\dfrac{1}{\mathbf{B}}\right)\right]} \quad (6\text{-}131)$$

In equation 6-131 a, b, c, and d are defined as

$$a = \left(1 + \frac{K_a}{\mathbf{A}}\right)\left[1 + \frac{\mathbf{C}}{K_c\left(1 + \frac{\mathbf{A}}{K_a}\right)}\right] \tag{6-132}$$

$$b = K_{b(a)}\left(1 + \frac{K_{ia}}{\mathbf{A}}\right) + d\left(1 + \frac{K_a}{\mathbf{A}}\right) + \frac{K_a\mathbf{C}\left(K_{b(c)} + \frac{K_{ia}K_{b(a)}}{K_a}\right)}{K_c\mathbf{A}} \tag{6-133}$$

$$c = dK_{b(a)}\left(1 + \frac{K_{ia}}{\mathbf{A}}\right)\left[1 + \frac{\mathbf{C}}{K_{ic}\left(1 + \frac{\mathbf{A}}{K_{ia}}\right)}\right] \tag{6-134}$$

$$d = \frac{K_{ic}K_{b(c)}}{K_c} \tag{6-135}$$

Two ratios determine the shape of the reciprocal plots: (1) $(K_{ic}K_{b(c)}/K_c)/(K_{ia}K_{b(a)}/K_a)$, which is the ratio of the apparent Michaelis constants for B with low levels of either C or A as the other substrate; and (2) K_{ic}/K_c, the ratio of the dissociation constant of C from EC and its Michaelis constant with B as the second substrate. If either ratio is unity, reciprocal plots are linear in the presence of C and the inhibition constants for the noncompetitive inhibition of C are the same as those given above for the random case.

When one ratio is greater than unity and the other is less than unity, reciprocal plots in the presence of C are concave downward. When both ratios are less than unity, reciprocal plots in the presence of C are concave upward but cannot have a minimum. When both ratios are above unity, reciprocal plots with C present are concave upward but can have a minimum (that is, partial substrate inhibition) if

$$\frac{K_{ic}}{K_c} > \left[1 + \frac{K_{ia}K_{b(a)}}{K_aK_{b(c)}}\right] \tag{6-136}$$

A 2/1 reciprocal plot has a linear asymptote at low **B**, and the slopes and intercepts of these asymptotes will be linear functions of **C**, so that the inhibition appears noncompetitive. The inhibition constant for the asymptote slopes is the same as K_{is} in equation 6-123 above, but the K_{ii} for the asymptote intercepts is given by equation 6-137:

$$K_{ii} = \frac{K_{ic}\left(1 + \frac{\mathbf{A}}{K_a}\right)}{1 + \frac{K_{ia}K_{b(a)}\left(1 - \frac{K_c}{K_{ic}}\right)}{K_a K_{b(c)}}} \tag{6-137}$$

Equation 6-137 can have a negative value if

$$\frac{K_c}{K_{ic}} > \left[1 + \frac{K_a K_{b(c)}}{K_{ia} K_{b(a)}}\right] \tag{6-138}$$

Thus the asymptote intercept will decrease with increasing **C** and become negative at high enough **C** values. This requires that the two ratios above are less than unity and the reciprocal plots in the presence of C are concave upward without a minimum. If equation 6-138 above is an equality, the asymptote intercept will not vary with **C** even though the true intercepts increase linearly with **C** and reciprocal plots are concave upward. When both ratios are above unity, equation 6-137 gives only positive values of K_{ii}, although the reciprocal plots are concave upward and may have a minimum.

The above approach was attempted by Rudolph and Fromm with the thio-NAD analogue as an alternate substrate for NAD with ethanol concentration varied in the presence of liver alcohol dehydrogenase [reaction followed at 342 nm, the isosbestic point for thio-NAD and its reduced form (B 9, 4660)]. These authors reported concave upward reciprocal plots with partial substrate inhibition in the presence of thio-NAD, but the asymptote intercepts decreased with increased thio-NAD concentration, which is not predicted by the equations above. Thus some other interactions must have caused the partial substrate inhibition and these equations remain to be verified in an actual case.

Double Inhibition

On the basis of the above discussion, it can be seen that corroborative methods would be very useful for the interpretation of observed inhibition patterns. For example, a noncompetitive inhibition pattern by an analogue of B versus B in a steady-state ordered mechanism may indicate combination of I_b either with E and EA or with EA and EQ, with the most reasonable being the latter. If one could demonstrate the existence of an EBQ complex, this would constitute supportive evidence consistent with combination with EQ. In addition, knowledge of the true dissociation constant for the EBQ complex would allow calculation of the amount of EQ present in the steady state and thus the amount of limitation of the overall reaction by release of Q from the EQ complex. An initial velocity method capable of demonstrating whether two inhibitors are capable of being bound to enzyme at the same time or whether they are mutually exclusive was developed by Yonetoni and Theorell (ABB *106*, 243). The above example of an EBQ complex will be used to demonstrate the method.

The ordered mechanism with an EBQ complex is written below (equation 6-139) to show that, under initial velocity conditions, Q is a competitive dead-end inhibitor versus A. (The product Q can be considered dead-end since the EQ complex is not competent in the absence of added P.)

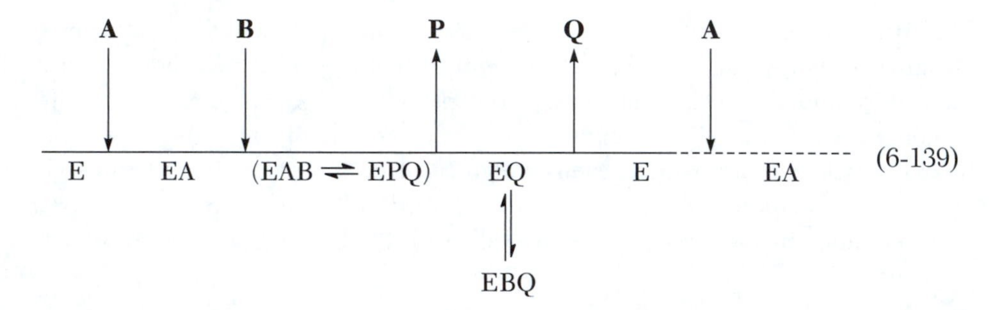

At a level of **B** sufficient to give substrate inhibition and with **A** fixed equal to or below its K_a, and allowing for the possible formation of an EB complex to complete the thermodynamic cycle, a portion of mechanism 6-139 can be rewritten as shown in mechanism 6-140. In order to generalize the theory, the following equation uses the general descriptors I and J for B and Q, respectively.

$$\begin{array}{ccc} E & \overset{K_i}{\rightleftharpoons} & EI \\ K_j \updownarrow & & \updownarrow \alpha K_j \\ EJ & \underset{\alpha K_i}{\rightleftharpoons} & EIJ \end{array} \qquad (6\text{-}140)$$

In mechanism 6-140, v and v_0 will be the inhibited and uninhibited initial rates, respectively; I and J are two inhibitors that are suspected of forming a ternary complex with enzyme; K_i and K_j represent the apparent dissociation constants for EI and EJ complexes, respectively; and α is an interaction constant that estimates the influence of one bound inhibitor on the binding of the other. The overall rate expression describing mechanism 6-140 is equation 6-141:

$$\frac{1}{v} = \frac{1}{v_0}\left(1 + \frac{\mathbf{I}}{K_i} + \frac{\mathbf{J}}{K_j} + \frac{\mathbf{IJ}}{\alpha K_i K_j}\right) \qquad (6\text{-}141)$$

Since the equilibrium constant for forming EIJ from E, I, and J is independent of the pathway, the effect of I on J binding must be equal to the effect of J on I binding. Data are obtained at constant nonsaturating **A**, with the initial rate measured as a function of **I**, at different fixed **J** including zero levels of both. Assuming that all complexes are observed, a Dixon plot similar to that shown in Figure 6-11 will be obtained.

In Figure 6-11, the line obtained at zero **J** reflects the formation of the EI complex, and if EI did not form, the line would have a slope of zero. Data on the

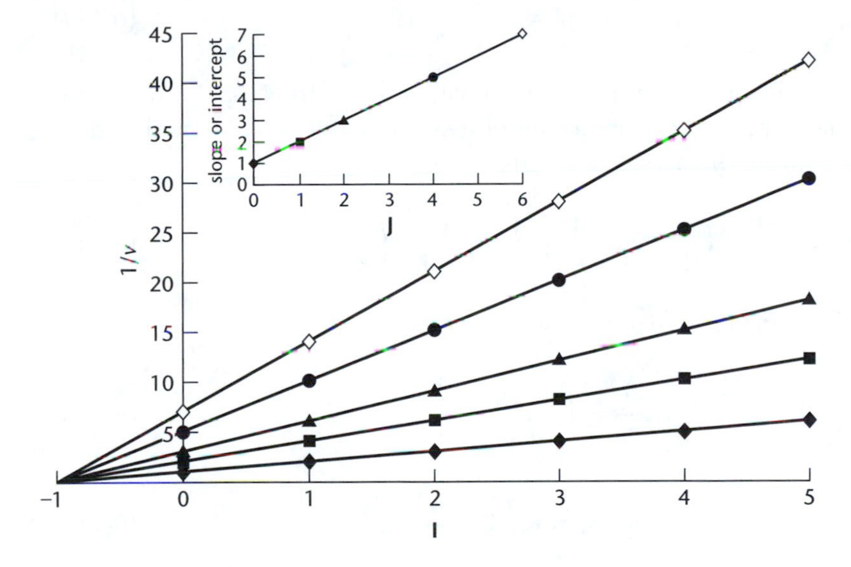

Figure 6-11. Dixon plot of $1/v$ versus **I** as a function of a second inhibitor J. Units on the ordinate and abscissa are minutes/millimolar and millimolar^{-1}, respectively. The concentrations of J used are as follows: 0 mM, ◆; 1 mM, ■; 2 mM, ▲; 4 mM, ●; and 6 mM, ◇. Inhibition by I alone is observed at $\mathbf{J} = 0$, while inhibition by J alone is observed in the intercept replot. The inset shows the interaction between the two inhibitors. Parameters used in the plot are as follows: $v = 1$ mM/min, $K_i = K_j = 1$ mM, and $\alpha = 1$.

ordinate, zero **I** but increasing **J**, reflect the EJ complex, and if EJ did not form, all lines would intersect on the ordinate. On the other hand, if both EI and EJ form but the EIJ complex does not—that is, I and J are mutually exclusive—the lines would be parallel. Lines with increasing slope indicate that the EIJ complex is allowed, and a quantitative analysis allows estimation of α. Rearranging equation 6-141 gives the following, along with expressions for slope and intercept terms. (Slope and intercept replots are shown as insets in Figure 6-11.) As can be seen, all terms can be estimated graphically.

$$\frac{v_0}{v} = 1 + \left(\frac{1}{K_i} + \frac{\mathbf{J}}{\alpha K_i K_j}\right)\mathbf{I} + \frac{\mathbf{J}}{K_j} \tag{6-142}$$

$$\text{slope} = \frac{1}{K_i} + \left(\frac{1}{\alpha K_i K_j}\right)\mathbf{J} = \left(\frac{1}{K_i}\right)\left(1 + \frac{\mathbf{J}}{\alpha K_j}\right) \tag{6-143}$$

$$\text{intercept} = 1 + \left(\frac{1}{K_j}\right)\mathbf{J} \tag{6-144}$$

It should be noted that the parameters in equation 6-141 are apparent ones. For example, if one considers the interaction of two competitive inhibitors in a one-substrate mechanism, the actual rate equation is

$$\frac{1}{v} = \left(\frac{K_a}{V}\right)\left[1 + \frac{\mathbf{I}}{K_i} + \frac{\mathbf{J}}{K_j} + \frac{\mathbf{IJ}}{\alpha K_i K_j}\right]\left(\frac{1}{\mathbf{A}}\right) + \frac{1}{V} \tag{6-145}$$

Comparing this equation to 6-140 shows that

$$\frac{1}{v_0} = \left(\frac{1}{V}\right)\left(1 + \frac{K_a}{\mathbf{A}}\right) \tag{6-146}$$

$$\text{app}K_i = K_i\left(1 + \frac{\mathbf{A}}{K_a}\right) \tag{6-147}$$

$$\text{app}K_j = K_j\left(1 + \frac{\mathbf{A}}{K_a}\right) \tag{6-148}$$

$$\text{app}\alpha = \frac{\alpha}{(1 + \mathbf{A}/K_a)} \tag{6-149}$$

but $(\text{app}K_i)(\text{app}\alpha) = \alpha K_i$ and $(\text{app}K_j)(\text{app}\alpha) = \alpha K_j$.

As noted above, an α value of infinity gives a parallel pattern and shows mutually exclusive binding of I and J. A finite α value shows that both inhibitors can be bound at the same time, with α values above unity showing some interference and values below unity showing synergistic binding. This method was first used with competitive inhibitors of NAD and liver alcohol dehydrogenase (ABB *106*, 243). *o*-Phenanthroline gave α values of 1.0, 0.5, and 0.3 with ADPR, ADP, and AMP, showing that it occupied only the nicotinamide part of the NAD binding site. ADPR, ADP, and AMP gave α values of ∞ among themselves, as expected.

When $1/v$ is plotted versus **I** at different **J** levels, the crossover point from equation 6-141 is at $-\alpha K_i$, and when $1/v$ is plotted versus **J** at different **I** levels, it is at $-\alpha K_j$. The vertical coordinate of the crossover point is $(1/v_o)(1 - \text{app}\alpha)$. If equation 6-145 is the rate equation, the horizontal coordinate of the crossover point is still $-\alpha K_i$, since the product of appK_i and appα is αK_i. But the vertical crossover point is now

$$\left(\frac{1}{V}\right)\left(1+\frac{K_a}{\mathbf{A}}\right)\left[1-\frac{\alpha}{1+\mathbf{A}/K_a}\right] \tag{6-150}$$

A pattern crossing on the horizontal axis corresponds to $\alpha = (1 + \mathbf{A}/K_a)$, while if $\alpha = 1$, the pattern crosses at $1/V$. In practice, one determines αK_i or αK_j from crossover points and uses K_i and K_j values from separate competitive inhibition patterns to evaluate α.

Double inhibition has been used for the NAD-malic enzyme, which has a steady-state random kinetic mechanism (B *23*, 5446). Oxalate is a noncompetitive inhibitor versus malate. One possibility for the noncompetitive inhibition is combination of oxalate with the E–NAD and E–NADH complexes. A double inhibition experiment with NADH and oxalate as inhibitors at low concentrations of NAD^+ and malate indicated the ternary E–NADH–oxalate complex can form (B *25*, 227).

Another example is the elucidation of the reason for UC substrate inhibition by α-ketoglutarate in the isocitrate dehydrogenase reaction. Double inhibition by α-ketoglutarate and NADP gave a parallel pattern, ruling out formation of an EBQ complex. This likely is a rare example of combination of the α-ketoglutarate as an imine with a lysine in the central complex (JBC *249*, 2928).

Slow Binding and Tight Binding Inhibition

Until now we have assumed that equilibrium between an inhibitor and an enzyme is attained rapidly enough that it is complete by the time initial rates are measured. This is true for most reversible inhibitions, but in some cases there is a lag in attaining full inhibition and the initial rate slows down until the final steady-state rate is achieved.

Slow Binding Inhibition. Two mechanisms have been considered for slow binding inhibition. Mechanism A, equation 6-151, is one in which the conversion of E to EI is slow.

$$\begin{array}{ccccc} \mathrm{E} & \underset{k_2}{\overset{k_1\mathbf{A}}{\rightleftharpoons}} & \mathrm{EA} & \xrightarrow{k_7} & \mathrm{E} + \text{products} \\ k_4 \upharpoonleft\downharpoonright k_3\mathbf{I} & & & & \\ \mathrm{EI} & & & & \end{array} \quad (6\text{-}151)$$

The initial velocity of the reaction, if it is initiated by the addition of enzyme, is given by equation 6-152, where $K = (k_2 + k_7)/k_1$:

$$v_0 = \frac{V\mathbf{A}}{K + \mathbf{A}} \quad (6\text{-}152)$$

The steady-state velocity, after E and EI come to equilibrium, is given by equation 6-153, where $K_i = k_4/k_3$:

$$v_{ss} = \frac{V\mathbf{A}}{K\left(1 + \frac{\mathbf{I}}{K_i}\right) + \mathbf{A}} \quad (6\text{-}153)$$

The equation for the actual rate at any time is given by equation 6-154, where

$$v = v_{ss} + (v_0 - v_{ss})e^{-kt} \quad (6\text{-}154)$$

$$k = k_4 + \frac{k_3\mathbf{I}}{\left(1 + \frac{\mathbf{A}}{K}\right)} \quad (6\text{-}155)$$

Note that, for mechanism 6-151, k is a linear function of $\mathbf{I}$ and k_4 is determined from the vertical intercept of a plot of k versus $\mathbf{I}$ or from $k_4 = kv_{ss}/v_0$, while k_3 is the slope of equation 6-155 multiplied by $(1 + \mathbf{A}/K)$. The horizontal intercept of the plot is $K_i(1 + \mathbf{A}/K)$.

The time course for product formation is given by equation 6-156:

$$\frac{\mathbf{P} = v_{ss}t + (v_0 - v_{ss})(1 - e^{-kt})}{k} \tag{6-156}$$

The asymptote to the curve generated by equation 6-156 at high t values is given by equation 6-157:

$$\mathbf{P} = v_{ss}t + \frac{(v_0 - v_{ss})}{k} \tag{6-157}$$

If the reaction is started by addition of enzyme, the vertical intercept of this asymptote allows one to calculate k. If enzyme and inhibitor are preincubated and the reaction is started with substrate, the initial velocity is given by equation 6-158:

$$v_0 = \frac{\mathbf{VA}}{(K + \mathbf{A})\left(1 + \dfrac{\mathbf{I}}{K_i}\right)} \tag{6-158}$$

The rate then increases as competition with substrate decreases the amount of EI at equilibrium with free E. Equations 6-154 to 6-157 still apply, but now the asymptote intercept of equation 6-156 is negative. The same k value will be obtained as in the experiment where enzyme was added last.

The inhibition of enolase by the trianion of phosphonoacetohydroxamate shows slow binding inhibition, fitting mechanism 6-151 (B 23, 2779). Initial velocities did not vary with inhibitor, but the steady-state ones did. The value of k_3 depended on pH, increasing a factor of 10 per pH unit up to pH 9 (the pK of the hydroxamate is 10.2), with k_4 pH-independent at $10^{-2}\,s^{-1}$ at high Mg^{2+}. The calculated value of K_i for the fully ionized inhibitor in the presence of saturating Mg^{2+} was 15 pM.

A more general and more common mechanism of slow binding inhibition is mechanism 6-159:

$$\begin{array}{ccccc} E & \underset{k_2}{\overset{k_1\mathbf{A}}{\rightleftharpoons}} & EA & \xrightarrow{k_7} & E + \text{products} \\ k_3\mathbf{I} \downarrow\uparrow k_4 & & & & \\ EI & \underset{k_6}{\overset{k_5}{\rightleftharpoons}} & EI^* & & \end{array} \tag{6-159}$$

In this mechanism it is assumed that k_3 and k_4 are fast, so that the initial equilibration of E and EI is complete before initial velocities are measured. The isomerization of EI to EI* is then slow, with $k_4 \gg k_5$ (for the general case where k_4 and k_5 are similar, see BBA *1298*, 78).

In mechanism 6-159, the initial velocity is given by equation 6-160. Note that unlike in mechanism 6-151, the inhibitor does affect v_0.

$$v_0 = \frac{V\mathbf{A}}{K\left(1 + \frac{\mathbf{I}}{K_i}\right) + \mathbf{A}} \tag{6-160}$$

The steady-state velocity is given by equation 6-161, where $K^*_i = k_4k_6/[k_3(k_5 + k_6)]$:

$$v_{ss} = \frac{V\mathbf{A}}{K\left(1 + \frac{\mathbf{I}}{K_i^*}\right) + \mathbf{A}} \tag{6-161}$$

The value of k in equations 6-155 to 6-158 is now given by equation 6-162:

$$k = k_6 + \frac{k_5 \frac{\mathbf{I}}{K_i}}{1 + \frac{\mathbf{A}}{K} + \frac{\mathbf{I}}{K_i}} \tag{6-162}$$

and

$$\frac{k_5}{k_6} = \frac{K_i}{K_i^*} - 1 \tag{6-163}$$

The value of k is no longer a linear function of **I** but rather a hyperbolic function of **I**, ranging from k_6 at low **I** to $(k_5 + k_6)$ at high **I**. The hyperbolic variation of k with **I** and the initial velocities that also vary with **I** distinguish mechanism 6-159 from 6-151. In practice all slow binding cases probably are represented by mechanism 6-159, but if the initial rapidly reversible complex is too weak it will not be detected, and the data will appear to fit mechanism 6-151.

The inhibition of yeast hexokinase by LuATP fits mechanism 6-159 (B *22*, 5507). At pH 8, $K_i = 0.9\ \mu\text{M}$, $K^*_i = 0.015\ \mu\text{M}$, $k_5 = 13\ \text{min}^{-1}$, and $k_6 = 0.21\ \text{min}^{-1}$. Other lanthanide ATP complexes also gave slow binding inhibition, with k_5/k_6 ratios of 62 for LuATP, 25 for YbATP, and 11 for TmATP.

The nitro analogue of isocitrate, when in its carbanionic form (pK 9.46), gave slow binding inhibition of aconitase (B *19*, 2358). The extrapolated values for the fully ionized inhibitor were $K_i = 72$ nM and $K^*_i = 0.7$ nM. At pH 8, $k_5 = 7\ \text{min}^{-1}$ and $k_6 = 0.06\ \text{min}^{-1}$. The non-ionized form of the inhibitor also showed slow binding inhibition, with 10^3-fold higher values for K_i and K^*_i.

Nitropropionate gave very strong slow binding inhibition of isocitrate lyase (B *21*, 4420), and the carbanion form is the active inhibitor (pK 9.0). The initial K_i value was 0.25 mM, but k_5/k_6 in this case was very large and K^*_i was calculated as 1.5 nM. The value of k_5 at pH 8 was $16\ \text{min}^{-1}$, while the value of k_6 was 14,000-fold less ($1.1 \times 10^{-3}\ \text{min}^{-1}$). The half-life for recovery of activity by enzyme preincubated with excess inhibitor and then diluted into an assay mixture was 14.5 hours!

A more complete list of systems that exhibit slow binding inhibition together with a full theoretical discussion is given in AdvEnz *61*, 201.

Effect of a Reversible Inhibitor on a Slow Binding Inhibitor. An interesting effect of the presence of a non-slow binding reversible inhibitor in a solution of a slow binding inhibitor was reported in AB *161*, 438. While the half-time for the initial burst in the presence of a slow binding inhibitor decreases from $0.693/k_6$ to $0.693/(k_5 + k_6)$ as the inhibitor concentration increases from zero to infinity, the presence of a classical inhibitor in constant ratio to the slow binding one increases the limit at high inhibitor concentrations by dividing k_5 in the above expression by $(1 + xK_i/K_j)$, where x is the ratio of classical and slow binding inhibitor concentrations and K_i and K_j are their inhibition constants for the initial inhibition before the slow binding phase.

One will thus not obtain the correct parameters for the slow binding inhibitor. This will be important if one is using a racemate and only one enantiomer shows slow binding inhibition or if one is testing a mixture of antibiotics, only one of which shows slow binding inhibition. The result of the contamination is too low a calculated value of k_5; the value of k_6 will be correct.

Tight Binding Inhibition. It has been implicitly assumed until now that the K_i values for inhibitors were sufficiently higher than $\mathbf{E_t}$ that formation of an EI complex did not lower the concentration of inhibitor in solution. When this is not true, one must consider the distribution of inhibitor, as well as enzyme, when deriving the rate equation. This situation arises with very tightly bound inhibitors, such as methotrexate with dihydrofolate reductase, or when one uses a very slow alternate substrate or makes a very slow mutant enzyme. If a mutant has lost

10^3-fold in activity, one may have to use 10 μM enzyme to observe the rate, and an inhibitor with a K_i value of 1 μM will act as a tight binding one, even though with the wild-type enzyme it is not when the enzyme is used at 10^{-8} M.

For a simple model such as that shown in equation 6-164, the conservation equation for enzyme is given in equation 6-165:

$$\mathrm{EI} \underset{}{\overset{k_i}{\rightleftharpoons}} \mathrm{E} \underset{k_2}{\overset{k_1\mathbf{A}}{\rightleftharpoons}} \mathrm{EA} \xrightarrow{k_3} \mathrm{E} + \text{products} \tag{6-164}$$

$$\mathbf{E_t} = \mathbf{E} + \mathbf{EA} + \mathbf{EI} \tag{6-165}$$

and $v = k_3\mathbf{EA}$. In the steady state $\mathbf{E}$ is equal to $\mathbf{EA}(K_a/\mathbf{A})$, where K_a is $(k_2 + k_3)/k_1$. The conservation equation for I is given in equation 6-166:

$$\mathbf{I} = \mathbf{I}_t - \mathbf{EI} \tag{6-166}$$

where $\mathbf{EI} = \mathbf{E}(\mathbf{I}/K_i)$, and substituting the expression for $\mathbf{I}$ gives $\mathbf{EI}$ as $\mathbf{E}(\mathbf{I_t} - \mathbf{EI})/K_i$. Solving for $\mathbf{EI}$ and substituting $\mathbf{E}$ from above gives

$$\mathbf{EI} = \frac{(\mathbf{EA})\mathbf{I_t}}{\dfrac{K_i\mathbf{A}}{K_a} + \mathbf{EA}} \tag{6-167}$$

Substituting equation 6-167 and the expression for $\mathbf{E}$ above into equation 6-165 gives equation 6-168:

$$\mathbf{E_t} = \mathbf{EA}\left(1 + \frac{K_a}{\mathbf{A}} + \frac{\mathbf{I_t}}{\dfrac{K_i\mathbf{A}}{K_a} + \mathbf{EA}}\right) \tag{6-168}$$

Solving for $\mathbf{EA}$ and replacing it with v/k_3 gives equation 6-169:

$$v^2 + vk_3\mathbf{A}\left(\frac{K_i}{K_a} + \frac{\mathbf{I_t} - \mathbf{E_t}}{K_a + \mathbf{A}}\right) - \frac{(k_3\mathbf{A})^2 K_i \mathbf{E_t}}{K_a(K_a + \mathbf{A})} = 0 \tag{6-169}$$

A more general derivation, which is valid for more complicated mechanisms as well, gives equation 6-170:

$$v^2 + v(\text{num})\left(\frac{1}{\sum \dfrac{\Delta_i}{K_i}} + \frac{\mathbf{I_t} - \mathbf{E_t}}{(\text{denom})}\right) - \frac{(\text{num})^2\mathbf{E_t}}{(\text{denom})\sum \dfrac{\Delta_i}{K_i}} = 0 \tag{6-170}$$

where (num) is the numerator of the rate equation ($V/\mathbf{E_t}$ or k_{cat} times the substrate concentrations), (denom) is the denominator of the rate equation in the absence of inhibitors, and each Δ_i is a denominator term representing an enzyme form that the inhibitor combines with. In our example of competitive inhibition, (num) $= k_3\mathbf{A}$, (denom) $= K_a + \mathbf{A}$, and $\Delta_i = K_a$. For a noncompetitive inhibition, $\sum(\Delta_i/K_i) = K_a/K_{is} + \mathbf{A}/K_{ii}$. This formulation is readily expanded to two substrate cases.

Data obeying equation 6-170 can be plotted several ways. A plot of v versus $\mathbf{E_t}$ at a fixed level of $\mathbf{I_t}$, Figure 6-12, is concave upward with an asymptote crossing the horizontal axis where $\mathbf{E_t} = \mathbf{I_t}$. This plot works well only when the $\mathbf{E_t}$ and $\mathbf{I_t}$ levels used exceed K_i by a factor greater than 20. For ratios less than this it is too difficult to determine the asymptote.

A plot of v versus $\mathbf{I_t}$ at different $\mathbf{E_t}$ levels, Figure 6-13, also gives concave upward curves that decrease from the uninhibited velocity with differing initial slopes and

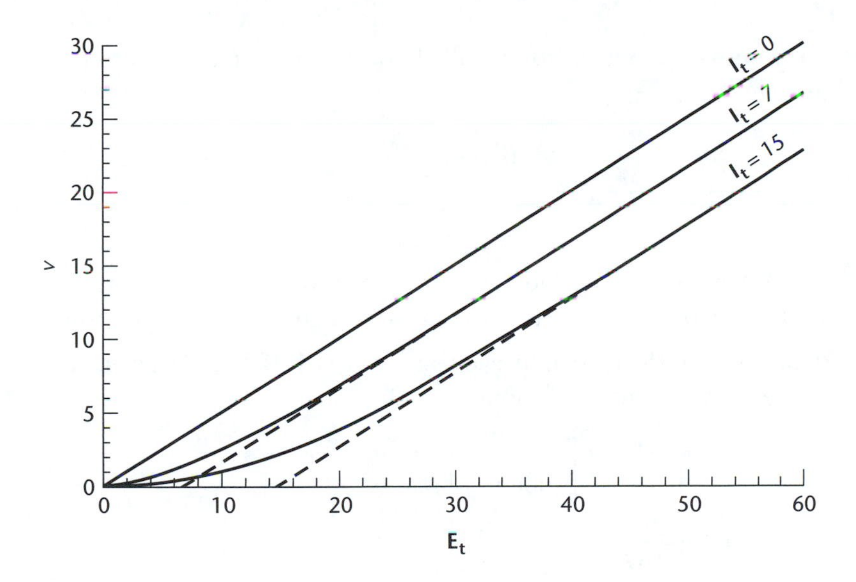

Figure 6-12. Plot of initial rate versus $\mathbf{E_t}$ in the absence ($\mathbf{I_t} = 0$) and presence of a tight binding inhibitor according to equation 6-169. All kinetic and rate constants were set equal to 1, as was the concentration of A. Concentrations of I_t are as indicated. Extrapolation of the linear portion of the curves in the presence of inhibitor to $v = 0$ gives $\mathbf{I_t} = \mathbf{E_t}$.

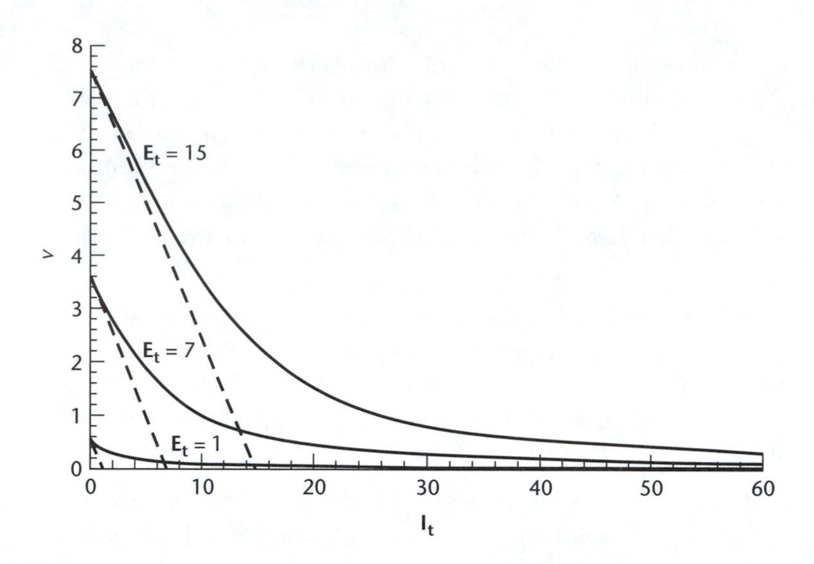

Figure 6-13. Data in Figure 6-12 plotted as initial rate versus $\mathbf{I_t}$.

approach a zero asymptote. A tangent to the initial portion of the curve intersects the horizontal axis at

$$\mathbf{I_t} = \mathbf{E_t} + \frac{(\text{denom})}{\sum \frac{\Delta_i}{K_i}} \tag{6-171}$$

A plot of $1/v$ versus $\mathbf{I_t}$ has a similar problem, and the intercept of the asymptote on the horizontal axis is given by equation 6-171 with a minus sign for the second term.

Equation 6-170 can be transformed into a linear form (BJ *127*, 321) given by equation 6-172, where $v_0 = (\text{num})\mathbf{E_t}/(\text{denom})$:

$$\frac{\mathbf{I_t}}{\left(1 - \frac{v}{v_0}\right)} = \left[\frac{(\text{denom})}{\sum \frac{\Delta_i}{K_i}}\right]\left(\frac{v_0}{v}\right) + \mathbf{E_t} \tag{6-172}$$

A plot of $\mathbf{I_t}/[1 - (v/v_0)]$ versus (v_0/v) has a vertical intercept of $\mathbf{E_t}$. If the experiment is carried out at different levels of enzyme, a series of parallel lines is obtained. The problem with this plot is that v_0 must be very accurately known, or the plot will appear curved. Also, the variables are not separated, which makes least-squares fitting problematic.

The best approach is to fit the data when $\mathbf{E_t}$ and $\mathbf{I_t}$ are varied to the solution of the quadratic equation 6-173, where appK_i is equal to (denom)/$\sum(\Delta_i/K_i)$:

$$v = \left[\frac{v_0}{2\mathbf{E_t}}\right]\left[-(\text{app}K_i + \mathbf{I_t} - \mathbf{E_t}) + [(\text{app}K_i + \mathbf{I_t} - \mathbf{E_t})^2 + 4(\text{app}K_i)\mathbf{E_t}]^{1/2}\right] \quad (6\text{-}173)$$

Note that $\mathbf{E_t}$ determined from a fit to this equation is only the active enzyme, which may be less than the actual amount of protein present if some is inactive. The value of appK_i in the mechanism represented by equation 6-164 is $K_i(1 + \mathbf{A}/K_a)$.

The pharmaceutical industry often uses I_{50} values, which are the levels of inhibitor needed to give 50% inhibition under a specified set of conditions. For tight binding inhibitors, I_{50} values vary with the enzyme level, and from equation 6-172 and the expression for appK_i in equation 6-173, the I_{50} is given by equation 6-174:

$$I_{50} = \frac{\mathbf{E_t}}{2} + \text{app}K_i \quad (6\text{-}174)$$

Thus it is necessary to specify enzyme as well as substrate levels when I_{50} values are reported for tight binding inhibitors.

If one varies the substrate concentration at different levels of a tight binding inhibitor, reciprocal plots are curved near the vertical axis, concave downward, but have linear asymptotes. The plots will intersect on the vertical axis for a competitive inhibitor, but unless very high substrate levels are used, the pattern will look noncompetitive. The rate equations can be obtained by substituting the expression for appK_i into equation 6-173 and making the appropriate substitutions for v_0, (denom), and Δ_i.

A more complete discussion of tight binding inhibition is given in ME *63*, 437.

Slow Tight Binding Inhibition. When an inhibitor shows tight binding inhibition because its K_i is nanomolar or below, it will also very likely show slow binding behavior. This is readily seen by remembering that the bimolecular rate constant for addition of an inhibitor to the enzyme is limited by diffusion to $10^9\ \text{M}^{-1}\ \text{s}^{-1}$ and is often less than this. If the inhibitor has a K_i of 10^{-11} M, the rate constant for dissociation cannot exceed $10^{-11} \times 10^9 = 0.01\ \text{s}^{-1}$ and will probably be less. If a level of inhibitor 10 times its K_i is used, the apparent first-order rate constant for formation of EI will be $0.1\ \text{s}^{-1}$, and the half-life for attaining equilibrium will be $0.69/0.1 = 6.9$ s. At a level equal to K_i, the half-life will be 69 s.

Thus an inhibitor with this low a K_i value must show slow binding behavior, in addition to being a tight binding inhibitor.

The equations for slow tight binding inhibitors are complex, and the reader should consult ME 63, 437.

Some Practical Considerations

How does one choose the concentration of inhibitor? For product inhibition, the K_m values for the products can be used as an initial estimate of K_i. However, generally the concentration of the varied substrate is fixed at its K_m and the concentration of inhibitor is increased. Data are then plotted as $1/v$ versus **I**. The intersection on the abscissa of this plot is the apparent K_i. (If the inhibitor is competitive versus the substrate fixed at K_m, the true K_i is half the estimated apparent K_i value.) A full inhibition pattern is then obtained by varying the concentration of substrate at four different concentrations of I (0, K_i, $2K_i$, and $4K_i$, where K_i is the value estimated as above). The levels of **I** are chosen to give a 2-fold change in the slope each time **I** is increased.

Curvature may sometimes be observed in the time courses used to estimate initial rates in the presence of the product inhibitor. Possible reasons for the curvature include the reaction approaching equilibrium in the presence of the product or very tight binding by the product inhibitor. The former can be overcome by changing to a more favorable pH (if the equilibrium is pH-dependent, as for example the alcohol dehydrogenase reaction), or by removing the other product as it is formed, by use of another enzymic (or nonenzymic) coupled reaction (for example, the phosphofructokinase reaction can be studied in the direction of F6P formation by removing ATP by use of hexokinase).

When a coupled enzyme assay is utilized to monitor a reaction, it is always important to rule out inhibition of the coupling enzymes by the added inhibitor. This can be carried out very easily by ensuring that there is a 2-fold increase in the reaction rate as the concentration of the enzyme being studied is doubled. The test should be carried out under conditions where the maximum amount of inhibition is observed. For example, in the case of a competitive inhibitor, the test should be carried out at the highest concentration of inhibitor and lowest concentration of the varied substrate.

7

PRE-STEADY-STATE AND RELAXATION KINETICS

The majority of this book has dealt with situations where the enzyme concentration is less than that of the substrate, so that a steady state exists during the reaction. Steady-state kinetic studies can be used to determine the order of addition and release of reactants and products and where the major rate-limiting steps are in a mechanism. However, one gets only limited information on isomerization of transitory intermediates after the substrates add. Isotope effects permit the determination of partition ratios for intermediates, and they and pH profiles give evidence of the chemistry that is taking place. To determine the intrinsic rate constants and to characterize intermediates usually requires pre-steady-state experiments.

We will first consider the theory of pre-steady-state treatment, specifically the most common cases encountered experimentally. For a more thorough treatment of reversible and irreversible first order processes, of more complex systems, and of practical uses of such methods, see Kinetics and Mechanism by J.W. Moore and R.G. Pearson (3rd edition, John Wiley & Sons, New York, 1981) and Kinetic Analysis of Macromolecules edited by K. Johnson (Oxford University Press, New York, 2003) and the review article by Fisher (ACR *38*, 157).

When carrying out pre-steady-state kinetic experiments, one normally arranges, when possible, to work under pseudo-first-order conditions. In this case, time courses can be treated as simple first-order processes, the sum of first-order processes, or consecutive first-order processes. Theory will first be presented for cases where reactant concentration is much greater than that of the enzyme, followed by a consideration of single-turnover experiments, and equilibrium perturbation and relaxation.

Reactant Concentration in Excess of Enzyme Concentration

Time courses can be treated as pseudo-first-order if $\mathbf{A} \gg \mathbf{E_t}$, and we will consider several common cases. The steady-state approximation no longer applies because the concentrations of enzyme species are not negligibly small. In some cases, the time courses reflect the approach to steady state.

Irreversible First-order Reactions

The treatment of irreversible first-order processes has been presented in Chapter 2. In brief, the first-order conversion of A to B, where $\mathbf{A} \gg \mathbf{E_t}$ is given in equation 7-1:

$$\mathrm{A} \xrightarrow{k_1} \mathrm{B} \tag{7-1}$$

The rate equation, $-\mathrm{d}\mathbf{A}/\mathrm{d}t = k_1\mathbf{A}$, is shown in integrated form in equation 7-2 when the initial concentration of A is $\mathbf{A_0}$:

$$\mathbf{A_t} = \mathbf{A_0}\mathrm{e}^{-k_1 t} \tag{7-2}$$

This simple case will be observed when dealing with an equilibrium that is far toward B. This could be a chemical step or some other step along the reaction pathway.

Reversible First-order Reactions

Allowing for reversibility in the first-order process increases the complexity of the overall process. Consider the following approach to equilibrium starting with A.

$$\mathrm{A} \underset{k_2}{\overset{k_1}{\rightleftharpoons}} \mathrm{B} \tag{7-3}$$

In this case the differential equation in terms of $\mathbf{A}$ is given by equation 7-4:

$$-\frac{\mathrm{d}\mathbf{A}}{\mathrm{d}t} = k_1\mathbf{A} - k_2\mathbf{B} \tag{7-4}$$

At any time

$$\mathbf{A_0} = \mathbf{A} + \mathbf{B} \tag{7-5}$$

where $\mathbf{A_0}$ is the initial concentration of A. Substitution of equation 7-5 into equation 7-4 gives

$$-\frac{d\mathbf{A}}{dt} = k_1\mathbf{A} - k_2[\mathbf{A_0} - \mathbf{A}] \tag{7-6}$$

Integration, with $\mathbf{B} = 0$ and $\mathbf{A} = \mathbf{A_0}$ when $t = 0$, then gives

$$\mathbf{A} = \left[\frac{\mathbf{A_0}}{k_1 + k_2}\right]\left[k_1 e^{-(k_1+k_2)t} + k_2\right] \tag{7-7}$$

where $\mathbf{A}$ is the concentration of A at any time t. In equation 7-7 the exponential term is

$$k_{obs} = k_1 + k_2 \tag{7-8}$$

The limit of equation 7-7 at $t = 0$ is $\mathbf{A_0}$, while at $t = \infty$ it is

$$\mathbf{A_{eq}} = \frac{\mathbf{A_0}k_2}{k_1 + k_2} \tag{7-9}$$

In order to obtain estimates of the microscopic rate constants, the concentrations of A and B at equilibrium must be known.

An example of the above would be the reversible association of enzyme and substrate.

$$\mathrm{E} \underset{k_2}{\overset{k_1\mathbf{A}}{\rightleftharpoons}} \mathrm{EA} \tag{7-10}$$

At a fixed concentration of substrate A, the process would be pseudo-first-order in $\mathbf{E}$, and under conditions where $\mathbf{E} = \mathbf{E_0}$ and $\mathbf{EA} = 0$, equation 7-7 applies, substituting $\mathbf{E}$ for $\mathbf{A}$ and $\mathbf{E_0}$ for $\mathbf{A_0}$, $k_{obs} = k_2 + k_1\mathbf{A}$. Measurement of k_{obs} as a function of the concentration of A will allow an estimate of k_1 and k_2, as shown in Figure 7-1.

Consecutive First-order Reactions

In the case of consecutive rate processes it is important to measure the time dependence of the concentrations of all components. We will consider the simple irreversible case first, followed by the reversible case.

$$\mathrm{A} \xrightarrow{k_1} \mathrm{B} \xrightarrow{k_2} \mathrm{C} \tag{7-11}$$

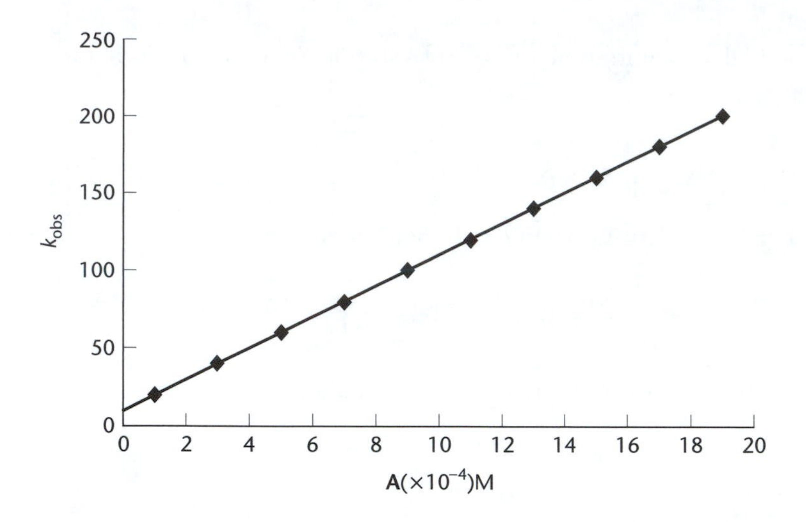

Figure 7-1. Plot of k_{obs} versus **A** for a reversible process such as that shown in equation 7-10. Values of 10^5 M^{-1} s^{-1} and 10 s^{-1} were used for k_1 and k_2, respectively.

In this case B is a transient intermediate. One first writes the differential equations as above and then integrates each to obtain expressions for **A**, **B**, and **C** as shown below. This assumes that, at $t = 0$, $\mathbf{A} = \mathbf{A_0}$ and $\mathbf{B} = \mathbf{C} = 0$.

$$\mathbf{A} = \mathbf{A_0} e^{-k_1 t} \tag{7-12}$$

$$\mathbf{B} = \frac{\mathbf{A_0} k_1}{k_2 - k_1}\left[e^{-k_1 t} - e^{-k_2 t}\right] \tag{7-13}$$

$$\mathbf{C} = \mathbf{A_0}\left[1 + \frac{1}{k_1 - k_2}\left(k_2 e^{-k_1 t} - k_1 e^{-k_2 t}\right)\right] \tag{7-14}$$

If $k_1 > k_2$, B is formed with rate constant k_1 and decomposes with rate constant k_2, while if $k_2 > k_1$, B is formed with rate constant k_2 and decomposes with rate constant k_1. A plot of concentration versus time for each of these cases is shown in Figure 7-2. Note that very little B accumulates under the condition where $k_2 > k_1$. When the reaction is A + B → C → D, with **B** in excess to maintain pseudo-first-order conditions, increasing **B** will also increase the buildup of C. At high enough concentrations of B, app$k_1 > k_2$, that is, $k_1\mathbf{B} > k_2$.

In the case of reversible consecutive first-order reactions, the treatment is complex but can be solved in a manner similar to that discussed above. In the case

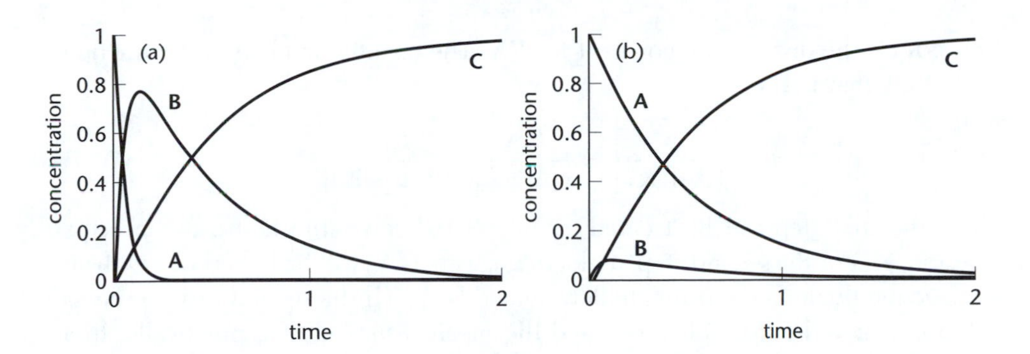

Figure 7-2. Dependence of **A**, **B**, and **C** on time according to mechanism 7-11 and equations 7-12 to 7-14. The plot on the left (panel a) is for $k_1 > k_2$, with $k_1 = 20\ s^{-1}$ and $k_2 = 2\ s^{-1}$. The plot on the right (panel b) is for $k_1 < k_2$, with $k_1 = 2\ s^{-1}$ and $k_2 = 20\ s^{-1}$. The rate of decrease of A is much more rapid and the maximum amount of B produced is much greater in the left plot, but the curves for C are identical in both cases.

of a reversible second-order process followed by a reversible first-order process, an absolute solution is not possible. There is, however, the possibility of treating the overall process as rapid pre-equilibrium or stationary model, and this has been treated in detail (JBC *250*, 4048). If the second-order and first-order processes are on different time scales, one can treat them as independent. For example, consider the following mechanism, which depicts association of enzyme and substrate followed by a conformational change.

$$\mathrm{E} \underset{k_2}{\overset{k_1\mathbf{A}}{\rightleftharpoons}} \mathrm{EA} \underset{k_4}{\overset{k_3}{\rightleftharpoons}} \mathrm{E'A} \tag{7-15}$$

If the second step in mechanism 7-15 is slow, which is a relatively common occurrence, then both $k_{obs\,1}$ and $k_{obs\,2}$, the first and second observed rate constants, will increase with increasing **A**, as a result of a build-up in the concentration of EA. In this case, if the starting material is E, then the treatment is the same as that discussed for equation 7-10 above for $k_{obs\,1}$. The second observed rate is

$$k_{obs\,2} = k_4 + k_3\left[\frac{\mathbf{A}}{\mathbf{A} + K_d}\right] \tag{7-16}$$

where K_d is the dissociation constant for EA. The term in brackets is the fraction of **E** + **EA** that is **EA**:

$$\left[\frac{\mathbf{A}}{\mathbf{A}+K_d}\right]=\frac{k_1\mathbf{A}}{k_1\mathbf{A}+k_2}=\frac{\mathbf{EA}}{\mathbf{EA}+\mathbf{E}} \tag{7-17}$$

Thus, the first step can be treated as an isolated reversible first-order process (see above). For the second step, a plot of k_{obs} versus $\mathbf{A_0}$ will be hyperbolic with an intersection point on the ordinate determined by k_4. (If the second-order process is too fast, it will not be observed and the mechanism reduces, practically, to a reversible first-order process as shown in mechanism 7-3 above.) If $k_4 \ll k_3$, the hyperbolic function will pass through zero. There are numerous examples of both cases in the literature, and two will be used to illustrate. A recently published example of one where k_4 is of the same order of magnitude as k_3 involves the transfer of phosphate from the response regulatory protein SLN1 to the phosphotransfer protein YPD1 in the yeast osmoregulatory system (Figure 7-3; B *44*, 377).

In this case k_3 (phosphoryl transfer to YPD1), k_4 (reverse phosphoryl transfer), and K_d (the dissociation constant for SLNI~P:YPD1 were about 30 s^{-1}, 10 s^{-1}, and 2 μM, respectively.

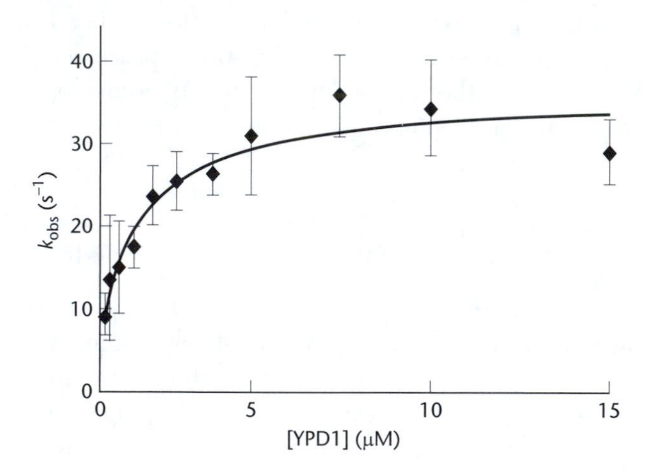

Figure 7-3. Dependence of the rate of phosphoryl transfer from phosphorylated form of the response regulatory protein SLN-1 to the phosphotransfer protein YPD1, components of a signal transduction pathway in yeast. The fitted parameters are $K_d = 1.4$ μM (the dissociation constant for the SLN-1~P–YPD1 complex), $k_3 = 29$ s^{-1}, and $k_4 = 7.5$ s^{-1} on the basis of a fit to equation 7-16.

The first half of the reaction catalyzed by the PLP-dependent *O*-acetylserine sulfhydrylase involves rapid formation of a Schiff base with the substrate, *O*-acetyl-L-serine (OAS), followed by a slow, practically irreversible elimination of acetate to give the Schiff base with α-aminoacrylate. This is an example of a case where $k_4 \ll k_3$. In this case, the hyperbola passes through zero, while the rate constant for the elimination and dissociation constant for the OAS Schiff base were 300 s^{-1} and 5 mM, respectively (Figure 7-4; B 35, 4776).

If the second step of a reversible consecutive reaction is faster than the first, for example, when a slow isomerization of E is required before A can bind:

$$\mathrm{E} \underset{k_2}{\overset{k_1}{\rightleftharpoons}} \mathrm{E}' \underset{k_4}{\overset{k_3\mathbf{A}}{\rightleftharpoons}} \mathrm{E}'\mathrm{A} \tag{7-18}$$

then the rate equation is

$$-\frac{\mathrm{d}\mathbf{E}}{\mathrm{d}t} = k_1\mathbf{E} - k_2\mathbf{E}' \tag{7-19}$$

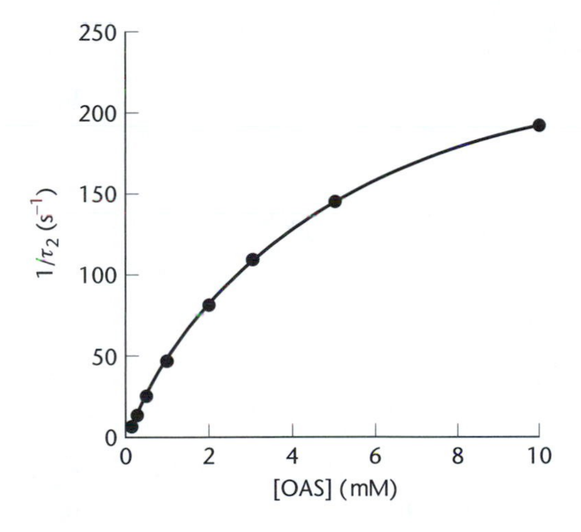

Figure 7-4. Dependence of rate on the concentration of *O*-acetyl-L-serine for the *O*-acetylserine sulfhydrylase reaction. In the first half of the sulfhydrylase-catalyzed Ping Pong reaction, OAS is converted to an α-aminoacrylate Schiff base with pyridoxal 5′-phosphate upon the elimination of acetate. Values of 4.5 mM and 300 s^{-1} are estimated for K_{ESB} (the dissociation constant for the initial Schiff base between PLP and OAS) and k_3 according to equation 7-16 when $k_4 \ll k_3$.

In this case:

$$\begin{aligned}\mathbf{E} = \mathbf{E_0} - (\mathbf{E}' + \mathbf{E}'\mathbf{A}) &= \mathbf{E_0} - \mathbf{E}'\left(1 + \frac{\mathbf{E}'\mathbf{A}}{\mathbf{E}'}\right) \\ = \mathbf{E_0} - \mathbf{E}'\left(1 + \frac{\mathbf{A}}{K_\mathrm{d}}\right) &= \mathbf{E_0} - \mathbf{E}'\left(\frac{K_\mathrm{d} + \mathbf{A}}{K_\mathrm{d}}\right)\end{aligned} \tag{7-20}$$

Substitution for E′ in equation 7-19, simplification, and integration of the final rate equation gives

$$k_{\mathrm{obs\,1}} = k_1\mathbf{A} + k_2\left[\frac{K_\mathrm{d}}{\mathbf{A} + K_\mathrm{d}}\right] \tag{7-21}$$

Thus, the observed rate constant decreases with increasing concentrations of A as a result of a decrease in the concentration of E′. The two cases, first step or second step fast in a two-step process, can thus be differentiated from one another.

Parallel First-order Reactions

Consider the following mechanism:

$$\begin{array}{ccc} \mathrm{A} & \xrightarrow{k_1} & \mathrm{B} \\ {\scriptstyle k_2}\downarrow & & \\ \mathrm{C} & & \end{array} \tag{7-22}$$

Integrated rate expressions for **A**, **B**, and **C** are given in equations 7-23 to 7-25 with the assumption that $\mathbf{A} = \mathbf{A_0}$ at $t = 0$, and $\mathbf{B} = \mathbf{C} = 0$.

$$\mathbf{A} = \mathbf{A_0}\mathrm{e}^{-(k_1 + k_2)t} \tag{7-23}$$

$$\mathbf{B} = \frac{\mathbf{A_0}k_1}{k_2 + k_1}\left(1 - \mathrm{e}^{-(k_1 + k_2)t}\right) \tag{7-24}$$

$$\mathbf{C} = \frac{\mathbf{A_0}k_2}{k_2 + k_1}\left(1 - \mathrm{e}^{-(k_1 + k_2)t}\right) \tag{7-25}$$

In this case, **A** decreases with a rate constant that is equal to the sum of the microscopic rate constants k_1 and k_2, while **B** and **C** must be formed with the same rate constant as for the disappearance of A. At $t = \infty$, **B** is equal to $\mathbf{A_0}k_1/(k_1 + k_2)$, while **C** is equal to $\mathbf{A_0}k_2/(k_1 + k_2)$. The ratio of the two is equal to k_1/k_2, and thus one obtains an estimate of the kinetic partitioning between the two pathways.

Burst in a Time Course

Observation of a burst in a time course is diagnostic of the significant build-up of an intermediate along the reaction pathway. It is the rate of establishment of the steady state. In the pre-steady-state, one often observes a burst of product formation. This happens when product P is liberated in the first step, but the second step is slower than the first. This can be illustrated with mechanism 7-26. [Note that mechanism 7-26 can also be applied to serine proteases where EQ is acylenzyme (JCEd *44*, 84).]

$$\mathrm{E} \underset{k_2}{\overset{k_1\mathbf{A}}{\rightleftharpoons}} (\mathrm{EA}) \xrightarrow[\downarrow \mathrm{P}]{k_3} \mathrm{EQ} \xrightarrow{k_5} \mathrm{E} + \mathrm{Q} \tag{7-26}$$

If we assume equilibrium binding, then K_a is $K_{ia}/(1+k_3/k_5)$. The steady-state k_{cat} is not k_3 but rather $k_3k_5/(k_3+k_5)$ and will be closer to k_5 than to k_3 if $k_3 > k_5$.

If $\mathbf{A} \gg \mathbf{E_t}$, the only conservation equation is

$$\mathbf{E_t} = \mathbf{E} + \mathbf{EA} + \mathbf{EQ} \tag{7-27}$$

Since

$$\mathbf{EA} = \mathbf{E}\left(\frac{\mathbf{A}}{K_{ia}}\right), \ \mathbf{EA} = \left(\frac{\mathbf{A}}{K_{ia}}\right)(\mathbf{E_t} - \mathbf{EA} - \mathbf{EQ}) \tag{7-28}$$

Solving this equation for **EA** gives

$$\mathbf{EA} = \left(\frac{\mathbf{A}}{K_{ia}}\right)\left(\frac{\mathbf{E_t} - \mathbf{EQ}}{1 + \left(\frac{\mathbf{A}}{K_{ia}}\right)}\right) \tag{7-29}$$

Then

$$\frac{\mathrm{d}(\mathbf{EQ})}{\mathrm{d}t} = k_3\mathbf{EA} - k_5\mathbf{EQ} \tag{7-30}$$

and substituting the value of **EA**, one gets

$$\frac{\mathrm{d}(\mathbf{EQ})}{\mathrm{d}t} = \frac{k_3\mathbf{E_t}}{1 + \left(\frac{K_{ia}}{\mathbf{A}}\right)} - \mathbf{EQ}\left[k_5 + \frac{k_3}{1 + \left(\frac{K_{ia}}{\mathbf{A}}\right)}\right] \tag{7-31}$$

Integration gives

$$\mathbf{EQ} = \frac{k_3\mathbf{E_t}(1 - e^{-\text{app}kt})}{\left[k_5\left(1 + \frac{K_{ia}}{\mathbf{A}}\right) + k_3\right]} \tag{7-32}$$

where

$$\text{app}k = \frac{(k_3 + k_5)\mathbf{A} + k_5K_{ia}}{K_{ia} + \mathbf{A}} \tag{7-33}$$

Then, $d\mathbf{P}/dt = k_3\mathbf{EA}$, and substituting from equations 7-29 and 7-32, one obtains

$$\frac{d\mathbf{P}}{dt} = \left[\frac{k_3\mathbf{E}_t}{1 + \left(\frac{K_{ia}}{\mathbf{A}}\right)}\right]\left\{1 - \frac{k_3(1 - e^{-\text{app}kt})}{k_5\left(1 + \frac{K_{ia}}{\mathbf{A}}\right) + k_3}\right\} \tag{7-34}$$

which integrates, with appropriate substitutions, to equation 7-35:

$$\mathbf{P} = k_{cat}\mathbf{E_t}\left[\frac{\mathbf{A}}{K_a + \mathbf{A}}\right]t + \left(\frac{k_{cat}\mathbf{A}}{k_5(K_a + \mathbf{A})}\right)^2 \mathbf{E_t}\left[1 - e^{-\text{app}kt}\right] \tag{7-35}$$

The steady-state rate is given by the first term, while the amplitude of the burst is

$$\left(\frac{k_{cat}\mathbf{A}}{k_5(K_a + \mathbf{A})}\right)^2 \mathbf{E_t} \tag{7-36}$$

It takes saturation with A and $k_3 \gg k_5$ to see a burst equal to $\mathbf{E_t}$ (Figure 7-5). The rate constant of the burst is given by equation 7-33 and varies from k_5 at low $\mathbf{A}$ to $(k_3 + k_5)$ at saturating $\mathbf{A}$, with K_{ia} being the level of A that gives the average of the two limiting values. Analysis of the burst as a function of $\mathbf{A}$ thus gives estimates of k_3 as well as K_{ia}. If k_3 is too high, however, one may not be able to determine it, since the limit of rates that can be accurately determined is 200 s^{-1} (up to 500 s^{-1} if the signal is strong) given a dead time of 3–4 ms. In this case, one can only determine a minimum value of k_3 and of K_{ia}. If, in mechanism 7-26, $k_3 < k_5$, a burst will not be observed (Figure 7-5).

The assumption has been made that, in mechanism 7-26, EA and E are in equilibrium; that is, the rate of equilibration is much faster than the rest of the

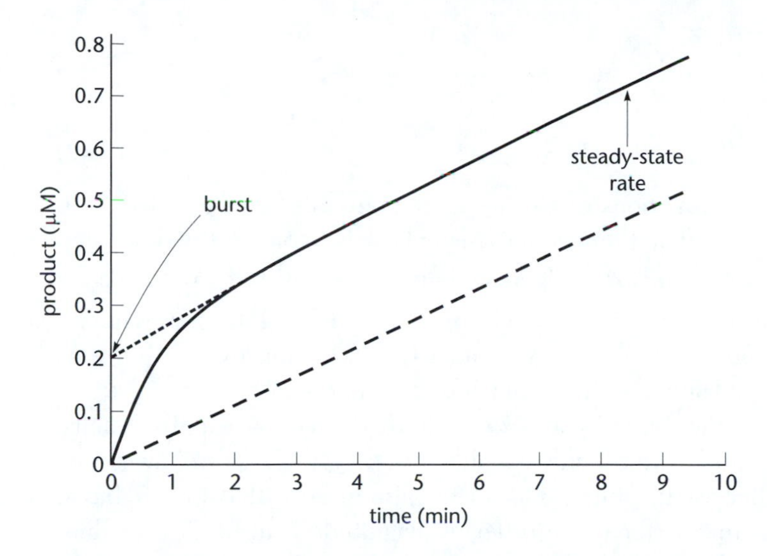

Figure 7-5. A burst in the pre-steady-state time course for the production of P, followed by a steady-state rate of production of P. Note that the amplitude of the burst is obtained by extrapolation of the steady-state rate to zero time. If $k_3 < k_5$, a burst will not be observed, as shown in the dashed line in the figure.

reaction. If equilibrium between E and EA is not established, an additional transient term appears in the rate equation as a second exponential. For most enzymes this transient cannot be observed. If equilibrium is not established, an accurate analysis of the burst phase, ignoring the later release of Q from EQ (or subtracting the steady-state rate), involves the following equations:

$$\mathbf{EQ} = \mathbf{E_t}\left(1 + C_\alpha e^{-k_\alpha t} - C_\beta e^{-k_\beta t}\right) \tag{7-37}$$

where

$$k_\alpha = \frac{p+q}{2} \qquad k_\beta = \frac{p-q}{2} \tag{7-38}$$

$$p = k_1\mathbf{A} + k_2 + k_3 \tag{7-39}$$

$$q = \left(p^2 - 4k_1k_3\mathbf{A}\right)^{1/2} \tag{7-40}$$

$$C_\alpha = \frac{k_\beta}{k_\alpha - k_\beta} \tag{7-41}$$

$$C_\beta = C_\alpha + 1 \tag{7-42}$$

Note that k_α and k_β are functions of all three rate constants in mechanism 7-26 (that is, $k_1\mathbf{A}$, k_2, and k_3) and neither can be assigned solely to one or the other step in the mechanism. Only if k_2 is zero do k_α and k_β correspond to $k_1\mathbf{A}$, and k_3.

The above equations apply when EQ itself is colored (E-NADH, for example) or to the initial portion of a burst when **P** and **EQ** will be approximately equal. The term in k_α causes a lag in the burst, which is often not seen because $C_\alpha \ll C_\beta$. The initial portion of the burst is also difficult to determine experimentally. However, when $k_1\mathbf{A}$ and k_3 are roughly equal and $k_3 \approx k_2$, the lag can have a rate constant ~6 times that of the burst and an amplitude 20% that of the burst. Values of $k_1\mathbf{A}$ and k_3 that differ by an order of magnitude lead to C_α/C_β values of 0.1 or lower, as do values of k_2/k_3 that exceed 2. As k_2/k_3 decreases below unity, the amplitude of the lag increases, but the two rate constants approach each other and the apparent rate constant for the burst (for a graph of these relationships, see ZN *44a*, 445) from a single exponential fit can be low by as much as 8%.

For more complex mechanisms than mechanism 7-26, one may see in a pre-steady-state study more than the two exponentials predicted by equation 7-37, as one intermediate changes into another. By recording at more than one wavelength, one can often tell what is happening. Fisher has pioneered this approach with glutamate dehydrogenase (B *41*, 11284), and by combining absorbance and fluorescence changes with measurement of proton release he has been able to identify iminoglutarate, carbinolamine, and α-ketoglutarate intermediates in the glutamate dehydrogenase reaction and tell when the active site is open and closed. He also uses isotope effects on the pre-steady-state time courses to help sort out what is happening. This will be discussed in Chapter 9.

Reactant Concentration Comparable to Enzyme Concentration

When enzyme and substrate concentrations are comparable, which may be necessary with very slow mutant enzymes requiring very high enzyme concentration to observe a rate, the steady-state assumption no longer holds. For a simple mechanism such as that shown in equation 7-43:

$$\mathrm{E} \underset{k_2}{\overset{k_1\mathbf{A}}{\rightleftharpoons}} (\mathrm{EA}) \xrightarrow{k_3} \mathrm{E} + \text{products} \tag{7-43}$$

the differential equation for EA is

$$\frac{d(\mathbf{EA})}{dt} = k_1(\mathbf{E})(\mathbf{A}) - (k_2 + k_3)(\mathbf{EA}) \tag{7-44}$$

and

$$\mathbf{E_t} = \mathbf{E} + \mathbf{EA} \qquad \mathbf{A_t} = \mathbf{A} + \mathbf{EA} \tag{7-45}$$

Substituting from equations 7-45 into equation 7-44:

$$\frac{d(\mathbf{EA})}{dt} = k_1(\mathbf{E_t} - \mathbf{EA})(\mathbf{A_t} - \mathbf{EA}) - (k_2 + k_3)(\mathbf{EA}) \tag{7-46}$$

This equation can be integrated, but an analytical solution for ($\mathbf{EA}$) is not possible, so one cannot substitute this value into $v = k_3(\mathbf{EA})$ to get a usable rate equation. The result is that solutions of equation 7-46 can only be obtained by iterative numerical integration methods using a computer. With more complicated mechanisms, the situation is even more complex. Thus in practice one has to simplify the situation by making assumptions or changing the conditions of the experiment.

If we assume that $\mathbf{A} \gg \mathbf{E_t}$, then equation 7-46 is simplified to

$$\frac{d(\mathbf{EA})}{dt} = k_1\mathbf{A}(\mathbf{E_t} - \mathbf{EA}) - (k_2 + k_3)(\mathbf{EA}) \tag{7-47}$$

and this equation can be integrated to

$$\mathbf{EA} = \left[\frac{\mathbf{AE_t}}{K_a + \mathbf{A}}\right](1 - e^{-kt}) \tag{7-48}$$

where K_a is the Michaelis constant for A, and

$$k = k_1\mathbf{A} + k_2 + k_3 \tag{7-49}$$

Since $d\mathbf{P}/dt$ is equal to $k_3(\mathbf{EA})$, if one substitutes for $\mathbf{EA}$ and integrates, equation 7-50 is obtained:

$$\mathbf{P} = \left[\frac{V\mathbf{A}}{K_a + \mathbf{A}}\right]t + \left[\frac{V\mathbf{A}}{k(K_a + \mathbf{A})}\right](e^{-kt} - 1) \tag{7-50}$$

The first term is the steady-state rate after the lag, while the second term describes the lag prior to establishment of the steady state, with the amplitude of the lag being $V\mathbf{A}/k(K_a + \mathbf{A})$. In most initial velocity experiments, the lag takes place in the mixing time and can be ignored. In pre-steady-state experiments, however, this lag in product formation can readily be measured if the rate constant $k \leq 100\ s^{-1}$ but not if $k > 200\ s^{-1}$.

Temperature Jump

This is a useful method for studying the rate constants of binding and dissociation when the equilibrium is temperature-dependent. However, it is certainly not as useful as stopped-flow, where one can make the reaction pseudo-first-order for both forward and reverse reactions. Typically, temperature changes of 5–10 °C can be produced in less than 5 μs by capacitative discharge or by an infrared laser pulse (Z. Electrochem. 63, 652). For example, if A is the first substrate to add to an enzyme and a pulse of heat is applied to a reaction mixture at equilibrium:

$$\mathrm{E} \underset{k_2}{\overset{k_1\mathbf{A}}{\rightleftharpoons}} \mathrm{EA} \tag{7-51}$$

the system relaxes to the equilibrium for the new temperature. The rate constant can be calculated as follows:

$$\mathbf{A} = \mathbf{A_{eq}} + \Delta\mathbf{A} \tag{7-52}$$

$$\mathbf{E} = \mathbf{E_{eq}} + \Delta\mathbf{E} \tag{7-53}$$

$$\mathbf{EA} = \mathbf{EA_{eq}} + \Delta\mathbf{C} \text{ where } \Delta\mathbf{C} = -\Delta\mathbf{A} = -\Delta\mathbf{E} \tag{7-54}$$

and

$$\frac{d(\mathbf{EA})}{dt} = \frac{d(\Delta\mathbf{C})}{dt} = k_1(\mathbf{A})(\mathbf{E}) - k_2\mathbf{EA} \tag{7-55}$$

Substituting for **A, E**, and **EA** from equations 7-52 to 7-54:

$$\begin{aligned}\frac{d(\Delta\mathbf{C})}{dt} &= k_1[\mathbf{A_{eq}} + \Delta\mathbf{A}][\mathbf{E_{eq}} + \Delta\mathbf{E}] - k_2[\mathbf{EA_{eq}} + \Delta\mathbf{C}] \\ &= k_1[\mathbf{A_{eq}} - \Delta\mathbf{C}][\mathbf{E_{eq}} - \Delta\mathbf{C}] - k_2[\mathbf{EA_{eq}} + \Delta\mathbf{C}] \\ &= k_1\mathbf{A_{eq}}\mathbf{E_{eq}} - k_2\mathbf{EA_{eq}} - \Delta\mathbf{C}[k_1(\mathbf{A_{eq}} + \mathbf{E_{eq}}) + k_2] + k_1\Delta\mathbf{C}^2\end{aligned} \tag{7-56}$$

The first two terms cancel each other and the $\Delta\mathbf{C}^2$ term is small enough to ignore. Thus:

$$\frac{d(\Delta\mathbf{C})}{dt} = \Delta\mathbf{C}\left[k_1(\mathbf{A}_{\mathbf{eq}} + \mathbf{E}_{\mathbf{eq}}) + k_2\right] = \frac{\Delta\mathbf{C}}{\tau} \tag{7-57}$$

This equation integrates to

$$\Delta\mathbf{C} = \Delta\mathbf{C}_0 e^{-t/\tau} \tag{7-58}$$

where $\Delta\mathbf{C}_0$ is the initial perturbation from the final equilibrium at the new temperature, and thus the rate constant for the relaxation to the new equilibrium position is

$$1/\tau = k_1(\mathbf{E} + \mathbf{A}) + k_2 \tag{7-59}$$

where **E**, **A**, and **EA** are equilibrium concentrations. By determining this rate constant as a function of **A**, one can determine both k_1 and k_2, as long as **E** is known, by plotting $1/\tau$ versus **A**. The slope is k_1 and the intercept is $k_1\mathbf{E} + k_2$.

If the system under study involves a unimolecular reaction:

$$\mathrm{E} \underset{k_2}{\overset{k_1}{\rightleftharpoons}} \mathrm{E}' \tag{7-60}$$

the relaxation has a rate constant:

$$1/\tau = k_1 + k_2 \tag{7-61}$$

where $K_{eq} = k_1/k_2$. If the equilibrium constant is known, one can solve for the rate constants:

$$k_1 = \frac{k_{obs}}{1 + \left(\frac{1}{K_{eq}}\right)} \qquad k_2 = \frac{k_{obs}}{1 + K_{eq}} \tag{7-62}$$

The greatest limitation of the temperature jump method is that an equilibrium system is needed with significant concentrations of all species of reactant. In addition, there are problems with uneven heating and the resulting schlieren effects, which makes optical measurements at long times difficult. Therefore, longer time regimes are usually reserved for experiments involving flow methods as described above. Some of the first examples of the use of temperature jump were to study the binding of ligands to metmyoglobin (JBC *240*, 4312; JBC *241*, 2653) and the kinetics of the half-reactions of aspartate aminotransferase (B *6*, 1798).

Methods of Pre-Steady-State Analysis

We will first consider the techniques that allow spectra of chromophoric reaction components to be collected in the millisecond time regime, and this will be followed by a consideration of stopped-flow and rapid quench techniques.

Rapid Spectral Acquisition

There are two basic types of rapid-scanning instruments capable of collecting spectra rapidly. One type uses diode-array technology, and another makes uses of a rapidly spinning (1000 Hz) disc with slits at defined distances around its circumference that rapidly scan the wavelength range of interest. The diode-array spectrometer uses a high-intensity xenon lamp and integrates all of the light for the exposure time (usually 1–10 ms) to record a snapshot spectrum of the reacting solution. Therefore, it is not a practical approach for photosensitive systems. The diode-array instrument uses a high-intensity xenon lamp and works on the same principle of a conventional diode-array spectrophotometer but with a more rapid collection system. The instrument that makes use of the rapidly spinning disc is capable of rapidly scanning wavelengths and allows a spectrum to be collected every 16 μs (after the instrument dead time). The signal-to-noise is low at this rate, and first-level signal averaging is often used for every turn of the wheel to generate a spectrum every millisecond. Both types of instrument provide valuable information in that they allow one to survey the landscape for a given reaction, obtain an estimate of the spectrum for an intermediate (especially useful when multiple chromophoric species are present, as in the case of PLP-, flavin-, and heme-dependent enzymes), provide estimates of rate constants, and pinpoint wavelengths where more accurate single wavelength stopped-flow experiments can be carried out. This method of rapid spectral acquisition can be combined with singular value decomposition (SVD) matrix methods and global analysis approaches to derive spectra of intermediates in reactions. Such data give direct information about the chemistry of the reaction.

Single Wavelength Methods

These involve either continuous flow or stopped flow. In continuous flow, two solutions are mixed to start the reaction, and the absorbance or other property is measured at a fixed distance from the mixing point corresponding to a time that depends on volume and flow rate, which must be fast enough to avoid laminar flow (which results in an inhomogeneous solution). Modern instruments utilize very small volumes and permit short times to be measured. Different time points are measured by changing the flow rate or by moving the detector to a different distance from the mixing point. This method requires large amounts of enzyme and substrate and thus is only used when these are cheap and readily available.

Stopped Flow

A more common technique is called stopped flow (see Figure 7-6). In this case the mixed solution passes the detection point and enters a syringe, which hits a stopping switch to stop the drive mechanism. The solution in the cell when stop occurs has usually aged 1–3 ms after mixing. This is called the dead time of the instrument. The time course of absorbance or fluorescence change is then measured as the solution continues to age. With a fast driving mechanism the dead time of the apparatus can be several milliseconds, but this does not mean that mixing is complete. Even the best crossed grid mixers lead only to homogenization of the two solutions being mixed, and diffusion across the boundaries of the two phases has to complete the mixing process. In practice the solutions are not well mixed until at least 2 ms (longer if the solutions are viscous or of different density), and thus it is difficult to determine any rate constant accurately that exceeds 500 s^{-1} (particularly with a strong signal).

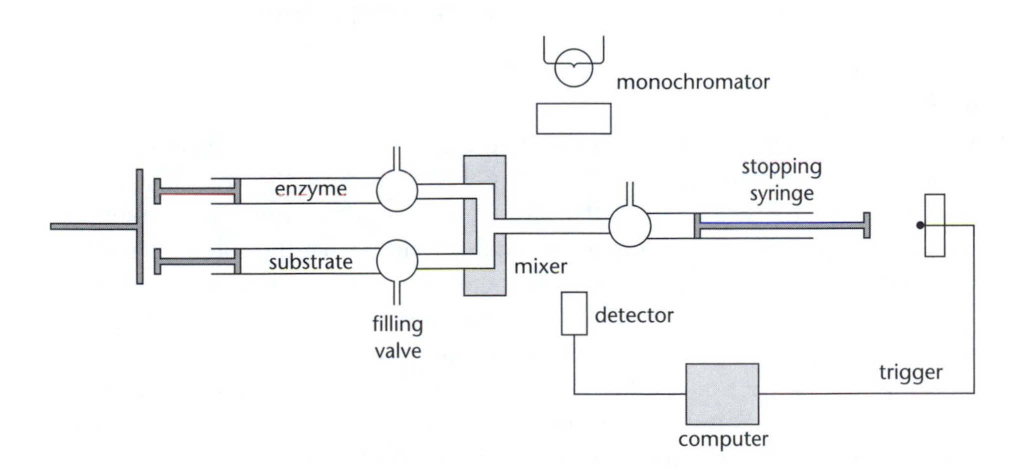

Figure 7-6. Schematic of a stopped-flow apparatus.

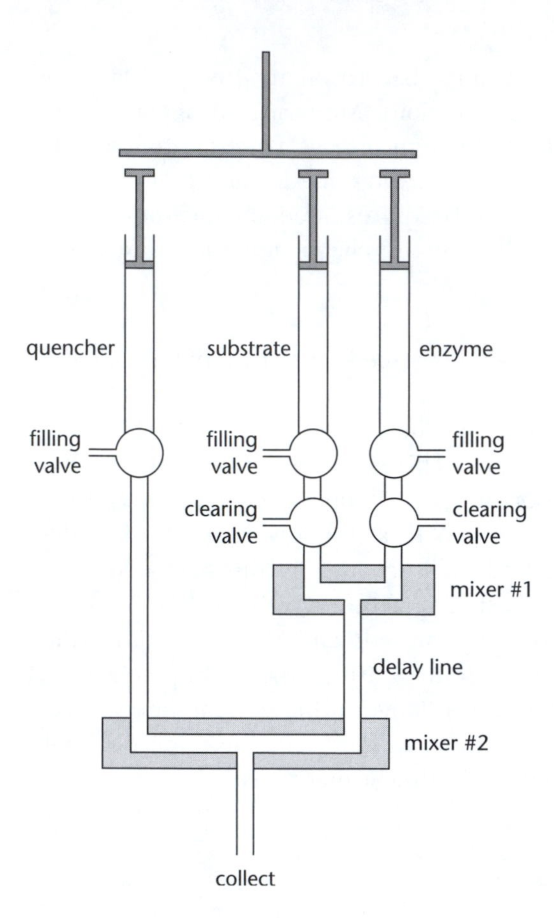

Figure 7-7. Schematic of a rapid quench apparatus.

Rapid Quench

When one cannot monitor absorbance or fluorescence changes, one can use rapid mix and quench methods (Figure 7-7). In this procedure the mixed solution passes through a delay line and is then mixed in turn with a quenching solution (acid, for example) or sprayed into a very cold solvent so that the solution freezes very quickly and the reaction stops. The quenched reaction mixtures are then analyzed at a convenient time by the appropriate technique. Because the reaction has been stopped (enzymatic reactions are very sensitive to pH and other denaturing agents that are employed to stop the reactions), separation methods can be used to isolate substrates, products, and intermediates from the quenched reaction mixture. The freezing method is often referred to as rapid cryogenic quenching and has been used to prepare samples for analysis by electron paramagnetic resonance (EPR), nuclear magnetic resonance (NMR), and

Mössbauer spectroscopy. In the rapid quench procedure, the reaction time is varied by adjusting the volume of the delay line and the flow rate. These quenching methods are often used in conjunction with radioisotopes, which permit very sensitive detection methods of various species in reactions. Phosphatases and kinases are frequently studied by these methods (see discussion of the yeast osmoregulatory system above).

Relaxation Methods

The methods described above involve rapid mixing of solutions, which is subject to problems of incomplete mixing, pressure surges as the mixing begins and the exit syringe hits the stop block, and cavitation. Such limitations make flow methods most practical only for events occurring between 1 ms and 1000 s. An alternate way to measure rapid changes involves relaxation methods such as temperature jump. In this technique a pulse of heat is applied to the reaction mixture at equilibrium by discharge of a condenser or use of an infrared laser. This changes the temperature of the solution very quickly, and if there is a ΔH value for the reaction of interest or if the reaction can be coupled to a second reaction with a significant ΔH (e.g., buffer), the system relaxes to a new equilibrium position in a first-order fashion. Another pulsed method is flash photolysis. In this method a short pulse of intense light (usually from a pulsed laser) on the sample will cause a photoreaction to occur. Such reactions can often be initiated in times as short as 10^{-11} s. Thus, very fast events can be monitored if a suitable photoreaction is available.

8

ISOTOPIC PROBES OF KINETIC MECHANISM

In the last few chapters, discussion has focused on the use of initial rate studies to obtain information on kinetic mechanism. The conventional means of determining the order of addition of reactants makes use of initial velocity studies in the absence and presence of product and dead-end inhibitors, as described in Chapters 5 and 6. These studies may not be conclusive, however, and information obtained from other techniques is often required to confirm conclusions based on initial rate studies and provide additional documentation of the kinetic mechanism. Complementary techniques include but are not limited to isotope exchange at equilibrium, isotope partitioning, positional isotope exchange, and rapid or pre-steady-state kinetics. In this chapter, some of these alternative methods that make use of isotopes will be introduced. The techniques discussed in this chapter are generally thought to be complementary to those discussed previously. Thus, although the results of methods discussed in this chapter overlap with those obtained from initial rate studies, new information is generated that extends knowledge of a given mechanism. In particular, these methods allow determination of the relative rates of steps in the mechanism. Isotope effects will be discussed separately in Chapter 9.

Isotopic Exchange

Most kinetic studies follow the conversion of substrates to products, or the net chemical reaction. The actual forward and reverse rates, especially near equilibrium, will be different, and the net chemical rate is the difference between them. The actual flux rates, however, can be measured with isotopes. Two types of experiment have been used. First, at equilibrium the rate of transfer of a labeled reactant to one or the other side of the reaction is measured. Radioactive reactants are normally used with ^{3}H, ^{14}C, or ^{32}P labels in positions that do not give an isotope effect, but stable isotopes such as deuterium, ^{13}C, ^{15}N, or ^{18}O can also be used with NMR or mass spectrometry for detection. In particular, a number of exchange studies have been carried out with ^{18}O.

The second type of experiment involves exchange of label from a product back into a substrate while the chemical reaction is proceeding in the forward direction. Both types of experiment are useful for determining kinetic mechanism and the relative rates of steps in the mechanism.

There are two ways to measure the rate of exchange. One is to determine the initial velocity of exchange (v^*) by taking samples over the first 10% or so of the time course to isotopic equilibrium; this is similar to using labeled substrates to determine initial velocities of the chemical reaction in a fixed time assay.

The second way to determine the rate of exchange will work only if the reaction is at equilibrium and involves measurements at a greater degree of approach to equilibrium and at final isotopic equilibrium. Since at chemical equilibrium the approach to isotopic equilibrium is a first-order process, the equation for the exchange rate is given by equation 8-1:

$$v^*_{\mathrm{A}\to\mathrm{P}} = -\left(\frac{1}{t}\right)\left[\frac{\mathbf{AP}}{\mathbf{A}+\mathbf{P}}\right]\ln(1-f) \tag{8-1}$$

where **A** and **P** are the concentrations of the two reactants between which exchange occurs, t is the time of the sample, and A originally has the label. f is the fractional approach to isotopic equilibrium, and $1-f$ is given in equation 8-2:

$$1-f = \frac{(\mathbf{A}^* - \mathbf{A}^*_\infty)}{(\mathbf{A}^*_0 - \mathbf{A}^*_\infty)} = \frac{(\mathbf{P}^*_\infty - \mathbf{P}^*)}{\mathbf{P}^*_\infty} \tag{8-2}$$

where the asterisk indicates the label and subscripts 0 or ∞ indicate zero time or final isotopic equilibrium. Note that in equation 8-2 the units can be anything convenient, such as counts per minute (cpm) per micromole. In equation 8-1, the units of $v^*_{\mathrm{A}\to\mathrm{P}}$ will be the concentration units of A and P divided by the time, t.

The first method of measuring v^* is faster but requires more label, as one only follows the first ~10% of the label transferred. The second method is more sparing of label but requires that one accurately know the final equilibrium distribution of label. While the final equilibrium distribution can be calculated from the levels of A and P used, it is more accurate to let the reaction reach isotopic equilibrium, which takes considerable time.

Isotopic Exchange at Equilibrium

Ping Pong Exchange Patterns. In a Ping Pong kinetic mechanism, one can observe isotopic exchange between the components of the Ping or the Pong in the absence of the reactants of the other. Thus in the aspartate aminotransferase reaction one observes aspartate–oxaloacetate exchange in the absence of glutamate and α-ketoglutarate, and exchange between the latter two in the absence of aspartate and oxaloacetate. Observation of such exchange is a major criterion for a Ping Pong mechanism. The reaction occurs at equilibrium, since the reaction that is occurring is

$$\mathrm{A} + \mathrm{E} \rightleftharpoons \mathrm{F} + \mathrm{P} \tag{8-3}$$

One can readily change the levels of A and P, since equilibrium is maintained by a change in the ratio of the concentrations of F and E, which are much lower than those of A and P. Unless one is working with a very slow mutant or very slow reactants, equilibrium is attained almost instantaneously.

When the level of A is varied at different levels of P, the basic rate equation is given as equation 8-4:

$$v^* = \frac{V^*_{\max\,\mathrm{A}\rightarrow\mathrm{P}}\mathbf{AP}}{K_{\mathrm{ia}}\mathbf{P} + K_{\mathrm{ip}}\mathbf{A} + \mathbf{AP}} \tag{8-4}$$

where

$$V^*_{\max\,\mathrm{A}\rightarrow\mathrm{P}} = \frac{V_1 K_{\mathrm{ia}}}{K_{\mathrm{a}}} = \frac{V_2 K_{\mathrm{ip}}}{K_{\mathrm{p}}} \tag{8-5}$$

A plot of $1/v^*$ versus $1/\mathbf{A}$ at different concentrations of P exhibits a series of parallel lines. This is similar to an initial velocity pattern where the initial rate is measured at different concentrations of A and different fixed levels of B, with the exception that the denominator of the rate equation contains dissociation constants, rather than Michaelis constants because the reaction is at chemical equilibrium. Measurement of such an exchange pattern is, in fact, the best way to determine the dissociation constants of the reactants in a Ping Pong mechanism.

The reason exchange patterns for Ping Pong mechanisms are parallel is that the slopes of the reciprocal plots represent the apparent first-order rate constants for exchange at very low **A**. When **A** is low, most of the enzyme is E, and any level of P is sufficient to convert any F that forms back into E and A. Thus the slope is not a function of **P**.

When B and Q are the exchange partners, the rate equation is similar:

$$v^* = \frac{V^*_{\text{max B}\rightarrow\text{Q}}\mathbf{BQ}}{K_{\text{ib}}\mathbf{Q} + K_{\text{iq}}\mathbf{B} + \mathbf{BQ}} \tag{8-6}$$

where

$$V^*_{\text{max B}\rightarrow\text{Q}} = \frac{V_2 K_{\text{iq}}}{K_{\text{q}}} = \frac{V_1 K_{\text{ib}}}{K_{\text{b}}} \tag{8-7}$$

The maximum velocities of the exchange reactions are related to the maximum velocities of the chemical reactions as follows:

$$\frac{1}{V^*_{\text{max A}\rightarrow\text{P}}} + \frac{1}{V^*_{\text{max B}\rightarrow\text{Q}}} = \frac{1}{V_1} + \frac{1}{V_2} \tag{8-8}$$

This relationship *must* hold for a Ping Pong mechanism, and failure of the equality for the acetate kinase reaction showed that the mechanism was in fact sequential and not Ping Pong (JBC *251*, 6775). Acetate kinase shows a fast MgATP/MgADP exchange because the enzyme reversibly phosphorylates one of its own carboxyl groups in the absence of acetate. The initial velocity pattern for the acetate kinase reaction with acetyl-P and MgADP is nearly parallel, and this along with the MgATP/MgADP exchange fooled early workers into thinking the mechanism was Ping Pong. However, the enzyme shows no acetate/acetyl-P exchange if it is treated with charcoal to remove traces of nucleotides. Moreover, the initial velocity pattern in the direction of formation of acetyl-P is intersecting and the reaction occurs with inversion of stereochemistry at phosphorus, evidence for a sequential kinetic mechanism.

Substrate inhibition can occur in Ping Pong mechanisms (Chapter 5), and this can complicate interpretation of exchange kinetics. Given the similarity in chemical structure of all reactants in Ping Pong mechanisms (for example, aspartate aminotransferase and nucleotide diphosphate kinase reactions), it is not uncommon for P to have some affinity for E as well as F. This will result in substrate inhibition of exchange by P that is competitive versus A, and thus the exchange pattern will no longer look parallel. Likewise, if A has some affinity for F as well as E, it will give substrate inhibition that is competitive versus P. Competitive substrate inhibition by P or A of the A to P exchange is described by equations 8-9 and 8-10:

$$v^* = \frac{V^*_{\max\,A\to P}\mathbf{AP}}{K_{ia}\mathbf{P}\left(1+\frac{\mathbf{P}}{K_{Ip}}\right)+K_{ip}\mathbf{A}+\mathbf{AP}} \tag{8-9}$$

$$v^* = \frac{V^*_{\max\,A\to P}\mathbf{AP}}{K_{ia}\mathbf{P}+K_{ip}\mathbf{A}\left(1+\frac{\mathbf{A}}{K_{Ia}}\right)+\mathbf{AP}} \tag{8-10}$$

This type of substrate inhibition exhibited in the exchange patterns was seen with nucleoside diphosphate kinase (B 8, 633). If both combinations, EP and FA, can occur, one can observe double competitive substrate inhibition of the A to P exchange and the expression in this case is given by equation 8-11:

$$v^* = \frac{V^*_{\max\,A\to P}\mathbf{AP}}{K_{ia}\mathbf{P}\left(1+\frac{\mathbf{P}}{K_{Ip}}\right)+K_{ip}\mathbf{A}\left(1+\frac{\mathbf{A}}{K_{Ia}}\right)+\mathbf{AP}} \tag{8-11}$$

Similar results could also be observed for the B to Q exchange, with the $K_{ib}\mathbf{Q}$ term modified by $(1+\mathbf{Q}/K_{Iq})$ in the case of substrate inhibition by Q binding to F and the $K_{iq}\mathbf{B}$ term modified by $(1+\mathbf{B}/K_{Ib})$ in the case of substrate inhibition by B. Both terms would be modified in the case of double competitive substrate inhibition.

When the partial reaction of a Ping Pong mechanism involves three reactants, such as shown in mechanism 8-12:

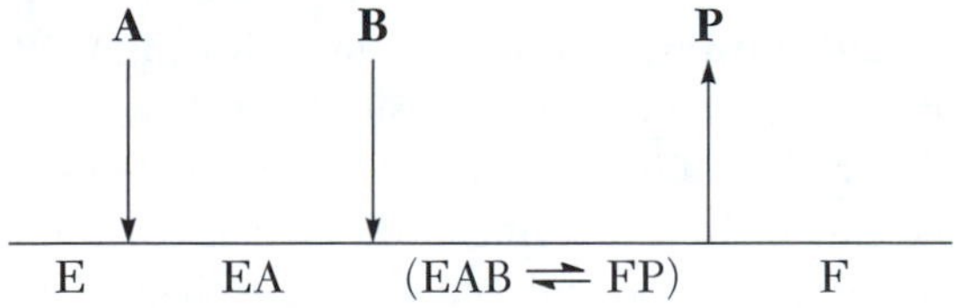

$$\begin{array}{ccccc} & \mathbf{A}\downarrow & \mathbf{B}\downarrow & \mathbf{P}\uparrow & \\ \text{E} & \text{EA} & (\text{EAB} \rightleftharpoons \text{FP}) & & \text{F} \end{array} \tag{8-12}$$

the order of addition of A and B can be readily determined by measuring the isotopic exchange pattern of P with either A or B as a function of the concentrations of A and B with the level of P held constant. Typically A and B are MgATP and an acid, and P is $MgPP_i$. One measures $MgPP_i$–MgATP exchange and varies the levels of acid at different fixed levels of MgATP. There are three possible results:

(1) Reaction is ordered, with A (MgATP) adding first. The pattern is intersecting, with uncompetitive substrate inhibition by B (acid). A high level of B causes substrate inhibition by displacing the E + A to EA equilibrium to the right by combining with EA and converting it to the central complex. Since the exchange involves the release of A from EA and the rate of release of A is near zero when the level of EA is near zero, the exchange is inhibited by high levels of B. The inhibition is seen only on the intercepts when A is varied, however, since at very low levels of A where the slopes are evaluated, the E + A + B to P + F equilibrium lies far to the left and the level of EA is so low that even a high level of B cannot tie up enzyme as central complexes.

(2) Reaction is ordered, with A (acid) adding first. Since the exchange occurs only between EA and F, the pattern will be equilibrium ordered. That is, the apparent dissociation constant of A approaches zero (actually the concentration of enzyme) as the level of B is raised. Thus the denominator of the rate equation for exchange will not contain a **B** term.

(3) The combination of A and B is random. The exchange pattern will be intersecting without substrate inhibition unless either A or B combines in the other's site, in which case the substrate inhibition will be competitive, not uncompetitive. This technique was first used with an asparagine synthetase, where MgATP and aspartate add randomly (JBC *244*, 4122).

With multisite Ping Pong mechanisms, one must be cautious in interpreting substrate inhibition of isotopic exchange. For example, with pyruvate carboxylase, bicarbonate will show uncompetitive substrate inhibition of exchanges between MgATP and either phosphate or MgADP, since bicarbonate can keep carboxybiotin out of the site and prevent it from reacting with MgADP and phosphate. The effects of inhibitors on exchanges in multisite Ping Pong mechanisms can also be unusual. An inhibitor combining at one site can activate exchange at another site by preventing the carrier from binding at the first site. An inhibitor that traps the carrier at one site will be a powerful inhibitor of exchange at other sites, since exchange requires the carrier to participate.

Exchange at Equilibrium in Sequential Mechanisms. In a sequential mechanism all reactants must be present to see isotopic exchange at equilibrium.

To maintain equilibrium, one must thus vary the concentrations of more than one reactant at the same time. While one could raise the level of one substrate and lower the level of another, the usual protocol is to vary in constant ratio the levels of one substrate and one product. For a dehydrogenase such as alcohol dehydrogenase, one can measure exchange between NAD^+ and NADH or between alcohol and aldehyde. One could also measure exchange between alcohol and NADH, but this involves the hydride transferred during the reaction and there will be a sizable isotope effect. There are no common atoms between NAD and aldehyde, so there is no way to measure exchange between them. The usual protocol involves monitoring the NAD–NADH and alcohol–aldehyde exchanges, using labels in places not expected to cause isotope effects.

The usual experiments involve varying together in constant ratio the levels of NAD^+ and NADH, or the levels of alcohol and aldehyde, maintaining the other reactants at constant concentrations. If one varies the levels of NAD^+ and aldehyde, or NADH and alcohol together, the formation of an E–NAD–aldehyde or E–NADH–alcohol dead-end complex will cause substrate inhibition and confuse the analysis. In a random mechanism, such an experiment does demonstrate the presence of such dead-end complexes.

If the mechanism is ordered, the expected exchange patterns are shown in Figure 8-1. When the levels of A and Q are varied together, the A–Q and B–P exchanges give linear reciprocal plots that intersect the horizontal axis at the same point as long as the concentrations of B and P are the same for both plots. The apparent K_m decreases from K_{ia} (when $1/v^*$ is plotted versus $1/\mathbf{A}$) or K_{iq} (when $1/v^*$ is plotted versus $1/\mathbf{Q}$) at low levels of B and P to zero at very high levels of B and P. The slope of the plot for the B–P exchange is $K_{ia}K_b/(V_1\mathbf{B})$ when the plot is versus $1/\mathbf{A}$ or $K_pK_{iq}/(V_2\mathbf{P})$ when it is versus $1/\mathbf{Q}$. The intercept of the plot for the B–P exchange when it is plotted versus $1/\mathbf{A}$ is $K_pK_{iq}/(V_1K_{ip}/K_q)$ at high levels of B and P and increases linearly with $1/\mathbf{B}$ as the B and P levels are reduced. The ratio of the slopes and intercepts of the plots for the A–Q and B–P exchanges is given by $(1+K_a\mathbf{B}/K_{ia}K_b+K_q\mathbf{P}/K_pK_{iq})$. Thus at low levels of B and P the two plots coincide, but they become farther apart as the B and P levels are increased.

When the levels of B and P are varied in constant ratio, the reciprocal plots for the A–Q and B–P exchanges exhibit parallel lines, but the A–Q exchange shows complete substrate inhibition. The substrate inhibition occurs because high levels of B and P at equilibrium reduce the concentrations of EA and EQ to near zero, and the exchange, which requires the release of A from EA and of Q from EQ, is prevented. The slopes of the reciprocal plots are $(K_b/V_1)[1+(K_{ia}/\mathbf{A})(1+\mathbf{Q}/K_{iq})]$ when plotted versus $1/\mathbf{B}$. The intercept of the reciprocal plot for the B–P

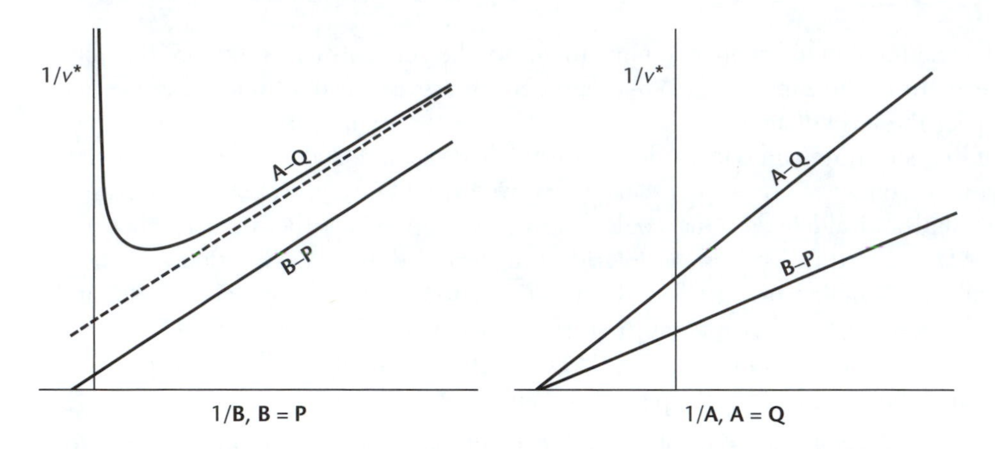

Figure 8-1. Isotope exchange patterns for an ordered mechanism. Left, the inner reactant pair, B and P, are varied in constant ratio, with $\mathbf{A} = \mathbf{Q} = 1$. Right, the outer pair, A and Q, are varied in constant ratio with $\mathbf{B} = \mathbf{P} = 1$. Exchange rates are calculated for a mechanism with all rate constants and $K_{eq} = 1$. This gives $K_{ia} = K_{iq} = K_b = K_p = 1$, $K_{ib} = K_{ip} = 2$, and $K_a = K_q = V_1/\mathbf{E_t} = V_2/\mathbf{E_t} = 0.5$.

exchange is independent of the levels of A and Q and is $K_pK_{iq}/(V_1K_{ip}/K_q)$ or $K_{ia}K_b/(V_2K_a/K_{ib})$, which is $1/\mathbf{E_t}$ times the sum of the reciprocals of the net rate constants for dissociation of B and P from the central complexes. The intercept is also equal to the sum of the vertical coordinates of the crossover points in the initial velocity patterns in forward and reverse directions (see Chapter 5), and when divided by $(1/V_1 + 1/V_2)$, it gives the value of R that is useful for determining the location of the rate-limiting step in the slower direction of the reaction.

The intercept of the asymptote to the plot for the A–Q exchange when B and P levels are varied together is greater than that for the B–P exchange by the ratio:

$$1 + \left[\frac{K_{ip}K_q}{K_pK_{iq}}\right]\left(1 + \frac{\mathbf{A}}{K_{ia}} + \frac{\mathbf{Q}}{K_{iq}}\right)\left[\frac{K_a}{\mathbf{A}} + \frac{V_1K_q}{V_2\mathbf{Q}}\right] \tag{8-13}$$

and also provides information about the rate-limiting steps. For a Theorell–Chance mechanism, the B–P reciprocal plot passes through the origin and this ratio is thus infinity. There is no substrate inhibition of the A–Q exchange in a Theorell–Chance mechanism, since there is no central complex for B and P to push the enzyme into. These experiments are thus very sensitive tests for the presence of central complexes.

For an equilibrium ordered mechanism, however, the two reciprocal plots coincide and again there is no substrate inhibition of the A–Q exchange. The substrate inhibition is most prominent for ordered mechanisms that do not approximate either Theorell–Chance or equilibrium ordered.

When the mechanism is not completely ordered, a degree of randomness alters the patterns in Figure 8-1. First, the apparent K_m values for the plots when A and Q levels are varied together do not coincide. With liver alcohol dehydrogenase and cyclohexanol as substrate, the K_m for the NAD–NADH exchange was 44 times that for the cyclohexanol–cyclohexanone exchange when NAD and NADH levels were varied in constant ratio (JBC *247*, 946). The higher value was assigned to the dissociation constant of the nucleotides from the central complexes, while the much lower value for the cyclohexanol–cyclohexanone exchange indicates that only a low degree of saturation with the nucleotides is sufficient to give half of the maximum exchange rate. Thus catalysis on the enzyme is faster than release of cyclohexanol and cyclohexanone from the enzyme.

When cyclohexanol and cycohexanone levels were varied in constant ratio, there was only partial substrate inhibition of the NAD–NADH exchange, indicating randomness. Further, the maximum exchange rate for the cyclohexanol–cyclohexanone exchange when their concentrations were varied in constant ratio, while large, was not infinite, showing that there are central complexes in the mechanism, although their steady-state level is low and the mechanism is fairly close to the one that Theorell and Chance postulated for liver alcohol dehydrogenase (JBC *247*, 946).

A classic case where catalysis is much faster than reactant release in a random mechanism is NADP–isocitrate dehydrogenase. $NADP^+$, NADPH, and isocitrate all dissociate from the enzyme much more slowly than they undergo the chemical reaction. The CO_2–isocitrate exchange was usually faster than the NADP–NADPH exchange, but when isocitrate and α-ketoglutarate levels were varied together the apparent K_m for the NADP–NADPH exchange was so much lower than that for the CO_2–isocitrate exchange that the nucleotide exchange became faster at low α-ketoglutarate and isocitrate levels (JBC *249*, 2920). When $NADP^+$ and NADPH levels were varied together, the apparent K_m for the nucleotide exchange was 15 times that for the CO_2–isocitrate exchange, so the nucleotide exchange was always slower.

In a rapid equilibrium random mechanism, all exchanges should go at the same maximum rate under all conditions. In most cases, however, reactant release partly limits at least one exchange. For kinases, the sugar–sugar phosphate exchange is usually slower than the MgATP–MgADP exchange by a factor of 1.5–36 (B *7*, 566; B *16*, 2176; JBC *239*, 3645). There will not normally be substrate inhibition of an exchange when the concentrations of similar molecules are varied together (MgATP and MgADP, or sugar and sugar–P), but substrate inhibition is likely when the levels of unlike molecules are varied together as the result of forming abortive complexes, for example, E–MgADP–sugar.

Isotope Exchange Not at Equilibrium

Most of the published isotope exchange studies have been carried out at equilibrium, but a number of useful experiments have involved measuring exchange from a product back into a substrate while the reaction is proceeding in the forward direction. For example, in the reaction catalyzed by glucose-6-phosphatase

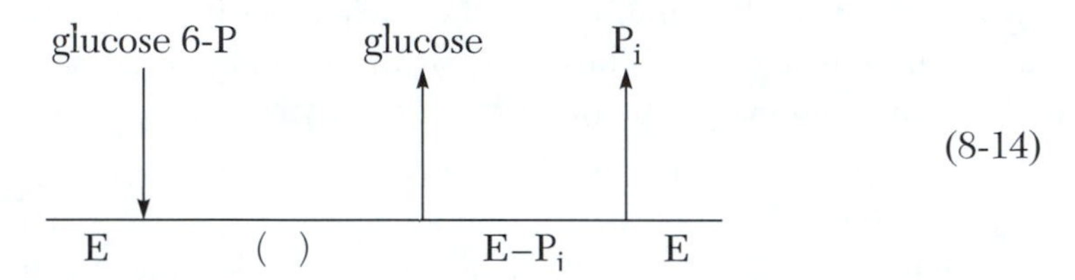

(8-14)

exchange was observed from glucose, but not from phosphate, into glucose-6-P when only one of the products was added at a time (JACS *82*, 947). The ratio of the rates of glucose to glucose-6-P exchange and the forward chemical reaction is given in equation 8-15:

$$\frac{v^*}{v} = \frac{[\text{glucose}]}{K_{is}} \tag{8-15}$$

where K_{is} is the slope inhibition constant for the noncompetitive product inhibition by glucose versus glucose 6-phosphate.

The observation of glucose to glucose 6-phosphate exchange shows that there is an appreciable level of $E–P_i$ present in the steady state and that glucose is released prior to phosphate. It does not necessarily indicate that phosphate cannot be released from the central complex, rather that release of glucose is much faster. In this case the chemistry involves formation of a covalent phosphoenzyme, which is then hydrolyzed, so product release is truly ordered. (The mechanism is really Ping Pong with water as the second substrate.)

An enzyme where product release is random is 5-aminolevulinate synthase, which catalyzes the reaction of succinyl-CoA and glycine to give CO_2, 5-aminolevulinate (ALA), and CoA. $^{14}CO_2$ exchanges into glycine only when both substrates are present, so the mechanism is sequential despite a nearly parallel initial velocity pattern (JBC *253*, 8872). The amount of exchange is doubled by the addition of CoA, tripled by the addition of ALA, and increased 5 times in the presence of both CoA and ALA. Thus CO_2 is the first product, with CoA and ALA released in random and partly rate-limiting fashion. ALA is released somewhat faster than CoA from the E–CoA–ALA complex, but release of CoA is significant, since its addition doubles the exchange rate.

Oversaturation and Iso Mechanisms

An Iso mechanism occurs when the form of free enzyme that combines with the substrate (E) is different from the one that is generated by product release (E′), mechanism 8-16. Thus, isomerization of E′ to E is a necessary part of the reaction cycle, and if this is a relatively slow step, the kinetics of the system are modified. The last product released becomes noncompetitive versus the first substrate, rather than competitive as in non-Iso mechanisms:

$$\mathrm{E} \underset{k_2}{\overset{k_1\mathbf{A}}{\rightleftharpoons}} \mathrm{EA} \underset{k_4}{\overset{k_3}{\rightleftharpoons}} \mathrm{E'P} \underset{k_6\mathbf{P}}{\overset{k_5}{\rightleftharpoons}} \mathrm{E'} \underset{k_8}{\overset{k_7}{\rightleftharpoons}} \mathrm{E} \qquad (8\text{-}16)$$

Many enzymes have two groups acting as general acid and general base during the reaction, and typically one group must be protonated and the other unprotonated. In most of these cases, however, the group that is protonated is different in E than in E′. Thus, if the proton shift between the groups can be made rate-limiting, one can demonstrate an Iso mechanism (B *14*, 4515). If the catalytic groups involve nitrogen or oxygen, the proton shift will be fast unless a conformation change in the protein accompanies it, and thus Iso mechanisms resulting from such proton shifting are not common.

Proline racemase, however, uses two cysteine thiols as general acid and general base and one must be protonated for combination of L-proline and the other one protonated for combination of D-proline (B *14*, 4515). Sulfhydryl groups do not show rapid proton release because they are so weakly hydrogen-bonded in water (hydroxide reacts with RSH at about 5% the rate seen with oxygen or nitrogen acids). As a result, interconversion of the protonation states of the two cysteines in proline racemase occurs with a rate constant of $10^5\ s^{-1}$, while the turnover numbers are $1600\ s^{-1}$. Under nonsaturating conditions, or even at substrate levels giving nearly the maximum rate, the free enzyme isomerization does not limit the rate. However, if the substrate concentration is raised to a high enough level and the time course to equilibrium is measured, most of the enzyme will be tied up as substrate or product complexes, and the conversion of E′ to E can be made rate limiting. This phenomenon is called oversaturation and exhibits a slower approach to equilibrium as the substrate concentration is raised above the saturation point (B 25, 2529).

The kinetics of proline racemase were studied by use of optical rotation as a monitor, starting the reaction with one isomer and proceeding to equilibrium. The time courses were normalized by plotting $(\mathbf{A}-\mathbf{P})/\mathbf{A_0}$ versus $t/\mathbf{A_0}$, which produces curves starting at unity on the vertical axis and becoming zero at infinite time (B 25, 2529). Below saturation, the initial slopes of these curves became more negative as $\mathbf{A_0}$ was increased, and as saturation was reached the initial slopes became identical. As $\mathbf{A_0}$ is increased further, however, the initial slopes remained the same, but the rate of the later portion of the time course decreased so that the $t/\mathbf{A_0}$ value to give 0.5 on the vertical scale increased. This phenomenon is called oversaturation and is illustrated by the proline racemase reaction in Figure 8-2.

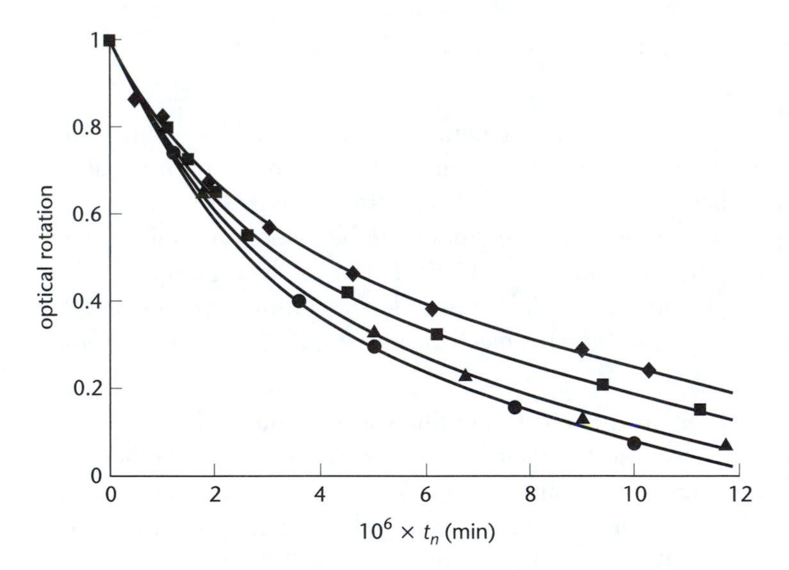

Figure 8-2. Demonstration of oversaturation in the proline racemase reaction. The initial concentrations of proline were 15.6 mM (•), 31.7 mM (▲), 94.8 mM (■), and 189 mM (◆). The optical rotation is plotted against t_n, which is defined as $t\mathbf{E_t}/c$, where t is time, $\mathbf{E_t}$ is the enzyme concentration (19.4 nM in these studies), and c is the initial proline concentration. Experiments were carried out at pH 8 and 37 °C.

What is happening is readily understood by examining Figure 8-3, which shows the free energy profile for proline racemase. The energy level for the bimolecular step where A adds to E depends on the level of A, as shown on the left. At an infinite level of A the energy level is equal to the top of the first barrier, as the reaction of E + A will go at an infinite rate (the difference between the energy levels on a free energy diagram is proportional to the reciprocal of the rate constant, or in this case the apparent rate constant, $k_1\mathbf{A}$). When **A** is equal to the dissociation constant for EA (K_s), the energy level is equal to that of the EA complex and also for that of E′P and E′ + P. Under these conditions, the barrier for the interconversion of E′ and E lies below that for converting EA and E′P, so the isomerization of free enzyme is not rate-limiting. At substrate and product concentrations marked C_P, however, the two barriers become equal and the E′ to E conversion contributes 50% of the rate limitation. C_P is called the "peak switch concentration" and is 125 mM for proline racemase (the K_m values are 2–3 mM). When the substrate and product concentrations are raised above C_P, the E′ to E isomerization becomes rate-limiting, rather than the conversion of EA to E′P, and this is the oversaturation region.

As noted in Chapter 6, in an Iso mechanism product inhibition by the last product is noncompetitive versus the first substrate, but where only one substrate and

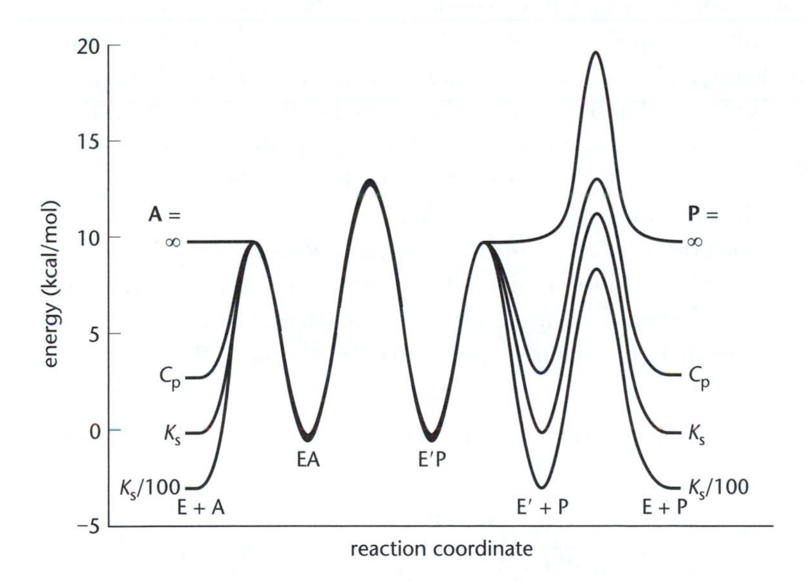

Figure 8-3. Free energy profiles for the proline racemase reaction as a function of reactant concentrations.

product are involved, one cannot add enough of the product to see the effect without causing a reversal of the reaction. Thus the method described above for proline racemase of looking at time courses to equilibrium as a function of initial substrate concentration was used to show the Iso mechanism. Northrop has given more details of how to analyze such data and the reader should consult his article (ME *249*, 211).

Proline racemase shows another interesting phenomenon. When the reaction was followed in D_2O, the optical rotation overshot zero by 20% and then eventually returned to zero. The reaction initially produces product containing a deuterium, and at the point where the optical rotation first reaches zero one has largely unlabeled substrate and deuterated product. The overshoot is thus an equilibrium perturbation. The equations for this system were presented by Cleland (1977, pages 252–256, Chapter 9) and allow one to determine the isotope effect in the system. Similar overshoots in D_2O have been seen with mandelate racemase (B *30*, 9255) and other similar enzymes. Observation of an overshoot establishes a two-base mechanism for the reaction, as opposed to a single-base mechanism.

Countertransport of Label

A very definitive experiment to demonstrate an Iso mechanism is the demonstration of countertransport. In fact this experiment is obligatory to determine the type of oversaturation being observed. With proline racemase (B *25*, 2538), 16 mM each of labeled D- and L-proline (about 6 K_m) were incubated with enzyme until equilibrium was achieved and then 183 mM unlabeled L-proline was added. This level of substrate ties up all of the E present and generates E′ + unlabeled D-proline. The E′ form can then return to E either by the isomerization reaction or by reacting with labeled D-proline to give labeled L-proline. As a result there is countertransport of labeled D-proline in the opposite direction of the net chemical conversion from L to D. The proportion of labeled L-proline transiently increased (in this case to 65%) and then decreased with time to 50%. Countertransport provides proof of an Iso mechanism and further defines the type of oversaturation.

The fractional excess of label in substrate A at any point in the reaction is given by equation 8-17:

$$\frac{\mathbf{A}^* - \mathbf{P}^*}{\mathbf{A}_0^* + \mathbf{P}_0^*} = \left[\frac{\mathbf{A} - \mathbf{P}}{\mathbf{A}_0 + \mathbf{P}_0}\right]\left(1 - \left[\frac{\mathbf{A} - \mathbf{P}}{\mathbf{A}_0 - \mathbf{P}_0}\right]^x\right) \tag{8-17}$$

where asterisks are for labeled species and the subscript “0” indicates the initial values.

$\mathbf{A_0^*} = \mathbf{P_0^*}$ but $\mathbf{A_0} \gg \mathbf{P_0}$ because of the large excess of unlabeled A added at the start. The fractional excess is zero both at the beginning of the reaction, where $\mathbf{A}-\mathbf{P}=\mathbf{A_0}-\mathbf{P_0}$, and at the end, where $\mathbf{A}=\mathbf{P}$. The point of maximum fractional excess label in A occurs at

$$\mathbf{A} - \mathbf{P} = \frac{\mathbf{A_0} - \mathbf{P_0}}{(1+x)^{1/x}} \tag{8-18}$$

and the fractional excess at this point is given by the following expression:

$$\left[\frac{\mathbf{A}^* - \mathbf{P}^*}{\mathbf{A_0^*} + \mathbf{P_0^*}}\right]_{\text{max}} = \left[\frac{\mathbf{A_0} - \mathbf{P_0}}{\mathbf{A_0} + \mathbf{P_0}}\right]\left[\frac{x}{(1+x)^{(1+1/x)}}\right] \tag{8-19}$$

In these equations, x is the ratio of $(\mathbf{A_0}+\mathbf{P_0})$ to the concentration at which the rate-limiting step changes from the conversion of EA to $E' + P$ to the conversion of E' to E (C_P). With a value of 125 mM for C_P with proline racemase, proof was obtained that the mechanism is one where EA to $E' + P$ limits V_{max}, and the E' to E conversion is rate-limiting only in the oversaturation region above C_P. The alternative will be discussed below.

In order for the countertransport experiment to work, the initial reactant concentrations must be high enough for x to be at least 0.3 if the maximum fractional excess is to be ~0.1. With proline racemase x was 1.6, and the maximum fractional excess label was 0.3. It is also required that $\mathbf{A_0} \gg \mathbf{P_0}$ to see the maximum effect; a value of 17 was used with proline racemase.

Another enzyme that exhibits oversaturation and an Iso mechanism is fumarase. Countertransport was clearly observed when a pulse of fumarate was added, while very little countertransport was seen when malate was added. Results obtained with fumarase have been discussed by Northrop (ME *249*, 211).

Most enzymes having Iso mechanisms will exhibit behavior similar to proline racemase, in that C_P will be well above the Michaelis constants and very high reactant concentrations will be required to demonstrate oversaturation and for E' to E conversion to become rate-limiting. There is one enzyme that shows the alternative pattern, carbonic anhydrase, which catalyzes a reaction in two parts. First, zinc-bound hydroxide attacks CO_2 to give bicarbonate. The reaction is fast and is not rate-limiting at saturation with substrates. The second step is conversion of zinc-bound water (which replaced bicarbonate) to zinc-bound hydroxide. This reaction requires transfer of a proton to buffer in solution and is the rate-limiting step at saturation with substrates. In this reaction, countertransport can be shown at levels of substrates above K_m (that is, C_P is close to the K_m levels) since the barrier for zinc-bound water to return to zinc-bound hydroxide by reaction with bicarbonate is lower than for loss of a proton to solution. Simpson and Northrop were able to measure this countertransport by adding 50 mM bicarbonate to an equilibrated mixture of ^{13}C-containing bicarbonate and CO_2 and by rapid sampling of the CO_2 via a membrane by using an isotope ratio mass spectrometer (ABB *352*, 288). Oversaturation in carbonic anhydrase will occur only at substrate concentrations well above the K_ms and C_P where E–CO_2 and E–bicarbonate become the dominant species present (ME *308*, 3).

Positional Isotopic Exchange (PIX)

This method measures the rate of internal isotopic exchange in a substrate. The result can be expressed as an absolute rate or, more usually, as a ratio to the overall reaction rate. Thus if MgATP transfers its γ-phosphate to an acceptor in a reversible manner and the next step is not too fast (either because the third substrate is omitted or because product release is sufficiently slow), ^{18}O in the $\beta-\gamma$ bridge will move to the β-nonbridge position as the result of PIX. The exchange occurs if the β-phosphate of MgADP is free to rotate during the reaction. Positional isotope exchange does not always occur (the biotin-containing carboxylases are an example), and in these cases the rotation of the β-phosphate is prevented, possibly because there are two Mg^{2+} ions involved in the reaction. But when PIX is observed, it clearly indicates the chemistry that takes place in the active site and directly measures the partition ratio of a key intermediate complex.

The method was developed by Rose to show that γ-glutamyl-P was an intermediate in the glutamine synthetase reaction in the absence of ammonia (JBC *251*, 5881). The exchange was measured by mass spectrometry, but a simpler procedure (although requiring more material) is to use ^{31}P NMR to follow the exchange. ATP containing ^{18}O in all four oxygens of the γ-phosphate is readily synthesized by reaction of ADP-morpholidate with [$^{18}O_4$]phosphate made from PCl_5 and $H_2{}^{18}O$. The NMR chemical shift for a phosphorus atom is shifted upfield by ^{18}O substitution, with the shift a function of bond order. When PIX occurs in this labeled ATP, a new peak forms 0.02 ppm downfield of each peak of the γ-phosphate doublet, corresponding to the species containing only three ^{18}O atoms. The final equilibrium distribution will be ⅔ with ^{16}O in the bridge and ⅓ with ^{18}O in the bridge.

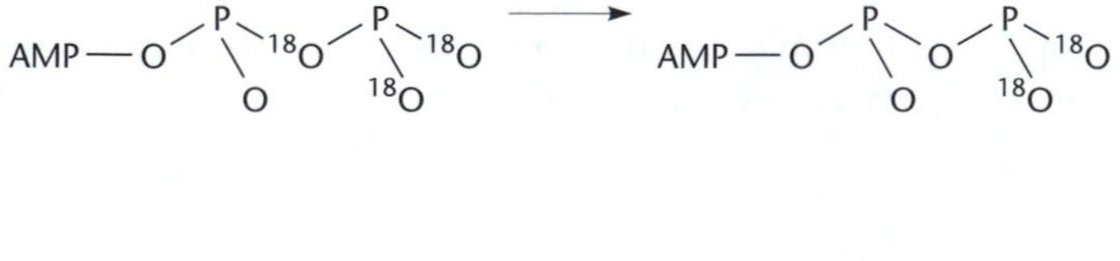

Since PIX is usually carried out in a system undergoing the forward reaction, the level of substrate is continuously decreasing. The rate of PIX is given by

$$v_{\text{PIX}} = \left[\frac{x}{\ln(1-x)}\right]\left[\frac{\mathbf{A_0}}{t}\right]\ln(1-f) \tag{8-20}$$

where $\mathbf{A_0}$ is the original substrate concentration, t is the time of measurement, f is the fractional approach to isotopic equilibrium at time t, and x is the fractional change in the concentration of A at this time. As x approaches 0 (substrate is not used up), $x/\ln(1-x)$ approaches unity, and equation 8-20 is simplified.

The PIX method is a partitioning experiment. In mechanism 8-21:

$$\mathrm{E} \underset{k_2}{\overset{k_1\mathbf{A}}{\rightleftharpoons}} \mathrm{EA} \underset{k_4}{\overset{k_3}{\rightleftharpoons}} \mathrm{EX} \xrightarrow{k_5} \mathrm{E} + \text{products} \tag{8-21}$$

X is an intermediate that can undergo the rotation needed to give PIX.

The ratio of the rate of PIX to that of the forward reaction is the ratio of the net rate constant for return of EX to E + A, $[k_4k_2/(k_2+k_3)]$, to k_5:

$$\frac{v_{\text{PIX}}}{v_{\text{chem}}} = \frac{k_2k_4}{k_5(k_2+k_3)} \tag{8-22}$$

Clearly the lower the value of k_5, the greater the degree of PIX, while a higher value for k_5 will suppress PIX. If k_5 is large, the addition of a product may reduce the proportion of EX that undergoes product release and thus induce PIX that is not seen in the absence of product. This was the case with argininosuccinate lyase (B *23*, 1791). Argininosuccinate was made enzymatically with ^{15}N only in the bridge position, originally from L-aspartate:

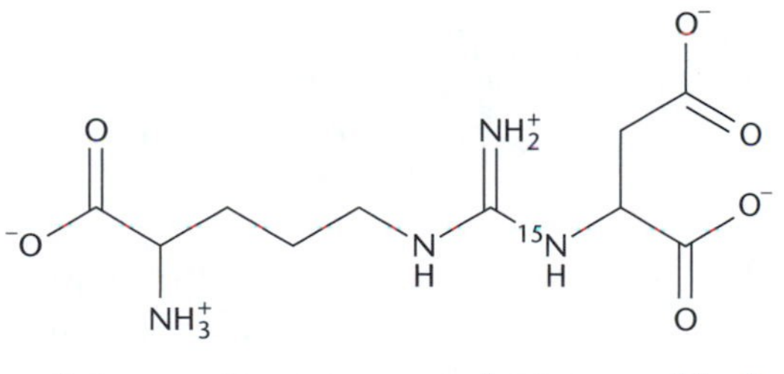

The cleavage of argininosuccinate was made irreversible by the addition of arginase to convert the arginine product to urea and ornithine. The PIX was measured by ^{15}N NMR, since the chemical shifts of the nitrogens at the bridge and nonbridge positions differ, and they differ from that of urea.

When the experiment was first carried out no PIX was observed, but addition of the product fumarate induced PIX. The mechanism for breakdown of EX in mechanism 8-21 is now

$$\begin{array}{ccc} EX & \xrightarrow{k_5} & EP + Q \\ {\scriptstyle k_{10}\mathbf{P}}\uparrow\downarrow{\scriptstyle k_9} & & \\ EQ & \xrightarrow{k_{11}} & E + Q \end{array} \tag{8-23}$$

where P is fumarate and Q is arginine. Equation 8-22 becomes

$$\frac{v_{\mathrm{PIX}}}{v_{\mathrm{chem}}} = \frac{(k_2k_4/k_2 + k_3)}{k_5 + k_9k_{11}/(k_{11} + k_{10}\mathbf{P})} \tag{8-24}$$

If $k_9 = 0$, that is, Q is released before P, then $v_{\mathrm{PIX}}/v_{\mathrm{chem}}$ is independent of the level of P added. If $k_5 = 0$, P is released first, and $v_{\mathrm{PIX}}/v_{\mathrm{chem}}$ is a linear function of **P** with a value of $k_2k_4/[(k_2 + k_3)k_9]$ at zero **P** and a slope of $k_2k_4k_{10}/[(k_2 + k_3)k_9k_{11}]$. If both k_5 and k_9 are finite (random release of products), $v_{\mathrm{PIX}}/v_{\mathrm{chem}}$ is a hyperbolic function of **P**, with a value of $k_2k_4/[(k_2 + k_3)(k_5 + k_9)]$ at zero **P** and a value of $k_2k_4/[(k_2 + k_3)k_5]$ at infinite **P**. With argininosuccinate lyase, the ratio of $v_{\mathrm{PIX}}/v_{\mathrm{chem}}$ as a function of fumarate concentration was hyperbolic, with a value of <0.15 at zero fumarate and 1.8 at infinite fumarate. Thus fumarate is released much faster than arginine, and the guanidinium group of arginine can rotate in the active site.

Another use of PIX was demonstrated with carbamoyl-P synthetase (B8, 3170). Carbamoyl-P was synthesized from $[^{18}O_4]$phosphate and KOCN to give bridge-labeled material:

$$H_2N{-}C(=O){-}^{18}O{-}P(={}^{18}O)({}^{18}O^-)({}^{18}O^-)$$

When this was incubated with MgADP and enzyme, bridge to nonbridge PIX was followed by observing the ^{31}P NMR chemical shift. A peak 0.02 ppm downfield of the original one appeared, and now the equilibrium level of exchange is 50%, since a carboxyl group has only two oxygen atoms.

When PIX is observed in the first substrate of an ordered mechanism, high levels of the second substrate will suppress the exchange, just as they do in an isotope exchange experiment, and for the same reason. High levels of B simply keep A trapped on the enzyme as EAB, so no release of A is possible. This approach was used with UDP-glucose pyrophosphorylase, where MgUTP adds before glucose-1-P and $MgPP_i$ is released before UDPG (B *26*, 6465). The PIX in UTP was linearly suppressed by increasing levels of glucose-1-P, and the PIX in UDPG was suppressed by $MgPP_i$. If the mechanism is partly random, PIX will be decreased, but not to zero, by high levels of the second substrate.

In a Ping Pong mechanism where the aim is to measure PIX in substrate A in the absence of substrate B, one must add finite amounts of P to permit F to be reconverted to E:

$$\mathrm{E}k_2 \underset{k_2}{\overset{k_1\mathbf{A}}{\rightleftharpoons}} \mathrm{EA} \underset{k_4}{\overset{k_3}{\rightleftharpoons}} \mathrm{FP} \underset{k_6\mathbf{P}}{\overset{k_5}{\rightleftharpoons}} \mathrm{F} \tag{8-25}$$

Excess P also dilutes out the labeled P formed by isotopic exchange and prevents apparent PIX resulting from back-reaction of labeled P that has been released. By comparing the rate of PIX to the rate of formation of unlabeled A, one can calculate $v_{\mathrm{PIX}}/v_{\mathrm{chem}}$. The correction for the level of labeled P that has dissociated and recombined is given in JBC *262*, 12092. These authors showed that in this system

$$\frac{k_5}{V_2/\mathbf{E_t}} \geq \frac{v_{\mathrm{A}\rightarrow\mathrm{P}}}{v_{\mathrm{PIX}}} \tag{8-26}$$

With galactose-1-P uridylyltransferase, they measured β-nonbridge to β-bridge ^{18}O PIX in UDPG with a 4–12-fold excess of glucose-1-P and found that glucose-1-P was released at least 3.4 times faster than $V_2/\mathbf{E_t}$. This type of experiment should be very useful in determining the relative rates of reactant release in Ping Pong mechanisms. The experiment can also be carried out at a high concentration of B, which will generate E. In this case the correction for P need not be considered. However, the concentration of A will change with time, and the experiment must be carried out under conditions where significant A still remains.

For a comprehensive discussion of the PIX method, see Raushel & Villafranca (CRB *23*, 1).

Determination of Stickiness

A sticky substrate is one that reacts through the first irreversible step (usually product release) faster than it dissociates from the enzyme. The stickiness ratio,

S_r, is thus the ratio of the net rate constant for reaction of the initial ES complex through the first irreversible step and the rate constant for substrate dissociation. Stickiness affects pH profiles and the degree to which isotope effects on V/K are expressed (see Chapters 10 and 9), and mechanistic analysis is greatly simplified if one uses a nonsticky substrate or eliminates the stickiness so that substrate binding is at equilibrium in the steady state. Two methods have been used for determining stickiness: the isotope partition method and the effect of viscosity on V/K values.

Isotope Partitioning

This method measures partitioning in a single turnover and can be used for any substrate that can form a binary complex with the enzyme without undergoing the overall reaction. Its use is thus limited to enzymes with two or more substrates, but it works for any substrate in a random mechanism and for all but the last substrate to add in an ordered one. The method was introduced by Alton Meister (JBC 238, 1179), who used it in an attempt to detect the presence of a glutamyl phosphate intermediate in the glutamine synthetase reaction. The potential of the method, however, was not recognized until Irwin Rose applied it to the yeast hexokinase reaction in an attempt to trap ^{14}C-glucose in the E–^{14}C-glucose binary complex (JBC 255, 7569). The method determines whether a substrate is "sticky," that is, remains bound to enzyme once a binary enzyme–substrate complex is diluted into a solution containing the other substrate(s). The following treatment will consider trapping from a binary complex, but experiments can be carried out for all substrates that add to enzyme with the exception of the last one to combine in an ordered mechanism.

Experimental

(1) A pulse solution composed of the buffered radioactive (EA*) binary complex is prepared. Thus a knowledge of, or estimate of, the dissociation constant of interest must be known. Optimally, the final concentration of E_t must be such that isotope trapping on the order of 5% could be measured, that is, $\mathbf{E} \sim 100$–1000 μM). The concentrations of E_t and A* (of known specific radioactivity) are adjusted such that all or the large majority of E is present as EA* and the concentration of free A* is low. If this cannot be accomplished, A^*_{free} and $\mathbf{EA}^*$ must be known so that correction can be made for free A* (see below). The concentration of enzyme is high, but the pulse volume is low, so relatively small amounts of protein are required per experiment. The volume of the pulse solution

is small ($<50\ \mu L$), delivered as accurately as possible, for example, by use of a Hamilton syringe.

(2) The pulse solution is diluted into a rapidly stirring chase solution containing the same buffer as in the pulse solution, a large excess of the unlabeled reactant present in the pulse solution, and any other reactants, activators, etc., required for reaction. The volume of the chase solution is much larger than that of the pulse solution (>5 mL), so that when the pulse solution is diluted into the chase solution, the final concentration of A^* is 1000-fold lower than that of unlabeled A. Under these conditions, there is very little possibility that A^* can effectively compete with A for binding to E.

(3) After dilution of the pulse into the chase, the reaction is terminated after a very brief time period that is the same from one experiment to the other, for example, 3 s. The product that contains the label present in A^* is then isolated and counted. On the basis of the known specific radioactivity, the molar amount of P^* formed is determined. Time after mixing is not critical since a very high dilution is used for the final concentration of A^* compared to A, and thus dissociation of A^* from EA^*B is for all intents and purposes irreversible. Nonetheless, a control is carried out to correct for reaction of the diluted A^* in the chase as discussed below. The reaction is normally terminated with something that will immediately denature the enzyme, for example, acid, base, or organic solvent. If acid or base is used to denature, and the solution is to be titrated to neutrality prior to separation of P^*, for example, chromatographically, neutralization should not be carried out until just before application to the column to be certain that the enzyme does not regain activity. If product P^* is not stable to acid or base treatment, some other means should be found to terminate activity. As an example, for kinases, a high concentration of EDTA can be used to chelate Mg^{2+}.

(4) As a control, the above experiment is repeated with no A^* in the pulse solution but with the same amount of A^* originally present in the pulse solution included in the chase solution. The control corrects for the small amount of combination of E and A^* in the chase solution to produce P^*. This control must be carried out for each level of the second substrate present in the chase.

(5) Separate experiments and controls are then carried out in which the concentration of the second substrate B in the chase is varied.

Data Analysis

The amount of radioactive product formed in micromoles at each concentration of B is corrected for the control reaction carried out at each B concentration, and the corrected value is plotted in double reciprocal fashion versus the concentration of B, as shown in Figure 8-4.

In order to obtain estimates of the kinetic parameters for the trapping experiment, data can be fitted to

$$\mathbf{P}^* = \frac{\mathbf{P}^*_{\text{max}}\mathbf{B}}{K'_b + \mathbf{B}} \tag{8-27}$$

Note that one can graphically estimate $\mathbf{P}^*_{\text{max}}$ as the reciprocal of the ordinate intercept and K'_b as the reciprocal of the abscissa intercept.

Theory

Consider the following mechanism:

$$\begin{array}{ccccc} EA^* & \underset{k_4}{\overset{k_3\mathbf{B}}{\rightleftharpoons}} & EA^*B & \xrightarrow{k_5} & E + P^* \\ \downarrow k_2 & & \downarrow k_7 & & \\ E + A^* & & EB + A^* & & \end{array} \tag{8-28}$$

where an asterisk indicates the label. Any A^* that dissociates is diluted by the large excess of unlabeled A in the solution, and only that which is converted to EA^*B

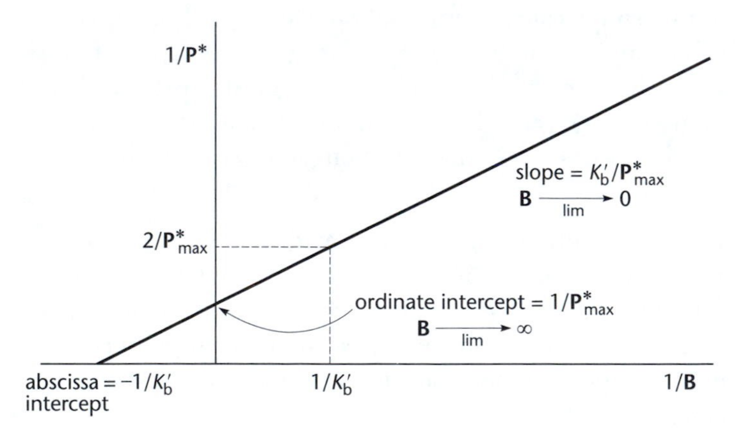

Figure 8-4. Double reciprocal plot of **P*** and **B**. The graphical estimation of kinetic constants is shown.

and on to P* contributes to the formation of P* in the first turnover. Subsequent turnovers involve only the highly diluted A*, and their contribution is removed by subtracting the control.

The basic experiment is repeated at different levels of B and the maximum level of P* and the trapping constant (apparent K_m) determined from a reciprocal plot of $1/\mathbf{P}^*$ versus $1/\mathbf{B}$. The ratio of k_7/k_5, which is the reciprocal of the stickiness ratio for A in the presence of saturating B, is given by

$$\frac{1}{S_r} = \frac{k_7}{k_5} = \frac{\mathbf{E_t}}{\text{app}\mathbf{P}^*_{max}(1 + K_{ia}/\mathbf{A})} - 1 \tag{8-29}$$

where K_{ia} is the dissociation constant of A from EA, $\mathbf{E_t}$ is the concentration of enzyme active sites, and app$\mathbf{P}^*_{max}$ the level of P* formed at saturating B, as determined from the reciprocal plot of $1/\mathbf{P}^*$ versus $1/\mathbf{B}$. Note that it is k_5, and not $V/\mathbf{E_t}$, or k_{cat}, that is compared to k_7. $V/\mathbf{E_t}$ is often limited by release of the second, rather than the first product, so the ratio of $k_7/(V/\mathbf{E_t})$ will be larger than k_7/k_5.

If the mechanism is ordered, $k_7 = 0$, and app$\mathbf{P}^*_{max}(1 + K_{ia}/\mathbf{A}) = \mathbf{E_t}$. In practice, it is difficult to tell whether k_7 is truly zero, since $\mathbf{E_t}$ values are usually somewhat uncertain. When the method was applied to hexokinase with glucose as first substrate, Rose found $k_7/k_5 < 0.05$ (JBC *249*, 5763). Likewise in the reverse direction, glucose-6-P appeared not to dissociate from the ternary complex (B *21*, 1295). By contrast, $k_7/k_5 = 0.53$ for fructokinase with fructose as first substrate (B *16*, 2176).

The rate at which A* dissociates from EA* relative to $V/\mathbf{E_t}$ is

$$\frac{\left(\frac{K'_b}{K_b}\right)\mathbf{E_t}}{\text{app}\mathbf{P}^*_{max}(1 + K_{ia}/\mathbf{A})} \geq \frac{k_2}{V/\mathbf{E_t}} \geq \frac{K'_b}{K_b} \tag{8-30}$$

where K'_b is the trapping constant (i.e., apparent K_m) from the reciprocal plot of $1/\mathbf{P}^*$ versus $1/\mathbf{B}$ and K_b is the Michaelis constant for the chemical reaction of B at saturating A. The comparison *is* with $V/\mathbf{E_t}$ [it really is a comparison of k_2 with $V/K_b\mathbf{E_t}$]. If the mechanism is ordered and $k_7 = 0$, both sides of equation 8-30 are equal, giving an exact solution for $k_2/(V/\mathbf{E_t})$. In a random mechanism, $k_2/(V/\mathbf{E_t})$ equals the left side expression of equation 8-30 if B is not sticky and cannot be trapped in an isotope partitioning experiment. It equals K'_b/K_b if B is quite sticky. If both A and B are sticky enough to be trapped:

$$\frac{k_2}{V/\mathbf{E_t}} = \frac{\left(\frac{K'_b}{K_b}\right)(1 + S_{ra})(1 + S_{rb})}{S_{ra}S_{rb} + S_{ra} + S_{rb}} \tag{8-31}$$

where S_{ra} and S_{rb} are stickiness ratios for A and B from equation 8-29. When equation 8-30 was used for hexokinase, Rose found $k_2/(V/\mathbf{E_t})$ to be 0.3, while Viola found 1.4 (the enzyme preps were different). In the back-reaction, the value for glucose-6-P was 165 (B *21*, 1295), showing that glucose-6-P is very sticky from the ternary complex but not very sticky at all from its binary complex with the enzyme. With fructokinase, where k_7/k_5 was 0.53, the value for $k_2/(V/\mathbf{E_t})$ lay between 80 and 130. Since ATP could not be trapped, however, the correct value is 130.

When K_{ia} is not less than $\mathbf{E_t}$, the results are very dependent on the accuracy with which K_{ia} is known. The proper procedure then is to vary both $\mathbf{A}^*$ and $\mathbf{B}$ in the isotope partitioning experiments. The amount of P* formed will then be given by

$$\mathbf{P}^* = \frac{\mathbf{P}^*_{max}\mathbf{A}^*\mathbf{B}}{(K_{ia} + \mathbf{A})(K'_b + \mathbf{B})} \tag{8-32}$$

Since the denominator is factored, it is easy to get an excellent fit to this equation and thus to obtain accurate values for $\mathbf{P}^*_{max}$, K_{ia}, and K'_b. Equation 8-29 then becomes

$$\frac{1}{S_r} = \frac{\mathbf{E_t}}{\mathbf{P}^*_{max}} - 1 \tag{8-33}$$

and the left term in equation 8-30 is $(K'_b/K_b)(\mathbf{E_t}/\mathbf{P}^*_{max})$. This method was used for hexokinase by Viola and deserves to be used more often, especially when there is doubt about the accuracy of K_{ia} values determined from initial velocity patterns (B *21*, 1295).

In a Ping Pong mechanism, an EA* complex will react to give F and P* to the extent determined by the equilibrium constant for the half-reaction, and after dilution with unlabeled A, the reaction will not reverse because of the dilution of P* and because F will be reconverted to E by the added B. Thus the level of P* formed should be independent of the level of B added, and the zero value of K'_b proves the mechanism is Ping Pong. This should be a useful method for establishing or eliminating a Ping Pong mechanism in cases where the initial velocity pattern is parallel, but no isotopic exchange of P back into A is observed because of a high equilibrium constant. It has not yet been applied, however.

Theoretical Limits

If one observes no trapping, there are several possible explanations.

(1) EA does not form; that is, the mechanism is ordered with B adding to enzyme prior to A.

(2) The rate constant for dissociation of EA^*B to EB and A^*, k_7, is much greater than the rate constant for turnover of EA^*B to give P^*, k_5.

(3) The rate constant for dissociation of EA^* to E and A^*, k_2, is much greater than $(V/K_b\mathbf{E_t})\mathbf{B}$ even at the highest **B** used experimentally. One can estimate the minimum **B** required to give trapping of A^* as P^*. The effective dissociation constant for A, K_{ia}, is equal to k_2/k_1 on the basis of mechanism 8-28 above. An estimate of the maximum on-rate constant, k_1, for A binding to enzyme is $<10^9\ M^{-1}\ s^{-1}$ (Fersht, 1999). An estimate for k_2 is thus obtained as $[K_{ia}\ (M)][10^9\ (M^{-1}\ s^{-1})]$. The amount of A^* trapped by B will also depend on the rate constant for conversion of EA^* and B to EQ and P^* and the concentration of B, that is, $V/K_b\mathbf{E_t}(M^{-1}\ s^{-1})\mathbf{B}(M)$. The minimal **B** required to give trapping is thus given in equation 8-34:

$$\mathbf{B}(M) \geq \frac{[K_{ia}\ (M)][10^9\ (M^{-1}\ s^{-1})]}{V/K_b\ (M^{-1}\ s^{-1})} \tag{8-34}$$

Variation of V/K with Viscosity

The second method of determining stickiness involves determining the effect of viscosity on V/K values. Molecules such as sucrose or glycerol raise the microviscosity and thus slow down diffusion. For example, 30% sucrose has a relative viscosity of ~ 3 and reduces diffusion-limited rate constants by this factor. Polymers like Ficoll, however, increase macroviscosity but do not affect microviscosity and thus do not slow down diffusion.

In a simple mechanism:

$$E \underset{k_2}{\overset{k_1\mathbf{A}}{\rightleftharpoons}} EA \xrightarrow{k_3} E + \text{products} \tag{8-35}$$

k_1 is limited by the rate at which E and A diffuse together, and k_2 is limited by the rate at which they diffuse apart. If viscosity does not affect unimolecular rate constants on the enzyme, however, k_3 will not be sensitive to viscosity (see below for problems with this assumption). If A is not sticky (that is, $k_2 \gg k_3$), viscosity will not affect V/K, which is $k_1k_3\mathbf{E_t}/(k_2+k_3)$, since k_1 and k_2 are equally affected by viscosity. If A is very sticky ($k_3 \gg k_2$), however, $V/K = k_1\mathbf{E_t}$ and should be inversely proportional to viscosity. In the general case:

$$\frac{K}{V} = \frac{\eta_{rel}}{k_1\mathbf{E_t}} + \frac{k_2/k_3}{k_1\mathbf{E_t}} \tag{8-36}$$

where η_{rel} is relative viscosity. The horizontal intercept of a plot of *K/V* versus η_{rel} is $-(k_2/k_3)$, or the reciprocal of S_r, the stickiness ratio, while the slope is $1/(k_1\mathbf{E_t})$, from which k_1 can be determined if $\mathbf{E_t}$ is known.

To compare different substrates, a useful plot is the ratio of *V/K* at a relative viscosity of 1 to that at η_{rel}, plotted versus η_{rel}:

$$\frac{(V/K)_{\eta_{rel}=1}}{(V/K)_{\eta_{rel}}} = \frac{\eta_{rel}}{1+(k_2/k_3)} + \frac{k_2/k_3}{1+(k_2/k_3)} = \frac{\eta_{rel}S_r}{S_r+1} + \frac{1}{S_r+1} \tag{8-37}$$

This plot has a slope of unity if S_r is large but zero for a nonsticky substrate. The value of S_r is determined from

$$S_r = \frac{\text{slope}}{1-\text{slope}} \tag{8-38}$$

or from the ratio of slope to vertical intercept.

This method seems to work well with simple hydrolytic enzymes. Kirsch used the method with chymotrypsin to show S_r values of 0, 0.1, and 0.67 for three substrates (B *21*, 1302). Similarly, S_r values of 0, 0.03, 0.43, and 1.3 were found for substrates of β-lactamase (B *23*, 1275). In these studies, sucrose was the viscosogen and gave no effects on *V*.

Phosphotriesterase catalyzes the hydrolysis of triesters of the type $(EtO)_2P(O)$-O-R, where R is a phenol. When the effects of viscosity on *V/K* values were determined with a series of substrates where the p*K* of the leaving group (phenol) varied from 4.1 to 8.56, it was found that the slope of equation 8-37 was $\sim$0.96 for leaving groups with p*K*s of 7 or less (B *30*, 7438). Leaving groups with higher p*K*s gave lower slopes, reaching 0.005 with a p*K* of 8.56. Thus, *V/K* is diffusion-limited with S_r values of $\sim$24 for faster substrates, but S_r is less than 0.01 for a substrate with 2.5% the value of the *V/K* of the better substrate.

When the effects of viscosity on *V*, rather than *V/K*, were measured, the slope of equation 8-37 was 0.32–0.37 for fast substrates but decreased for ones with a leaving group p*K* above 7. The slope was only 0.005 for a p*K* of 8.56. The value of $\sim$0.35 for good substrates shows that $V/\mathbf{E_t}$ is largely limited by unimolecular steps that are not viscosity-dependent and only partly by the rate of diffusion of product from the active site. With slow substrates, the chemistry becomes the rate-limiting step and $V/\mathbf{E_t}$ is not sensitive to viscosity.

For more complex enzymes, however, viscosity can have more complex effects. Thus if product release limits *V/K*, viscosity will decrease the value even if the substrate is not sticky. One or more of the viscosogens may bind in the active site and act as inhibitors (and thus one should always use two or more separate viscosogens). This is particularly a problem with enzymes operating on sugars or similar molecules. The viscosogen or viscosity may affect the rates of conformation changes prior to or following catalysis and thus may affect k_3 in our simple model. Sucrose affects the *V*/*K* for adenosine with adenosine deaminase (B *26*, 3027), although other studies clearly show that adenosine is not sticky (B *26*, 7378). With malic enzyme, sucrose or glycerol decreased K_m for malate and increased *V*, so that *V*/*K* increased with viscosity (B *27*, 2934). The ^{13}C isotope effect on decarboxylation was decreased, presumably as the result of changes in the rate of conformation changes.

It is thus important to use this method with care and always to use slow, nonsticky substrates as controls. The advantage of the method is that it works for the last substrate to add to the enzyme in an ordered mechanism, or when there is only one substrate. These are cases where isotope partitioning does not work and where the initial velocity patterns do not distinguish rapid equilibrium binding from highly committed binding. An estimation of the stickiness of substrates can also be obtained from the analysis of pH profiles and the pH variation of isotope effects. A discussion of these applications will be deferred until Chapters 9 and 10.

9

ISOTOPE EFFECTS AS A PROBE OF MECHANISM

Isotope effects are perhaps one of the most powerful tools available to the mechanistic enzymologist because of the amount and different types of information one can obtain. In this chapter, an overview of isotope effects applied to the determination of the kinetic and chemical mechanisms of enzymatic reactions will be provided.

An isotope effect can be defined as a change in the rate or equilibrium constant of a reaction upon substitution of a heavy atom for a light one at or adjacent to the position of bond cleavage in a molecule undergoing reaction. Isotopes provide an excellent tool for the study of reaction because they are isosteric and isoelectronic, and thus nonperturbing. However, isotope effects reflect changes in vibrational frequencies of reactants as they are converted to products in the rate-determining transition states.

The application of isotope effects to enzyme-catalyzed reactions has provided another probe of kinetic mechanism yielding both qualitative and quantitative information. In addition, isotope effects have been used to obtain information on the regulatory kinetic mechanism of enzymes, that is, the steps along the reaction pathway (chemical interconversion, reactant release, etc.) affected by bound allosteric modulators. In the first part of this chapter, theory will be developed for the determination of kinetic mechanism from isotope effects with a discussion of advantages and shortcomings of the technique. In the second part of the chapter, theory will be developed for the use of isotope effects in the elucidation of chemical mechanism. First, nomenclature will be briefly discussed followed by a presentation of the methods utilized for the measurement of isotope effects.

Types of Isotope Effects

There are two types of isotope effects termed primary and secondary. If a heavy atom is substituted for a light one in the bond undergoing cleavage a primary kinetic isotope effect is observed on the rate of the reaction, expressed as the ratio of the rates with light and heavy atoms, k_L/k_H, where L and H refer to light and heavy, respectively. If a heavy atom is substituted for a light one in a bond that does not undergo bond cleavage but is bonded to or in a position remote to an atom that does undergo bond cleavage, a secondary kinetic isotope effect is observed on the rate of the reaction. If the heavy atom is attached to the one undergoing bond cleavage it is termed α, while if it is in a position attached to an atom adjacent to the one undergoing bond cleavage it is termed β. Isotope effects can also be observed on the equilibrium constant for a given reaction, and these may be primary or secondary.

Nomenclature

Isotope effects are given as the ratio of the rates with light and heavy atom substitutions, for example, k_H/k_D. In the following discussion, isotope effects will be abbreviated with a leading superscript to identify the heavy atom (B *14*, 2644; B *20*, 1790). Leading superscripts D, T, 13, 14, 15, and 18 correspond to deuterium, tritium, ^{13}C, ^{14}C, ^{15}N, and ^{18}O isotope effects. For example, k_H/k_D and $k_{C\text{-}12}/k_{C\text{-}13}$ are ^{D}k and ^{13}k, respectively, while α- and β-secondary deuterium kinetic isotope effects are $^{\alpha\text{-}D}k$ and $^{\beta\text{-}D}k$, respectively. Isotope effects on equilibrium constants are likewise given as $^{D}K_{eq}$, $^{13}K_{eq}$, etc., again representing ratios of K_{eq} for molecules containing the light and heavy isotopes. In the case of observed isotope effects in enzyme-catalyzed reactions, the limiting macroscopic rate constants are used, for example, ^{D}V and $^{D}(V/K)$.

Measurement of Kinetic Isotope Effects

Direct Comparison of Initial Rates

Of the three ways to measure an isotope effect, this is the least accurate, but it permits the measurement of isotope effects on both V and V/K for the substrate whose concentration is varied, while other methods do not allow measurement of isotope effects on V. The method is straightforward; one simply compares double reciprocal plots of the initial rate versus substrate concentration with labeled and unlabeled substrates, as in Figure 9-1. The ratio of slopes with labeled and unlabeled substrates gives the isotope effect on *V/K*, while the ratio of vertical intercepts gives the isotope effect on *V*. Direct comparison is useful only for isotope effects of 5% or larger, for example, primary deuterium or larger secondary deuterium isotope effects. A few attempts have been made to determine ^{18}O or other heavy atom isotope effects by direct comparison, but the errors are large.

Accurate measurement of the *V/K* isotope effect requires that the substrate concentrations be precisely known, and thus stock solutions of both labeled and unlabeled substrates should be calibrated by an enzymatic end-point analysis. The error in the isotope effect will depend on the precision of this calibration. By contrast, the purity of the substrate is not a factor, and the $^{D}(V/K)$ value will be correct as long as the substrate concentrations are accurately known. A competitive inhibitor causes an effect only on *V*, not on *V/K*, so it is possible, for example, to use a D, L-deuterated substrate and a pure L unlabeled one to determine $^{D}(V/K)$ as long as an enzymatic analysis sensitive only to the L isomer is used to calibrate the substrate concentrations.

The *V* isotope effect, however, is not sensitive to the accuracy with which the substrate concentrations are known but is sensitive to the presence of inhibitors.

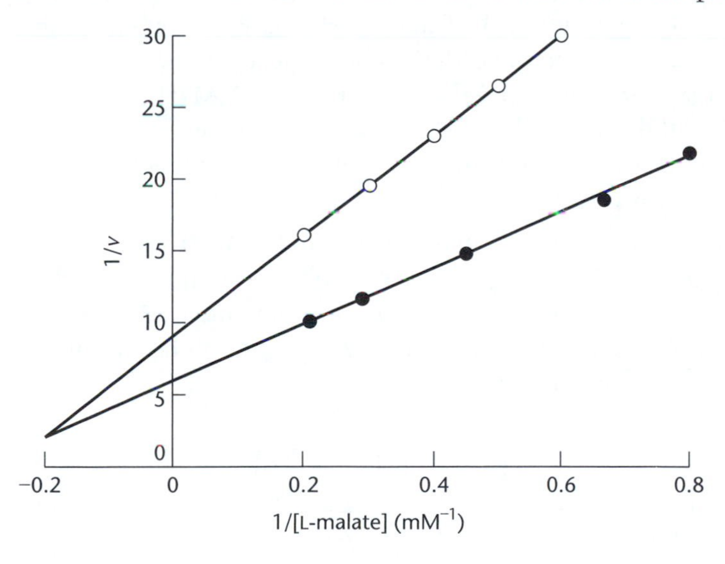

Figure 9-1. Deuterium isotope effects with malic enzyme: (○) L-malate-2-*d*; (•) L-malate-2-*h*. The concentration of NAD^+ was 1.2 mM and the experiment was performed at pH 9.6.

Thus use of a D, L-deuterated substrate and an L unlabeled one will give an apparent V isotope effect if the D isomer has affinity for the enzyme. This is described in equation 9-1 (B *19*, 3153)

$$\text{app}^{\mathrm{D}}V = \frac{V_{\mathrm{H}}\left(1 + \dfrac{r_{\mathrm{D}}K_{\mathrm{D}}}{K_{\mathrm{is}}}\right)}{V_{\mathrm{D}}\left(1 + \dfrac{r_{\mathrm{H}}K_{\mathrm{H}}}{K_{\mathrm{is}}}\right)} = {}^{\mathrm{D}}V\left[\frac{\left(\dfrac{K_{\mathrm{is}}}{K_{\mathrm{H}}}\right) + {}^{\mathrm{D}}(V/K)(r_{\mathrm{D}})}{\left(\dfrac{K_{\mathrm{is}}}{K_{\mathrm{H}}}\right) + r_{\mathrm{H}}}\right] \tag{9-1}$$

where the r values are ratios of inhibitor to substrate concentrations for deuterated and unlabeled substrate solutions, K_{is} is the slope inhibition constant of the inhibitor, and K_H and K_D are the apparent Michaelis constants of the unlabeled and deuterated substrates whose concentrations are varied.

One must be careful in picking which substrate concentration to vary when determining isotope effects in a two- or three-substrate reaction. Tritium isotope effects, as will be noted below, are ones on the apparent V/K for the labeled substrate, and if one wants to compare tritium and deuterium isotope effects, one must vary the concentrations of deuterated and unlabeled versions of this molecule with similar concentrations of the other substrates in order to make the comparison. If one is interested only in the deuterium isotope effects, however, one can vary the concentration of any substrate. For example, with lactate dehydrogenase one can determine the isotope effect on $V/K_{pyruvate}$ by varying the pyruvate concentration at fixed levels of either NADH or A-side deuterated NADH (NADD). Varying the concentration of the last substrate to add (as pyruvate here) gives the greatest expression of the isotope effect on the catalytic reaction, which is usually what is desired.

One must be careful with impurities in the NADD and NADH in such an experiment. The apparent ${}^{D}(V/K_{pyruvate})$ is sensitive to impurities in the nucleotide, since a competitive inhibitor decreases the amount of E–NADD or E–NADH present. The equation for apparent ${}^{D}(V/K_{pyruvate})$ will be similar to equation 9-1 and is given below as equation 9-2:

$$\text{app}\ {}^{\mathrm{D}}\!\left(\frac{V}{K_{\mathrm{pyruvate}}}\right) = \frac{{}^{\mathrm{D}}\!\left(\dfrac{V}{K_{\mathrm{pyruvate}}}\right)\left(r_{\mathrm{D}} + \dfrac{K_{\mathrm{is}}}{K_{\mathrm{ia}}}\right)}{\left(\dfrac{K_{\mathrm{is}}}{K_{\mathrm{ia}}}\right) + r_{\mathrm{H}}} \tag{9-2}$$

where K_{ia} is the dissociation constant of E–NADH. The apparent ${}^{D}V$ value is also affected by impurities in the dinucleotide solution, with equation 9-1 applying and

K_H and K_D being the K_m values of NADH and NADD, respectively. If pyruvate contains a competitive inhibitor, but the dinucleotides are pure, a correct value of $^D(V/K_{pyruvate})$ is obtained, but the apparent ^DV is given by equation 9-1 with $r_D = r_H$ and K_H and K_D as the K_m values of pyruvate with NADH and NADD, respectively.

Equilibrium Perturbation

The next most accurate method for determining an isotope effect is equilibrium perturbation. In this method a reaction mixture is set up at concentrations expected to be at equilibrium and with a single labeled reactant. Enzyme is then added and the reaction is followed until isotopic equilibrium is reached. The method was discovered by Michael Schimerlik using malic enzyme, which catalyzes the oxidative decarboxylation of malate to CO_2 and pyruvate with $NADP^+$ as the oxidant (B *14*, 5347). When 2-deuterated malate and unlabeled NADPH were used, the absorbance first decreased and then returned to near the starting point, as in Figure 9-2. The effect is caused by slower reaction of deuterated malate in the forward reaction than reaction of unlabeled NADPH in the reverse direction. After isotopic mixing is complete, the rates become equal and the reaction returns to the equilibrium position.

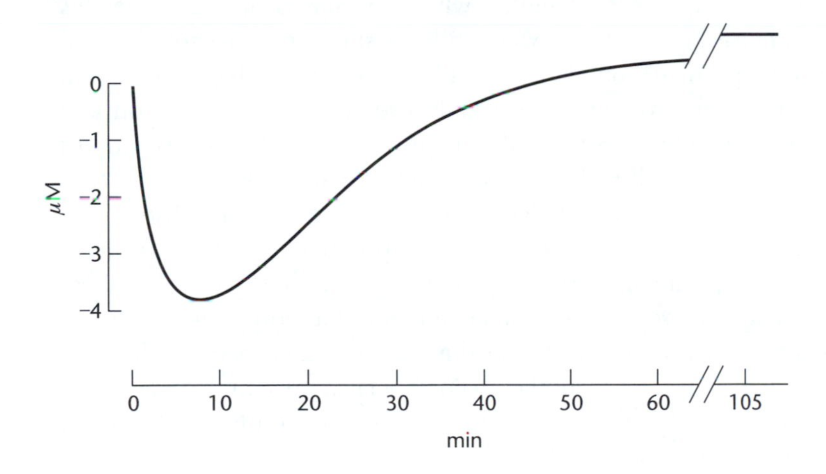

Figure 9-2. Equilibrium perturbation caused by 0.4 mM L-malate-2-*d* with malic enzyme. The concentration of the other perturbant, unlabeled NADPH, was 0.08 mM, while nonperturbant concentrations were NADP, 0.l mM; CO_2, 3.8 mM; and pyruvate, 3.8 mM. The experiment was carried out at pH 7.

If care is taken, this method can determine isotope effects as small as 1–2%, and it gives the same ^{13}C isotope effect of 3% on the malic enzyme reaction as that determined by the much more accurate isotope ratio method (see below). The reaction must be reversible, and the experiment must be carried out at chemical equilibrium. One also must have a suitable method for following the reaction, such as absorbance or circular dichroism. There are two methods for carrying out the experiment. In the two-pot method, two solutions are prepared, one containing the substrates and the other containing the products, with a labeled reactant in one or the other. Aliquots of these solutions are then mixed so that chemical equilibrium is expected, and enzyme is added. If the reaction does not return to the starting point, the mixing ratio of the two solutions is changed until a satisfactory perturbation returning to the starting point is observed.

In the one-pot method, a single solution containing all reactants is made up and enzyme is added to a portion of it. If the reaction does not return to the starting point, a small amount of one reactant is added so that the reaction will be at equilibrium and the experiment is repeated.

Excellent temperature control must be maintained in equilibrium perturbation experiments, as most enzymatic reactions have fairly high temperature coefficients. The cuvette should not be handled when enzyme is added, since this generates a pulse of heat in the quartz, which diffuses slowly into the solution and produces a spurious perturbation. Use an adder–mixer to add enzyme! As a control for this and other problems, one ideally uses the two-pot method and perturbs unlabeled substrates versus unlabeled products and labeled substrates versus labeled products as well as labeled substrates versus unlabeled products and unlabeled substrates versus labeled products. One should observe no perturbation in the first two cases and equal and opposite perturbations in the latter two. Doing the experiment in this way eliminates another insidious artifact caused by the presence in a reactant of a small amount of a more rapidly or more slowly reacting alternative substrate. Since this contaminant reaches chemical equilibrium at a different rate than the major species present, it can cause spurious perturbations. Carrying out the four combinations outlined above will detect this and other problems.

The equation that describes the size of the observed perturbation is as follows:

$$\frac{A_{\max} - A_0}{A'_0} = \alpha^{-1/(\alpha - 1)} - \alpha^{-\alpha/(\alpha - 1)} = (\alpha - 1)\alpha^{-\alpha/(\alpha - 1)} \qquad (9\text{-}3)$$

In equation 9-3, α is an apparent isotope effect that is related to the actual isotope effects in the two directions of the reaction by

$$^{D}(\text{Eq.P.})_{\text{labeled side}} = \frac{\alpha(^{D}K_{eq} + K)}{(1 + K)} \quad (9\text{-}4)$$

$$^{D}(\text{Eq.P.})_{\text{unlabeled side}} = \frac{\alpha(1 + K/^{D}K_{eq})}{(1 + K)} \quad (9\text{-}5)$$

where $K = \mathbf{P_{eq}/A_{eq}}$ for the unlabeled molecules and $^{D}K_{eq}$ is the equilibrium isotope effect for reaction in the direction of P formation. The initially labeled reactant is A and P is the unlabeled reactant, and these molecules, between which label is exchanged, are called the perturbants. Note that K in equations 9-4 and 9-5 is not equal to K_{eq} if there are reactants other than A and P in the reaction. In addition, the actual concentrations of A and P one uses in the experiment will differ from those determined by K, because $K_{eq\ H}$ and $K_{eq\ D}$ are usually not identical and when the experiment is complete partially deuterated perturbants are present on both sides of the reaction.

In equation 9-3, $A_{max} - A_0$ is the size of the perturbation measured from the average of the absorbance value at the beginning and end of the perturbation experiment (if they differ by $\leq 10\%$). A_0' is equal to the reciprocal of the sum of reciprocal concentrations of the two perturbants, with a correction added for low levels of any other reactants that are not perturbants (see ME *64*, 104 for these corrections).

For small isotope effects, α is approximately equal to $1 + 2.72(A_{max} - A_0)/A_0'$, so that a 1% perturbation corresponds to almost a 4% isotope effect. (The 2.72 in the expression for α is the value of e.) For larger isotope effects one must solve equation 9-3 for α, and this can be done either by a computer program (ME 64, 104) or by use of a table of values of the magnitude of the perturbation versus α (Cleland, 1976, page 274–278).

The time to reach the maximum perturbation is given by

$$t_{max} = \frac{\left[\frac{\alpha}{\alpha - 1}\right](\ln \alpha)}{k} \quad (9\text{-}6)$$

or at small values of α

$$t_{max} = \frac{(\alpha + 1)}{2k} \quad (9\text{-}7)$$

where k is the rate constant for final approach to equilibrium when all reactants are unlabeled. The value of t_{max} is thus only slightly greater than $1/k$, the time needed for an unlabeled reaction mixture to get to 63% of the equilibrium position. The level of enzyme should be adjusted to give the maximum perturbation between 2 and 10 min, since the reaction must be followed for at least $10t_{max}$ in order to determine the final equilibrium position.

The isotope effect measured by equilibrium perturbation is similar to one on *V*/*K* but is not always identical to it, and thus we write it as ^{D}(Eq.P.) rather than $^{D}(V/K)$. The equation for ^{D}(Eq.P.) has the same terms for the intrinsic isotope effect, the commitments, and the equilibrium isotope effect (see below), but the commitments are calculated for the perturbants rather than for the varied or labeled substrate and the first product released. This can lead to very different isotope effects when measured by equilibrium perturbation and by direct comparison, and this will be illustrated below.

Internal Competition

The most sensitive method for measuring isotope effects is the internal competition method. The method involves following changes in the isotope ratio between labeled and unlabeled molecules as the reaction proceeds. It is the method one has to use for ^{14}C or ^{3}H isotope effects, since these isotopes are not available or practical to use in carrier-free form. It is also the method used with the natural abundances of ^{13}C (1.1%), ^{15}N (0.37%), or ^{18}O (0.20%) as the label, with an isotope ratio mass spectrometer employed for detection. With stable isotopes one is not limited to trace labels, and it is possible, for example, to use mixtures of deuterated and unlabeled substrates with isotope ratios close to unity; the accuracy is then limited by the precision of the mass spectrometer used for analysis.

The accuracy of measuring isotope effects with ^{14}C or ^{3}H depends on the care and precision of counting, and thus it is difficult to determine changes in specific activities of less than 1%. The isotope ratio mass spectrometer, on the other hand, determines isotope effects with a precision of 0.02%. It is limited in practice to measuring mass ratios in CO_2, N_2, SO_2, or CH_3Cl, but this is sufficient for ^{13}C, ^{15}N, ^{34}S, or ^{37}Cl isotope effects. Most ^{18}O isotope effects have to be determined by the remote label method (see below).

Isotope effects determined by internal competition are on *V*/*K* for the labeled substrate. Considering a mixture of tritiated and unlabeled substrates, the rate equation for the unlabeled substrate is

$$v_{H} = \frac{V_{H}\mathbf{H}}{K_{H}\left(1 + \frac{\mathbf{T}}{K_{T}}\right) + \mathbf{H}} \tag{9-8}$$

where **H** and **T** represent the concentrations of the unlabeled and tritiated substrate, V_H and K_H are kinetic parameters for the unlabeled substrate, and K_T is the K_m for the tritiated substrate. It is clear that the tritiated species will act as a competitive inhibitor, and its inhibition constant will be its K_m (the definition of a K_m is the apparent dissociation constant in the steady state). A similar equation can be written for the tritiated substrate:

$$v_T = \frac{V_T\mathbf{T}}{K_T\left(1+\frac{\mathbf{H}}{K_H}\right)+\mathbf{T}} \tag{9-9}$$

where V_T is for the tritiated substrate. Now the unlabeled substrate acts as a competitive inhibitor. These equations can be rearranged by dividing each by the respective K_m value:

$$-\frac{d\mathbf{H}}{dt} = v_H = \frac{(V/K)_H\mathbf{H}}{\left(1+\frac{\mathbf{T}}{K_T}+\frac{\mathbf{H}}{K_H}\right)} \tag{9-10}$$

$$-\frac{d\mathbf{T}}{dt} = v_T = \frac{(V/K)_T\mathbf{T}}{\left(1+\frac{\mathbf{H}}{K_H}+\frac{\mathbf{T}}{K_T}\right)} \tag{9-11}$$

Note that the denominators of equations 9-10 and 9-11 are identical, as has to be the case for reactions taking place in the same solution (the denominator of a rate equation expresses the distribution of the enzyme among its possible forms). If equation 9-10 is divided by equation 9-11:

$$\frac{d\mathbf{H}}{d\mathbf{T}} = \frac{(V/K)_H\mathbf{H}}{(V/K)_T\mathbf{T}} \text{ or } \left(\frac{d\mathbf{T}}{\mathbf{T}}\right){}^{T}(V/K) = \frac{d\mathbf{H}}{\mathbf{H}} \tag{9-12}$$

which integrates to

$$\ln\left(\frac{\mathbf{T}}{\mathbf{T}_0}\right)^{T}(V/K) = \ln\left(\frac{\mathbf{H}}{\mathbf{H}_0}\right) \text{ or } {}^{T}(V/K) = \frac{\ln\left(\frac{\mathbf{H}}{\mathbf{H}_0}\right)}{\ln\left(\frac{\mathbf{T}}{\mathbf{T}_0}\right)} = \frac{\log\left(\frac{\mathbf{H}}{\mathbf{H}_0}\right)}{\log\left(\frac{\mathbf{T}}{\mathbf{T}_0}\right)} \tag{9-13}$$

since the ratio of natural logarithms or logarithms to the base 10 is the same. Equation 9-13 is valid regardless of whether the labeled compound is present in only a small amount (as is the case for tritium) or at a level similar to that of the unlabeled substrate.

If the labeled compound is present in low concentration, the fractional reaction is defined as

$$f = 1 - \frac{\mathbf{H}}{\mathbf{H}_0} = \frac{\mathbf{H}_0 - \mathbf{H}}{\mathbf{H}_0} \text{ and } 1 - f = \frac{\mathbf{H}}{\mathbf{H}_0} \tag{9-14}$$

The isotope ratio in the product at fractional reaction f is

$$R_\mathrm{p} = \frac{\mathbf{T}_0 - \mathbf{T}}{\mathbf{H}_0 - \mathbf{H}} \tag{9-15}$$

while the ratio in residual substrate at f is

$$R_\mathrm{s} = \frac{\mathbf{T}}{\mathbf{H}} \tag{9-16}$$

and the initial ratio in substrate is

$$R_0 = \frac{\mathbf{T}_0}{\mathbf{H}_0} \tag{9-17}$$

With these definitions, we now have

$$\frac{\mathbf{T}}{\mathbf{T}_0} = 1 - \frac{fR_\mathrm{p}}{R_0} = \frac{(1-f)R_\mathrm{s}}{R_0} \tag{9-18}$$

Substituting from equations 9-14 and 9-18 into equation 9-13 gives

$${}^{\mathrm{T}}(V/K) = \frac{\log(1-f)}{\log\left(1 - fR_\mathrm{p}/R_0\right)} = \frac{\log(1-f)}{\log\left[(1-f)(R_\mathrm{s}/R_0)\right]} \tag{9-19}$$

Thus the isotope effect is one on V/K, and it can be determined from the fractional reaction and the isotope ratios in initial substrate and either product or residual substrate at f. If isotope ratios are determined in both residual substrate and product, one obtains two independent determinations of the isotope effect, and if they agree, one has presumably a valid result. If it is difficult to measure isotope ratios in the substrate, one can convert it totally (this means 100%) to product and use this for determination of R_o (this is commonly done to circumvent problems with contaminants that copurify with the labeled substrate).

Determination of the V/K isotope effect by following the isotope ratio in residual substrate is practical only for isotope effects smaller than 2, since for large primary tritium isotope effects almost all of the tritium stays behind in the substrate. For example, at $f=0.5$, R_s/R_o is 1.87 for an isotope effect of 10, 1.92 for a value of 18, and 2.0 for an effect of infinity. Even for $f=0.9$, the values are 7.94, 8.8, and 10.

For small isotope effects, however, one can readily use either R_p or R_s values to determine the isotope effect. If only R_p values are used, f values of 0.2–0.3 are best, while if R_s values are used, 0.5–0.7 is best. If both are being determined, an f value of 0.5 is a good compromise. Figure 9-3 shows the variation of isotope ratios with f values during a reaction. It is important to determine f accurately, as the error in the isotope effect often is determined by the error in the f value, rather than in the isotope ratios, especially if the latter are determined with the isotope ratio mass spectrometer.

Use of the internal competition method requires that a reaction go to completion and not be reversible, since reversal lowers the apparent isotopic discrimination. The usual approach is to use coupling enzymes to remove a product other than the one being determined so that reversal cannot occur. If there is only one product, one can remove it by coupling enzymes and analyze the eventual form for its mass ratio, or one can use R_s values for determining the isotope effect after

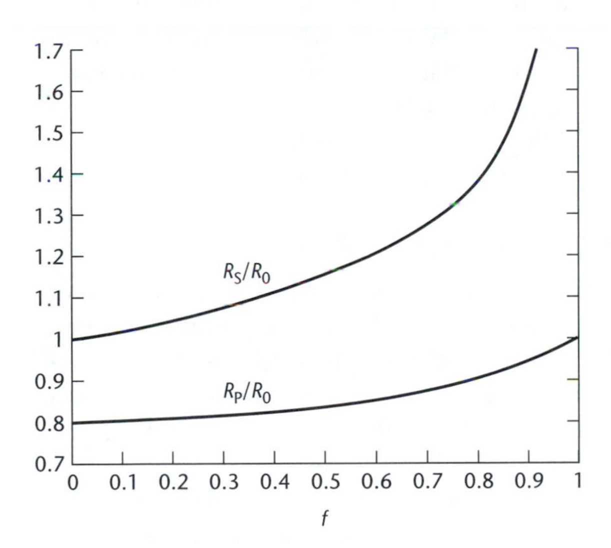

Figure 9-3. Observed isotope ratios in substrate (R_s) and product (R_p) as a function of fractional reaction f and the initial isotope ratio in the substrate (R_0). A $^D(V/K)$ value of 1.25 is assumed.

the product is coupled away. Equations for a reversible case have been published by Cleland (CRB *13*, 385).

If comparable concentrations of labeled and unlabeled substrates are used in an internal competition experiment, such as for a deuterium isotope effect, equation 9-19 becomes more complicated. Now if X_H and X_D are initial mole fractions of unlabeled and deuterated substrates:

$$f = \frac{\mathbf{H}_0 + \mathbf{D}_0 - \mathbf{H} - \mathbf{D}}{\mathbf{H}_0 + \mathbf{D}_0} \quad R_s = \frac{\mathbf{D}}{\mathbf{H}} \tag{9-20}$$

$$R_0 = \frac{\mathbf{D}_0}{\mathbf{H}_0} = \frac{X_D}{X_H} \quad R_p = \frac{\mathbf{D}_0 - \mathbf{D}}{\mathbf{H}_0 - \mathbf{H}} \tag{9-21}$$

$$^{D}(V/K) = \frac{\log\left[1 - \frac{f}{\left(X_H + \frac{X_D R_p}{R_0}\right)}\right]}{\log\left[1 - \frac{f}{\left(X_D + \frac{X_H R_0}{R_p}\right)}\right]} = \frac{\log\left[\frac{(1-f)}{\left(X_H + \frac{X_D R_s}{R_0}\right)}\right]}{\log\left[\frac{(1-f)}{\left(X_D + \frac{X_H R_0}{R_s}\right)}\right]} \tag{9-22}$$

This approach would be useful when the deuterated substrate also contains a ^3H label and the nondeuterated substrate contains ^{14}C, with the $^3H/^{14}C$ ratio in product or residual substrate versus initial substrate used to measure the discrimination. In order to get sufficient ^3H labeling it may be necessary to have D_0 and H_0 be nearly equal. If isotope effects are less than 1% (for example, ^{37}Cl ones, where R_0 is ~0.32), equations 9-19 and 9-22 give the same answer. But for an isotope effect of 2.4 with $X_H = X_D$ and $f = 0.375$, the use of R_p/R_0 in equation 9-19 gives 2.26 and the use of R_s/R_0 gives 7.28!

A novel method has been developed by Singleton for the measurement of small isotope effects using the natural abundance of heavy atoms in the substrate (JACS *117*, 9357). The method makes use of NMR to measure the increase in heavy atom in the substrate as the reaction nears completion (Figure 9-3). By ^2H and ^{13}C NMR, the isotope effects at every hydrogen and carbon in isoprene were measured in the Diels-Alder reaction of isoprene and maleic anhydride, after the isoprene remaining in a reaction taken to 98.9% completion was isolated.

Remote Label Method. As noted above, the isotope ratio mass spectrometer accepts in practice only small molecular weight gases such as CO_2 or N_2. If there is only a single nitrogen in a molecule, the sample can be sealed in a quartz tube with CuO and heated to 850 °C, which converts everything to CO_2, H_2O, and N_2. After cooling, the tube is cracked on a vacuum line and the N_2 is separated for analysis. Ammonia is readily isolated by steam distillation and converted by alkaline hypobromite to N_2. The Kjeldahl method will produce ammonia from amino acids and amides but is less satisfactory in practice than the CuO oxidation.

^{13}C isotope effects on decarboxylation reactions are readily studied by isolating the CO_2 on a vacuum line, and in some cases other carbon atoms are readily converted to CO_2. Thus C-1 of glucose can be converted to CO_2 by the combined action of hexokinase, glucose-6-P dehydrogenase, and 6-P-gluconate dehydrogenase. Periodate converts carboxyl groups of molecules like glycerate to CO_2. I_2 in dimethyl sulfoxide converts anhydrous formate or oxalate salts to CO_2 and in this case one can determine both ^{13}C and ^{18}O isotope effects, since the absence of water means there is no oxygen exchange with the solvent.

For most ^{18}O isotope effects, however, and for many ^{13}C or ^{15}N isotope effects, no suitable degradation is available and one has to use the remote label method. We will illustrate it with measurement of isotope effects on reactions of *p*-nitrophenyl acetate (pNPA) (JACS *116*, 5045):

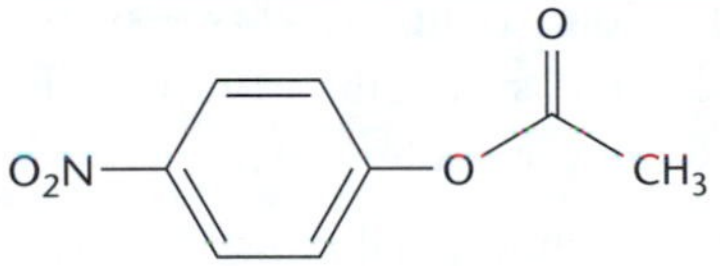

To determine the primary ^{18}O isotope effect in the phenolic oxygen, two samples of pNPA are prepared. One contains ^{18}O in the phenolic oxygen and ^{15}N in the nitro group, both at as high a degree of labeling as possible. The second contains ^{14}N (nitrogen depleted in ^{15}N, which can be purchased as $^{14}NH_4{}^{14}NO_3$), but no other labels. The two samples are mixed to give the ^{15}N natural abundance ratio so that the nitro group contains 0.37% ^{15}N, but every ^{15}N is accompanied by an ^{18}O in the phenolic oxygen. The reaction (hydrolysis or reaction with another nucleophile) is carried out and the *p*-nitrophenol released, as well as the residual substrate, is separated and oxidized by CuO as described above to give N_2 for analysis in the isotope ratio mass spectrometer. A sample of the initial substrate is oxidized in the same way to give the R_0 value. The isotope ratio mass spectrometer is designed to compare mass ratios of unknowns relative to a tank standard that contains the natural abundance of the label. Operating near natural abundance also minimizes the effect of contamination from air.

The isotope effect determined in this way is the product of isotopic discrimination caused by ^{18}O and ^{15}N substitution. The experiment is then repeated with pNPA containing only natural abundance of isotopes, and this determines the ^{15}N isotope effect alone. The $^{18}O,^{15}N$ value is divided by the ^{15}N one to give the desired ^{18}O isotope effect. To determine the isotope effects in other positions of pNPA, the double-labeled material contained, in addition to ^{15}N in the nitro group, ^{18}O in the carbonyl group or ^{13}C at C-1 of acetate or three deuteriums in the methyl group of acetate. In each case the double-labeled material was mixed with ^{14}N-labeled pNPA to give material with 0.37% ^{15}N in the nitro group. Thus it was possible to determine five isotope effects on reactions of pNPA (primary and secondary ^{18}O, primary ^{13}C, β-secondary deuterium, and secondary ^{15}N). In this case there is an isotope effect in the nitro group, as the ^{15}N value provides an estimate of the degree of electron delocalization into the nitro group in the transition state (JACS *116*, 5045).

There are a number of convenient remote labels. Nitro groups are readily inserted into phenols or into *m*-nitrobenzyl alcohol. Amide nitrogens are useful remote labels, as is the exocyclic amino group of adenine, since it is readily generated from ammonia and chloropurine riboside and removed with adenosine deaminase. This permits remote labeling of ATP, NAD, and similar molecules. C-1 of glucose can be used, since either ^{13}C or ^{12}C can be introduced by synthesis into this position, and this atom as noted above is readily converted to CO_2. In practice one can usually measure any isotope effect in any position of a molecule by the use of a suitable remote label.

The degree of labeling in a remote labeling experiment will not be 100%, and certain corrections need to be made to the observed isotope effects when this method is used. The equation for calculating the isotope effect in the discriminating position (for a single heavy atom) is

$$^{j}(V/K) \text{ per atom} = 1 + \frac{(S-1)}{\left[1 - \frac{S(1-y)}{i}\right]} \tag{9-23}$$

where

$$S = \sqrt[i]{\frac{\frac{P}{R}}{1 - Q\left(\frac{P}{R} - 1\right)}} \tag{9-24}$$

In equations 9-23 and 9-24, i equals the number of discriminating atoms (1, 2, or 3); j reflects the nature of the discriminating atoms (^{15}N, ^{18}O, etc.); q represents the nature of remote label (^{15}N, ^{18}O); P is equal to $^{q,j}(V/K)$, the observed isotope effect with remote label [for example, $^{15,18,18}(V/K)$]; R is the isotope effect in the remote-labeled position determined with natural-abundance material; Q is equal to $(1-b)z/bx \approx z/b$, the degree to which light material in the remote-labeled mixture is depleted below natural abundance; b is the fraction of double-labeled material in the remote labeled mixture; z is the fraction of heavy label present in the remote-labeled position of light material used for the remote-labeled mixture, x is the fraction of heavy label in the remote labeled position of the double-labeled material used for mixing, and y is the fraction of heavy discriminating label in the double-labeled material (or if two or three discriminating atoms are present, the fraction containing *all* atoms labeled, for example, 90% labeling with $i=2$ gives $y=0.81$).

For the details of how one carries out studies with the isotope ratio mass spectrometer, see Weiss, 1991 (in Cook 1991, page 291–311) and JACS *116*, 5045.

Types of Isotope Effects

Equilibrium Isotope Effects. When an isotopic atom is more stiffly bonded in substrate than in product, or vice versa, there will be an equilibrium isotope effect on the reaction. The value will be normal, or greater than unity, when the substrate is more stiffly bonded and inverse, or less than unity, when the product is more stiffly bonded. Thus $^{D}K_{eq\ forward} = 1/^{D}K_{eq\ reverse}$ where $^{D}K_{eq} = K_{eq\ H}/K_{eq\ D}$. Equilibrium isotope effects are readily measured for freely reversible reactions, either by direct measurement of K_{eqH} and K_{eqD} or by measuring mass ratios in reactant and product at equilibrium and taking their ratios (used for ^{13}C and ^{15}N equilibrium isotope effects). The ^{18}O isotope effect on the ionization of any carboxyl, phosphoryl, phenol, or alcohol group can be measured via NMR by following the separation of ^{13}C or ^{31}P signals during titration of a 50/50 mix of ^{16}O- and ^{18}O-containing molecules. If the ^{16}O pK is lower than the ^{18}O one, the small upfield shift of the ^{18}O species will increase to a maximum at the average of the pKs and then decrease, and the magnitude of the changes allows calculation of the isotope effect on the pK (JCSCC 745). A similar approach can determine the ^{15}N isotope effect on the ionization of an amino group.

For accurate direct comparison of K_{eqH} and K_{eqD}, one makes up reaction mixtures with either deuterated or unlabeled reactants calculated to be close to equilibrium and varies the concentration of one reactant to bracket the equilibrium position. One then adds enzyme and measures how far the reaction goes to reach equilibrium (easily done for pyridine nucleotide-dependent dehydrogenases, for example, by measuring absorbance at 340 nm). One then plots absorbance versus [product]/[reactant] ratio and estimates K_{eq} as the point where the line crosses zero.

It is possible to calculate equilibrium isotope effects from IR and Raman spectra if one has a full force field for substrate and product. Since the stiffness of bonding is determined by local stretching, bending, and torsional vibrations involving the isotopic atom, the effects are local and thus one can calculate equilibrium isotope effects by reference to a table of fractionation factors, which are equilibrium isotope effects relative to a standard (H_2O for deuterium, tritium, or ^{18}O; aqueous CO_2 for ^{13}C and ^{14}C; aqueous NH_3 for ^{15}N).

For example, to estimate the $^{D}K_{eq}$ for the hydrogen transferred in a reaction of a secondary alcohol with NAD^+ to give NADH, one takes the ratio of the fractionation factors in Table 9-1 for a secondary alcohol (1.16) and NADH (0.98) to give a value of 1.18. Tritium equilibrium isotope effects are obtained by raising the deuterium isotope effect to the 1.442 power. The fractionation factor of [H]DO is higher than that of [H]HO, so that if one is comparing equilibrium constants in H_2O and D_2O for reactions such as that catalyzed by fumarase, for which water is a product, the apparent K_{eq} in D_2O must be multiplied by 0.945 since K_{eq} for the reaction $D_2O + H_2O \rightleftharpoons 2HDO$ is 3.78, not 4.00 as one would expect. In mixed H_2O–D_2O solvents, one must use an intermediate value for this parameter. The values in the table predict secondary as well as primary equilibrium isotope effects. Thus, 0.87 for NAD^+ divided by 0.98 for NADH gives a value of 0.89, the $^{D}K_{eq}$ observed experimentally for the hydrogen at C4 of the nicotinamide ring of NAD^+ in dehydrogenase reactions.

Deuterium fractionation factors are mainly sensitive to atoms bonded to the carbon bonded to the isotopic hydrogen, and replacement of another H on this same carbon by S, C, N, or O increases the fractionation factor by 1.03, 1.10, 1.15, or 1.18. However, due to hyperconjugation, a CH_2 or CH_3 group adjacent to a carbonyl carbon has a fractionation factor 5% less than if the carbonyl group were not present.

Table 9-1 includes a number of useful fractionation factors. For more complete lists, see ZN *44a*, 337, and CJC 77, 967 for ^{13}C, ^{15}N, and ^{18}O fractionation factors. For ^{14}C equilibrium isotope effects, the [^{13}C equilibrium isotope effect minus 1] is multiplied by 1.9 and added to 1. The CJC article also includes rules for estimating ^{13}C fractionation factors other than those included in the table.

Table 9-1. Fractionation Factors in Aqueous Solution

Structure	Examples (position)	Value
	(Deuterium Fractionation Factors are Relative to H_2O) Nonexchangeable	
H-C≡C-[H]	acetylene	0.64
$(-COPO_3^{2-})$=CH[H]	phosphoenolpyruvate (C3)	0.81
[H]COO^-	formate	0.80
-CO-CH_2[H]	pyruvate	0.84
C-C[H]=C	NAD^+ (C4), Fumarate	0.87
C-CH_2-[H]	lactate (C3)	0.88
-C[H]=O	acetaldehyde	0.83
-CO-CH[H]-C	α-ketoglutarate (C3)	0.93
C-CH[H]-C	malate (C3), NADH (C4)	0.98
-CO-CH[H]OH	dihydroxyacetone-P (unhydrated)	0.99
C-(C[H]S-C)-C	dUMP adduct to thymidylate synthase	1.01
C-C[H]OH	ethanol	1.04
-CO-C[H]OH-C	keto sugars	1.10
C-C[H]NH_3^+-C	amino acids	1.13
C-C[H]OH-C	malate (C2), 2-propanol (C2)	1.16
C-(C[H]-NH_2^+)-C	proline	1.17
C-C[H]$OPO_3^=$-C	2-phosphoglycerate (C2)	1.19
C-C[H]-$(OH)_2$	aldehyde hydrates (C1)	1.14
C-C[H]-(OH,OC)	sugar hemiacetals (C1)	1.24
	Exchangeable	
$H_2O[H]^+$	hydrated proton	0.69
$H_aO(H_b\text{-}OH_c)_3^-$	hydroxide overall	0.43-0.48
	H_a	~1.25
	H_b	~0.70
	H_c	1.00
-CH_2-S[H]	thiols	0.45
	alcohols, acids, amines	0.9-1.1
(C_2-COH)-O[H]	hydrated aldehydes	1.23
(-C-O···[H]···O-C-)	low barrier H-bond	0.3-0.5
(-C-O···[H]···N-C-)	low barrier H-bond	0.4-0.5
	^{13}C Fractionation Factors Relative to Aqueous CO_2	
[C]O_2	CO_2 gas	1.001
[C]O_2(OH)	bicarbonate	1.009

(Continued)

Table 9-1. Continued

Structure	Examples (position)	Value
	^{13}C Fractionation Factors Relative to Aqueous CO_2 (Cont.)	
$[C]O_3^{=}$	carbonate	1.008
NH_2-$[C]O_2^-$	carbamate	1.011
NH_2-[C]=O-$OPO_3^{=}$	carbamoyl phosphate	1.013
C-[C]-OO^-	oxaloacetate (C4)	1.0075
	isocitrate (center carbon)	1.003
	6-phosphogluconate (C1)	0.996
	malate (C4)	0.988
$[C]H_4$	methane	0.944
CH_3-$[C]H_3$	ethane	0.956
CH_2=$[C]H_2$	ethylene	0.954
C-CH_2-$[C]H_3$	propane (C1)	0.960
C-$[C]H_2$-CH_3	propane (C2)	0.966
$[C]H_3OH$	methanol	0.963
$[C]H_3NH_2$	methylamine	0.955
$[C]H_3NH_3^+$	methylamine, protonated	0.959
	^{15}N Fractionation Factors Relative to Aqueous NH_3	
$[N]H_3$	NH_3 gas	0.994
$[N]H_4^+$	ammonium ion	1.0192
C-(CH$[N]H_2$)-C	amino acid, unprotonated	1.022
C-(CH$[N]H_3^+$)-C	amino acid, protonated	1.0333
C-$[N]H_2^+$-C	secondary amine, protonated	1.038
C-(C=NH_2^+)-C	imine, protonated	1.0192
C-C=O-$[N]H_2$	amide	1.005
	^{18}O Fractionation Factors Relative to H_2O	
$C[O]_2$	CO_2 gas	1.0412
$H_2[O]$	H_2O gas	0.991
$H_3[O]^+$	hydrated Proton	1.023
$[O]H^-$	hydroxide	0.960
C-CH[O]H-C	malate	1.033
C-C=[O]-C	α-ketoglutarate	1.031
$HC[O]_2H$	formic acid	1.0345
$HC[O]_2^-$	formate	1.0124
$CH_3[O]H$	methanol	1.021
C-$CH_2[O]H$	ethanol	1.029
CH_3-[O]-CH_3	diethyl ether	1.043

For example, replacing H on methane with C or O raises the fractionation factor by 1.011 or 1.019, while replacing H with NH_2 or NH_3^+ raises it by 1.012 or 1.016. With larger molecules, the effects are smaller because the changes make up less of the overall stiffness of bonding. Thus, the factor for replacing H with C drops from 1.013 to 1.010 when a C of ethane becomes the center carbon of propane. The CJC article also has a list of known $^{13}K_{eq}$ values for various reactions. Of particular interest is the value of 1.001 for ionization of a carboxyl group.

Similar effects are seen with ^{15}N fractionation factors. Thus, the fractionation factor of NH_4^+ is 1.0192, while that for a protonated amino group is larger than that for the unprotonated amine by 1.0167. Replacing H with C to give a protonated secondary amine causes a further increase of only 1.0045. With amines, deuteration raises ^{15}N fractionation factors, with the effect being 0.4–0.5% for each extra deuterium on one side of the equation. Thus, the fractionation factor for NH_4^+ relative to NH_3 is 1.0192, while the $^{15}K_{eq}$ for ND_4^+ going to ND_3 in D_2O is 1.0246. The deprotonation of pyridinium ion has a $^{15}K_{eq}$ of 1.0211 in water but 1.025 in D_2O. A protonated imine ($=NH_2^+$) has the same fractionation factor as an ammonium ion, but in D_2O the value is 1% lower than that of ND_4^+, since there are two extra deuteriums in the latter molecule.

Not many ^{18}O fractionation factors have been determined. However, calculated values for small molecules suggest empirical rules for determining other fractionation factors. Thus, replacing H with CH_3 on the isotopic oxygen raises the fractionation factor by 1.022, while replacing H with CH_3 on a carbon bonded to the oxygen raises it by 1.008, and a second such replacement raises it by an additional 1.004. The $^{18}K_{eq}$ values for dissociation of various molecules are 2% (the value of the isotope effect minus 1 times 100) for a carboxyl group, 1.5% for a phenol, 1.9% for phosphate, and 1.46% for a phosphate monoester. In D_2O these values are multiplied by 1.006.

Equations for Isotope Effects. Application of isotope effects to the study of uncatalyzed and enzyme-catalyzed reactions differs in one major respect. In the case of the uncatalyzed reaction, the chemical (isotope-sensitive) step is usually rate-limiting (mechanism 9-25), while in the case of the enzyme-catalyzed reaction, the isotope-sensitive step is seldom the only rate-limiting step along the reaction pathway (mechanism 9-26).

$$A \xrightarrow{k_{chem}} B \tag{9-25}$$

The physical steps, substrate binding and enzyme conformation changes, often limit or contribute to rate limitation overall:

$$\mathrm{E} \underset{k_2}{\overset{k_1\mathbf{A}}{\rightleftharpoons}} \mathrm{EA} \underset{k_4}{\overset{k_3\mathbf{B}}{\rightleftharpoons}} \mathrm{EAB} \underset{k_6}{\overset{k_5}{\rightleftharpoons}} \mathrm{E'AB} \underset{k_8}{\overset{k_7}{\rightleftharpoons}} \mathrm{E'PQ} \underset{k_{10}}{\overset{k_9}{\rightleftharpoons}} \mathrm{EPQ} \xrightarrow{k_{11}} \mathrm{EQ} \xrightarrow{k_{13}} \mathrm{E} \quad (9\text{-}26)$$

Thus, although the measured isotope effect, ^{D}k, can be directly interpreted in terms of transition-state structure for the uncatalyzed reaction, the rate expressions for enzyme-catalyzed reactions are usually much more complex, with k_7 and k_8 isotope-sensitive. This may preclude interpretation of the observed isotope effects in terms of transition-state structure, unless the observed isotope effect is equal to the intrinsic isotope effect or the intrinsic isotope effect ($^{D}k_7$ or $^{D}k_8$) can be calculated. However, the observed isotope effect in an enzyme-catalyzed reaction will depend on which substrate is varied and on the kind of kinetic mechanism. Application of isotope effects to determination of kinetic mechanism is based on the theory presented below.

The isotope effect for V/K_b in mechanism 9-26 is

$$^{D}\left(\frac{V}{K_b}\right) = \frac{^{D}k_7 + c_f + c_r{}^{D}K_{eq}}{1 + c_f + c_r} \quad (9\text{-}27)$$

where $c_f = (k_7/k_6)(1 + k_5/k_4)$ and $c_r = (k_8/k_9)(1 + k_{10}/k_{11})$. The constants c_f and c_r are called commitments and are the ratio of the rate constant for the catalytic step to the net rate constant for release of a reactant from the enzyme. For isotope effects determined by direct comparison of initial velocities, c_f is for the varied substrate, even if it is not the labeled one. For isotope effects determined by the internal competition method, c_f is for the labeled substrate. For both types of experiment, c_r is for the first irreversible step, normally release of the first product, although if that product is present it will be for a later step. For isotope effects determined by equilibrium perturbation, c_f and c_r are calculated for the perturbants between which the label is exchanged. The difference in the isotope effects obtained by these two methods is illustrated for isocitrate dehydrogenase. $^{D}(V/K_{isocitrate})$ is unity since isocitrate is very sticky and c_f is large, while c_r is for CO_2 and is very small. But $^{D}(\mathrm{Eq.P.})_{isocitrate}$ is 1.15, since c_r for NADPH is even larger than c_f for isocitrate, and one sees $^{D}K_{eq}$ as the observed isotope effect (B *20*, 1797).

The isotope effect on V in mechanism 9-26 is

$$^{D}V = \frac{^{D}k_7 + c_{Vf} + c_r{}^{D}K_{eq}}{1 + c_{Vf} + c_r} \tag{9-28}$$

where c_r is the same as that for $^{D}(V/K)$ above, but c_{Vf} is given by equation 9-29:

$$c_{Vf} = \left[\frac{k_5 k_7}{k_5 + k_6}\right]\left[\frac{1}{k_5} + \left(\frac{1}{k_9}\right)\left(1 + \frac{k_{10}}{k_{11}}\right) + \frac{1}{k_{11}} + \frac{1}{k_{13}}\right] \tag{9-29}$$

c_{Vf} is the sum of the ratios of the rate constant for the isotope-sensitive step (multiplied by the proportion of enzyme in the EAB* form at equilibrium prior to this step) to each unimolecular net rate constant that may limit $V/\mathbf{E_t}$. This includes steps prior to the catalytic step that form EAB* (k_5), as well as steps after it [$k_9 k_{11}/(k_{10} + k_{11})$, k_{11}, and k_{13}].

Calculation of Dissociation Constant. The Michaelis constant for a reaction is the apparent dissociation constant in the steady state and is usually not equal to the thermodynamic dissociation constant, which represents binding at equilibrium. A useful comparison of deuterium isotope effects on V and V/K can be used to compute the dissociation constant for a substrate if there is only one isotope-sensitive step in the mechanism (only deuterium isotope effects will be large enough to measure the isotope effect on V by direct comparison experiments) (JACS *107*, 1058).

To make this comparison we subtract 1 from each side of equations 9-27 and 9-28:

$$^{D}\left(\frac{V}{K_b}\right) - 1 = \frac{^{D}k_7 - 1 + c_r(^{D}K_{eq} - 1)}{1 + c_f + c_r} \tag{9-30}$$

$$^{D}V - 1 = \frac{^{D}k_7 - 1 + c_r(^{D}K_{eq} - 1)}{1 + c_{Vf} + c_r} \tag{9-31}$$

Note that the numerators are now the same. If we take the ratio of equations 9-31 and 9-30:

$$\frac{(^{D}V - 1)}{^{D}(V/K_b) - 1} = \frac{1 + c_f + c_r}{1 + c_{Vf} + c_r} \tag{9-32}$$

In mechanism 9-26, expressions for $V/\mathbf{E_t}$ and $V/K_b\mathbf{E_t}$ in terms of commitment factors are given in equations 9-33 and 9-34:

$$\frac{V}{\mathbf{E_t}} = \frac{\left(\dfrac{k_5 k_7}{k_5 + k_6}\right)}{1 + c_{Vf} + c_r} \tag{9-33}$$

$$\frac{V}{K_b \mathbf{E_t}} = \frac{\left(\dfrac{k_3 k_5 k_7}{k_4 k_6}\right)}{1 + c_f + c_r} \tag{9-34}$$

Then

$$K_b = \frac{V/\mathbf{E_t}}{V/K_b\mathbf{E_t}} = \frac{\left(\dfrac{k_4 k_6}{k_3(k_5 + k_6)}\right)(1 + c_f + c_r)}{1 + c_{Vf} + c_r} \tag{9-35}$$

and

$$\frac{(1 + c_f + c_r)}{(1 + c_{Vf} + c_r)} = K_b\left(\frac{k_3(k_5 + k_6)}{k_4 k_6}\right) \tag{9-36}$$

But the dissociation constant of B from EAB and E′AB is

$$K_d = \frac{k_4 k_6}{k_3(k_5 + k_6)} \tag{9-37}$$

Thus from equations 9-32, 9-36, and 9-37 we have

$$\frac{{}^{D}V - 1}{{}^{D}(V/K_b) - 1} = \frac{K_b}{K_d} \tag{9-38}$$

The difference between K_b and K_d is determined by the differences in c_f and c_{Vf}. When k_4 and/or k_6 are small (a sticky substrate that reacts to give products faster than it dissociates), c_f can be larger than c_{Vf}, ${}^{D}V$ larger than ${}^{D}(V/K_b)$, and K_b larger than K_d. On the other hand, if a step following the isotope-sensitive one is slow, c_{Vf} can be larger than c_f, ${}^{D}(V/K_b)$ larger than ${}^{D}V$, and K_b less than K_d. This is common with dehydrogenases, where release of the second product, the nucleotide, limits $V/\mathbf{E_t}$. K_b is smaller than K_d, since only a small degree of saturation with the substrate suffices to keep the early part of the mechanism faster than the later rate-limiting step.

Isotope Effects on More Than One Step. The equations given above for isotope effects on V/K and V assume that only one step in the reaction is isotope-sensitive. But there are cases where more than one step will show an isotope effect. One example is where there is a binding isotope effect (see below). When an amino acid is a substrate for an aminotransferase, there will be a primary ^{15}N isotope effect on the initial transimination and on the final C–N cleavage, as well as secondary ^{15}N isotope effects on other steps as the fractionation factor of the nitrogen changes. In these cases one must consider the possibility of isotope effects on all steps of the mechanism.

$$\mathrm{E} \underset{k_2}{\overset{k_1\mathbf{A}}{\rightleftharpoons}} \mathrm{EA} \underset{k_4}{\overset{k_3}{\rightleftharpoons}} \mathrm{EA}' \underset{k_6}{\overset{k_5}{\rightleftharpoons}} \mathrm{EA}'' \underset{k_8}{\overset{k_7}{\rightleftharpoons}} \mathrm{EP} \xrightarrow{k_9} \mathrm{E} + \mathrm{P} \tag{9-39}$$

V/K for mechanism 9-39 can be written

$$\frac{V}{K\mathbf{E_t}} = \frac{\dfrac{k_1k_3k_5k_7k_9}{k_2k_4k_6k_8}}{1 + \left(\dfrac{k_9}{k_8}\right)\left[1 + \left(\dfrac{k_7}{k_6}\right)\left(1 + \left(\dfrac{k_5}{k_4}\right)\left(1 + \dfrac{k_3}{k_2}\right)\right)\right]} \tag{9-40}$$

If an isotope effect on each rate constant is assumed:

$$^{x}\left(\frac{V}{K}\right) = \frac{\left\{\begin{array}{l} {}^{x}K_{eq1}{}^{x}K_{eq3}{}^{x}K_{eq5}{}^{x}K_{eq7}{}^{x}k_9 + \left(\dfrac{k_9}{k_8}\right){}^{x}K_{eq1}{}^{x}K_{eq3}{}^{x}K_{eq5}{}^{x}k_7 \\ \quad + \left(\dfrac{k_7k_9}{k_6k_8}\right){}^{x}K_{eq1}{}^{x}K_{eq3}{}^{x}k_5 + \left(\dfrac{k_5k_7k_9}{k_4k_6k_8}\right){}^{x}K_{eq1}{}^{x}k_3 \\ \quad + \left(\dfrac{k_3k_5k_7k_9}{k_2k_4k_6k_8}\right){}^{x}k_1 \end{array}\right\}}{1 + \left(\dfrac{k_9}{k_8}\right)\left[1 + \left(\dfrac{k_7}{k_6}\right)\left(1 + \left(\dfrac{k_5}{k_4}\right)\left(1 + \dfrac{k_3}{k_2}\right)\right)\right]} \tag{9-41}$$

where x defines the nature of the isotope effect (15 for ^{15}N; D for deuterium, etc.). In equation 9-41 each numerator term contains a kinetic isotope effect multiplied by the equilibrium isotope effects for each preceding step in the mechanism. The pattern can be extended for reactions of any numbers of steps.

Determination of Intrinsic Isotope Effects

It is clear that unless commitments are small and the isotope-sensitive step is rate-limiting, one needs a way to determine the intrinsic isotope effect in order for observed isotope effects to be useful. Two methods have been used for this. The first involves comparison of deuterium and tritium isotope effects on *V*/*K*. The second uses the effects of deuteration on ^{13}C or other heavy atom isotope effects to solve for all intrinsic isotope effects and commitments in the system.

Northrop's Method

This method (B *14*, 2644) assumes that either there is no reverse commitment or the equilibrium isotope effect is unity. Then one defines a single commitment or sum of commitments as *c* and

$$^{D}\left(\frac{V}{K}\right) = \frac{^{D}k + c}{1 + c} \quad \text{and} \quad ^{T}\left(\frac{V}{K}\right) = \frac{^{T}k + c}{1 + c} \tag{9-42}$$

Subtracting 1 from both sides of both equations gives

$$^{D}\left(\frac{V}{K}\right) - 1 = \frac{^{D}k - 1}{1 + c} \quad \text{and} \quad ^{T}\left(\frac{V}{K}\right) - 1 = \frac{^{T}k - 1}{1 + c} \tag{9-43}$$

and the ratio of the two is thus

$$\frac{^{D}(V/K) - 1}{^{T}(V/K) - 1} = \frac{^{D}k - 1}{^{T}k - 1} \tag{9-44}$$

Since $^{T}k = (^{D}k)^{1.44}$ (JACS *80*, 5885), one can rewrite equation 9-44 as

$$\frac{^{D}(V/K) - 1}{^{T}(V/K) - 1} = \frac{^{D}k - 1}{(^{D}k)^{1.44} - 1} \tag{9-45}$$

Equation 9-45 has experimental quantities on the left side and only one unknown on the right side. Solution of this equation requires either a computer program or reference to tables of $[^{D}(V/K) - 1]/[^{T}(V/K) - 1]$ versus ^{D}k (Cleland, 1977, page 280–283). Northrop's method depends on the tritium isotope effect being equal to the deuterium isotope effect raised to the 1.44 power. This is fairly accurate for intrinsic isotope effects greater than 1.5 or less than 0.7, but for values between 0.8 and 1.2 the relationship does not hold in many cases (JACS *127*, 3294).

If there is a finite reverse commitment and $^{D}K_{eq}$ is not unity, equation 9-45 will not give an exact answer. Approximate answers are then obtained by dividing $^{D}(V/K)$ by $^{D}K_{eq}$ and $^{T}(V/K)$ by $^{T}K_{eq}$ [which is $(^{D}K_{eq})^{1.44}$] to give the isotope effects in the reverse direction. When these are used in equation 9-45, one obtains a value for ${}^{D}k_{rev}$. The true value of ${}^{D}k_{for}$ then lies between the one given by equation 9-45 with isotope effects measured in the forward direction and ${}^{D}k_{rev}{}^{D}K_{eq}$. The true value of ${}^{D}k_{rev}$ then is between the value given by equation 9-45 with the calculated *V/K* isotope effects in the reverse direction and ${}^{D}k_{for}/{}^{D}K_{eq}$. With malic enzyme, this approach gave limits of 5–8 for ${}^{D}k$ in the forward direction and 4–6.5 in the back reaction (B *16*, 571).

Multiple Isotope Effect Method

This method uses the effect of deuteration on a ^{13}C or other heavy atom isotope effect to determine limits for intrinsic isotope effects. If there is no reverse commitment, or if one uses both primary and secondary deuteration, one can determine the true values of the intrinsic isotope effects (B *21*, 5106). The isotope effects measured must be on the same step and deuteration will make this step more rate-limiting and thus increase the size of observed ^{13}C isotope effects (or cause no change if there are no commitments). The equations involved are

$$^{D}\left(\frac{V}{K}\right) = \frac{{}^{D}k + c_f + c_r{}^{D}K_{eq}}{1 + c_f + c_r} \qquad (9\text{-}46)$$

$$^{13}\left(\frac{V}{K}\right)_H = \frac{{}^{13}k + c_f + c_r{}^{13}K_{eq}}{1 + c_f + c_r} \qquad (9\text{-}47)$$

$$^{13}\left(\frac{V}{K}\right)_D = \frac{{}^{13}k + \dfrac{c_f}{{}^{D}k} + \dfrac{c_r{}^{13}K_{eq}{}^{D}K_{eq}}{{}^{D}k}}{1 + \dfrac{c_f}{{}^{D}k} + \dfrac{c_r{}^{D}K_{eq}}{{}^{D}k}} \qquad (9\text{-}48)$$

Note that when ${}^{13}(V/K)_D$ is measured, the commitments are reduced by the intrinsic deuterium isotope effects (${}^{D}k$ in the forward direction and ${}^{D}k/{}^{D}K_{eq}$ in the reverse direction). Equations 9-46 to 9-48 contain four unknowns (${}^{D}k$, ${}^{13}k$, and the two commitments; the equilibrium isotope effects can be measured or calculated). To solve the three equations, one assumes ratios for c_f/c_r ranging from 0 to ∞ and solves the equations simultaneously (B *21*, 5106). Solutions that give negative values or totally ridiculous values for the intrinsic isotope effects can be discarded, and in most cases this gives reasonable limits for the possible solutions.

If c_r is negligible, one needs only the multiple isotope effect of primary deuteration on ${}^{13}(V/K)$, as there are then only the two intrinsic isotope effects and c_f to determine. This was the case with prephenate dehydrogenase, where ^{13}C isotope effects of 1.0033 and 1.0103 with unlabeled and deuterated substrate and a deuterium isotope effect on V/K of 2.34 gave $^{13}k = 1.0155$, $^{D}k = 7.3$, and $c_f = 3.7$ (B *23*, 6263). The increased ^{13}C isotope effect upon deuteration but the small intrinsic ^{13}C isotope effect show that the reaction is concerted but with an asynchronous transition state, with well advanced C–H cleavage but much less C–C cleavage.

If one measures a secondary isotope effect on the same step as well as a ^{13}C one with a secondary deuterated substrate, one has two more equations:

$$^{\text{sec-D}}\left(\frac{V}{K}\right) = \frac{^{\text{sec-D}}k + c_f + c_r\,^{\text{sec-D}}K_{eq}}{1 + c_f + c_r} \tag{9-49}$$

$$^{13}\left(\frac{V}{K}\right)_{\text{sec-D}} = \frac{^{13}k + \dfrac{c_f}{^{\text{sec-D}}k} + \dfrac{c_r\,^{13}K_{eq}\,^{\text{sec-D}}K_{eq}}{^{\text{sec-D}}k}}{1 + \dfrac{c_f}{^{\text{sec-D}}k} + \dfrac{c_r\,^{\text{sec-D}}K_{eq}}{^{\text{sec-D}}k}} \tag{9-50}$$

This then gives five equations for $^{13}(V/K)_H$, $^{13}(V/K)_{\text{pri-D}}$, $^{13}(V/K)_{\text{sec-D}}$, $^{\text{pri-D}}(V/K)$, and $^{\text{sec-D}}(V/K)$. Since there are only five unknowns (^{13}k, $^{\text{pri-D}}k$, $^{\text{sec-D}}k$, c_f, and c_r; the equilibrium isotope effects are known or determinable separately), the equations can be solved simultaneously for the five unknowns.

A study of glucose-6-P dehydrogenase involved the use of all five equations to solve for the parameters (B *21*, 5106; B *23*, 5471). The study was carried out in both D_2O and H_2O, and the results are in Table 9-2.

Note that ^{13}k is very well determined, while $^{\text{pri-D}}k$ is somewhat less so. The value of $^{\text{sec-D}}k$ is not significantly different from unity, although since the equilibrium isotope effect is 0.89, any value of unity or above represents a real isotope effect. Although the individual values of c_f and c_r are not well determined, their sum *is* well determined. In D_2O, ^{13}k was not changed, but $^{\text{pri-D}}k$ was significantly reduced. $^{\text{sec-D}}k$ also appeared to decrease, although the difference, while probably real, is not significant. The surprising result was that the commitments increased, with their sum being double the value in water.

In this reaction there are three hydrogens in motion in the transition state. The primary hydrogen is being transferred from C-1 of glucose-6-P to NADP. The hydrogen at the 4-position of the nicotinamide ring of NADP is moving from an in-plane to an out-of-plane position in NADPH, and the proton on the 1-OH group of glucose-6-P is being transferred to the aspartate general

Table 9-2. Intrinsic Isotope Effects and Commitments for Glucose-6-P Dehydrogenase and Liver Alcohol Dehydrogenase[a]

Parameter	Glucose-6-P dehydrogenase H_2O	Glucose-6-P dehydrogenase D_2O	Liver alcohol dehydrogenase H_2O
^{13}k	1.041 ± 0.002	1.044 ± 0.004	1.026 ± 0.002
$^{pri\text{-}D}k$	5.3 ± 0.3	3.7 ± 0.3	3.85 ± 0.39[b]
$^{sec\text{-}D}k$	1.054 ± 0.035	1.00 ± 0.04	1.10 ± 0.03
c_f	0.75 ± 0.26	1.9 ± 0.6	0.48 ± 0.33
c_r	0.49 ± 0.27	0.65 ± 0.40	0.13 ± 0.21
$c_f + c_r$	1.24 ± 0.14	2.5 ± 0.3	0.61[c]

[a]The primary deuterium isotope effect is for the hydride transfer, and the α-secondary one is for the 4-hydrogen of the nicotinamide ring of the nucleotide.
[b]Product of primary and α-secondary isotope effects in benzyl alcohol, which was dideuterated.
[c]Error not calculated.

base on the enzyme. Coupled hydrogen motions of this type involve tunneling, and the first deuterium substitution in the system reduces the tunneling, and leads to smaller isotope effects on subsequent deuterium substitution. Thus in D_2O, where the 1-hydroxyl is an OD group, the other deuterium isotope effects are reduced.

The striking thing about these results is that the commitments are doubled in D_2O. The effect of the D_2O solvent is clearly greater on the rate constants for the conformation changes that precede and follow hydride transfer than on the rate of the latter. One must be cautious about interpreting the effects of D_2O on enzymatic reactions unless the intrinsic isotope effects and commitments are fully determined as in this case!

The only other example of using five isotope effects to solve for the intrinsic values was with liver alcohol dehydrogenase with dideuterated benzyl alcohol as the substrate (B 23, 5471). These results are also listed in Table 9-2. Note that in this case also the α-secondary deuterium isotope effect at the 4-position of the nicotinamide ring of NAD is above unity, although the equilibrium isotope effect is 0.89. This is the rule in other dehydrogenases as well where the motion of an α-secondary hydrogen is part of the reaction coordinate motion. We will discuss this coupled motion effect in more detail later when we deal with the use of isotope effects to determine transition-state structure.

Multiple Isotope Effects in Stepwise Mechanisms

When deuterium and ^{13}C isotope effects are not on the same step, as is the case with malic enzyme with NADP$^+$ as the nucleotide substrate, deuteration of the substrate decreases the observed ^{13}C isotope effect by making the deuterium-sensitive step more rate-limiting. In this case $^{13}(V/K)_H$ and $^{13}(V/K)_D$ are not independent (B *21*, 5106) but are related to $^D(V/K)$ by equation 9-51 if the deuterium-sensitive step comes first

$$\frac{^{13}(V/K)_H - 1}{^{13}(V/K)_D - 1} = \frac{^D(V/K)}{^DK_{eq}} \tag{9-51}$$

and by equation 9-52 if the ^{13}C-sensitive step comes first, as in the back reaction for malic enzyme:

$$\frac{^{13}(V/K)_H - {}^{13}K_{eq}}{^{13}(V/K)_D - {}^{13}K_{eq}} = {}^D(V/K) \tag{9-52}$$

Equations 9-51 and 9-52 are really the same equation, but one is expressed in terms of the parameters in the forward direction and the other in terms of those for the back reaction. Since $^DK_{eq}$ is often different from unity (1.18 for malic enzyme), the experimental data should fit one equation and not the other. This was the case with malic enzyme with NADP$^+$, where the first equation gave $1.21 \pm 0.05 = 1.25 \pm 0.03$, but the second one gave $1.20 \pm 0.05 \neq 1.47 \pm 0.03$ (B *21*, 5106). Thus to no surprise, malate is oxidized to oxaloacetate and then decarboxylated to pyruvate.

Intermediate Partitioning

The intrinsic isotope effects in the malic enzyme reaction were determined by adding oxaloacetate and NADPH and following the partitioning of oxaloacetate to pyruvate and malate (B *24*, 944). The formation of malate was followed at 340 nm, while the disappearance of oxaloacetate was followed at 281.5 nm, the isosbestic point of NADP$^+$ and NADPH. If there is no reverse commitment to the decarboxylation step, which is likely, the mechanism can be modeled according to mechanism 9-53:

$$\mathrm{E} \underset{k_2}{\overset{k_1\mathbf{B}}{\rightleftharpoons}} \mathrm{EAB} \underset{k_4}{\overset{k_3}{\rightleftharpoons}} \mathrm{E^*AB} \underset{k_6}{\overset{k_5}{\rightleftharpoons}} \mathrm{ERX} \xrightarrow{k_7} \mathrm{ERQ} + \mathrm{CO_2} \tag{9-53}$$

where EA is E-NADP, B is malate, R is NADPH, Q is pyruvate, and X is oxaloacetate. In this mechanism k_5 and k_6 involve hydride transfer and will show a deuterium isotope effect with 2-deuterated malate, while k_7 involves decarboxylation and will give a ^{13}C isotope effect. The partition ratio of added oxaloacetate is then given by equation 9-54:

$$\frac{[\text{pyruvate}]}{[\text{malate}]} = r_H = \left(\frac{k_7}{k_6}\right)\left[1 + \left(\frac{k_5}{k_4}\right)\left(1 + \frac{k_3}{k_2}\right)\right] \quad (9\text{-}54)$$

and r_H is the forward commitment factor for the decarboxylation step. Thus since

$$^{13}\left(\frac{V}{K}\right)_H = \frac{^{13}k_7 + r_H}{1 + r_H} \quad (9\text{-}55)$$

one can solve for $^{13}k_7$ as follows:

$$^{13}k_7 = {}^{13}(V/K)_H + r_H[^{13}(V/K)_H - 1] \quad (9\text{-}56)$$

The same equation should apply if $^{13}(V/K)_D$ and r_D (the partition ratio measured with A-side deuterated NADPH) are used, and in fact values of 1.044 ± 0.002 and 1.045 ± 0.004 were obtained with an r_H value of 0.47 and an r_D value of 0.81. The values of r_H and r_D are not independent but are related by

$$^{D}(V/K) = {}^{D}K_{eq}\left(\frac{r_D + 1}{r_H + 1}\right) \quad (9\text{-}57)$$

The value of r_H can be broken into two parts: (k_7/k_6), which is the reciprocal of the reverse commitment for hydride transfer, c_{rH}, and $(k_5/k_4)(1 + k_3/k_2)$, which is the forward commitment for hydride transfer, c_{fH}. The equation for the deuterium isotope effect is thus

$$^{D}(V/K) = \frac{^{D}k_5 + c_{fH} + c_{rH}{}^{D}K_{eq}}{1 + c_{fH} + c_{rH}} \quad (9\text{-}58)$$

and the equation for $^{T}(V/K)$ is similar except that $^{D}k_5$ and $^{D}K_{eq}$ are both raised to the 1.44 power (JACS *80*, 5885). We can then define

$$T_r = \frac{[^{T}K_{eq} - {}^{T}(V/K)]}{r_H} \qquad D_r = \frac{[^{D}K_{eq} - {}^{D}(V/K)]}{r_H} \quad (9\text{-}59)$$

The equations for $^{D}(V/K)$ and $^{T}(V/K)$ can be combined with r_H to get

$$\frac{^{T}(V/K) - 1 - T_r}{^{D}(V/K) - 1 - D_r} = \frac{(^{D}k_5)^{1.44} - {}^{T}(V/K) + T_r}{^{D}k_5 - {}^{D}(V/K) + D_r} \tag{9-60}$$

This equation can be solved by Newton's method of successive approximations for $^{D}k_5$ and then

$$c_{fH} = \frac{^{D}k_5 - {}^{D}(V/K) + D_r}{^{D}(V/K) - 1 - D_r} \tag{9-61}$$

and

$$c_{rH} = \frac{1 + c_{fH}}{r_H} \tag{9-62}$$

For malic enzyme, these equations gave $^{D}k_5 = 5.7 \pm 0.3$, $c_{fH} = 3.3 \pm 0.4$, and $c_{rH} = 10 \pm 1$.

When the nucleotide is changed from NADP (redox potential –0.320 V) to acetylpyridine-NADP (redox potential –0.258 V), which makes the reaction 2 orders of magnitude more spontaneous, the malic enzyme reaction becomes concerted rather than stepwise [deuteration increases the ^{13}C isotope effect (B *30*, 5755; B *33*, 2667; B *36*, 1141)]. However, oxaloacetate when added with acetylpyridine-NADPH still partitions to both malate and pyruvate, with an r_H value of 9.9. When combined with the $^{13}(V/K)_H$ value of 1.0037 ± 0.0006, this r_H value gives 1.040 ± 0.009 for $^{13}k_7$, which is not significantly different from the values calculated from the intermediate partitioning with NADPH. This isotope effect is thus similar to that when oxaloacetate is an actual intermediate. As we will discuss below, the ^{13}C isotope effect on the concerted reaction of malate and acetylpyridine-NADP is lower than this (1.012; B *33*, 2667).

Reactant Dependence of Isotope Effects

Substrate Dependence of Isotope Effects in Bireactant Sequential Mechanisms

In an ordered kinetic mechanism, the $V/K\mathbf{E_t}$ for the first substrate bound is the on-rate constant k_1, which will not be sensitive to substitution of deuterium for protium and thus the isotope effect on V/K_a will be unity. An exception is when an isotope effect is observed on substrate binding (B *28*, 3619; ME *308*, 301). The lack of an isotope effect on V/K_a is the basis for the theory for determination of kinetic mechanism, specifically the order of addition of substrates. Consider the following minimal scheme for a sequential mechanism:

$$\begin{array}{ccccc} & k_1\mathbf{A} \nearrow\!\!\swarrow k_2 & \mathrm{EA} & k_3\mathbf{B} \searrow\!\!\nwarrow k_4 & \\ \mathrm{E} & & & & \mathrm{EAB} \xrightarrow{k_9} \mathrm{EQ} \xrightarrow{k_{11}} \mathrm{E} \\ & k_8\mathbf{B} \searrow\!\!\nwarrow k_7 & \mathrm{EB} & k_6\mathbf{A} \nearrow\!\!\swarrow k_5 & \end{array} \qquad (9\text{-}63)$$

In mechanism 9-63, only k_9 is sensitive to isotopic substitution. The rate constant k_9 is not a simple microscopic rate constant but is the net rate constant for conversion of EAB to EQ and can contain contributions from conformation changes, the bond-breaking step(s), and release of the first product. Unless the bond-breaking step is the slowest step, the isotope effect on k_9 will be lower than that on the bond-breaking step. Mechanism 9-63 is sufficient for a consideration of kinetic mechanism since k_9 will not depend on the concentration of substrates. Thus, the equations used for the isotope effects on V and V/K will include only an external forward commitment. The expression for ${}^{D}k_9$ will include the internal part of the forward commitment as well as all of the reverse commitment.

The expression for V based on mechanism 9-63 is given below:

$$V = \frac{k_9 k_{11}\mathbf{E_t}}{k_9 + k_{11}} \qquad (9\text{-}64)$$

The expression for the isotope effect on a kinetic parameter is the ratio of the expression obtained with protium-labeled substrate compared to that obtained with deuterium-labeled substrate. The expression for the isotope effect on V is

$${}^{D}V = \frac{{}^{D}k_9 + \dfrac{k_9}{k_{11}}}{1 + \dfrac{k_9}{k_{11}}} \qquad (9\text{-}65)$$

where ${}^{D}V$ is V_H/V_D and ${}^{D}k_9$ is k_{9H}/k_{9D}. Since V is obtained under conditions of saturating substrates, the isotope effect on V will be independent of whether a random or ordered kinetic mechanism is considered. The isotope effect on V will be maximal when k_9 is small compared to k_{11}. Only when second product release is rapid compared to the net rate constant for the bond-breaking step (k_9) will ${}^{D}V$ be equal to ${}^{D}k_9$. (It should be noted that ${}^{D}k_9$ will not necessarily be equal to the intrinsic deuterium isotope effect on the chemical step; k_9 is a net rate constant.) The ratio k_9/k_{11} is called the ratio to catalysis, c_{Vf} (B *14*, 2644).

Isotope effects on the *V/K* values for the individual substrates will depend on the order of addition of reactants, with the differences reflected in the expressions for the forward commitment factor for (1) the varied substrate in the case of a deuterium isotope effect or (2) the labeled substrate when the competitive method is used. The general expression for the *V/K* isotope effect in mechanism 9-63 is given in equation 9-66, where c_f is the external part of the forward commitment.

$$^{D}(V/K) = \frac{^{D}k_9 + c_f}{1 + c_f} \tag{9-66}$$

The forward commitment factor is the ratio of k_9 to the net rate constant for dissociation of the varied substrate in a direct comparison experiment or the labeled substrate in an internal competition experiment. In mechanism 9-63 the forward commitment (c_f) is given in equations 9-67 and 9-68 for A and B as the substrates varied, respectively:

$$c_{fA} = \frac{k_9}{k_5 + \dfrac{k_2 k_4}{k_2 + k_3\mathbf{B}}} \tag{9-67}$$

$$c_{fB} = \frac{k_9}{k_4 + \dfrac{k_5 k_7}{k_7 + k_6\mathbf{A}}} \tag{9-68}$$

Differences in sequential kinetic mechanisms are essentially the result of differences in the magnitudes of rate constants in mechanism 9-63. For example, if k_5, k_6, k_7, and k_8 are equal to zero, the mechanism becomes ordered with A adding before B. Several of the commonly encountered sequential kinetic mechanisms will be discussed below with respect to their discrimination by isotope effects obtained on V/K_a and V/K_b.

Ordered Mechanisms (k_5, k_6, k_7, and $k_8 = 0$). In a steady-state ordered kinetic mechanism, the off-rate constant for the first substrate, k_2, is not much greater than $V/\mathbf{E_t}$. The expression for c_f with **A** varied, equation 9-67, is then

$$c_{fA} = \frac{k_9(k_2 + k_3\mathbf{B})}{k_2 k_4} \tag{9-69}$$

but with **B** varied, c_{fB} in equation 9-68 becomes k_9/k_4. Although the value of $^{D}(V/K_b)$ is independent of **A**, the value of $^{D}(V/K_a)$ will depend on **B**.

The expression for c_{fA} in equation 9-69 will be equal to k_9/k_4 as **B** approaches zero, and $^D(V/K_a)$ will become equal to $^D(V/K_b)$. As **B** increases, k_3**B** will become much greater than k_2, c_{fA} approaches infinity, and the value of $^D(V/K_a)$ will be unity at infinite **B**. The value of unity is a result of A becoming trapped on the enzyme by B at high concentrations. As a result, once A is bound it is committed to form product rather than dissociate to regenerate free reactant and enzyme, and thus no isotope effect is seen. The level of B giving a value of $[^D(V/K_b)+1]/2$ is $K_{ia}K_b/K_a$. An example of the above behavior is observed with liver alcohol dehydrogenase with the secondary alcohol cyclohexanol as a substrate (B *20*, 1790). In this case, the value of $^D(V/K_{cyclohexanol})$ is 3, while the isotope effects on V and V/K_{NAD} are, within error, unity. Data are consistent with an ordered kinetic mechanism with NAD^+ adding prior to cyclohexanol. In addition, the isotope effect of 3 on $V/K_{cyclohexanol}$ indicates significant rate limitation by the hydride transfer step at limiting alcohol, while the value of unity on V suggests slow release of NADH at saturating reactant concentration.

In an equilibrium ordered kinetic mechanism, k_2 is much greater than $V/\mathbf{E_t}$ and k_3**B** at any **B**. Under these conditions, the expression for c_{fA} with **A** varied reduces to k_9/k_4, independent of **B**, as for the steady-state mechanism. With **B** varied, however, c_{fB} is also k_9/k_4 and the isotope effect will be constant, that is, $^D(V/K_a) = {}^D(V/K_b)$ whichever substrate is varied and at any concentration of the fixed substrate (as long as the condition $k_2 \gg k_3\mathbf{B}$ is satisfied).

On the basis of isotope effects alone, it is not possible to distinguish this mechanism from a rapid equilibrium random one. The equilibrium ordered mechanism, however, gives a distinctive initial velocity pattern that intersects on the ordinate when **B** is varied and intersects to the left of the abscissa when **A** is varied, with the replot of slope versus 1/**B** passing through the origin (Chapter 5). If a dead-end EB complex is present, the isotope effects will not be affected. The initial velocity pattern will no longer intersect on the ordinate with **B** varied, nor will the slope replot pass through the origin when **A** is varied. As a result, the equilibrium ordered mechanism with a dead-end EB complex cannot be distinguished from a random mechanism in which k_2, $k_7 > V/\mathbf{E_t}$ (see below).

Random Mechanisms (all rate constants of mechanism 9-63 apply). In the case of a steady-state random mechanism k_2, k_4, k_5, and k_7 are not much greater than k_9. With **A** or **B** varied and the fixed substrate maintained at low levels (**B**/K_{ib} or **A**/K_{ia} approach zero), the expression for the forward commitment factor, equation 9-67, is given by equation 9-70:

$$c_{f(\text{A or B})} = \frac{k_9}{k_4 + k_5} \tag{9-70}$$

Under these conditions, the predominant enzyme form is E when V/K is measured and both pathways for formation of EAB are operative. The magnitude of the observed $^D(V/K)$ will depend on the relative values of k_4 and k_5 compared to k_9.

With **A** varied and **B** fixed, the value of $^D(V/K_a)$ may depend on **B**. As **B** approaches infinity equation 9-67 reduces to k_9/k_5, and thus the observed isotope effect will be determined by the values of k_9 and k_5, that is, the off-rate constant for B from EAB, and k_9. If $k_4 < k_5$, that is, B is sticky in the EAB complex, the isotope effect will be independent of **B**, while if $k_4 \geq k_5$, the isotope effect will decrease as **B** increases, with c_{fA} eventually becoming k_9/k_5. The mechanism is symmetrical, and the same logic can be used with **B** varied and **A** at several fixed levels. As **A** approaches infinity, c_{fB} in equation 9-68 reduces to k_9/k_4. The observed isotope effect will be independent of **A** as it is increased if $k_5 < k_4$ (A is sticky in EAB), while the isotope effect will decrease as **A** is varied if $k_5 > k_4$ (A is not sticky in EAB).

As a result, one obtains both qualitative and semiquantitative information from the measured isotope effects. Finite values of the isotope effects on V/K_a (**B** saturating) and V/K_b (**A** saturating) indicate that the kinetic mechanism is random. The values of the two isotope effects provide information on the preference of the two pathways, EAB to EA or EAB to EB. If, on the other hand, $^D(V/K_a) = {}^D(V/K_b)$, there are two possibilities: either neither substrate is sticky and the kinetic mechanism is rapid equilibrium random (k_4, $k_5 > k_9$) or both substrates are equally sticky and the kinetic mechanism is steady-state random ($k_4 = k_5 \leq k_9$). Both possibilities have been reported.

The NAD-malic enzyme from *Ascaris suum* catalyzes the oxidative decarboxylation of L-malate using NAD^+ as the oxidant to give pyruvate and CO_2. Deuterium isotope effects are finite with $^DV = 2$ and $^D(V/K_{NAD}) = {}^D(V/K_{malate}) = 1.6$ (B 25, 227; B 33, 2096). The equal isotope effects on V/K_{NAD} and V/K_{malate} indicate the kinetic mechanism is random, while the greater value of the isotope effect on V suggests both reactants are somewhat sticky. Isotope partitioning with radiolabeled NAD^+ and radiolabeled malate in separate experiments confirm the equal stickiness of the two reactants (B *27*, 212). Sheep liver 6-phosphogluconate dehydrogenase catalyzes the oxidative decarboxylation of 6-phosphogluconate using $NADP^+$ as the oxidant, a reaction very similar to that catalyzed by malic enzyme. In this case ${}^{D}V = {}^{D}(V/K_{NAD}) = {}^{D}(V/K_{malate}) = 1.6$, suggesting a rapid equilibrium random kinetic mechanism (B *37*, 12596), in agreement with initial velocities in the absence and presence of product and dead-end inhibitors (ABB *336*, 215).

A modification of the steady-state random mechanism is one in which substrates are sticky in the ternary complex but bind in rapid equilibrium to the binary complexes; that is, k_2 and $k_7 \gg V/\mathbf{E_t}$ but k_4 and k_5 are less than or not much greater than $V/\mathbf{E_t}$. When **A** is varied at any finite **B**, k_2 will exceed $k_3\mathbf{B}$, and when **B** is varied at any finite **A**, k_7 will exceed $k_6\mathbf{A}$, and equations 9-67 and 9-68 give $k_9/(k_4+k_5)$ as $c_{f(\mathrm{A\ or\ B})}$. The full value of ${}^{D}k_9$ will not be seen unless A or B are not sticky from the ternary complex. For this mechanism, partitioning of EAB will include EQ, EA, and EB whether **A** or **B** is varied and whatever the level of the other substrate. One needs additional information, for example, binding constants for the EA and EB complex, to distinguish between this mechanism and those discussed above. Nonetheless, the conclusions concerning ordered versus random kinetic mechanisms still apply.

Ping Pong Mechanisms. Although initial velocity and isotope exchange studies will usually already have been used to show that the mechanism appears Ping Pong, isotope effects are also useful. Consider the following mechanism:

$$\mathrm{E} \underset{k_2}{\overset{k_1\mathbf{A}}{\rightleftharpoons}} \mathrm{EA} \xrightarrow{k_3} \mathrm{F} \underset{k_6}{\overset{k_5\mathbf{B}}{\rightleftharpoons}} \mathrm{FB} \xrightarrow{k_7} \mathrm{E} \qquad (9\text{-}71)$$

In mechanism 9-71, only the rate constant k_3 is sensitive to isotopic substitution. It will usually be true that only one of the two half-reactions in a Ping Pong mechanism is sensitive to isotopic substitution unless (1) there is a significant amount of internal return of deuterium to a nonexchangeable covalent position (which may be true for some of the aminotransferases where the α-proton could be transferred to C4′ of pyridoxamine 5′-phosphate) or (2) a position is monitored in which the label is not potentially eliminated in the first half-reaction (as is the case for secondary isotope effects or some heavier atom isotope effects). We will consider only the case in which one of the two half-reactions is isotope-sensitive since this is the most common case encountered. In mechanism

9-71 the constants k_3 and k_7 are not simple microscopic rate constants but are the net rate constants for conversion of either EA to F or FB to E and contain contributions from conformation changes, the bond-breaking step(s), and release of the product. However, only the bond-breaking step is sensitive to isotopic substitution; and thus, unless the bond-breaking step is the slowest step, the isotope effect on k_3 will be lower than the actual (intrinsic) isotope effect on the bond-breaking step. The following brief discussion will hold whether one considers a classical one-site or a nonclassical two-site Ping Pong kinetic mechanism.

The expressions for V, V/K_a, and V/K_b are given in equations 9-72 to 9-74:

$$\frac{V}{\mathbf{E_t}} = \frac{k_3}{1 + \frac{k_3}{k_7}} \tag{9-72}$$

$$\frac{V}{K_a \mathbf{E_t}} = \frac{k_1 k_3}{k_2 + k_3} = \frac{\frac{k_1 k_3}{k_2}}{1 + \frac{k_3}{k_2}} \tag{9-73}$$

$$\frac{V}{K_b \mathbf{E_t}} = \frac{k_5 k_7}{k_6 + k_7} = \frac{\frac{k_5 k_7}{k_6}}{1 + \frac{k_7}{k_6}} \tag{9-74}$$

The expressions for the isotope effects on V and V/K_a are given in equations 9-75 and 9-76. The isotope effect on V/K_b will be equal to unity, since it does not include k_3.

$${}^{D}V = \frac{{}^{D}k_3 + \frac{k_3}{k_7}}{1 + \frac{k_3}{k_7}} \tag{9-75}$$

$${}^{D}\left(\frac{V}{K_a}\right) = \frac{{}^{D}k_3 + \frac{k_3}{k_2}}{1 + \frac{k_3}{k_2}} \tag{9-76}$$

If it is not known whether the kinetic mechanism is sequential or Ping Pong, the isotope effects on the V/Ks for A and B will not distinguish between a steady-state ordered and a Ping Pong mechanism. However, if initial velocity patterns already implicate a Ping Pong mechanism, the isotope effects will provide information on the relative rates of the two half-reactions. If the first half-reaction completely limits the overall reaction ($k_7 \gg k_3$) then ${}^{D}V$ will exhibit the full isotope effect on k_3 (${}^{D}k_3$) and ${}^{D}(V/K_a)$ will be equal to ${}^{D}V$ if A is not sticky, that is, if $k_2 > k_3$. If the value of ${}^{D}V < {}^{D}(V/K_a)$, the second half-reaction contributes to rate limitation (if it is unity, the second half-reaction is rate-limiting). An estimate of k_3/k_7 can be obtained by using ${}^{D}(V/K_a)$ as an estimate of ${}^{D}k_3$ and substituting into equation 9-75. If ${}^{D}V > {}^{D}(V/K_a)$, A is sticky, and an estimate of k_3/k_2 can be obtained by using ${}^{D}V$ as an estimate of ${}^{D}k_3$ and substituting into equation 9-76.

Aspartate aminotransferase is a PLP-dependent enzyme that catalyzes the transfer of the α-amino group of L-aspartate to α-ketoglutarate to give L-glutamate. The deuterium isotope effects on V and $V/K_{glutamate}$ are 1.9 and 3.8, respectively, for the mitochondrial isozyme, suggesting the second half-reaction contributes to rate limitation at saturating substrate concentrations (B *28*, 3815). Isotope effects measured for the cytosolic enzyme, however, are 1.4 on both V and $V/K_{aspartate}$, suggesting the aspartate half-reaction is limiting overall.

Substrate Dependence of Isotope Effects in Terreactant and Higher Order Mechanisms

The simplest way to treat a terreactant mechanism is to measure the initial rate with one reactant varied while the other two are maintained at saturating concentration. Consider the following random terreactant kinetic mechanism:

$$
\begin{array}{ccccccccc}
 & & EA & \rightleftharpoons & EAB & & & & \\
 & & & & & \overset{k_5\mathbf{C}}{\underset{k_6}{\searrow\!\!\nwarrow}} & & & \\
E & \rightleftharpoons & EB & \rightleftharpoons & EAC & \underset{k_8}{\overset{k_7\mathbf{B}}{\rightleftharpoons}} & EABC & \xrightarrow{k_{11}} & \text{products} \\
 & & & & & \overset{k_9\mathbf{A}}{\underset{k_{10}}{\nearrow\!\!\swarrow}} & & & \\
 & & EC & \rightleftharpoons & EBC & & & &
\end{array}
\qquad (9\text{-}77)
$$

(E is connected reversibly to EA, EB and EC; EA to EAB and EAC; EB to EAB and EBC; EC to EAC and EBC.)

In equation 9-77, only the rate constants for formation of the Michaelis complex, EABC, and the net rate constant for product formation, k_{11}, are given. Three experiments can thus be carried out with **A** and **B** saturating and **C** varied; with **A** and **C** saturating and **B** varied; and with **B** and **C** saturating and **A** varied. Under these conditions, the mechanism reduces to the following:

$$\mathrm{EAB} \underset{k_6}{\overset{k_5\mathbf{C}}{\rightleftharpoons}} \mathrm{EABC} \xrightarrow{k_{11}} \text{products} \tag{9-78}$$

$$\mathrm{EAC} \underset{k_8}{\overset{k_7\mathbf{B}}{\rightleftharpoons}} \mathrm{EABC} \xrightarrow{k_{11}} \text{products} \tag{9-79}$$

$$\mathrm{EBC} \underset{k_{10}}{\overset{k_9\mathbf{A}}{\rightleftharpoons}} \mathrm{EABC} \xrightarrow{k_{11}} \text{products} \tag{9-80}$$

The deuterium isotope effect on V/K_c, V/K_b, and V/K_a would then be measured. The c_f terms corresponding to equations 9-78 to 9-80 are k_{11}/k_6, k_{11}/k_8, and k_{11}/k_{10}, respectively. If finite values are obtained for all three V/K values, evidence is obtained for a random mechanism and an estimate of the relative rates of the three pathways can be obtained. If any of the isotope effects is equal to unity, c_f is large and the data suggest the pathway does not exist. For example, a finite isotope effect on V/K_c but values of unity for the other two suggests an ordered mechanism with C adding last. The mechanism can be further defined by setting **C** equal to its K_m and one of the other two reactants (e.g., **B**) at saturation and varying the third (**A**), and then repeating this experiment with the other reactant (e.g., **A**) at saturation and varying the other (e.g., **B**). Thus, isotope effects are measured on V/K_a and V/K_b, respectively. If finite isotope effects are obtained in both cases, the data indicate a random addition of A and B. If one of the two isotope effects is unity, for example, $^{D}(V/K_a)$, the data suggest ordered addition of A, B, and C. A quadreactant or higher order mechanism can be studied by a similar approach.

Product Dependence of Isotope Effects

Addition of a product will affect the magnitude of the reverse commitment factor and thus the magnitude of the observed isotope effect. In this section we will briefly discuss the effects of added product on isotope effects in sequential and Ping Pong kinetic mechanisms (B 32, 1795).

In the case of sequential kinetic mechanisms the following overall mechanism can be used:

$$\begin{array}{ccccccccc} & \overset{k_1\mathbf{A}}{\underset{k_2}{\nearrow\!\!\swarrow}} & \mathrm{EA} & \overset{k_3\mathbf{B}}{\underset{k_4}{\searrow\!\!\nwarrow}} & & & \overset{k_{11}}{\underset{k_{12}\mathbf{P}}{\nearrow\!\!\swarrow}} \ \mathrm{EQ} & \overset{k_{13}}{\underset{k_{14}\mathbf{Q}}{\searrow\!\!\nwarrow}} & \\ \mathrm{E} & & & & \mathrm{EAB} \underset{k_{10}}{\overset{k_9}{\rightleftharpoons}} \mathrm{EPQ} & & & & \mathrm{E} \\ & \overset{k_6}{\underset{k_5\mathbf{B}}{\nwarrow\!\!\searrow}} & \mathrm{EB} & \overset{k_8}{\underset{k_7\mathbf{A}}{\nearrow\!\!\swarrow}} & & & \overset{k_{18}\mathbf{Q}}{\underset{k_{17}}{\nwarrow\!\!\searrow}} \ \mathrm{EP} & \overset{k_{16}\mathbf{P}}{\underset{k_{15}}{\nearrow\!\!\swarrow}} & \end{array} \qquad (9\text{-}81)$$

In mechanism 9-81, k_9 and k_{10} represent the isotope-sensitive steps, and all other rate constants have their usual meaning.

Ordered Kinetic Mechanisms. The bottom pathway (k_5, k_6, k_7, k_8, k_{15}, k_{16}, k_{17}, and k_{18}) in mechanism 9-81 is not present, and under initial rate conditions with a single product added, either **P** or **Q** will be zero. Isotope effects are observed on V and V/K_b only (see above). We will assume finite isotope effects greater than $^{D}K_{eq}$ on both kinetic parameters, prior to the addition of P. Addition of Q will have no effect on the measured isotope effect since **A** is saturating when V and V/K_b are measured and Q and A compete for E. Derivation of the expressions for $^{D}(V/K_b)$ and ^{D}V on the basis of mechanism 9-81 with **Q** maintained at zero gives

$$^{D}\left(\frac{V}{K_b}\right) = \frac{^{D}k_9 + \dfrac{k_9}{k_4} + \left(\dfrac{k_{10}}{k_{11}}\right)\left(1 + \dfrac{k_{12}\mathbf{P}}{k_{13}}\right)^{D}K_{eq}}{1 + \dfrac{k_9}{k_4} + \left(\dfrac{k_{10}}{k_{11}}\right)\left(1 + \dfrac{k_{12}\mathbf{P}}{k_{13}}\right)} \qquad (9\text{-}82)$$

$$^{D}V = \frac{^{D}k_9 + k_9\left[\dfrac{1}{k_{13}} + \left(\dfrac{1}{k_{11}}\right)\left(1 + \dfrac{k_{12}\mathbf{P}}{k_{13}}\right)\right] + \left(\dfrac{k_{10}}{k_{11}}\right)\left(1 + \dfrac{k_{12}\mathbf{P}}{k_{13}}\right)^{D}K_{eq}}{1 + k_9\left[\dfrac{1}{k_{13}} + \left(\dfrac{1}{k_{11}}\right)\left(1 + \dfrac{k_{12}\mathbf{P}}{k_{13}}\right)\right] + \left(\dfrac{k_{10}}{k_{11}}\right)\left(1 + \dfrac{k_{12}\mathbf{P}}{k_{13}}\right)} \qquad (9\text{-}83)$$

As the concentration of P increases, the c_r term $(k_{10}/k_{11})(1 + k_{12}\mathbf{P}/k_{13})$ increases and the measured value of $^{D}(V/K_b)$ decreases to $^{D}K_{eq}$. This will allow assignment of ordered product release with P released prior to Q. In the case of ^{D}V, however, both c_{Vf} and c_r increase and the expression at infinite **P** is thus

$$^{D}V = \frac{^{D}K_{eq} + \dfrac{k_9}{k_{10}}}{1 + \dfrac{k_9}{k_{10}}} \qquad (9\text{-}84)$$

and the value of ${}^{D}V$ will be between unity and ${}^{D}K_{eq}$ dependent on the value of the equilibrium constant, on enzyme, for the chemical step. A value closer to unity will be obtained when the equilibrium favors the EPQ complex, while a value of ${}^{D}K_{eq}$ will be obtained when the equilibrium favors the EAB complex.

Random Kinetic Mechanisms. All rate constants in mechanism 9-81 apply. As long as k_9 contributes to overall rate limitation and both substrates are not very sticky, finite isotope effects will be observed on all parameters, V, V/K_a, and V/K_b; see above. Qualitatively identical results will be observed whether **A** is varied at saturating **B** or vice versa. The product dependence of the isotope effects should be measured by use of the largest of the effects on V/K_a or V/K_b. Below we will assume the largest isotope effect is observed on V/K_b. The expression for ${}^{D}(V/K_b)$ with **Q** at zero and **P** varied is given in equation 9-85:

$$ {}^{D}\left(\frac{V}{K_b}\right) = \frac{{}^{D}k_9 + \dfrac{k_9}{k_4} + \dfrac{k_{10}\,{}^{D}K_{eq}}{k_{17} + \dfrac{k_{13}}{\left(\dfrac{\mathbf{P}}{K_{ip}} + \dfrac{k_{13}}{k_{11}}\right)}}}{1 + \dfrac{k_9}{k_4} + \dfrac{k_{10}}{k_{17} + \dfrac{k_{13}}{\left(\dfrac{\mathbf{P}}{K_{ip}} + \dfrac{k_{13}}{k_{11}}\right)}}} \tag{9-85} $$

As was true for the ordered mechanism, the c_r term in ${}^{D}(V/K_b)$ contains **P**, but the form of the expressions differs because of the two pathways for release of product. At zero **P**, c_r is $k_{10}/(k_{11}+k_{17})$, with k_{11} and k_{17} reflecting release of P and Q, respectively, from EPQ. As the concentration of P increases, however, the pathway for product release is forced to release of Q, and c_r becomes k_{10}/k_{17}. The expression for ${}^{D}(V/K_b)$ with **P** at zero and **Q** added is identical to equation 9-84, with the exception that the c_r term is given by equation 9-86.

$$ c_r = \frac{k_{10}}{k_{11} + \dfrac{k_{15}}{\left(\dfrac{\mathbf{Q}}{K_{iq}} + \dfrac{k_{15}}{k_{17}}\right)}} \tag{9-86} $$

As the concentration of Q increases, the pathway for product release is forced to release of P, and c_r becomes k_{10}/k_{11}. Thus, if there is a preference for release of P or Q from EPQ, the isotope effects will differ at saturating concentrations of the added product, for example, a larger isotope effect obtained with saturating P than with saturating Q would indicate a more rapid release of Q than P from EPQ.

The expression for ${}^{D}V$ with **Q** at zero and **P** varied is given in equation 9-87:

$$
{}^{D}V = \frac{{}^{D}k_9 + \dfrac{k_9\left[1 + \dfrac{k_{17}}{k_{15}} + \dfrac{1}{(\mathbf{P}/K_{ip} + k_{13}/k_{11})}\right]}{k_{17} + \dfrac{k_{13}}{(\mathbf{P}/K_{ip} + k_{13}/k_{11})}} + \dfrac{k_{10}{}^{D}K_{eq}}{k_{17} + \dfrac{k_{13}}{(\mathbf{P}/K_{ip} + k_{13}/k_{11})}}}{1 + \dfrac{k_9\left[1 + \dfrac{k_{17}}{k_{15}} + \dfrac{1}{(\mathbf{P}/K_{ip} + k_{13}/k_{11})}\right]}{k_{17} + \dfrac{k_{13}}{(\mathbf{P}/K_{ip} + k_{13}/k_{11})}} + \dfrac{k_{10}}{k_{17} + \dfrac{k_{13}}{(\mathbf{P}/K_{ip} + k_{13}/k_{11})}}} \quad (9\text{-}87)
$$

As is true for the ordered mechanism, the c_r and c_{Vf} terms contain **P**. At zero and infinite **P**, respectively, the expressions for ${}^{D}V$ are given in equations 9-88 and 9-89:

$$
{}^{D}V_{P=0} = \frac{{}^{D}k_9 + \dfrac{k_9\left[1 + \dfrac{k_{17}}{k_{15}} + \dfrac{k_{11}}{k_{13}}\right]}{k_{17} + k_{11}} + \dfrac{k_{10}{}^{D}K_{eq}}{k_{17} + k_{11}}}{1 + \dfrac{k_9\left[1 + \dfrac{k_{17}}{k_{15}} + \dfrac{k_{11}}{k_{13}}\right]}{k_{17} + k_{11}} + \dfrac{k_{10}}{k_{17} + k_{11}}} \quad (9\text{-}88)
$$

$$
{}^{D}V_{P=\infty} = \frac{{}^{D}k_9 + k_9\left(\dfrac{1}{k_{15}} + \dfrac{1}{k_{17}}\right) + \dfrac{k_{10}{}^{D}K_{eq}}{k_{17}}}{1 + k_9\left(\dfrac{1}{k_{15}} + \dfrac{1}{k_{17}}\right) + \dfrac{k_{10}}{k_{17}}} \quad (9\text{-}89)
$$

The isotope effect on V will thus decrease to some finite value that will be different than the value of ${}^{D}(V/K_b)$ and will depend on the relative rates of the two pathways for product release. Again the dependence of the isotope effect on **Q** will be qualitatively similar. The expressions for c_{Vf} and c_r are given in equation 9-90:

$$
c_r = \frac{k_{10}}{k_{11} + \dfrac{k_{15}}{\dfrac{\mathbf{Q}}{K_{iq}} + \dfrac{k_{15}}{k_{17}}}} \qquad c_{Vf} = \frac{k_9\left[1 + \dfrac{k_{11}}{k_{13}} + \dfrac{1}{(\mathbf{Q}/K_{iq} + k_{15}/k_{17})}\right]}{k_{11} + \dfrac{k_{15}}{(\mathbf{Q}/K_{iq} + k_{15}/k_{17})}} \quad (9\text{-}90)
$$

Thus, the pathway for dissociation of products—k_{17}, k_{15}, or k_{11}, k_{13} that is favored will give the larger isotope effect, for example, if P dissociation from EPQ is preferred, the value of ^{D}V will be larger at saturating **Q** than at saturating **P**.

Ping Pong Kinetic Mechanisms. An example of a Ping Pong Bi Bi mechanism is given schematically in equation 9-91.

$$E \underset{k_2}{\overset{k_1\mathbf{A}}{\rightleftharpoons}} EA \underset{k_4}{\overset{k_3}{\rightleftharpoons}} EP \underset{k_6\mathbf{P}}{\overset{k_5}{\rightleftharpoons}} F \underset{k_8}{\overset{k_7\mathbf{B}}{\rightleftharpoons}} FB \underset{k_{10}}{\overset{k_9}{\rightleftharpoons}} EQ \underset{k_{12}\mathbf{Q}}{\overset{k_{11}}{\rightleftharpoons}} E \qquad (9\text{-}91)$$

In mechanism 9-91, k_3 and k_4 are isotope-sensitive while k_9 and k_{10} are usually not, and all other rate constants have their usual meaning.

In the absence of added products, an isotope effect is usually obtained on only the first half-reaction $[^{D}(V/K_b) = 1]$, unless the label is maintained through the second half-reaction, for example, in the case of a secondary kinetic isotope effect. The presence of added Q will have no effect on the isotope effect, since it competes with A and only changes the amount of E that A can bind to. However, addition of P can generate a finite isotope effect on V/K_b. The expression for the isotope effects on V, V/K_a, and V/K_b are given in equations 9-92 to 9-94:

$$^{D}V = \frac{^{D}k_3 + k_3\left[\frac{1}{k_5} + \frac{1}{k_{11}} + \left(\frac{1}{k_9}\right)\left(1 + \frac{k_{10}}{k_{11}}\right)\right] + \left(\frac{k_4}{k_5}\right)^{D}K_{eq,1}}{1 + k_3\left[\frac{1}{k_5} + \frac{1}{k_{11}} + \left(\frac{1}{k_9}\right)\left(1 + \frac{k_{10}}{k_{11}}\right)\right] + \frac{k_4}{k_5}} \qquad (9\text{-}92)$$

$$^{D}\left(\frac{V}{K_a}\right) = \frac{^{D}k_3 + \frac{k_3}{k_2} + \left(\frac{k_4}{k_5}\right)^{D}K_{eq,1}}{1 + \frac{k_3}{k_2} + \frac{k_4}{k_5}} \qquad (9\text{-}93)$$

$$^{D}\left(\frac{V}{K_b}\right) = \frac{\left[1 + \left(1 + \frac{^{D}K_{eq,1}}{K_{eq,1}}\right)\frac{\mathbf{P}}{K_{ip}}\right]}{\left[1 + \left(1 + \frac{1}{K_{eq,1}}\right)\frac{\mathbf{P}}{K_{ip}}\right]} \qquad (9\text{-}94)$$

where $K_{eq,1}$ is k_3/k_4 and $^{D}K_{eq,1}$ is $(^{D}k_3/^{D}k_4)$. There is no **P** term in the expression for V/K_a since it is obtained at saturating **B**. Addition of P, however, may give a finite isotope effect on V/K_b if a significant equilibrium isotope effect is observed, as long as $K_{eq,1}$ is not large, that is, the equilibrium for the first half-reaction is not too far to the right.

The above theory has only been applied to a few systems. Cook and Cleland (B *20*, 1790) used the equilibrium perturbation technique (see above) to study the yeast and liver alcohol dehydrogenase reactions with the secondary alcohols 2-propanol and cyclohexanol, respectively, as substrates. With deuterated 2-propanol and NADH as perturbants in the yeast alcohol dehydrogenase reaction, increasing the concentration of the acetone product causes the isotope effect to decrease from a value of 3.2 to 1.18 (the equilibrium isotope effect) at infinite acetone, consistent with an ordered release of acetone prior to NADH. Similar experiments were also carried out with liver alcohol dehydrogenase with deuterated cyclohexanol and NADH as perturbants. In this case the isotope effect decreased from a value of 2.9 at low cyclohexanone to 1.5 at infinite ketone concentration, consistent with a random release of ketone and NADH.

Studies have also been carried out by the noncompetitive method (B *32*, 1795) for the lactate and yeast alcohol dehydrogenase reactions with NADD as the labeled reactant. In the case of lactate dehydrogenase, ${}^{D}(V/K_{pyruvate})$ decreases from a value of 1.9 at zero to 1.16 at infinite ($>$500 mM) lactate, while ${}^{D}V$ decreases from 1.75 to 0.93. Given an equilibrium isotope effect of 0.85 in the direction of pyruvate reduction, the data support a random release of lactate and NAD^+. Similar results are obtained for the yeast alcohol dehydrogenase-catalyzed reduction of acetaldehyde. The ${}^{D}(V/K_{acetaldehyde})$ decreases from a value of 2.8 at zero to 1.8 at infinite ($>$300 mM) ethanol. The value of ${}^{D}V$ is unity in this case, consistent with a slow off-rate constant for the last product released. These data are consistent with a random release of NAD^+ and ethanol. In both of the above cases the pathway in which the dinucleotide is released last is greatly preferred given the very high concentrations of lactate and ethanol used to observe the alternative pathway. The technique is very sensitive and needs to be more widely applied to systems where there is a question as to the order of addition.

Isotope Effects as a Probe of Regulatory Mechanism

In Chapter 5, theory for the kinetic mechanism of regulation was presented. An extension of this theory makes use of the measurement of isotope effects as a function of allosteric regulator (B *21*, 113).

As discussed in Chapter 5, allosteric effectors can have an effect on either V or V/K or on both, and as a result, on ^{D}V, $^{D}(V/K)$, or both. For consistency, we use mechanism 9-95, the minimal mechanism for a unireactant enzyme or a bireactant enzyme with one reactant saturating:

$$\mathrm{EA} \underset{k_4}{\overset{k_3\mathbf{B}}{\rightleftharpoons}} \mathrm{EAB} \underset{k_6}{\overset{k_5}{\rightleftharpoons}} \mathrm{E^*AB} \underset{k_8}{\overset{k_7}{\rightleftharpoons}} \mathrm{E^*PQ} \underset{k_{10}}{\overset{k_9}{\rightleftharpoons}} \mathrm{EPQ} \xrightarrow{k_{11}} \mathrm{EQ} \xrightarrow{k_{13}} \qquad (9\text{-}95)$$

For mechanism 9-95, k_3 and k_4 represent the addition and release of B; k_5, k_6, k_9 and k_{10} represent conformation changes in the central reactant and product complexes; k_7 and k_8 represent forward and reverse catalytic (isotope-sensitive) steps; while k_{11} and k_{13} represent the off rates for P and Q, respectively. (In the case of a unireactant mechanism EA, EAB, EPQ, and EQ would be replaced by E, EA, EP, and E, respectively.) Expressions for ^{D}V and $^{D}(V/K_b)$ are given in equations 9-96 and 9-97, respectively:

$$^{D}V = \frac{^{D}k_7 + \dfrac{k_7\left[\dfrac{1}{k_5} + \left(\dfrac{1}{k_9}\right)\left(1 + \dfrac{k_{10}}{k_{11}}\right) + \dfrac{1}{k_{11}} + \dfrac{1}{k_{13}}\right]}{1 + \dfrac{k_6}{k_5}} + \left(\dfrac{k_8}{k_9}\right)\left(1 + \dfrac{k_{10}}{k_{11}}\right)^{D}K_{eq}}{1 + \dfrac{k_7\left[\dfrac{1}{k_5} + \left(\dfrac{1}{k_9}\right)\left(1 + \dfrac{k_{10}}{k_{11}}\right) + \dfrac{1}{k_{11}} + \dfrac{1}{k_{13}}\right]}{1 + \dfrac{k_6}{k_5}} + \left(\dfrac{k_8}{k_9}\right)\left(1 + \dfrac{k_{10}}{k_{11}}\right)} \qquad (9\text{-}96)$$

$$^{D}\left(\frac{V}{K_b}\right) = \frac{^{D}k_7 + \left(\dfrac{k_7}{k_6}\right)\left(1 + \dfrac{k_5}{k_4}\right) + \left(\dfrac{k_8}{k_9}\right)\left(1 + \dfrac{k_{10}}{k_{11}}\right)^{D}K_{eq}}{1 + \left(\dfrac{k_7}{k_6}\right)\left(1 + \dfrac{k_5}{k_4}\right) + \left(\dfrac{k_8}{k_9}\right)\left(1 + \dfrac{k_{10}}{k_{11}}\right)} \qquad (9\text{-}97)$$

An allosteric activator or inhibitor can affect V, V/K, or both (see Chapter 5). Depending on the step(s) affected, the isotope effects on V, V/K, or both can be changed.

As stated in Chapter 5, the most common effect of allosteric effectors is on the rate constants for the conformation changes preceding and following the chemical steps, while the least common effect is on the rate constants for the chemical steps. As a result, an increase in the isotope effects on V and V/K in the presence of an allosteric inhibitor likely indicates either a decrease in the rate of the chemical step, k_7, or an increase in the reverse rates of the conformational changes, k_6 and/or k_{10}. An activator, however, will likely give an increase in the isotope effects on both parameters with an increase in the net rate constant for release of the first product, that is, a decrease in the magnitude of the reverse commitment factor, $(k_8/k_9)(1+k_{10}/k_{11})$. When ${}^DV$ is increased or decreased with no effect on V/K, the most likely effect is on the off-rate constant for the last product, k_{13}, while an increase in ${}^D(V/K)$ with no effect on V is most likely a result of a change in the off-rate constant for the reactant, k_4.

There are a few examples of isotope effects in regulated systems in the literature. The glutamate dehydrogenase reaction is affected by ADP and GTP (B *21*, 113). ADP increases V by about 3-fold but decreases the $V/K_{glutamate}$ and V/K_{NADP} by 1.5-fold and 3-fold, respectively. GTP, on the other hand decreases V but increases the V/K for glutamate. These effectors thus likely exert their effect on the net rate constants for product and reactant release, as suggested in Chapter 5. ADP thus increases the net off rates for glutamate and NADP, giving a decrease in the amount of ternary complex and thus a decrease in the rate under V/K conditions. When reactants are saturating, however, the rate is increased as a result of an increase in the rate of release of products (the biggest effect here is likely on NADPH release). GTP has the opposite effect, decreasing the net off rate for glutamate and NADP and increasing the amount of ternary complex under V/K conditions but decreasing the off rate for products at saturating reactant concentrations. In agreement with this suggestion, ADP increases the deuterium isotope effects on V and V/K from values of about 1.05 in the absence of the allosteric effector to 1.3 in its presence. The allosteric inhibitor, GTP, has exactly the oppposite effect in that the isotope effects on V and V/K decrease to unity.

Isotope Effects as a Probe of Chemical Mechanism

The major kinetic tools for determining the chemical mechanisms of enzymatic reactions are isotope effects and pH profiles. We have alluded to the use of isotope effects to distinguish between stepwise and concerted reactions in our discussion of the determination of intrinsic isotope effects (see above). When a C–H bond is cleaved in one part of the substrate and another isotope effect occurs in another position, deuteration of the C–H bond allows one to tell whether the reaction is concerted or stepwise. As noted earlier, the malic enzyme reaction with NAD(P)$^+$ as substrate is stepwise because the observed ^{13}C isotope effect at C-4 is reduced when the hydrogen at C-2 is replaced with deuterium (B *21*, 5106). Furthermore, the ^{13}C isotope effects with unlabeled and deuterated malate and the deuterium isotope effect on V/K_{malate} satisfy the equation expected when the deuterium-sensitive step comes first. These data indicate the presence of an oxaloacetate intermediate in the reaction.

When thio-NAD(P)$^+$ or acetylpyridine-NAD(P)$^+$, which have more positive redox potentials, are used as substrates, however, the ^{13}C isotope effect is larger with deuterated malate and the reaction is now concerted (B *30*, 5755; B *33*, 2667; B *36*, 1141). With acetylpyridine-NADP$^+$ the intrinsic deuterium isotope effect with the chicken liver malic enzyme is now 3.4, lower than the value of 5.7 estimated with NADP$^+$. Data suggest an earlier transition state with the more redox positive dinucleotide, as expected from the 2 orders of magnitude larger equilibrium constant. However, the ^{13}C intrinsic isotope effect is 1.012; the small value shows that the transition state is asynchronous, with C–H cleavage further advanced than C–C cleavage. This change in mechanism results from the increased equilibrium constant for the oxidation. With 3-fluoromalate the equilibrium constant is reduced by the inductive effect of the fluorine and in this case the reaction remains stepwise even with acetylpyridine-NADP$^+$ (B *37*, 18026).

Other oxidative decarboxylation reactions that are stepwise are those catalyzed by 6-phosphogluconate (B *26*, 6257; B *37*, 12596) and isocitrate dehydrogenases (B *27*, 2934).

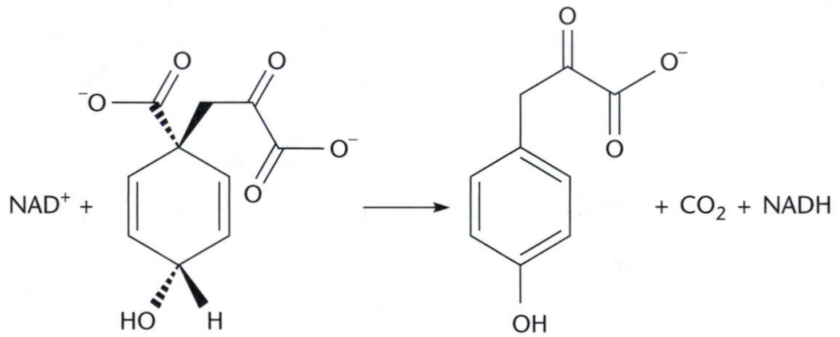

Prephenate dehydrogenase, on the other hand, which converts prephenate to *p*-hydroxyphenylpyruvate, has a concerted mechanism (B *23*, 6263). With a substrate lacking the keto group but showing a similar V_{max}, the ^{13}C isotope effect was 1.0033 with unlabeled and 1.0103 with 4-deuterated substrate, and the $^{D}(V/K)$ was 2.34. These numbers give $^{D}k = 7.3$, $^{13}k = 1.0155$, and $c_f = 3.7$. The transition state is asynchronous, with C–C cleavage only just beginning but C–H cleavage well advanced. The reaction violates the Woodward–Hoffman rules, but the energy of aromatization is so high that it takes place anyway. If one removes one double bond from the ring, the substrate is simply oxidized to a ketone and does not decarboxylate. The enzyme is thus really a secondary alcohol dehydrogenase, and decarboxylation takes place spontaneously because the putative ketone intermediate is unstable.

Aspartate transcarbamoylase catalyzes the reaction between aspartate and carbamoyl phosphate to give carbamoyl-L-aspartate and inorganic phosphate.

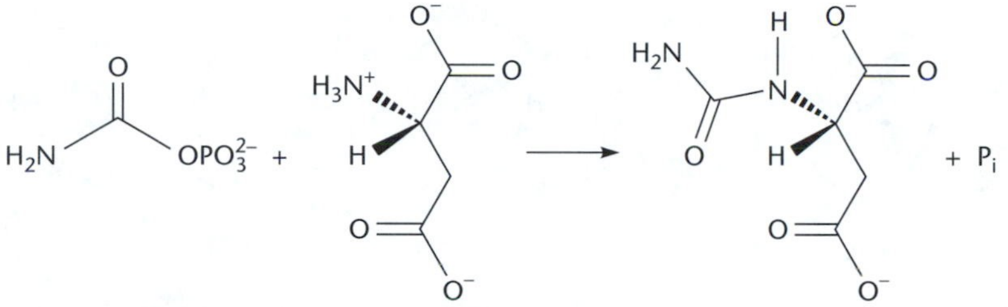

The enzyme exhibits a highly cooperative, sigmoid saturation curve for aspartate. In the presence of the allosteric activator ATP, the curve becomes a normal hyperbola, while addition of CTP causes it to become more sigmoid. The six regulatory subunits to which ATP and CTP bind are arranged like a barber pole between two catalytic triads, each with three active sites. The free enzyme exists largely in an inactive T state, with only a small level of the active R form present. Aspartate binds only to R, and since all six active sites change together from T to R, the binding of one aspartate increases the number of active sites available for further addition of aspartate. ATP binds selectively to R and CTP to the T form.

When the ^{13}C isotope effects were measured in carbamoyl-P in the absence of the allosteric modifiers, the value varied from 1.022 at very low aspartate to unity at very high aspartate, showing that the reaction is ordered, with carbamoyl-P adding first (see earlier section Ordered Kinetic Mechanism). Furthermore, the same curve was seen in the presence of ATP or CTP, showing that the properties of the R form do not depend on the allosteric modifiers, which simply change the amount of R present (B *31*, 6570).

The catalytic triads can be separated from the regulatory subunits and they then display normal Michaelis–Menten kinetics. The ^{13}C isotope effects in carbamoyl-P now vary from 1.024 at low aspartate to 1.004 at high aspartate. The mechanism has become partly random, and aspartate slows but does not prevent carbamoyl-P dissociation (B *31*, 6570). When the slow alternative substrate, cysteine sulfinate, where the β-carboxyl is now $-SO_2^-$, was used as a substrate, the ^{13}C isotope effect in carbamoyl-P was 1.039 and was independent of the cysteine sulfinate level (B *31*, 6570). Similarly, an H134A mutant showed an isotope effect of 1.04, which was independent of the aspartate level (B *31*, 6585). In these cases the chemistry is totally rate-limiting, and 1.04 is presumably the intrinsic ^{13}C isotope effect.

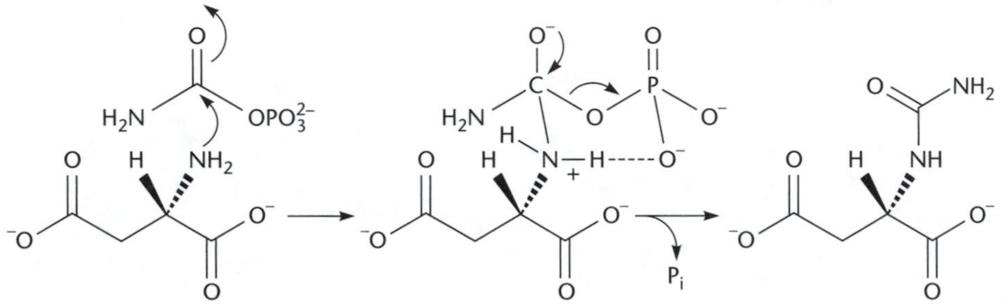

Most workers have assumed that aspartate attacked carbamoyl-P to form a tetrahedral intermediate, from which phosphate is released to give carbamoyl-aspartate; see above.

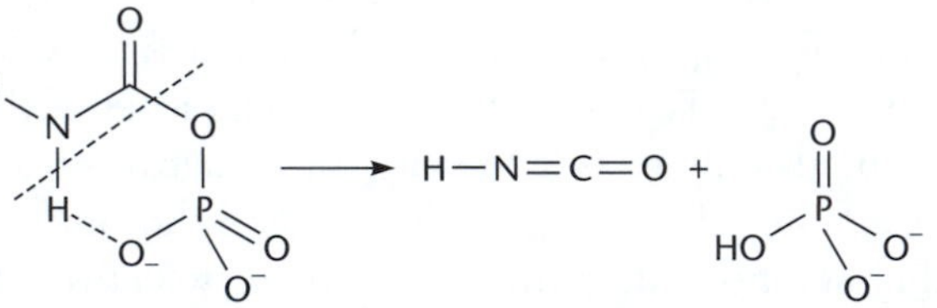

But the dianion of carbamoyl-P (the reactive form for the enzymatic reaction) decomposes nonenzymatically by internal general base catalysis and C–O cleavage to give cyanic acid, a good electrophile, which could certainly react with aspartate. To distinguish this mechanism from the one involving a tetrahedral intermediate, the ^{15}N isotope effects were determined in carbamoyl-P for its nonenzymatic breakdown and for the enzymatic reactions (JACS *114*, 5941). The nonenzymatic decomposition of carbamoyl-P dianion gave an ^{15}N isotope effect of 1.0114, which is consistent with a primary isotope effect resulting from N–H cleavage. By contrast, breakdown of the carbamoyl-P monoanion, which proceeds by P–O cleavage to give carbamic acid, gave 1.0028, a small secondary isotope effect probably resulting from a change from a N–C–O–P torsional vibration in carbamoyl-P, which is ^{15}N-sensitive (N has lower mass than P and thus moves more during the vibration), to a N–C–O–H vibration, which is not ^{15}N-sensitive, since the hydrogen undergoes most of the motion.

In the enzymatic reaction, the holoenzyme with low aspartate gave an ^{15}N isotope effect of 1.0014, while with cysteine sulfinate as substrate or with the H134A mutant with aspartate, the values were 1.0024 and 1.0027, respectively. Since the chemistry is rate-limiting for the latter two cases, the intrinsic ^{15}N isotope effect appears to be ~1.0025. This is consistent with formation of a tetrahedral intermediate, where only a small secondary isotope effect is expected from loss of amide resonance in the adduct, but not with formation of a cyanic acid intermediate, where a value similar to that for the nonenzymatic breakdown of carbamoyl-P dianion is expected.

Chorismate mutase catalyzes the isomerization of chorismate to prephenate.

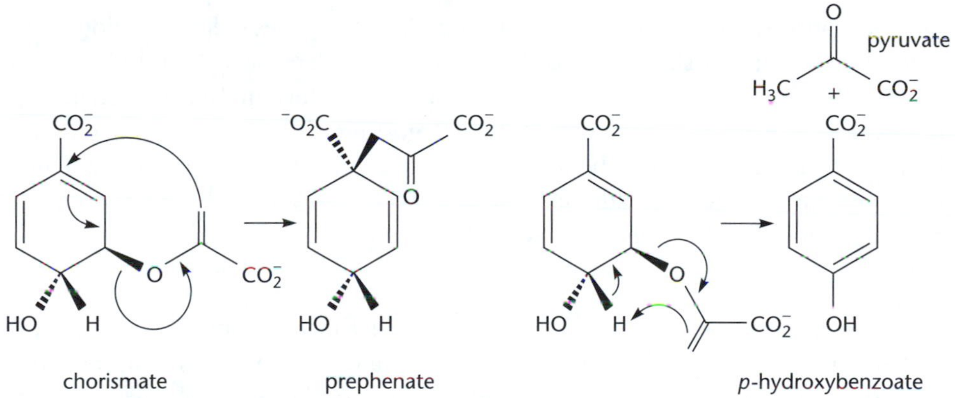

chorismate prephenate *p*-hydroxybenzoate pyruvate

The reaction proceeds in the absence of the enzyme at 60 °C, accompanied by a minor reaction in which *p*-hydroxybenzoate and pyruvate are the products. Isotope effects were measured by the remote label method, with C-10 as the remote label (it is readily decarboxylated in prephenate at low pH) (JACS *121*, 1758). The ^{18}O isotope effect in the ether oxygen was 4.5–5%, showing that C–O cleavage is well advanced in the transition state. The ^{13}C isotope effect at C-1 was ~0.5% normal, while that at C-9 of the side chain was 1.1–1.3%. These values show that the reaction is a concerted pericyclic one but with C–C formation lagging behind C–O cleavage.

The ^{18}O isotope effects in the nonenzymatic reaction were also 4–5%, while the ^{13}C isotope effect at C-1 was 1.2%. Thus the reaction has a similar transition state to the enzymatic one. In the side reaction that produces *p*-hydroxybenzoate, a different rotamer is involved, and C-9 of the side chain removes the 4-hydrogen, as shown by transfer of deuterium (JACS *127*, 12957).

L-Ribulose-5-P 4-epimerase converts L-ribulose-5-P to D-xylulose-5-P.

L-ribulose 5-phosphate

D-xylulose 5-phosphate

The mechanism of this reaction was established by degrading xylulose-5-P carbon by carbon and showing that the ^{13}C isotope effects at C-3 and C-4 were over 2% (B *39*, 4808). With a slow mutant the deuterium isotope effects at these carbons were only 4% and 19%, suggesting that these are secondary isotope effects and no C–H bond cleavage is involved in the reaction. Thus the mechanism involves an aldol cleavage to the enediolate of dihydroxyacetone and glycolaldehyde-P, followed by rotation of the aldehyde group and recondensation.

Oxalate decarboxylase catalyzes the conversion of the monoanion of oxalate to CO_2 and formate. Since both oxalate and formate as anhydrous salts can be oxidized to CO_2 by I_2 in dimethyl sulfoxide without exchanging the oxygens, one can, by analyzing mass ratios in formate, CO_2, and residual as well as initial oxalate, determine the ^{13}C and ^{18}O isotope effects going to both CO_2 and formate (JACS *125*, 1244). At pH 5.7, where the chemistry is rate-limiting, the ^{13}C isotope effects were 1.9% going to formate and 0.8% going to CO_2. The ^{18}O isotope effects were 1.0% normal going to formate and 0.7% inverse going to CO_2. The small ^{13}C isotope effect going to CO_2 suggests that decarboxylation is not the major rate-limiting step, while the larger isotope effects going to formate suggest that the C–O bond order in the end of oxalate going to formate is reduced in a step prior to decarboxylation, which is the first irreversible step.

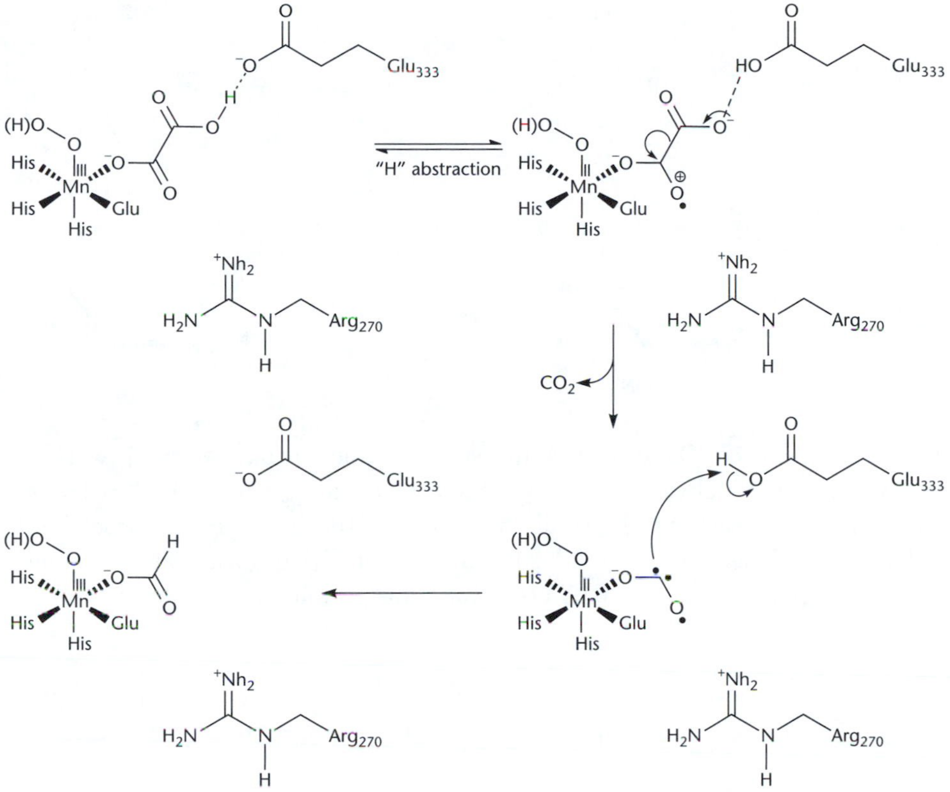

Anything happening after decarboxylation will not affect the *V*/*K* isotope effects determined in this study.

When appropriate equations for such a model were applied, it was possible to calculate the C–O bond order in the end of the intermediate going to formate as 1.16 from the ^{13}C isotope effects and 1.14 from the ^{18}O ones, with the intermediate decarboxylating 4 times faster than it returns to oxalate. The enzyme contains Mn^{2+} and requires catalytic oxygen, which is not used up during the reaction. The proposed mechanism calls for the Mn^{2+} and oxygen to form a Mn^{3+} superoxide, which extracts an electron from the end of oxalate that is coordinated to the metal ion, while a proton is removed by a general base from the other end. On the basis of the C–O bond order in the coordinated end of the intermediate, this carbon will have ~70% of a positive charge, which induces rapid decarboxylation to give a formate radical anion, which will subsequently pick up a proton and an electron to become formate.

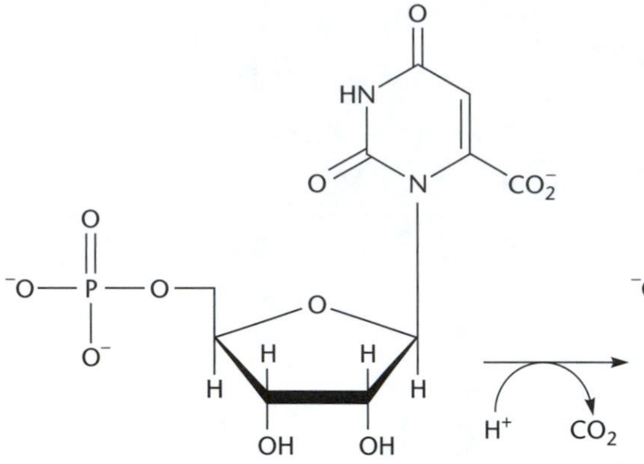

The mechanism of OMP decarboxylase (see above) has puzzled workers because there is no obvious stable intermediate resulting from decarboxylation. In nonenzymatic reactions of a similar type, N-1 of the ring becomes protonated so that the intermediate is an ylide. Protonation of N-1 gives an inverse equilibrium ^{15}N isotope effect of ~0.979 (the value for pyridine).

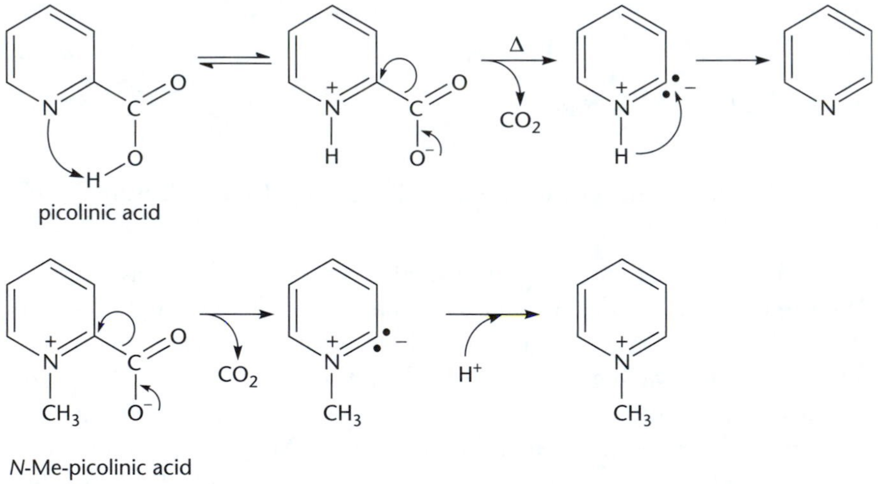

picolinic acid

N-Me-picolinic acid

Decarboxylation will then produce a small normal secondary kinetic ^{15}N isotope effect from loss of N–C–C bending and N–C–C–O torsional motions. With picolinic acid, where the proton has to shift from the carboxyl group to N-1 prior to decarboxyation, an overall inverse ^{15}N isotope effect of 0.993 (extrapolated to 25 °C) was observed, while with *N*-methylpicolinate, where N-1 already has a positive charge, a normal isotope effect of 1.007 was seen. The enzymatic reaction gave 1.0036, which shows that N-1 does not become positively charged prior to decarboxylation. The enzyme has a positively charged lysine next to C-6, so its charge stabilizes the transient carbanion intermediate and protonates it as soon as CO_2 moves away (B 39, 4569).

Equilibrium isotope effects can be used in some cases to determine chemical mechanisms of certain enzymatic reactions. A number of kinases show an ATPase activity when presented with an aldehyde analogue of their normal alcohol substrate. For example, glycerokinase phosphorylates L-glyceraldehyde in the 3-position in a fashion similar to glycerol, but D-glyceraldehyde induces an ATPase with a V_{max} 20% of that with glycerol (B 23, 5157). When D-glyceraldehyde is labeled with ^{18}O (by incubation in a small volume of $H_2{}^{18}O$ long enough for the exchange to occur by the hydration–dehydration equilibrium) and then added to a solution of glycerokinase and MgATP, the phosphate produced contains ^{18}O. Gyceraldehyde is 94% hydrated at 25 °C, and the equilibrium deuterium isotope effect on the hydration equilibrium is 1.37, so that there is less free aldehyde present by this ratio when glyceraldehyde is deuterated at C-1.

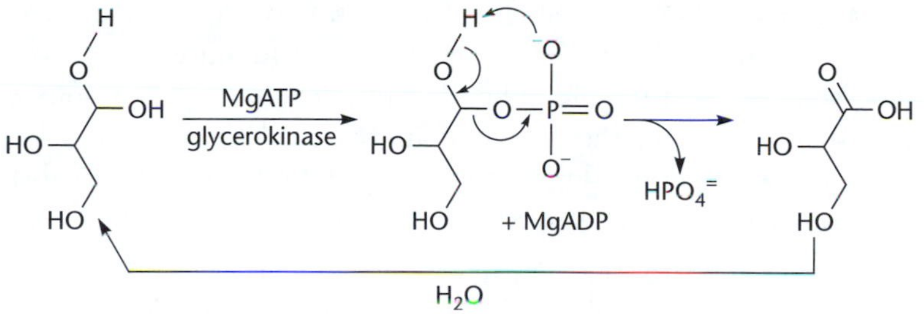

The $^{D}(V/K)$ for the ATPase reaction is unity, showing that the active substrate is the hydrated aldehyde (the value would be 1.37 if the aldehyde were the substrate). Thus MgATP phosphorylates the hydrate to give an unstable species that eliminates phosphate to give free aldehyde. Similar ATPase reactions involving phosphorylation of aldehyde hydrates are catalyzed by fructokinase, fructose-6-P kinase, and hexokinase.

Acetate kinase is inhibited by acetaldehyde, but there is no induced ATPase. The deuterium isotope effect on the inhibition constant is 0.87, which corresponds to the hydrate being the inhibitory form (expected value 0.89, versus 1.22 for the free aldehyde; acetaldehyde is only 60% hydrated at 25 °C). The failure of MgATP to phosphorylate the hydrate may result from the absence of a general base to remove the proton, as such a catalytic group is not needed for phosphorylation of acetate. Similarly, D-glyceraldehyde-3-P inhibits 3-P-glycerate kinase but does not induce an ATPase.

Isotope Effects as a Probe of Transition-State Structure

Isotope effects are one of the only practical ways to study transition-state structures of enzymatic reactions. Structure–reactivity relationships do not usually work because of steric problems. The interpretation of primary and secondary isotope effects differs. Secondary isotope effects, except when the bending motion of the α-secondary hydrogen is coupled into the reaction coordinate motion (see below), give direct evidence of transition-state structure, since they are determined solely by changes in the stiffness of bonding (a zero-point energy effect). The stiffness is sensitive to all vibrations that involve motion of the isotopic atom, with stretching ones the most important. But bending and torsional motions (if the isotopic atom is one of the end atoms and not much heavier than the other end atom) can also be major contributors.

If the stiffness of bonding of a secondary hydrogen in the transition state is between that in substrate and product, the secondary deuterium or tritium isotope effect will lie between unity and the equilibrium isotope effect. In a late (that is, productlike) transition state, the value will be closer to ${}^{D}K_{eq}$, while for an early (substratelike) transition state, the value will be closer to unity. If the bonding in the transition state is stiffer than in either substrate or product, the isotope effect will be inverse and outside the range of unity and ${}^{D}K_{eq}$. Conversely, if the hydrogen is more weakly bonded in the transition state than in either substrate or product, the isotope effect will be normal and outside the range of unity and ${}^{D}K_{eq}$. The same principles apply to ${}^{13}C$, ${}^{15}N$, and ${}^{18}O$ isotope effects, and in a number of cases the bending and torsional modes are important for the observed isotope effects (aspartate transcarbamoylase and OMP decarboxylase discussed above).

Primary isotope effects are not determined solely by the stiffness of bonding. At the transition state the vibration corresponding to the reaction coordinate motion does not have a restoring force and has an imaginary frequency, which depends on the curvature of the barrier. This imaginary frequency factor is temperature-independent, unlike the zero-point energy contribution, which extrapolates to unity at infinite temperature (equilibrium isotope effects and secondary ones, except where coupled hydrogen motions are involved, also extrapolate to unity at infinite temperature). Finally, this factor depends on the degree to which the atom involved is moving in the transition state. An example given below is that ${}^{13}C$ isotope effects for primary alcohol oxidation are much smaller than for secondary alcohol dehydrogenation, since the carbon is moving in the transition state in the latter case but the secondary hydrogen is moving in the case of primary alcohols.

With primary deuterium or tritium isotope effects where hydrogen is being transferred, the size of the primary isotope effect is small for early or late transition states but is at a maximum when the hydrogen is symmetrically placed at the transition state. We will give examples below. The ^{13}C isotope effect when a bond to carbon is broken depends on how early or late the transition state is. An early transition state gives a smaller number, and a late transition state a larger one, but the value does not go through a maximum, unlike the isotope effect in the hydrogen being transferred. These principles are readily illustrated by studies that involved changes in transition states as the redox potentials of nucleotide substrates were changed.

Formate Dehydrogenase. Isotope effects measured with formate dehydrogenase are shown in Table 9-3 (B *23*, 5479).

As the redox potential of the dinucleotide becomes more positive, the equilibrium constant becomes more favorable by 1–2 orders of magnitude. This causes the transition state to become earlier, as shown by the smaller ^{13}C isotope effects. The primary deuterium isotope effects become larger since the hydride being transferred is more nearly symmetrically arranged in the transition state. The transition state with NAD appears to be late because of the smaller primary deuterium isotope effect and the very large secondary deuterium isotope effect of 1.23, which arises from coupled motions of the hydrogens. That is, the out-of-plane bending of the secondary hydrogen at the 4-position of the NAD ring is

Table 9-3. Isotope Effects on the *V/K* Values for the Formate Dehydrogenase Reaction

Substrate	NAD	thio-NAD	acetylpyridine-NAD
Redox potential	−0.320 V	−0.285 V	−0.258 V
primary D in formate	2.17	2.60	3.32
secondary D (unlabeled formate)[a]	1.23	1.18	1.06
secondary D (deuterated formate)[a]	1.07	1.03	0.95
^{13}C in formate	1.042	1.038	1.036
^{18}O in formate	1.005	1.007	1.008

[a]Deuteration at C-4 of the pyridine ring of the nucleotide.

coupled into the reaction coordinate motion. As the transition state becomes earlier, the coupling is less and the secondary deuterium isotope effect is reduced. When the formate is deuterated, these secondary deuterium isotope effects are reduced halfway to the equilibrium isotope effect, which is 0.89 in this case, as the hydrogen is more stiffly bonded in NADH than in NAD.

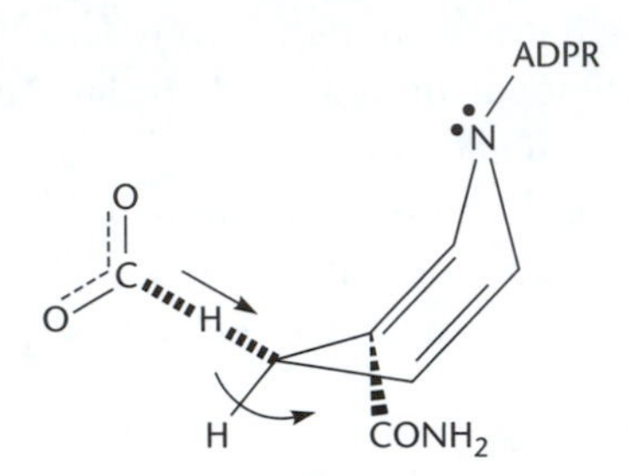

This is the result of eliminating the tunneling, which always seems to be involved with hydrogen motions.

The ^{18}O isotope effects vary in the expected direction but are not inverse as one might expect from the equilibrium isotope effect of 0.983. It is probable that the normal ^{18}O isotope effects seen in the enzymatic reaction come from a normal isotope effect on dehydration as formate enters the active site. While it is possible that some of the effect comes from motion of the oxygens in the transition state, the two oxygens have a mass of 32 and the carbon has a mass of 12, so most of the motion that makes the carboxyl group become linear in the CO_2 product is carbon motion. The I_2 oxidation of formate in DMSO (which has an early transition state) does give an inverse ^{18}O isotope effect of 0.9938, as expected.

Alcohol Dehydrogenase. A similar study was carried out on the oxidation of benzyl alcohol by liver alcohol dehydrogenase (Table 9-4) (B 23, 5471).

Table 9-4. Isotope Effects on the *V*/*K* Values for the Alcohol Dehydrogenase Reaction

Substrate	NAD	thio-NAD	acetylpyridine-NAD
Redox potential	−0.320 V	−0.285 V	−0.258 V
primary and secondary D[a]	4.05	4.77	6.25
^{13}C in benzyl alcohol	1.0254	1.0136	1.0115

[a] The benzyl alcohol was dideuterated.

The intrinsic deuterium isotope effects increase as the redox potential of the dinucleotide gets more positive, showing that the transition state with NAD is late and gets earlier with the increase in equilibrium constant.

The ^{13}C isotope effects show the opposite trend as expected, since earlier transition states involve less C–H cleavage.

transition-state structures for ^{13}C

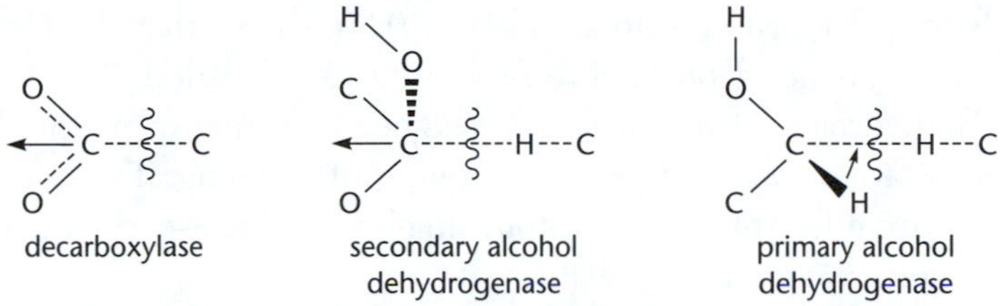

decarboxylase — C moves

secondary alcohol dehydrogenase — C moves

primary alcohol dehydrogenase — sec H moves

Note that the ^{13}C isotope effects are much smaller than those seen with formate or glucose-6-P dehydrogenases or with decarboxylations. The reason is that as the carbon changes from tetrahedral to trigonal during the oxidation, it does not have to move appreciably. The motion needed to bring about the change in geometry is made by the secondary hydrogen of the primary alcohol. In the oxidation of a secondary alcohol (as with glucose-6-P dehydrogenase) or in a decarboxylation, the carbon has to move to bring about the change in geometry, and thus is in motion in the transition state. This leads to much larger isotope effects.

Glutamate Mutase. An interesting use of isotope effects was used to study the formation of 5′-deoxyadenosine from adenosylcobalamin in the reaction catalyzed by glutamate mutase (B *43*, 2155). In the presence of glutamate the Co–C bond breaks, yielding a very short-lived 5′-deoxyadenosyl radical that abstracts a proton from glutamate, which then rearranges and retrieves a proton from the 5′-deoxyadenosine with subsequent re-formation of the Co–C bond.

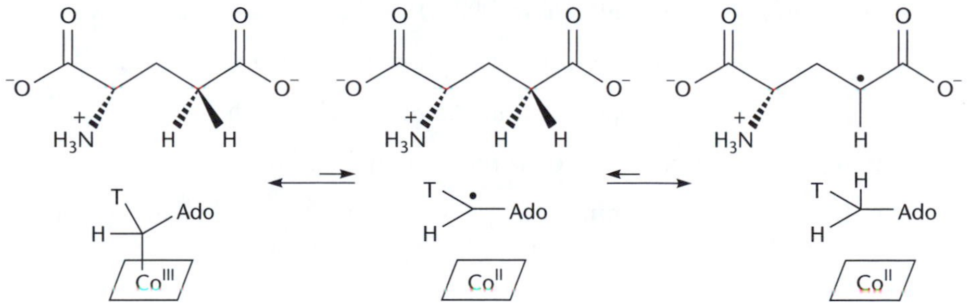

With pre-steady-state and rapid quench techniques, the formation of the 5′-deoxyadenosine intermediate was followed when a mixture of [5′-^{3}H]-deoxyadenosine and ^{14}C-labeled 5′-deoxyadenosine was mixed with glutamate and enzyme. The kinetic secondary tritium isotope effect on the initial rate was 0.76, while the equilibrium isotope effect on the formation of 5′-deoxyadenosine was 0.72. This suggests that the transition state is fairly late, as expected for an unfavorable reaction. But the highly inverse equilibrium isotope effect, which corresponds to a deuterium isotope effect of 0.80, shows that the hydrogens on the 5′-carbon of adenosylcobalamin are very weakly bonded. The fractionation factor of CH_2 hydrogens is normally 10% higher than that of methyl hydrogens, so the value 20% less shows a great weakening of bonding. This may represent a strong inductive effect of the Co^{3+} atom attached to the 5′-carbon to produce an effect like hyperconjugation.

Acyl and Phosphoryl Transfers. Isotope effects have proven very useful in probing transition-state structure for acyl and phosphoryl transfers. In studies of the reaction of *p*-nitrophenyl acetate with various nucleophiles, five different isotope effects were determined by the remote label method, by use of the nitrogen of the nitro group as the remote label (JACS *116*, 11256).

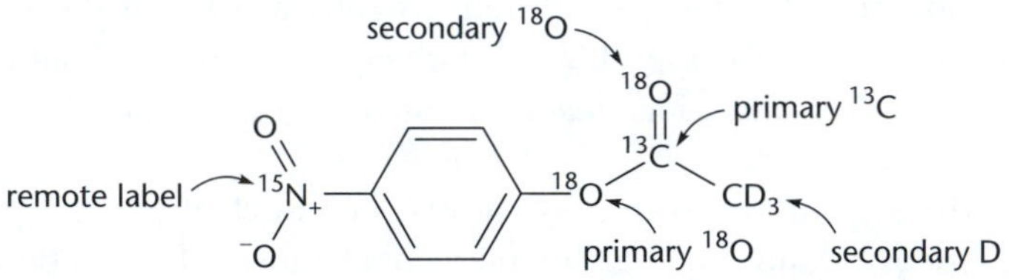

The isotope effects measured were as follows:

(1) The ^{15}k value in the nitro group, which determines the degree of electron delocalization into the nitro group (the equilibrium ^{15}N isotope for deprotonation of *p*-nitrophenol is 0.23%).

(2) The primary ^{18}O isotope effect in the phenolic oxygen.

(3) The secondary ^{18}O isotope effect in the carbonyl oxygen.

(4) The primary ^{13}C isotope effect in the carbonyl carbon.

(5) The β-secondary deuterium isotope effect in the methyl group (three deuteriums), which reports changes in hyperconjugation.

Table 9-5 shows the observed isotope effects in both enzymatic and nonenzymatic reactions. With oxygen nucleophiles the primary ^{18}O isotope effect is ~2%, with the exception of hydroxide, which appears to have a very early transition state, and the secondary ^{18}O isotope effect is 0.4–0.6%. The ^{15}N isotope effect is ~60% of the equilibrium isotope effect for ionization of *p*-nitrophenol with the exception of hydroxide because of its early transition state. The ^{13}C isotope effect, where measured, is 3–4%, as expected since all of the chemistry takes place at this carbon. The secondary deuterium isotope effects are several percent inverse, showing that hyperconjugation is reduced in the transition state. The important point, however, is that there is a sizeable primary ^{18}O isotope effect, showing that the reactions are concerted with a tetrahedral transition state but with no intermediate. Any tetrahedral intermediate would partition forward with release of *p*-nitrophenolate ion (p*K* 7) rather than backward with release of the nucleophiles with p*K*s above 9, and this would suppress the value of any primary ^{18}O isotope effect.

Attack by sulfur nucleophiles shows significant differences. The secondary ^{18}O isotope effect is twice as large as those observed with oxygen nucleophiles,

Table 9-5. Isotope Effects in Reactions of *p*-Nitrophenyl Acetate with Nucleophiles or Enzymes

Nucleophile (p*K*)	^{15}k	$^{18}k_{leaving}$	$^{18}k_{carbonyl}$	^{13}k	$^{D}k^{a}$
Hydroxide (15.7)	1.0002	1.0135	1.0039	1.038	0.956
phenolate (9.9)	1.0009	1.0182	1.0039	nd	0.962
$(CF_3)_2CHO^-$ (9.3)	1.0010	1.0210	1.0058	1.029	0.948
$HO(CH_2)_2S^-$ (9.5)	1.0001	1.0219	1.0119	nd	0.978
$^-S(CH_2)_2CO_2CH_3$ (9.3)	1.0003	1.0172	1.0117	nd	0.977
$CH_3O(CH_2)_2NH_2$ (9.7)	1.0011	1.0330	1.0064	1.028	0.968
NH_2OH, pH 6 (13.7)	1.0009	1.0310	1.0082	1.029	0.964
NH_2OH, pH 12	1.0011	1.0074	1.0008	1.034	0.952
Enzyme					
chymotrypsin	1.0011	1.0204	1.0065	1.030	0.982
carbonic anhydrase	1.0005	1.0255	1.0075	1.028	0.999
acid protease	0.9997	1.0141	1.007	1.036	0.986
papain	1.0011	1.0330	1.0064	1.034	0.995

[a]For three deuteriums in the methyl group.

there is less loss of hyperconjugation, and almost no electron delocalization into the nitro group occurs.

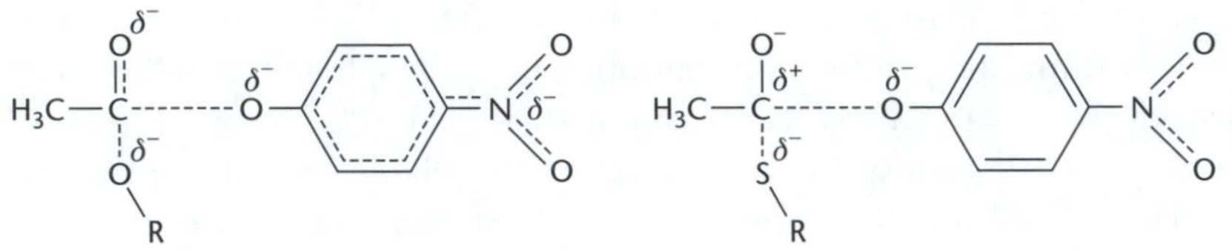

transition states with O and S nucleophiles

These effects suggest that the carbonyl C–O bond has almost become a single bond and the carbonyl carbon has retained a partial positive charge in the transition state. The difference in the transition-state structures with oxygen and sulfur nucleophiles of similar p*K* shows the power of isotope effects to detect transition-state structures.

The rapid reaction of hydroxylamine at pH 12 represents a concerted attack of the oxyanion (p*K* 13.7) with an early transition state, and the isotope effects are similar to those of other oxygen nucleophiles (JACS *119*, 6980).

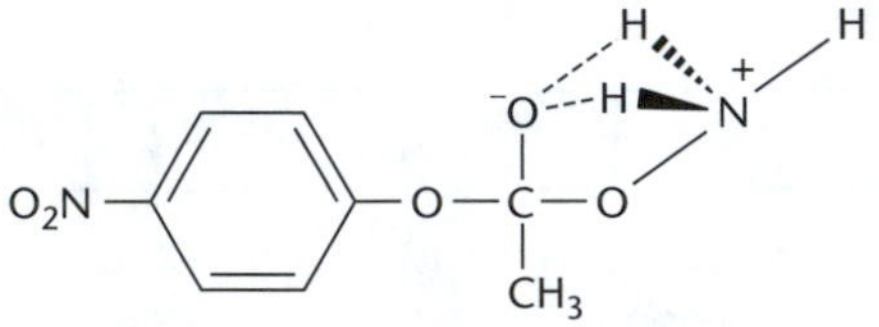

[Although one might think that it would be the amino group of hydroxylamine that attacks *p*-nitrophenyl acetate, Jencks showed that *O*-acetylhydroxylamine was the first product, followed by slow isomerization to the hydroxamate (JACS *80*, 4581, 4585).] At pH 6, however, the primary ^{18}O isotope effect is much larger and the reaction is 400 times slower than at pH 12. The reaction now involves attack by the neutral hydroxyl group, with formation of a zwitterionic intermediate by a proton shift from oxygen to nitrogen. Breakdown of the zwitterionic intermediate to give *p*-nitrophenol will be rate-limiting, as reversal to give hydroxylamine will be very fast. Thus the primary ^{18}O isotope effect is over 3%. The reaction of methoxyethylamine also involves a zwitterionic tetrahedral intermediate whose breakdown is rate-limiting, although now proton transfers must take place through hydrogen bonds between waters.

In the enzymatic hydrolysis reactions listed in Table 9-5, the ^{13}C isotope effects are all at least 3%, showing that the chemistry is rate-limiting (JACS *120*, 2703).

The primary ^{18}O isotope effects are all 2–3% with the exception of the acid protease, where at pH 3 the leaving group will be protonated, thus diminishing the value. This shows that the reactions are all concerted, similar to the nonenzymatic reactions. The secondary ^{18}O isotope effects show that the bond order of the carbonyl group is reduced in the transition state. The ^{15}N isotope effects show a moderate degree of electron delocalization into the nitro group except for the acid protease, where protonation of the leaving group prevents this.

The major difference in the enzymatic and nonenzymatic transition states comes in the β-deuterium isotope effects, which are less inverse with the enzymatic reactions. The β-secondary effects show that in the initial enzyme–substrate complex the carbonyl group is polarized by the enzyme, thus increasing hyperconjugation. Hyperconjugation will then be decreased as the transition state is approached, but from a higher value than with the nonenzymatic reactions.

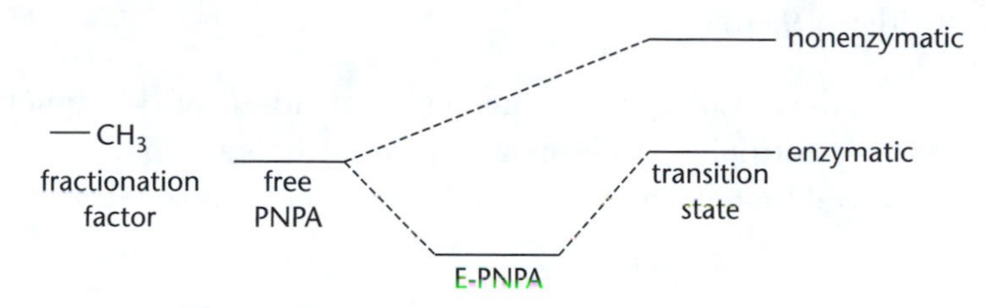

Since the isotope effect compares the transition state with the substrate in solution, not bound in the ES complex, the observed isotope effect is less inverse. The increased polarization of the carbonyl group by the enzyme makes the carbonyl carbon more electrophilic and thus helps catalyze the reaction.

Phosphoryl Transfer. ^{18}O isotope effects have been used extensively to study the transition-state structures of phosphoryl transfers (FASEB J 9, 1585).

Primary ^{18}O isotope effects in the leaving group have been used to show that the chemistry is rate-limiting. These are usually at least 2% unless the leaving group is protonated by the enzyme. Secondary ^{18}O isotope effects in nonbridge oxygens are then used to discern the structure of the transition-state structure.

For phosphate monoesters, for example, the bond order in the nonbridge oxygens of the substrate is 4/3. If the transition state is highly dissociative (that is, the PO_3 unit being transferred resembles metaphosphate), this bond order may be higher in the transition state and thus give an inverse ^{18}O isotope effect. How inverse depends on the actual bond order in the transition state, and in practice the values are only slightly inverse or normal. Both the nonenzymatic and enzymatic reactions of phosphate monoesters at neutral pH show dissociative transition states. It would appear that the axial bond orders in the transition state are 8–15% (FASEB J 9, 1585).

Table 9-6 shows the results of a number of studies of phosphotyrosine phosphatases with *p*-nitrophenyl phosphate or *m*-nitrophenyl phosphate as the substrate, as well as the nonenzymatic hydrolyses of these molecules.

For the enzymatic reactions of *p*-nitrophenyl phosphate, the primary isotope effects are smaller when the leaving group is protonated by the aspartate general acid, while the aspartate mutant enzymes, where *p*-nitrophenolate ion is released without protonation, show larger primary isotope effects; see above. There is

Table 9-6. Isotope Effects for Hydrolysis of Phosphate Monoesters

Enzyme	Primary $^{18}(V/K)$	Secondary $^{18}(V/K)$	Secondary $^{15}(V/K)$	Ref[a]
Enzymatic Cleavage with *p*-Nitrophenyl Phosphate as Substrate				
Rat PTP1	1.0142	0.998	1.000	a
same, D181N	1.028	1.0018	1.0019	a
Yersinia PTPase	1.0152	0.9998	1.000	a
same, D356A	1.0274	1.0007	1.0022	a
human VHR	1.012	1.0003	1.000	b
same, D92N	1.029	1.0019	1.0030	b
Cdc25A	1.036	0.9988	1.0030	c
Nonenzymatic Cleavage of PNPP				
monoanion	1.0106	1.0224	1.0005	d
dianion (H_2O)	1.0230	0.9993	1.0034	d
dianion (*t*-butanol)	1.0202	0.9997	1.0039	d
Enzymatic Cleavage with *m*-Nitrobenzyl Phosphate as Substrate				
Yersinia PTPase	0.9995	0.9999	0.9996	e
Cdc25A	1.0012	0.9983	1.0002	e
Nonenzymatic cleavage of MNBP				
monoanion	0.9982	1.0149	0.9999	e

[a]References: (a) B *34*, 13982; (b) B *35*, 7087; (c) JBC *277*, 11190; (d) JACS *116*, 5045; (e) B *43*, 8256.

also considerable electron delocalization into the nitro group, shown by the ^{15}N isotope effects (compare with the nonenzymatic reactions of the dianion). The secondary isotope effects are around unity, which is consistent with somewhat dissociative transition states. The reactions of *m*-nitrophenyl phosphate are surprising, since they show very small primary isotope effects in both the enzymatic and nonenzymatic reactions. The exception is the nonenzymatic cleavage of the monoanion, in which O–H cleavage causes the nonbridge isotope effect to become a primary one.

Phosphate diesters appear to have S_N2 mechanisms in which the sum of the axial bond orders in the transition state are approximately unity. With uridine 3′-*m*-nitrobenzyl phosphate, a slow substrate for ribonuclease, the primary ^{18}O isotope effect was pH-independent at 1.6%, while the ^{18}O isotope effect in the nonbridge oxygens was 0.5%.

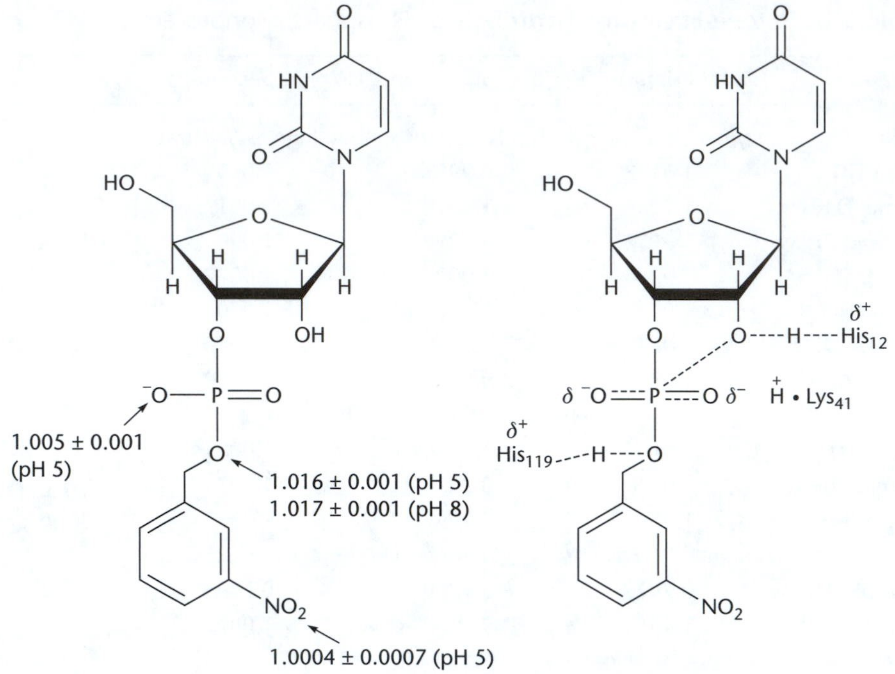

HO
O
HN
O
N
O
H H H H
O OH
$^{-}$O—P═O
1.005 ± 0.001 (pH 5)
O
1.016 ± 0.001 (pH 5)
1.017 ± 0.001 (pH 8)
NO_2
1.0004 ± 0.0007 (pH 5)

HO
O
HN
O
N
O
H H H H
δ^+
O----H----His_{12}
δ $^{-}$O═P═O δ^- $\overset{+}{H}$ • Lys_{41}
δ^+
His_{119}----H----O
NO_2

The small normal secondary isotope effect indicates a slightly associative reaction with a sum of the axial bond orders of 1.13–1.20 (JACS *119*, 2319).

These data support a concerted mechanism in which His-12 acts as a general base to activate the 2′-hydroxyl and His-119 acts as a general acid to protonate the leaving group. Protonation by His-119 lowers the primary isotope effect to the observed 1.6%.

A similar pattern was seen with kanamycin nucleotidyltransferase, which normally reacts MgATP with kanamycin to form kanamycin-4′-AMP and $MgPP_i$ (B *40*, 2972).

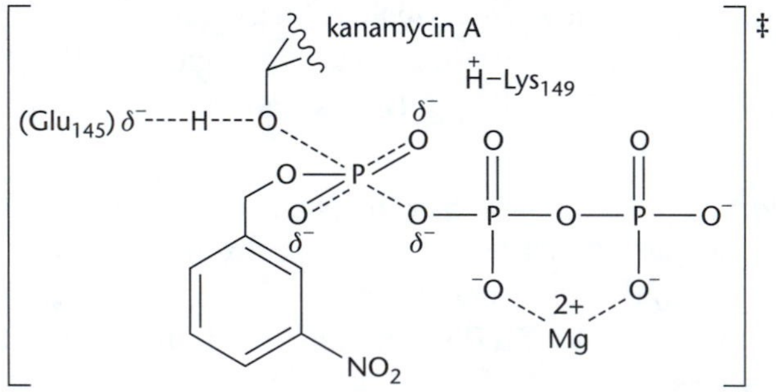

kanamycin A
$\overset{+}{H}$–Lys_{149}
(Glu_{145}) δ^-----H----O
δ^-
O
O O
O—P
O—P—O—P—O$^-$
O δ^- δ^-
$^{-}$O O^{-}
2+
Mg
NO_2
‡

In order to reduce commitments to catalysis, the substrate used was *m*-nitrobenzyl triphosphate, which reacts 2 orders of magnitude more slowly but with the same regiospecificity. The primary ^{18}O isotope effect in the α–β bridge was 1.6% and the secondary ^{18}O isotope effect in the α-nonbridge oxygens was 0.33%. This reaction also appears concerted with a slightly associative transition state.

In contrast to these two enzymes, the reaction of snake venom phosphodiesterase on 3,3-dimethylbutyl *p*-nitrophenyl phosphate gives a secondary ^{18}O isotope effect that is 1.6% inverse, showing that the enzyme has protonated a nonbridge oxygen in the transition state (JACS *113*, 5835).

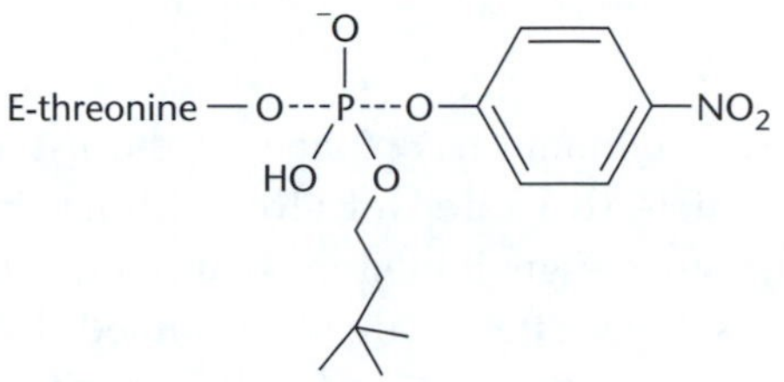

$^{-}$O
E-threonine — O---P---O—C$_6$H$_4$—NO$_2$
HO O

Acyclic phosphotriesters show associative transition states for their hydrolysis with the sum of the axial bond orders being well above unity. The transition state gets

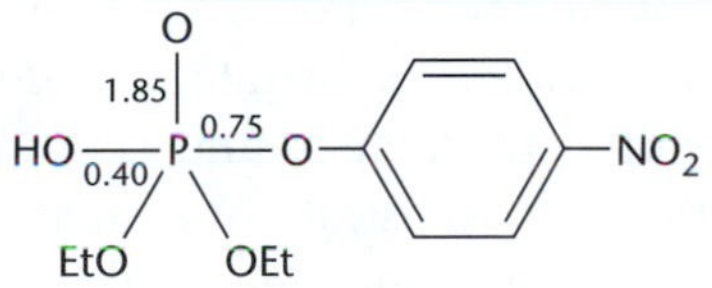

O
1.85
HO 0.40 P 0.75 O—C$_6$H$_4$—NO$_2$
EtO OEt

more associative as the p*K* of the leaving group increases. Table 9-7 shows a number of examples of both enzymatic (B *30*, 7449) and nonenzymatic reactions (JACS *123*, 9246).

For the alkaline hydrolysis of diethyl *p*-nitrophenyl phosphate, the ^{15}N isotope effect in the nitro group (1.0007) indicates about 25% P–O bond cleavage in the transition state, while the secondary ^{18}O isotope effect suggests that the P–O bond order to the nonbridge oxygen is ~1.85. This leaves a bond order to the attacking oxygen (presumably one of the hydrating waters) of hydroxide of ~0.40. The sum of the axial bond orders in the transition state is thus 1.15 in this case. For the molecules with higher p*K* leaving groups, the reactions appear more associative, since the increasing secondary ^{18}O isotope effects point to decreasing bond order to the nonbridge oxygen and thus more along the reaction coordinate.

Table 9-7. Isotope Effects for Phosphotriester Hydrolysis

Alkaline hydrolysis substrate, pK	Primary ^{18}k	Secondary ^{18}k
di-Et-*p*-nitrophenyl-P, 7.0	1.0060	1.0063
di-Et-*p*-carbamoylphenyl-P, 8.6	1.027	1.025
di-Et-choline-P, 13.9	1.041	1.033
di-Et-*m*-nitrobenzyl-P, 14.9	1.052	
Phosphotriesterase substrate, pK	primary $^{18}(V/K)$	secondary $^{18}(V/K)$
di-Et-*p*-nitrophenyl-P, 7.0	1.0020	1.0021
di-Et-*p*-carbamoylphenyl-P, 8.6	1.036	1.018

Glycosyl Ttransferases. Schramm has pioneered the use of isotope effects to determine the transition-state structures for glycosyl transfer reactions and used the resulting knowledge to design inhibitors that mimic the transition states (ME *308*, 301). The isotope effects are determined by the remote label method, with tritium or ^{14}C as the remote label. For tritium isotope effects in the ribose ring, 5′–^{14}C-labeled substrate is mixed with substrate tritiated in the desired position and the change in the ^{14}C-tritium ratio as the reaction proceeds is used to determine the isotope effect. For the ^{14}C isotope effect at C-1, one mixes 1′–^{14}C-substrate with 5′–tritiated substrate. For deuterium isotope effects, a substrate dual-labeled with deuterium in the position of interest and with tritium at C-5′ is mixed with 5′–^{14}C-substrate. A similar approach is used for the ^{15}N isotope effect in N-9 of the adenine ring or the ^{18}O isotope effect in the ribose ring.

The method thus allows for measurement of any isotope effect in any position of the substrate. With care in counting, isotope effects can be determined with a precision of 0.002. This is an order of magnitude less sensitive than the use of isotope ratio mass spectrometry but is sufficient for deuterium or tritium isotope effects and ^{14}C, ^{15}N, or ^{18}O effects of 1% or greater. Intrinsic ^{13}C isotope effects are given by $1+[^{14}(V/K) - 1]/1.9$. That is, a ^{13}C isotope effect of 1% corresponds to a ^{14}C effect of 1.9%. Intrinsic deuterium isotope effects are related to tritium ones by the Swain–Schaad relationship (see page 276 on Northrop's Method).

The results shown in the scheme with nucleoside hydrolase are typical of this method. Note a 4.4% ^{14}C isotope effect at C-1′ and a 2.6% ^{15}N one, showing that

C–N cleavage is taking place in the transition state.

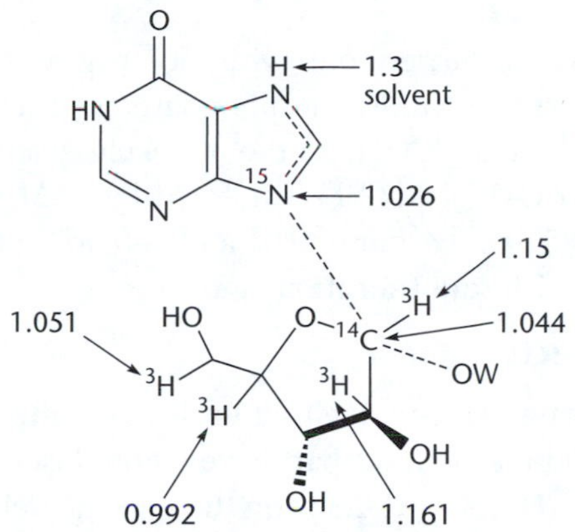

The tritium isotope effects at C-1′ and C-2′ show that there is considerable oxocarbenium ion character to the transition state, but modeling of the transition-state structure with the BEBOVIB program, which calculates isotope effects for assumed structures, showed that interaction with the water that attacks C-1′ was essential to give the observed isotope effects. The transition-state structure that reproduced the isotope effects is given in the scheme below.

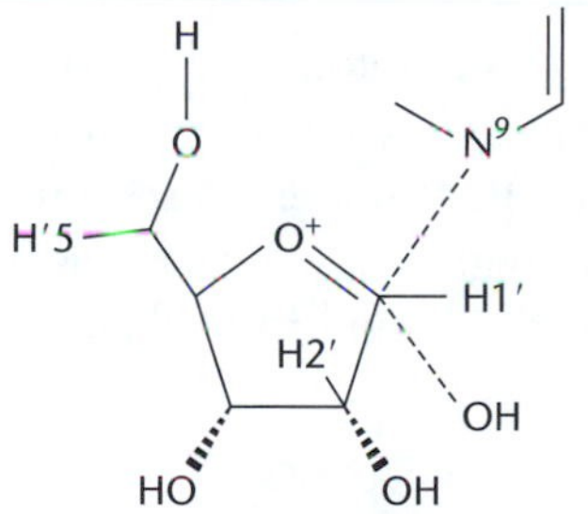

	expected KIE	observed KIE
^{15}N9	1.03	1.04
^{3}H1′	1.15	1.15
^{14}C1′	1.04	1.04
^{3}H2′	1.09	1.16
^{3}H5′	1.00*	1.05*

*KIE unexpected at this position unless enzyme-specific sp^3 distortion occurs

The C-1′–N-9 bond that is cleaved is 1.97 Å in the transition state versus 1.48 Å in inosine, while the bond between C-1′ and the attacking water is 3.0Å. Neither a fully dissociative mechanism with an oxocarbenium ion intermediate nor an S_N2 mechanism with 0.5 bond order to leaving group and attacking water reproduced the observed isotope effects.

Of interest is the tritium isotope effect of 5% at the C-5′ position of the ribose ring. This large effect was attributed to hydrogen bonding between the C-5′-hydroxyl and the ribose ring oxygen, and this geometry was later confirmed with an X-ray structure of the enzyme in complex with an inhibitor mimicking the transition state.

Similar studies have been carried out with other glycosyl transferases. The transition state for purine nucleoside phosphorylase, for example, shows a 1.77 Å distance to N-9 of the base and 3.0 Å to the attacking arsenate. In contrast, the attack of methionine on C-5′ of MgATP catalyzed by SAM synthetase is an S_N2 reaction, with an S to C-5′ bond order of 0.61 and C-5′ to tripolyphosphate leaving group bond order of 0.35 in the transition state.

Binding Isotope Effects

In our previous discussion we have focused on isotope effects that occur during the catalytic reaction and for the most part have ignored isotope effects caused by binding. However, the ^{18}O isotope effects on the formate dehydrogenase reaction (see pages 301–308) suggested a ~1% effect on binding of formate caused by removal of hydrating water. Similarly, the increased hyperconjugation in the methyl group of *p*-nitrophenyl acetate induced by binding to enzymes (see page 313) is a binding isotope effect. Equilibrium isotope effects on dissociation constants of aldehydes from kinases (see page 305) or from leucine aminopeptidase, where a value of unity showed that leucinal was bound as the hydrate (B *24*, 330), are also binding isotope effects.

Isotope effects on the bimolecular rate constant for the binding of PEP to PEP carboxylase or pyruvate kinase in the presence of saturating levels of the second substrate (bicarbonate or MgADP) were inverse (0.9943 with PEP carboxylase; 0.993 with pyruvate kinase). The *V/K* value for PEP under these conditions is just the bimolecular rate constant for binding. These data suggest that the bridging oxygen was more stiffly bonded in the bound form than in solution (B *34*, 2577).

The equilibrium isotope effect for binding of 1-^{18}O-oxamate to a lactate dehydrogenase–NADH complex was 0.984, showing that oxamate is more stiffly bonded in the active site than in solution. This is partly caused by an increased force constant for the O–C–C–(O,N) torsional modes when oxamate is bound (B *34*, 6050).

The binding of 4′-deuterated NAD to lactate dehydrogenase showed an equilibrium isotope effect of 1.10 ± 0.03, while the tritium isotope effect in this position was 1.085 ± 0.01 (corresponding to a deuterium isotope effect of 1.06). This 6–10% isotope effect upon deuteration indicates that the C-4′ hydrogen is less stiffly bound in the binary complex than in solution and appears to result from a movement of 0.3–0.5 of the positive charge at N-1′ to the 4′-position, which of course poises this carbon for receiving a hydride ion from the substrate (B *28*, 3619).

Lewis and Schramm have determined tritium isotope effects for the binding of glucose to brain hexokinase, either as a binary complex or in the presence of Mg–β,γ-CH_2-ATP. The remote label was ^{14}C at C-2 or C-6. The isotope effects are given in Table 9-8.

The normal isotope effects on binding come from formation of hydrogen bonds to the OH at that carbon, while inverse effects come from stiffer bonding. The large normal effects at C-4 and C-6 result from interaction with Asp657, the general base in the reaction. The inverse values at C-2 and C-5 are thought to result from steric compression by Ser603 and Asn683, respectively. The isotope effects at C-1, C-5, and C-6 change upon addition of the ATP analogue. The oxygen at C-6 is clearly more immobilized in the ternary complex, decreasing the size of the isotope effect, while the C-4 oxygen is not affected. The loosening of the C-5 hydrogen is attributed to relief of the steric compression generated by Asn683. The lower isotope effect at C-1 in the ternary complex suggests that the hydrogen bond of the 1-OH to Glu742 is weaker. Overall, this study shows the detail about binding and in particular hydrogen bonding that can be derived from careful measurements of binding isotope effects.

Transient-State Kinetic Isotope Effects

In steady-state kinetic studies, one determines the rate of formation of a product or disappearance of a substrate and one is not looking at intermediates in the

Table 9-8. Equilibrium Tritium Isotope Effects for Binding of Glucose to Brain Hexokinase[a]

Position	binary complex	ternary complex[b]
C-1	1.027	1.013
C-2	0.927	0.929
C-3	1.027	1.031
C-4	1.051	1.052
C-5	0.988	0.997
C-6 (both H)	1.065	1.034

[a] JACS *125*, 4672.

[b] With Mg–β,γ-CH_2-ATP.

reaction. The theory for isotope effects presented up to now in this chapter involves comparison of these rates for reactants containing light and heavy isotopes. In transient-state studies, however, one *is* looking at the intermediates, using multiwavelength spectroscopy or quench-flow methods. With the time courses of the several intermediates, one can study their interconversion, and isotope effects are quite useful for this.

Fisher (ABB *425*, 165) has developed the theory of transient-state isotope effects. He defines

$$tKIE_x(t) = \frac{\left(\frac{d[X_H]}{dt}\right)}{\left(\frac{d[X_D]}{dt}\right)} \tag{9-98}$$

for a deuterium isotope effect on the reaction of intermediate X, where *t*KIE is the ratio of rates for light and heavy isotopes at any time, *t*. The limit of this parameter at $t = 0$ is $tKIE^0$. There are three cases, depending on which step

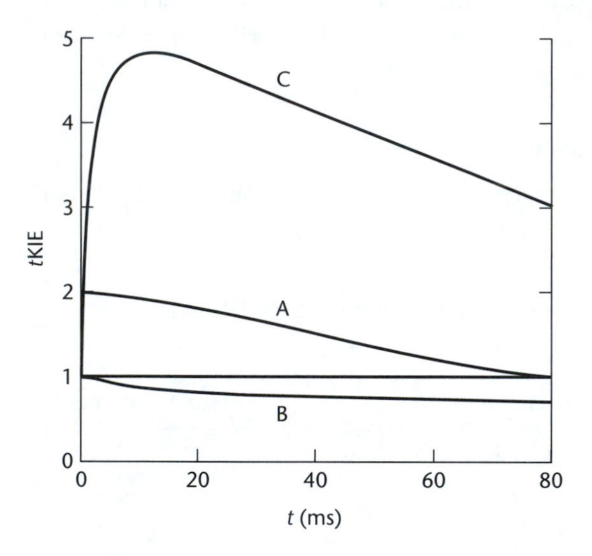

Figure 9-4. Transient kinetic isotope effect as a function of time for single isotopic substitution. (A) Only step(s) after the isotope-sensitive step contribute to the transient, case 1. (B) Only step(s) prior to the isotope-sensitive step contribute to the transient, case 2. (C) Step(s) before and after the isotope-sensitive step contribute to the transient, case 3.

is isotope-sensitive and which step gives the measured signal. We are assuming a normal isotope effect, ${}^{D}k$ (see Figure 9-4).

Case 1: The measured signal comes solely from an intermediate following (or produced by) the isotope-sensitive step. If there is only one isotope-sensitive step, the observed $t\mathrm{KIE}^0 = {}^{D}k$ at $t=0$ but decreases with time. If there are several isotope-sensitive steps, $t\mathrm{KIE}^0$ is the product of the ${}^{D}k$ values on the several steps.

Case 2: The measured signal comes from an intermediate preceding the isotope-sensitive step. In this case the observed $t\mathrm{KIE}^0$ is unity at $t=0$ and decreases with time.

Case 3: If there are isotope effects on steps both before and after the formation of the intermediate giving the measured signal, the $t\mathrm{KIE}^0$ at $t=0$ is unity, rises to a maximum value, and then decays with time.

The $t\mathrm{KIE}^0$ values at $t=0$ calculated for the various intermediates contain the information needed to evaluate intrinsic isotope effects on the various steps. Thus if the sequence of intermediates is

$$\mathrm{E} \xrightarrow{k_1\mathbf{A}} \mathrm{EA} \xrightarrow{k_3} \mathrm{EA'} \xrightarrow{k_5} \mathrm{EA''} \xrightarrow{k_7} \mathrm{EPQ} \tag{9-99}$$

the $t\mathrm{KIE}^0$ values at $t=0$ for the intermediates are given in equations 9-100 to 9-103:

$$\mathrm{EA} \qquad {}^{D}k_1 \tag{9-100}$$

$$\mathrm{EA'} \qquad {}^{D}k_1{}^{D}k_3 \tag{9-101}$$

$$\mathrm{EA''} \qquad {}^{D}k_1{}^{D}k_3{}^{D}k_5 \tag{9-102}$$

$$\mathrm{EPQ} \qquad {}^{D}k_1{}^{D}k_3{}^{D}k_5{}^{D}k_7 \tag{9-103}$$

This will be true even though the various steps are all reversible. Thus the intrinsic isotope effect for a given step is the $t\mathrm{KIE}^0$ at $t=0$ for that step divided by the $t\mathrm{KIE}^0$ for the preceding step.

To use these methods, one must be able to measure time courses for each intermediate, measure the slopes of the time courses, and extrapolate the ratios of the slopes back to $t=0$ to get the $t\mathrm{KIE}^0$ value. Fisher has used this approach successfully to study the kinetics of glutamate dehydrogenase. He gives a clear exposition of the application of this theory in *Enzymatic Mechanisms*; Frey, P.A., & Northrop, D.B., Eds.; IOS Press: Burke, VA, 1999; p. 264.

10

pH DEPENDENCE OF KINETIC PARAMETERS AND ISOTOPE EFFECTS

The pH dependence of kinetic parameters provides information on kinetic and chemical mechanism. The combination of pH and isotope effects can provide further information on chemical mechanism and the location of isotope-sensitive steps. In this chapter theory will be developed for the behavior of kinetic parameters as pH is varied, by use of unlabeled and deuterium (or heavier) labeled reactants.

pH–Rate Studies

The study of pH profiles allows one to tell the required state of protonation of groups required for binding and/or catalysis, and thus to detect acid–base catalysis and gain information on the chemical mechanism of an enzymatic reaction.

There are a number of things that must be considered when one carries out pH studies, and these are considered briefly below.

(1) *Stability of the enzyme over the pH range studied*. The stability of the enzyme as a function of pH can be determined as follows. Enzyme is

preincubated at a given pH and aliquots are withdrawn as a function of time and assayed at a pH where the enzyme is known to be stable, for example, pH 7. If the enzyme is stable over several minutes, this is normally sufficient since pH studies are usually done by the pH-jump method. For pH jump, enzyme is maintained at a pH where it is stable in dilute buffer, for example, 10 mM. Parameters of interest are then determined by diluting enzyme to start the reaction in a reaction mixture at the desired pH with 100 mM buffer. Since enzyme is usually diluted 100-fold, for example, 10 μL to 1 mL final volume, the final concentration of the buffer the enzyme is stored in is 0.1 mM, which will not contribute significantly to the final assay pH.

(2) *pH dependence of kinetic mechanism.* The kinetic mechanism can change as the pH is changed if the off-rate constants for reactant change with pH. This is known to occur in a number of cases, including the creatine kinase reaction where the mechanism is equilibrium ordered at low pH (JBC *248*, 8418), steady-state random at neutral pH (B *20*, 1204), and rapid equilibrium random at high pH (JBC *241*, 673). Similar changes in mechanism have been reported for dopamine β-hydroxylase (B *22*, 3096) and the cAMP-dependent protein kinase in the direction of phosphorylation of MgADP (B *32*, 6802). Thus, before pH studies are attempted, the initial velocity pattern in the absence of products and select diagnostic initial rate patterns should be carried out at the extremes of pH, for example, 5.5 and 9.5. In the case of an ordered mechanism, a diagnostic pattern would be the uncompetitive dead-end inhibition by an analogue of B with A as the varied substrate. In this way one can be certain that the mechanism is pH-invariant or if it does vary with pH the investigator knows how it changes. These studies will also provide preliminary estimates of the kinetic parameters at the extremes of pH values.

(3) *Effect of buffer on activity.* There are a number of examples in the literature of buffers inhibiting an enzyme reaction. The buffer normally acts as a competitive inhibitor of one or more of the reactants. Thus, the effect of changing buffer concentration on the reaction and the use of different buffers at the same pH will distinguish between a buffer and an ionic strength effect. The effect of the buffer should also be determined at the extremes of buffer capacity for a given buffer (p$K \pm 1$ pH unit), since the predominant protonation state of the buffer is different under these two conditions. For these studies, reactant concentrations should be maintained $\leq K_m$.

Buffers are used at a concentration sufficient to maintain the pH of the reaction constant at all concentrations of reactant used.The pH can be changed with single buffers at a range of p$K \pm 0.5$ pH unit or with a mixed buffer system.

Single Buffers with Overlap

For a single buffer system, the rate must be measured so that the buffer range of the two buffers overlaps and inhibitory buffer effects can be identified. Ionic strength should be maintained as nearly constant as possible by use of a background electrolyte. This can be accomplished either by calculating the amount of electrolyte required to maintain ionic strength low and constant or by adding a high concentration of electrolyte so that the concentration of species other than the electrolyte have no effect on ionic strength. The former would be used for an ionic strength-sensitive enzyme, while the latter would be used for an enzyme insensitive to ionic strength. A homologous buffer series that is mostly innocuous to enzymes is that developed by Good: 2-(*N*-morpholino)ethanesulfonic acid (Mes), piperazine-*N*,*N′*-bis (2-ethanesulfonic acid) (Pipes), *N*-(2-hydroxyethyl)piperazine-*N′*-2-ethanesulfonic acid (Hepes), 3-[tris(hydroxymethyl) methyl]aminopropanesulfonic acid (Taps), 2-(cyclohexylamino)ethanesulfonic acid (Ches), and 3-(cyclohexylamino)propanesulfonic acid (Caps) (B 5, 467).

Mixed Buffer Systems

A series of buffers is selected to cover the desired pH range, set at a concentration, as above, that will be higher than the highest reactant concentration, mixed together, and titrated to the desired pH value. If the investigator decides on 50 mM as the optimum buffer concentration and Mes, Pipes, and Hepes as the buffers, then a stock solution that contains 250 mM Mes, 250 mM Pipes, and 250 mM Hepes would be prepared and titrated to the correct pH. This mixture when diluted 1/5 would give the desired result. This method has the advantage that ionic strength is fairly constant.

The parameters plotted versus pH are as follows:

(1) The K_i, dissociation constant, for a competitive inhibitor, a metal ion, or a substrate other than the last one to add to the enzyme, or K_{act}, the activation constant for an activator. Data are plotted as pK_i or pK_{act}, that is, $\log_{10}(1/K_i)$ or $\log_{10}(1/K_{act})$, so that any decrease in the parameter is seen as a decrease in affinity.

(2) The $\log_{10}$ of the V/K for a reactant, the apparent first-order rate constant at low reactant concentration at a saturating concentration of all other

reactants. This parameter reflects protonation changes in free reactant and in the enzyme form with which it combines.

(3) The $\log_{10}$ of V, the rate at saturating concentrations of all reactants.

Since Michaelis constants are simply ratios of V and V/K and are thus not independent kinetic constants, their pH variation is complex and is not normally one of the profiles plotted. All of the above are log–log plots and should be plotted with equal vertical and horizontal scales. The reader should see the review article by Cleland (AdvEnz. *45*, 273) for a more complete treatment of pH studies.

pH Dependence of Equilibrium Dissociation Constant

The simplest pH profiles to interpret are those for the pH dependence of K_i, since they represent effects on binding only. Further, since K_i is an equilibrium dissociation constant, the p*K*s one sees in such profiles are actual p*K*s in the ligand or in the enzyme form with which it combines. It is thus very useful to determine true pK values from pK_i profiles for comparison with apparent values seen in log V/K or log V profiles.

When protonation or deprotonation of a group in inhibitor or enzyme decreases but does not eliminate binding, the following thermodynamic cycle applies:

$$\begin{array}{ccc} EH & \underset{}{\overset{\mathbf{I}/K_j}{\rightleftharpoons}} & EIH \\ \mathbf{H}/K_1 \updownarrow & & \updownarrow \mathbf{H}/K_2 \\ E & \underset{}{\overset{\mathbf{I}/K_i}{\rightleftharpoons}} & EI \end{array} \qquad (10\text{-}1)$$

In mechanism 10-1, **H** is hydrogen ion concentration, K_1 and K_2 are acid dissociation constants for EH and EIH, respectively, while K_i and K_j are dissociation constants for inhibitor from EI and EIH, respectively. The same behavior will be observed if I, and not E, becomes protonated to affect binding. The mechanism is applicable for a competitive inhibitor, since inhibition is obtained under conditions where A tends to zero, and thus all steps prior to addition of reactant, protonation of enzyme and binding of inhibitor, come to thermodynamic equilibrium. It is possible, however, that not all steps in mechanism 10-1 are observed, and several cases will be considered below.

When protonation of ligand or enzyme eliminates binding, the pK_i profile has a constant value at high pH and decreases with a slope of 1 below the pK, as shown in Figure 10-1.

Under these conditions the equation for the apparent K_i is

$$\text{app}K_i = K_{i\ \text{high pH}}\left(1 + \frac{\mathbf{H}}{K}\right) \tag{10-2}$$

or

$$\text{p}K_i = \text{p}K_{i\ \text{high pH}} - \log\left(1 + \frac{\mathbf{H}}{K}\right) \tag{10-3}$$

where **H** is $[H^+]$ and K is the acid dissociation constant of the group for which protonation eliminates binding. Tangents to the two linear portions of the curve intersect at the pK of the group being protonated. The vertical coordinate of the point corresponding to the pK lies 0.3 unit (log 2) below the high-pH plateau, and the curved region extends from one pH unit above the pK to one below it.

When deprotonation of a group prevents binding, the curve has a constant value at low pH and decreases with a slope of –1 at high pH, as shown in Figure 10-2.

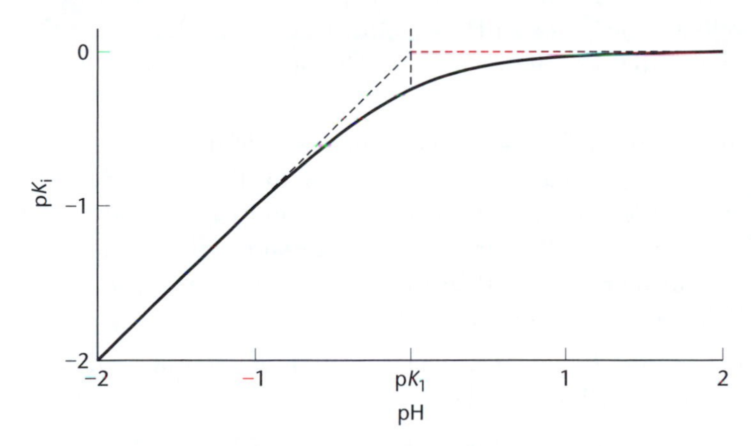

Figure 10-1. Plot of pK_i [log (1/K_i)] versus pH for the case where protonation eliminates binding affinity. Note that linear segments have slopes of 0 (pH-independent value of pK_i, high pH) and +1 (low pH), which intersect at the value of pK_1. pH is expressed relative to the pK of the group titrated.

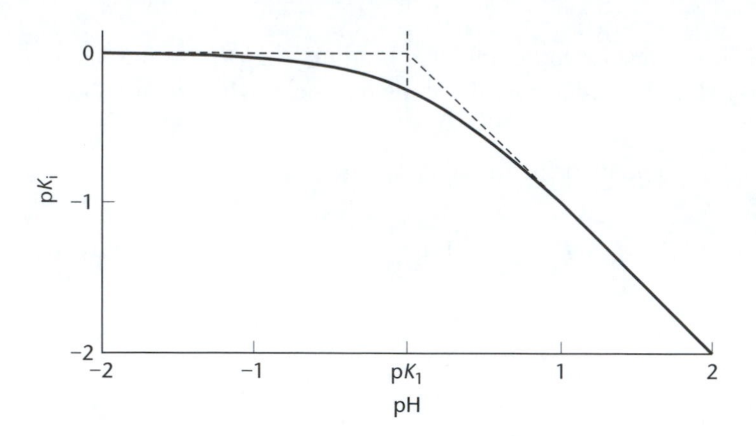

Figure 10-2. Plot of pK_i versus pH for the case where deprotonation eliminates binding affinity. The linear segments have slopes of 0 (pH-independent value of pK_i, low pH) and –1 (high pH), which intersect at the value of pK_1. pH is expressed relative to the pK of the group titrated.

The equation is now

$$\mathrm{p}K_\mathrm{i} = \mathrm{p}K_{\mathrm{i\,low\,pH}} - \log\left(1 + \frac{K}{\mathbf{H}}\right) \tag{10-4}$$

where K is the dissociation constant of the group. The plot curves as it decreases below the pH-independent region over 2 pH units, that is, the p$K \pm 1$. At the pK the curve is 0.3 units below the low-pH asymptote. The intersection of the two tangents to the curve is at the pK.

If the inhibitor has a finite affinity for both protonation states of the enzyme (or both protonation states of a ligand have significant affinity for the enzyme), the complete thermodynamic cycle shown in mechanism 10-1 must be considered. The inhibitor can bind more tightly to protonated or unprotonated enzyme (or protonated ligand can bind more tightly than the unprotonated form), that is, $K_\mathrm{j} < K_\mathrm{i}$ or vice versa. The latter case is shown in Figure 10-3.

Expressions can be written for K_1 and K_2, equation 10-5, and for K_i and K_j, equation 10-6, and are solved for **EI** and **EHI**:

$$K_1 = \frac{(\mathbf{E})(\mathbf{H})}{\mathbf{EH}} \quad \mathbf{EH} = \frac{(\mathbf{E})(\mathbf{H})}{K_1} \quad K_2 = \frac{(\mathbf{EI})(\mathbf{H})}{\mathbf{EIH}} \quad \mathbf{EIH} = \frac{(\mathbf{EI})(\mathbf{H})}{K_2} \tag{10-5}$$

$$K_\mathrm{i} = \frac{(\mathbf{E})(\mathbf{I})}{\mathbf{EI}} \quad \mathbf{EI} = \frac{(\mathbf{E})(\mathbf{I})}{K_\mathrm{i}} \quad K_\mathrm{j} = \frac{(\mathbf{EH})(\mathbf{I})}{\mathbf{EHI}} \quad \mathbf{EHI} = \frac{(\mathbf{EH})(\mathbf{I})}{K_\mathrm{j}} \tag{10-6}$$

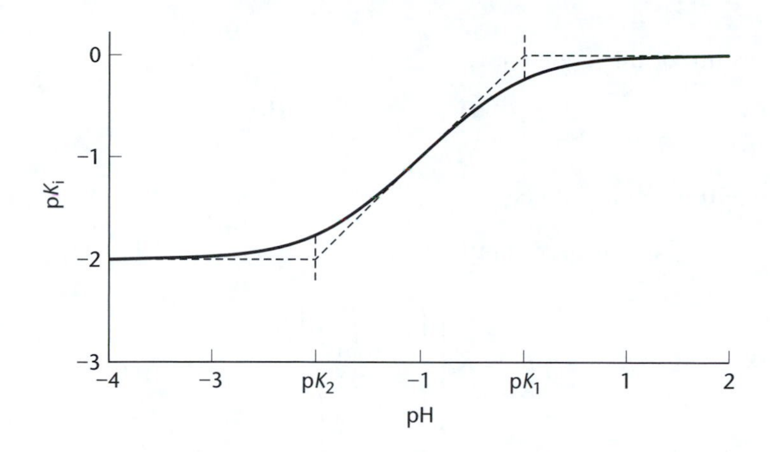

Figure 10-3. Plot of pK_i versus pH for the case where protonation decreases but does not eliminate binding. The linear segments have slopes of 0 [pH-independent value of pK_i (high) pH and pK_j (low) pH] and +1 (dashed line drawn through the midpoint of the change). The lines intersect at values of pK_1 (high pH) and pK_2 (low pH). pH is expressed relative to the pK of the group titrated.

An expression can then be written for appK_i (equation 10-7):

$$\text{app}K_i = \frac{(\mathbf{E} + \mathbf{EH})\mathbf{I}}{(\mathbf{EI} + \mathbf{EHI})} \tag{10-7}$$

Substituting for **EH**, **EI**, and **EHI** in equation 10-7 from equations 10-5 and 10-6 and rearranging gives equation 10-8:

$$\text{app}K_i = \frac{\left(1 + \dfrac{\mathbf{H}}{K_1}\right)}{\left(\dfrac{1}{K_i} + \dfrac{\mathbf{H}}{K_1 K_j}\right)} \qquad \text{app}K_i = K_i\left[\frac{\left(1 + \dfrac{\mathbf{H}}{K_1}\right)}{\left(1 + \dfrac{K_i\mathbf{H}}{K_1 K_j}\right)}\right] \tag{10-8}$$

and since $K_1K_j = K_2K_i$

$$\text{app}K_i = K_i\left[\frac{\left(1 + \dfrac{\mathbf{H}}{K_1}\right)}{\left(1 + \dfrac{\mathbf{H}}{K_2}\right)}\right] \tag{10-9}$$

and the negative logarithm of this equation is

$$pK_i = pK_{i\,\text{high pH}} + \log\left(1+\frac{\mathbf{H}}{K_2}\right) - \log\left(1+\frac{\mathbf{H}}{K_1}\right) \quad (10\text{-}10)$$

The limits of equation 10-9 are as follows:

$$\mathbf{H} \ll K_1, K_2 \quad \text{app}K_i = K_i \quad (10\text{-}11)$$

$$K_1 < \mathbf{H} < K_2 \quad \text{app}K_i = K_i\left[\frac{\left(1+\frac{\mathbf{H}}{K_1}\right)}{\left(1+\frac{\mathbf{H}}{K_2}\right)}\right] \quad (10\text{-}12)$$

These asymptotes cross at $\text{pH} = pK_1$. The final limit is

$$K_1K_2 \ll \mathbf{H} \quad \text{app}K_i = K_iK_2/K_1 = K_j \quad (10\text{-}13)$$

The middle and low asymptotes cross at $\text{pH} = pK_2$. Thus, all of the parameters in mechanism 10-1 can be calculated from a fit of data such as those shown in Figure 10-3 to equation 10-9.

The above treatment will always yield true pK values. It is the best way to obtain intrinsic pK values as long as one can find an inhibitor sensitive to the protonation state of enzyme groups or inhibitors. This may be used for any system at equilibrium, for example, K_a for metal ions or activators; K_i for substrates (normally determined as competitive product inhibition since K_i values from initial velocity patterns are usually not well-defined).

In Figure 10-3, unless the low-pH plateau is 3 or more log units below the high-pH one, there will not be a linear portion with slope of 1 between the two plateaus. Instead, there will simply be an inflection point. Note that the displacement of the pK caused by ligand binding is equal to the difference in high- and low-pH plateaus.

If the binding is tighter at low pH than at high pH, then the plot shown in Figure 10-4 is observed.

The following equation applies for this case:

$$pK_i = pK_{i\,\text{low pH}} - \log\left(1+\frac{K_1}{\mathbf{H}}\right) + \log\left(1+\frac{K_2}{\mathbf{H}}\right) \quad (10\text{-}14)$$

where again K_1 is the dissociation constant in the free species and K_2 is the value in the complex. The displacement of pK_1 to pK_2 equals the drop from the low-pH to high-pH plateaus. There will not be a linear segment with a slope of -1 unless pK_1 and pK_2 are 3 or more units apart.

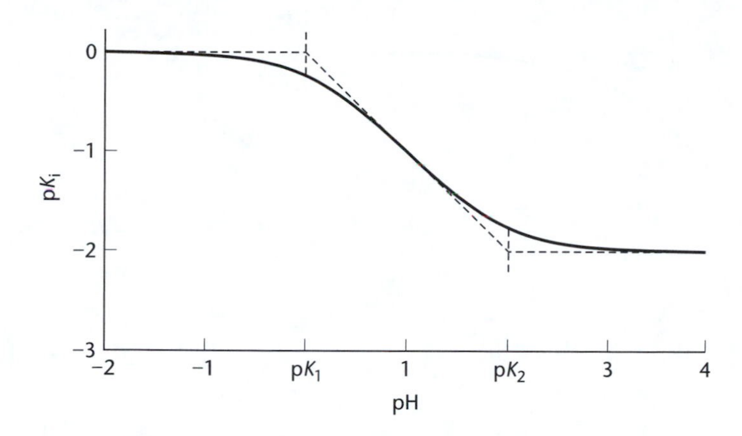

Figure 10-4. Plot of pK_i versus pH for the case where deprotonation decreases but does not eliminate binding. The linear segments have slopes of 0 [pH-independent value of pK_i (low) pH and pK_j (high) pH] and –1 (dashed line drawn through the midpoint of the change). The lines intersect at values of pK_1 (low pH) and pK_2 (high pH). pH is expressed relative to the pK of the group titrated.

When two groups ionize in the pH region of interest, the profiles become more complex. If both groups need to be unprotonated for binding, the profile will decrease with an ultimate slope of 2 at low pH (Figure 10-5).

Unless the pKs of the groups are at least 3 pH units apart, there will not be a linear region with a slope of 1. The intersection of the linear asymptote with slope of 2 and the high-pH plateau occurs at $(\mathrm{p}K_1 + \mathrm{p}K_2)/2$. The equation here is as follows:

$$\mathrm{p}K_\mathrm{i} = \mathrm{p}K_{\mathrm{i\ high\ pH}} - \log\left[1 + \left(\frac{\mathbf{H}}{K_1}\right)\left(1 + \frac{\mathbf{H}}{K_2}\right)\right] \tag{10-15}$$

where K_1 and K_2 are the acid dissociation constants of the two groups. One of the dissociation constants can be in the ligand and the other in the enzyme, or both can be in one or the other.

If the two groups must be protonated for binding, the profile decreases at high pH to an ultimate slope of –2. Again the intersection point of the asymptotes is at the average of the pK values and the equation is

$$\mathrm{p}K_\mathrm{i} = \mathrm{p}K_{\mathrm{i\ low\ pH}} - \log\left[1 + \left(\frac{K_1}{\mathbf{II}}\right)\left(1 + \frac{K_2}{\mathbf{II}}\right)\right] \tag{10-16}$$

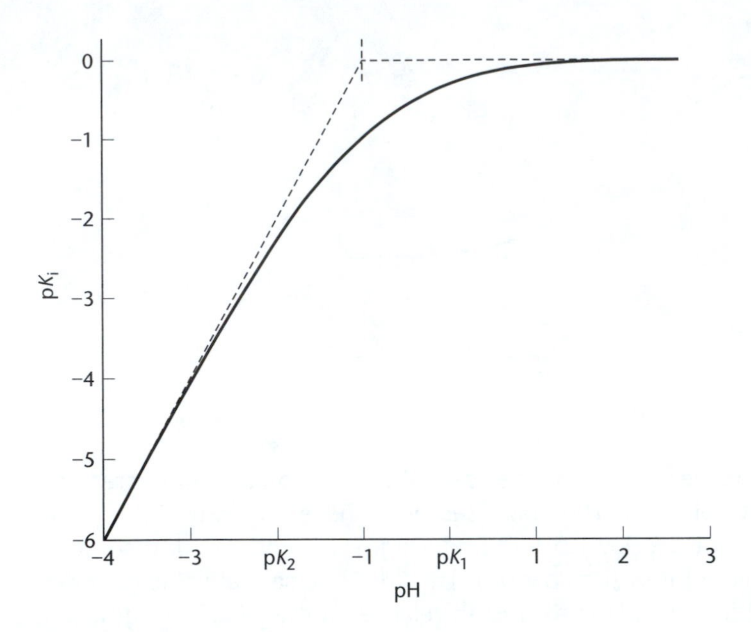

Figure 10-5. Plot of pK_i [log (1/K_i)] versus pH for the case where protonation of two groups with pK values 2 pH units apart eliminates binding affinity. Note that linear segments have slopes of 0 (pH-independent value of pK_i, high pH) and +2 (low pH), which intersect at (pK_1 + pK_2)/2. pH is expressed relative to the pK of the group titrated.

When one of two groups must be protonated for binding and the other unprotonated, the profile will decrease at both low and high pH (Figure 10-6).

Unless the two pKs are at least 3 pH units apart, there will not be a flat plateau at intermediate pH values. The intersection of the asymptotes will be at the average of the pK values of the two groups. If the pKs are more than a pH unit apart, equation 10-17 can be used to fit the data:

$$\mathrm{p}K_{\mathrm{i}} = \mathrm{p}K_{\mathrm{i0}} - \log\left(1 + \frac{\mathbf{H}}{K_1} + \frac{K_2}{\mathbf{H}}\right) \tag{10-17}$$

where K_1 and K_2 are acid dissociation constants of the two groups and pK_{i0} is the pH-independent value when the groups are correctly protonated. If the pK values are close together, however, the true equation, (equation 10-18) must be used:

$$\mathrm{p}K_{\mathrm{i}} = \mathrm{p}K_{\mathrm{i0}} - \log\left[\left(1+\frac{\mathbf{H}}{K_1}\right)\left(1 + \frac{K_2}{\mathbf{H}}\right)\right] = \mathrm{p}K_{\mathrm{i0}} - \log\left(1+\frac{K_2}{K_1} + \frac{\mathbf{H}}{K_1} + \frac{K_2}{\mathbf{H}}\right) \tag{10-18}$$

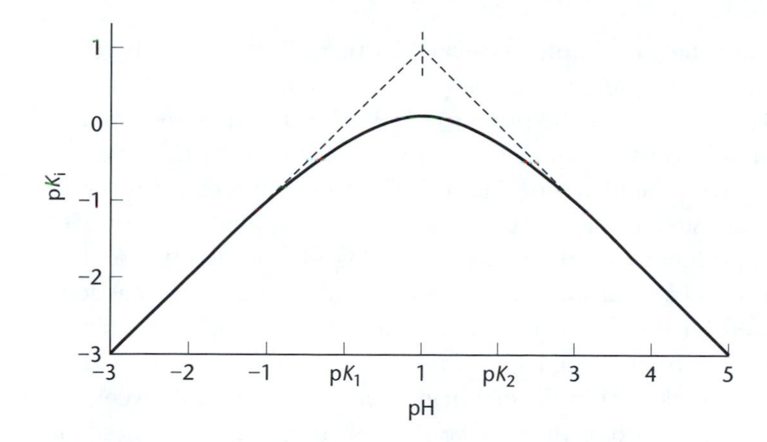

Figure 10-6. Plot of pK_i [log (1/K_i)] versus pH for the case where protonation and deprotonation both eliminate binding affinity. The linear segments have slopes of +1 and –1, which intersect at (pK_1 + pK_2)/2. pH is expressed relative to the pK of the group titrated.

If equation 10-17 is used to fit the data, the apparent pK values are

$$\text{app p}K_1 = \text{p}K_1 - \log\left(1 + \frac{K_2}{K_1}\right) \tag{10-19}$$

$$\text{app p}K_2 = \text{p}K_2 + \log\left(1 + \frac{K_2}{K_1}\right) \tag{10-20}$$

If $K_2/K_1 < 0.1$, the difference can be ignored, but if $K_2 = K_1$, the apparent pKs will appear 0.6 pH unit apart and cannot be closer than this. Computer programs fitting data to equation 10-17 must therefore not permit the fitted pKs to be closer than 0.6 pH unit. This is important, because the difference in the shape of a profile when the true pKs are equal or up to 0.5 pH unit apart is very small and errors in the key data points at the top of the profile can cause the fitted values of the pKs to appear to diverge or even to become closer together than 0.6 pH unit.

While interpreting the profiles that drop only at low or high pH is straightforward, this is not the case with ones that decrease both at low and high pH, especially if the pKs are not more than 2 or 3 pH units apart. It is clear that one group must be protonated and the other ionized for binding, but it is not clear which is which.

If the p*K*s are close together, the profile does not distinguish between the two possibilities. The intuitive interpretation is that pK_1 belongs to a group that must be ionized and pK_2 to one that must be protonated. But exactly the same profile will be seen if the opposite is true; that is, if the group with the low p*K* has to be protonated and the one with the higher p*K* ionized. This situation is called reverse protonation. The p*K*s are seen in their correct places in the profile, but the roles are reversed. Reverse protonation is quite common in log *V*/*K* profiles where the protonation states of the two catalytic groups are reversed for the back reaction relative to that required for the forward reaction. For example, enolase requires Lys345 to be neutral to act as a general base for proton removal and Glu211 to be protonated for protonation of –OH in the elimination of water from 2-P-glycerate (B *42*, 8298). This is reverse protonation, since the p*K*s of Glu211 and Lys345 are 7.4 and 8.8 in the V/K_{Mg} profile. For the backward reaction of P-enolpyruvate and water, normal protonation is required with Lys345 protonated and Glu211 ionized.

When faced with profiles of this type with p*K* values close to each other, one must use other means to determine the actual required protonation states, such as examination of the X-ray structure, construction of mutants lacking one group, or simple consideration of the chemistry involved. For example, Lys183 must be unprotonated, but Glu190 must be protonated for the 6-phosphogluconate dehydrogenase reaction (B *38*, 11231). This is required because the unprotonated form of Lys183 is required to accept a proton in the oxidation step, but must be protonated in the decarboxylation step in order to polarize the carbonyl of the 3-keto intermediate.

pH Dependence of *V*/*K*

Activity Lost at Low pH

V/*K* is the apparent first-order rate constant for reaction of enzyme and substrate at low levels of the latter. ($V/K\mathbf{E_t}$ or k_{cat}/K_m is an apparent second-order rate constant.) The rate expression for *V*/*K* starts with the combination of substrate and enzyme and includes all steps through the first irreversible one, which is usually either release of the first product (unless it is present) or an irreversible chemical step prior to this. The pH variation of *V*/*K* will show the required (or preferred) protonation states for binding and/or subsequent catalysis of groups in either the substrate or the enzyme form it combines with. Since catalysis usually requires a unique protonation state, wavelike profiles like those in Figures 10-3 and 10-4 are not usually seen, even if the substrate and enzyme can form an incorrectly protonated complex.

We will consider several models where protonation reduces activity. First, assume that substrate binds only to E and not to EH:

$$\begin{array}{c} \mathrm{EH} \\ \mathbf{H}/K_1 \updownarrow \\ \mathrm{E} \underset{k_2}{\overset{k_1\mathbf{A}}{\rightleftharpoons}} \mathrm{EA} \xrightarrow{k_3} \end{array} \tag{10-21}$$

Since only *V/K* will be discussed, steps after k_3, which includes all unimolecular steps up to and including the first irreversible step, will not be shown. Since protonation states are usually locked once the substrate binds, we will assume k_3 to be independent of pH. The equation for *V/K* in this mechanism is

$$\frac{V}{K} = \frac{k_1 k_3 \mathbf{E}_t}{(k_2 + k_3)\left(1 + \dfrac{\mathbf{H}}{K_1}\right)} \tag{10-22}$$

A log–log plot of equation 10-22 is identical to that shown in Figure 10-1 and the apparent p*K* is pK_1. Note that the substrate may be sticky ($k_3 > k_2$), but the correct p*K* is seen and should agree with one determined from the pK_i profile of a competitive inhibitor. The interconversion of E and EH is assumed to be fast and at equilibrium. *V* shows no pH variation in this mechanism, unless steps following k_3 are pH-dependent.

Similar behavior is seen if substrate only binds to EH, so that *V/K* decreases at high pH. The equation for *V/K* is similar to equation 10-22 but with a $(1 + K_1/\mathbf{H})$ term in the denominator, and the curve is similar to Figure 10-2. Triose-P isomerase shows this pattern, with *V/K* decreasing above a p*K* of 9, while *V* is pH-independent (BJ *129*, 311). Presumably the lysine, which interacts with the phosphate of the substrate, has its p*K* so elevated by interaction with the phosphate that it is not seen in the *V* profile.

If substrate and enzyme can form an EAH complex, *V* will be pH-dependent and the equation for *V/K* can be different, depending on the stickiness of the proton in EAH.

$$\begin{array}{ccc} \mathrm{EH} & \underset{k_8}{\overset{k_7\mathbf{A}}{\rightleftharpoons}} & \mathrm{EAH} \\ \mathbf{H}/K_1 \updownarrow & & k_6\mathbf{H} \updownarrow k_5 \\ \mathrm{E} & \underset{k_2}{\overset{k_1\mathbf{A}}{\rightleftharpoons}} & \mathrm{EA} \xrightarrow{k_3} \end{array} \tag{10-23}$$

This model assumes that k_5/k_6, that is, K_2, has a value greater than 10^{-7} M, so that the interconversion of EA and EAH involves reaction of a proton with EA, and donation of a proton from EAH to water to give EA. The interconversion of E and EH is assumed to be fast and at equilibrium. If pK_2 is above 7.3, the reaction will involve hydroxide and not H^+ (see page 340) and the equation for V/K will differ.

For the model in mechanism 10-23, the equation for V/K is

$$\frac{V}{K} = \frac{\left(\frac{k_1 k_3 \mathbf{E_t}}{k_2 + k_3}\right)\left(1 + \frac{\mathbf{H}}{K_\alpha}\right)}{\left(1 + \frac{\mathbf{H}}{K_1}\right)\left(1 + \frac{\mathbf{H}}{\beta K_\alpha}\right)} \tag{10-24}$$

where

$$K_\alpha = \left(\frac{k_1}{k_7}\right)\left(1 + \frac{k_8}{k_5}\right)K_1 \tag{10-25}$$

and

$$\beta = \left(1 + \frac{k_3}{k_2}\right) = 1 + S_r \tag{10-26}$$

The asymptotes of the low-pH portion of the curve with slope of 1 and the high-pH plateau cross at

$$\text{app p}K = \text{p}K_1 - \log\left(1 + \frac{k_3}{k_2}\right) = \text{p}K_1 - \log(1 + S_r) \tag{10-27}$$

but the shape of the profile near this pH may differ from that in Figure 10-1. Three rate constant ratios determine the overall shape:

(1) The stickiness of the substrate, given by $S_r = k_3/k_2$.

(2) The ratio of rates of release of the proton and the substrate from EAH (k_5/k_8).

(3) The ratio of rates of substrate combination with E and EH (k_1/k_7).

If the substrate is not sticky ($k_2 > k_3$), $\beta = 1$ and equation 10-24 reduces to one similar to equation 10-22 and the profile is given by Figure 10-1.

If the substrate is sticky, however, there can be three types of profile as discussed below.

(1) $(1+k_8/k_5)=k_7/k_1$. This occurs if $k_1=k_7$ and $k_5>k_8$ and is the most common case. The proton on EAH is released rapidly, and the rates of combination of the substrate with E and EH are similar. K_α in this case equals K_1, and equation 10-24 simplifies to

$$\frac{V}{K}=\frac{\left(\dfrac{k_1k_3\mathbf{E_t}}{k_2+k_3}\right)}{\left(1+\dfrac{\mathbf{H}}{\beta K_1}\right)} \tag{10-28}$$

The shape of the curve is similar to Figure 10-1, but the apparent p*K* is given by equation 10-29:

$$\text{app}\,\text{p}K=\text{p}K_1-\log\beta=\text{p}K_1-\log\left(1+\frac{k_3}{k_2}\right)=pK_1-\log\,(1+S_r) \tag{10-29}$$

The apparent p*K* is thus displaced to lower pH by the stickiness of the substrate, and comparison with the true p*K* from the pK_i profile of a competitive inhibitor, or the *V/K* profile with a slow, nonsticky substrate, allows calculation of S_r. There are a number of examples in the literature where the stickiness factor has been estimated (see for example B *35*, 6358; ABB *388*, 267; JBC *268*, 3407).

(2) $(1+k_8/k_5)>k_7/k_1$. This occurs when $k_1>k_7$, or when $k_5<k_8$ with k_7 not being larger than k_1. This is the next most likely case and occurs when the proton in the EAH complex is sticky and does not readily dissociate when the substrate is present. The crossover of the asymptotes is still given by equation 10-29 but the curve is flattened in the vicinity of the apparent p*K* and may have an actual hollow in it, as shown in Figure 10-7.

The initial decrease occurs at pK_1, but then the curve flattens out at pK_α before the curve decreases again at p$K_\alpha-\log\beta$ to become asymptotic to a line given by the following expression:

$$\log\left(\frac{V}{K}\right)=\log\left(\frac{k_1k_3\mathbf{E_t}}{k_2}\right)-\text{p}K_1+\text{pH} \tag{10-30}$$

The asymptote described in equation 10-30 intersects the asymptote of the high-pH plateau ($\log\,(V/K)=\log\,[k_1k_3\mathbf{E_t}/(k_2+k_3)]$) at p$K_1-\log\,(1+k_3/k_2)$.

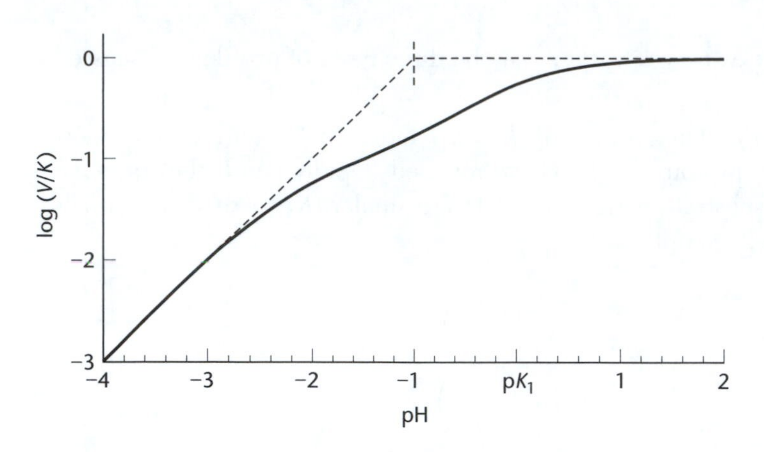

Figure 10-7. pH dependence of V/K exhibiting a "hollow" on the acidic limb of the profile. Intersection of the linear asymptotes is at $pK_1 - \log(1 + k_3/k_2)$. In this case $k_3/k_2 = 9$, while $k_8/k_5 = 9$, and $k_7/k_1 = 1$.

If k_5 actually is zero, the curve begins to decrease below pK_1, and since $K_\alpha = \infty$, the profile resembles Figure 10-1.

Creatine kinase shows a hollow in the $V/K_{\text{P-creatine}}$ pH–rate profile at 12 °C (B *20*, 1204). Activity in this case is lost at high pH. The profile shows less of a hollow at 25 °C and none at 35 °C. Fitting the data showed that phosphocreatine dissociates 2.5 times faster than the proton at 25 °C and 9.3 times faster at 12 °C, so the proton is much stickier at lower temperature. The value of k_3/k_2 was 11.3 at 12 °C, 4.0 at 25 °C, and ~2 at 35 °C, so the substrate is also more sticky at lower temperature. The catalytic reaction has the lowest ΔH value, the conformation change that permits substrate release has the next highest ΔH value, and the release of the proton has the highest ΔH value.

(3) $(1 + k_8/k_5) < k_7/k_1$. This case would occur only if $k_1 < k_7$, and k_5 is greater than k_8. This case has never been observed and is very unlikely to occur. The V/K profile would have a hump in the vicinity of pK_1 and then decrease at lower pH. Since $pK_\alpha > pK_1$ in this case, the curve begins to rise at pK_α and then turns over and drops as pK_1 and $(pK_\alpha - \log \beta)$ are reached (Figure 10-8).

Activity Lost at Low pH and pK₁ above pH 7.3

Protons react in aqueous solution with oxygen or nitrogen bases with rate constants of $\sim 4 \times 10^{10}\ M^{-1}\ s^{-1}$, while hydroxide reacts with the conjugate acids with a rate half this value (ACIEE *3*, 1). Therefore the pH at which the base reacts with a proton at the same rate as hydroxide carries out the reverse reaction is 7.3.

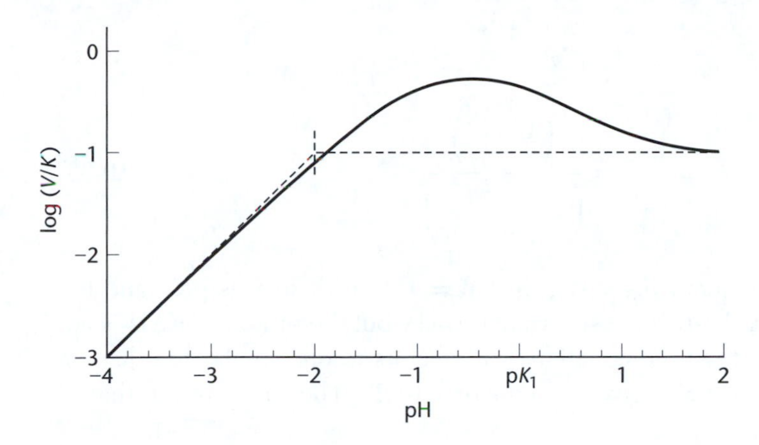

Figure 10-8. pH dependence of V/K exhibiting a "hump" on the acidic limb of the profile. The values for k_8/k_5, k_7/k_1, and β are 0.1, 10, and 10, respectively.

(The only exception to this rule is for RSH groups, which react with hydroxide 50 times more slowly than RS^- reacts with protons. The switch over for sulfhydryl groups will thus be at pH 7.85.) Below pH 7.3, the reaction involves protons reacting with the base and the acid donating a proton to water; this is the assumption we made in mechanism 10-23. But if a p*K* is above pH 7.3, hydroxide will react with the acid and the base will accept a proton from water. Thus mechanism 10-23 must be rewritten as shown below. The interconversion of E and EH is assumed to be fast and at equilibrium.

$$\begin{array}{ccc} \mathrm{EH} & \underset{k_8}{\overset{k_7\mathbf{A}}{\rightleftharpoons}} & \mathrm{EAH} \\ \mathbf{H}/K_1 \updownarrow & & k_6 \updownarrow k_5\mathbf{OH} \\ \mathrm{E} & \underset{k_2}{\overset{k_1\mathbf{A}}{\rightleftharpoons}} & \mathrm{EA} \xrightarrow{k_3} \end{array} \tag{10-31}$$

This changes the rate equation. Equation 10-24 still applies, but now

$$K_\alpha = \frac{K_1\left(\dfrac{k_1}{k_7}\right)}{\left(1+\dfrac{k_2}{k_6}\right)} \tag{10-32}$$

and

$$\beta = \frac{\left(1 + \frac{k_3}{k_2}\right)}{\left(1 + \frac{k_3}{k_2 + k_6}\right)} \tag{10-33}$$

If the substrate is not sticky ($k_3 < k_2$), then $\beta = 1$, the pK seen is pK_1, and the shape is that of Figure 10-1. If the substrate is sticky but the proton in EAH is not ($k_6 \gg k_2$) and $k_1 \approx k_7$, then the equation again factors to one similar to equation 10-22 and the apparent pK is given by equation 10-27. The shape is still that of Figure 10-1.

The critical ratios for this model are k_1/k_7 and $1 + k_2/k_6$. Equality leads to a simplified equation, as in the previous example. If $(k_1/k_7) < (1 + k_2/k_6)$, which corresponds to a sticky proton ($k_6 < k_2$), the curve at the apparent pK lies less than 0.3 below the high-pH plateau, but there is no hump. If $k_6 \ll k_2$, $\beta = 1$, and the apparent pK is pK_1.

If $(k_1/k_7) > (1 + k_2/k_6)$, which occurs when $k_1 > k_7$ and $k_6 > k_2$, there will be a flattening or a hollow in the profile similar to that in Figure 10-7. Unlike mechanism 10-23, the apparent pK given by the intersection of the two asymptotes of the curve in the general case is given by

$$\text{app p}K = \text{p}K_1 - \log\beta = \text{p}K_1 - \log\left[\frac{\left(1 + \frac{k_3}{k_2}\right)}{\left(1 + \frac{k_3}{k_2 + k_6}\right)}\right] \tag{10-34}$$

Thus the ratio of k_6 to k_2 determines the displacement of the apparent pK from pK_1. If $k_6 \gg k_2$, the denominator term drops out and the displacement is the same as that for mechanism 10-23 and is given by equation 10-27. But if $k_6 \ll k_2$, the numerator and denominator cancel and there is no displacement. Of course, if $k_3 < k_2$ (nonsticky substrate), there is also no displacement of the pK.

Alanine dehydrogenase shows a V/K profile that decreases both above and below the pH optimum of 9–10 (B *20*, 5655). Serine is a slow, nonsticky substrate and shows a pK_1 of 9.6, while alanine, a fast and clearly sticky substrate, shows the pK displaced to 8.7. The pK of alanine itself (pK_2) was displaced from 9.7 to 10.8. Thus both pKs are displaced outwards by 0.9 and 1.1 pH unit. Because of the high pH, EA and EAH in this case will be interconverted by reaction with hydroxide and water, so that mechanism 10-31 applies. The displacement of the pK shows that alanine is sticky, but the proton is not, and that $(k_1/k_7) \approx (1 + k_2/k_6)$ since no hollow was seen.

Activity Lost at High pH

When activity is lost at high pH, the V/K profile will depend first on whether the substrate can combine with deprotonated enzyme. If it cannot, V will be pH-independent and the correct pK will be seen even if the substrate is sticky. The equation is

$$\frac{V}{K} = \frac{k_1 k_3 \mathbf{E_t}}{\left(k_2 + k_3\right)\left(1 + \frac{K_1}{\mathbf{H}}\right)} \tag{10-35}$$

and the shape of the profile is similar to that in Figure 10-2.

If the substrate can combine with deprotonated enzyme, the following mechanism applies:

$$\begin{array}{ccccc} \mathrm{EH} & \underset{k_2}{\overset{k_1\mathbf{A}}{\rightleftharpoons}} & \mathrm{EAH} & \xrightarrow{k_3} & \\ \mathbf{H}/K_1 \updownarrow & & k_6 \updownarrow k_5\mathbf{OH} & & \\ \mathrm{E} & \underset{k_8}{\overset{k_7\mathbf{A}}{\rightleftharpoons}} & \mathrm{EA} & & \end{array} \tag{10-36}$$

It is assumed the pK is above 7.3, so that hydroxide and not a proton is involved in interconverting EA and EAH (see page 345 for the other option). The interconversion of E and EH is assumed to be fast and at equilibrium. The equation for V/K is

$$\frac{V}{K} = \frac{\left(\frac{k_1 k_3 \mathbf{E_t}}{k_2 + k_3}\right)\left(1 + \frac{K_\alpha}{\mathbf{H}}\right)}{\left(1 + \frac{K_1}{\mathbf{H}}\right)\left(1 + \frac{\beta K_\alpha}{\mathbf{H}}\right)} \tag{10-37}$$

where

$$K_\alpha = \frac{K_1\left(\frac{k_7}{k_1}\right)}{\left(1 + \frac{k_8}{k_6}\right)} \tag{10-38}$$

and

$$\beta = \frac{1}{\left(1 + \frac{k_3}{k_2}\right)} \tag{10-39}$$

The asymptotes to the low-pH plateau and the high-pH portion with a slope of –1 cross at

$$\text{app}\,pK = pK_1 - \log\beta = pK_1 + \log\left(1 + \frac{k_3}{k_2}\right) \tag{10-40}$$

The apparent pK is thus displaced to higher pH by a sticky substrate, and the shape of the profile is again determined by three ratios as follows:

(1) The stickiness of the substrate, given by $S_r = k_3/k_2$.

(2) The ratio of the rate constants for protonation of EA and release of substrate from EA (k_6/k_8).

(3) The ratio of the rate constants for substrate combination with EH and E (k_1/k_7).

If the substrate is not sticky ($k_3 < k_2$), then $\beta = 1$ and equation 10-37 reduces to one similar to equation 10-35 and the profile has the shape of Figure 10-2. With a sticky substrate, however, there are three possible shapes.

(1) $(1 + k_8/k_6) = (k_7/k_1)$. This corresponds to a nonsticky proton ($k_6 > k_8$) and $k_1 \approx k_7$ and is the most likely case. K_α equals K_1 and equation 10-37 becomes

$$\frac{V}{K} = \frac{\left(\frac{k_1 k_3 \mathbf{E_t}}{k_2 + k_3}\right)}{\left(1 + \frac{\beta K_1}{\mathbf{H}}\right)} \tag{10-41}$$

The shape of the curve is that in Figure 10-2, but the apparent pK is given by equation 10-40.

(2) $(1 + k_8/k_6) > (k_7/k_1)$. This occurs if $k_1 > k_7$, or when $k_6 < k_8$, with k_7 not larger than k_1. This is the next most likely case and corresponds to a sticky proton in EAH. Equation 10-40 still applies, but there will be flattening or a hollow in the profile as in Figure 10-9. This is the type of profile shown by $V/K_{\text{P-creatine}}$ for creatine kinase and discussed on page 340.

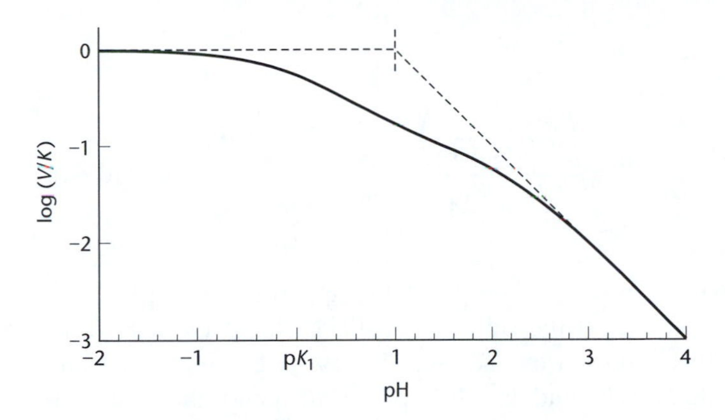

Figure 10-9. pH dependence of V/K exhibiting a "hollow" on the basic limb of the profile. Intersection of the linear asymptotes is at $\mathrm{p}K_1 + \log(1 + k_3/k_2)$, with $k_3/k_2 = 9$. The ratios k_8/k_6 and k_7/k_1 are 9 and 1, respectively.

The initial decrease occurs at $\mathrm{p}K_1$, but then the curve flattens out at $\mathrm{p}K_\alpha$ and then curves downward again at $\mathrm{p}K_\alpha + \log \beta$. If k_6 is actually zero, so that $K_\alpha = 0$, the curve decreases at $\mathrm{p}K_1$ but never flattens, so that the apparent pK is $\mathrm{p}K_1$ and the shape is that of Figure 10-2.

(3) $(1 + k_8/k_6) < (k_7/k_1)$. This unlikely case requires that $k_1 < k_7$ and $k_6 > k_8$. The profile would have a hump in the vicinity of $\mathrm{p}K_1$ and have a shape the mirror image of Figure 10-8.

Activity Lost at High pH and pK₁ Less Than 7.3

For an enzyme with a low pH optimum, a decrease in V/K may occur above an apparent pK that is below 7.3. In this case, the interconversion of EA and EAH will involve a proton and water, and not hydroxide. The interconversion of E and EH is assumed to be fast and at equilibrium. Mechanism 10-36 must thus be changed to

$$\begin{array}{ccccc} \mathrm{EH} & \underset{k_2}{\overset{k_1\mathbf{A}}{\rightleftharpoons}} & \mathrm{EAH} & \xrightarrow{k_3} & \\ \mathbf{H}/K_1 \Big\updownarrow & & k_6\mathbf{H} \Big\updownarrow k_5 & & \\ \mathrm{E} & \underset{k_8}{\overset{k_7\mathbf{A}}{\rightleftharpoons}} & \mathrm{EA} & & \end{array} \tag{10-42}$$

The equation for V/K is still equation 10-37, but now

$$K_\alpha = K_1\left(\frac{k_7}{k_1}\right)\left(1 + \frac{k_2}{k_5}\right) \tag{10-43}$$

and

$$\beta = \frac{\left(1 + \frac{k_3}{k_2 + k_5}\right)}{\left(1 + \frac{k_3}{k_2}\right)} \tag{10-44}$$

If the substrate is not sticky ($k_3 < k_2$), then $\beta = 1$, the apparent pK is pK_1, and the shape is that of Figure 10-2. If the substrate is sticky, the shape may vary. The critical ratios for this model are k_1/k_7 and $1 + k_2/k_5$. If they are equal (a nonsticky proton with $k_5 > k_2$ and $k_1 \approx k_7$), pK_α = pK_1, and the equation is simplified to one similar to equation 10-35, with the apparent pK equal to p$K_1 - \log \beta$.

If $(k_1/k_7) < (1 + k_2/k_5)$, which results from a sticky proton ($k_5 < k_2$), the curve at the apparent pK lies less than 0.3 units below the low-pH plateau, but there is no hump. If $k_5 \ll k_2$, $\beta = 1$ and the apparent pK is pK_1. If $(k_1/k_7) > (1 + k_2/k_5)$, which corresponds to $k_1 > k_7$ and $k_5 > k_2$, there will be a flattening or a hollow in the profile similar to that in Figure 10-9.

In the general case the apparent pK given by the intersection of the asymptotes to the curve is

$$\text{app p}K = \text{p}K_1 - \log \beta = \text{p}K_1 + \log \left[\frac{\left(1 + \frac{k_3}{k_2}\right)}{\left(1 + \frac{k_3}{k_2 + k_5}\right)} \right] \tag{10-45}$$

The ratio of k_5 to k_2 determines the displacement of the observed pK above pK_1. If $k_5 \gg k_2$, the denominator of the log term in equation 10-45 drops out and the displacement is the same as for mechanism 10-36, but if $k_5 \ll k_2$, the log term is zero and pK_1 is seen. If $k_3 < k_2$ (nonsticky substrate), pK_1 is also seen.

V/K Profile Decreases at both Low and High pH

pH–rate profiles for which the rate decreases at both low and high pH indicate two groups are required in a given protonation state for binding and/or catalysis, one protonated and the other unprotonated. One can consider the following simple unimolecular reaction mechanism in which only the free enzyme can be titrated. Mechanism 10-46 is used for simplicity and will give a pH-independent V.

If both E and EA can be protonated, mechanisms similar to equations 10-23 and 10-31 must be used.

$$
\begin{array}{c}
H_xEH_y \\
\mathbf{H}/K_x \Big\updownarrow \\
EH_y \underset{k_2}{\overset{k_1A}{\rightleftharpoons}} H_yEA \xrightarrow{k_3} H_xEP \xrightarrow{k_5} H_xE \overset{K'}{\rightleftharpoons} EH_y \\
\mathbf{H}/K_y \Big\updownarrow \\
E
\end{array}
\qquad (10\text{-}46)
$$

In mechanism 10-46, H_x and H_y reflect protons on two different acid-dissociable groups on enzyme, k_3 is the net rate constant for the chemical step, k_5 is the net rate constant for release of product and K' is an equilibrium constant for intramolecular proton transfer between the two acid-dissociable groups on enzyme. The rate of proton interchange will almost never contribute to rate limitation, but the case for proline racemase, where this step can be made to contribute to overall rate limitation, was discussed in Chapter 8.

The V/K pH–rate profile obtained in the case of the above mechanism will decrease at low and high pH, that is, it will have a bell shape, indicating the requirement for one group protonated and the other unprotonated. However, one cannot tell which group is protonated and which is unprotonated for activity, since both of these cases give a pH profile that is qualitatively identical but differs quantitatively. To show this we will assume pK values of 6 and 8 for groups H_xE and EH_y above. The pH dependence of the fraction of each of the groups calculated by use of the Henderson–Hasselbach equation, assuming the groups with pKs of 6 and 8 must be unprotonated and protonated, respectively, is shown in Figure 10-10. Note that when the group with a pK of 6 must be unprotonated and that with a pK of 8 must be protonated, 91% of the enzyme has the two groups correctly protonated between the two pK values at pH 7. A plot of log f_{EHy} versus pH, inset to figure, gives the expected bell-shaped curve with asymptotes of 1 and −1 and pK values of 6 and 8. When the two groups have opposite protonation states, that is, the group with a pK of 6 must be protonated and that with a pK of 8 must be unprotonated, only 9.1% of the enzyme is correctly protonated at pH 7. As shown in the inset to the figure, a bell-shaped curve is still generated with pK values of 6 and 8. Thus, one must be careful when assigning protonation states to groups when a bell-shaped profile is observed. In mechanism 10-46, the fraction of enzyme optimally protonated between the pK values will differ in the two reaction directions depending on the difference in the pK values

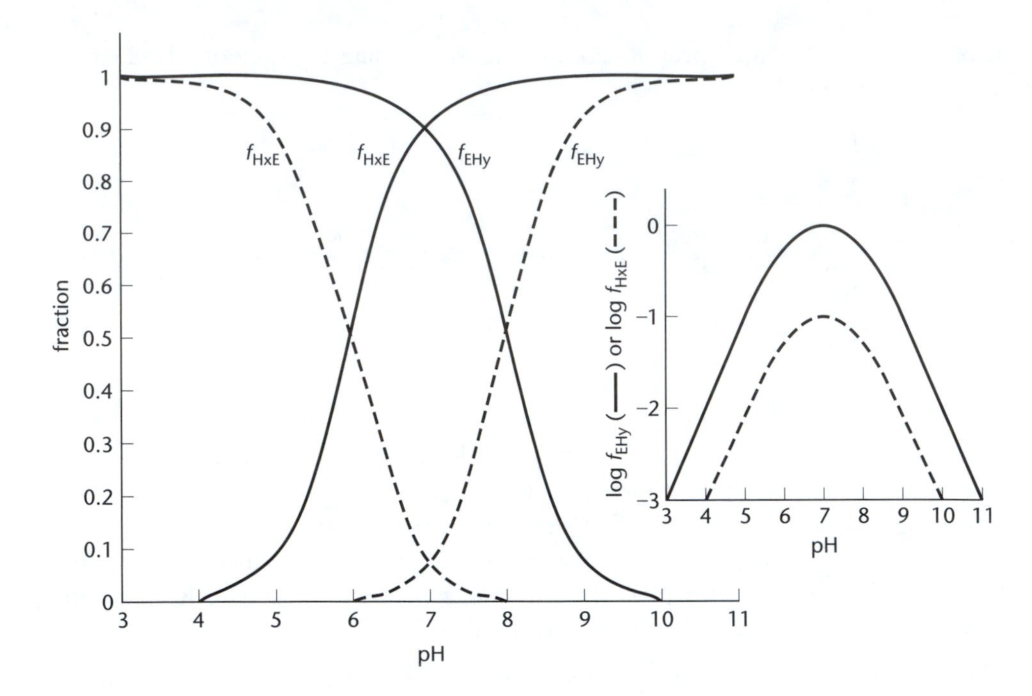

Figure 10-10. Plot of the fraction of each of two groups on enzyme required in opposite protonation states for reaction. The solid lines represent the proportion of free enzyme with H_y present (f_{EHy}) or H_x absent (f_{HxE}). The dashed lines represent the proportion of free enzyme with H_x present (f_{HxE}) or with H_y absent (f_{EHy}). The area under the curve formed by intersection of the two titration curves reflects the fraction of EH_y (solid line) and H_xE (dashed line). The inset shows the pH dependence of the log fraction of EH_y (solid line) and H_xE (dashed line), respectively.

and the amount of rate limitation of the catalytic pathway, k_3, in both reaction directions.

If the substrate is not sticky, the *V/K* profile will resemble Figure 10-6 and equations 10-17 to 10-20 will apply. If the substrate is sticky, however, one must use a more complete model that incorporates mechanisms 10-23 or 10-31 and 10-36 or 10-42. The rate equation is derived in the usual way by assuming the equilibration of E, EH, and EH_2 to be rapid but the equilibration of EA, EAH, EAH_2 not to be. If equilibration of EA, EAH, and EAH_2 is fast, the equation for *V/K* will be similar to equation 10-17 or 10-18, although the apparent p*K*s will be displaced outwards by log $(1+k_3/k_2)$. When protonic equilibration is not fast when substrate is bound, however, the same rules embodied in mechanisms 10-23, 10-31, 10-36, and 10-42 and their associated equations must be employed. The low- and high-pH asymptotes will be given by the same equations as for the mechanism when activity decreases only at low or high pH. For example, for an aspartic protease where the pH optimum is well below 7, the acid-side asymptote will be

$$\frac{V}{K} = \left(\frac{k_1 k_3 \mathbf{E_t}}{k_2}\right)\left(\frac{K_1}{\mathbf{H}}\right) \tag{10-47}$$

while the base-side asymptote will be

$$\frac{V}{K} = \frac{\left(\frac{k_1 k_3 \mathbf{E_t}}{k_2}\right)\left(\frac{\mathbf{H}}{K_2}\right)}{\left(1 + \frac{k_3}{k_2 + k_5}\right)} \tag{10-48}$$

where k_1, k_2, and k_3 are for reaction of A with EH to give EAH and then products, and k_5 is the rate constant for release of a proton from EAH. Note that the base-side asymptote has the extra term in the denominator because pK_2 is less than 7.3, and the model of mechanism 10-42 must be used.

What was said about reverse protonation in connection with profiles resembling Figure 10-6 applies to *V/K* profiles. It is a common phenomenon in *V/K* profiles and the case of enolase was mentioned above. The same situation applies to fumarase, where the protonation states of the two groups that remove the proton from malate to give the intermediate, and protonate the –OH group to allow it to leave as water, have to have reverse protonation to add water to fumarate. The *V/K* profiles, which have to have the same shapes and p*K* values, since K_{eq} is pH-independent and the Haldane relationship equates the ratio of the forward and reverse *V/K* values with K_{eq}, show p*K* values of 5.88 and 7.05 (JACS *85*, 2204). The pK_i profiles for competitive inhibitors, however, give equal p*K*s of 6.6 (JACS *82*, 5482). Thus the stickiness of the reactants has displaced the p*K* values by 0.72 and 0.45.[1] Cleland (AdvEnz *45*, 273) showed that these results are consistent with the observation that the proton removed from malate is not rapidly exchanged

[1]The p*K* values observed experimentally for free fumarase (*V/K* pH–rate profiles) are 6.3 and 6.9. However, these values were estimated from a fit of the data to an equation in which the denominator is $(1 + \mathbf{H}/K_1 + K_2/\mathbf{H})$. The denominator, assuming two groups, should be $(1 + \mathbf{H}/K_1)(1 + K_2/\mathbf{H})$. Equal (true) values of pK_1 and pK_2 will give apparent values that are separated by 0.6 pH unit. Thus, the observed values of 6.3 and 6.9 correspond to equal p*K* values of 6.6.

while substrate is bound, but that the group that protonates the –OH of malate does exchange rapidly.

Temperature studies (see page 352) suggest that the group with a p*K* of 5.88 in *V*/*K* is a carboxylate and the one with a p*K* of 7.05 may be a histidine. Furthermore, the p*K* of 5.88 is raised to 6.4 in the V_{malate} profile and to 7.0 in $V_{fumarate}$, while the temperature-dependent p*K* of 7.05 is raised to 9.0 in V_{malate} and decreased to 4.9 in $V_{fumarate}$. The elevation of the p*K* of 5.88 in both profiles and the small ΔH_{ion} suggest a carboxyl group that does not hydrogen-bond in the active site but whose p*K* is elevated by removal from aqueous solution. This is presumably the group that removes the proton of malate to give the aci-carboxylate intermediate. The 2 pH swings in the *V* profiles for the p*K* of 7.05 when malate or fumarate are bound is characteristic of a group that forms specific hydrogen bonds in the active site, which is what is expected of the group that protonates the hydroxyl of malate in the forward direction, and deprotonates water in the reverse reaction. This group may be histidine, although no structural proof for this is available at present. The chemical mechanism of fumarase can be diagrammed as shown in Figure 10-11.

V/*K* Profiles That Show Two p*K*s at Low or High pH

When there are two p*K*s on the acid or base side, the shapes of the *V*/*K* profile will be similar to Figure 10-5 or its mirror image. One must be careful about the results of substrate stickiness. If the first p*K* that causes a decrease in the profile at low pH results from the fact that a sticky substrate and enzyme can form both EA and EAH, then only this first p*K* is potentially displaced from its true value according to the equations given above. The second p*K* (or any subsequent one) in the profile is then seen at its true value, because the stickiness of the substrate is eliminated as one goes past the first p*K*. If the first p*K* results from the fact that only correctly protonated enzyme and substrate will combine (mechanism 10-21, or the equivalent one where A can combine but AH cannot), then the stickiness of the substrate is not eliminated below this p*K*, and the second p*K* may be displaced if it corresponds to a situation where both EA and EAH can form. Isocitrate dehydrogenase shows this behavior (B *20*, 1797; B *27*, 2934). $V/K_{isocitrate}$ decreases below a p*K* of 6.7, which is the third p*K* of isocitrate. Since only the trianion binds, however, decreasing pH below this p*K* does not eliminate the stickiness of isocitrate. The p*K* of the general base in the reaction is therefore perturbed to ~4.5 by the high stickiness of isocitrate at neutral pH. This p*K* is 5.7 in the pK_i profile of oxalylglycine, a competitive inhibitor (the shape of the profile is that of Figure 10-3). The stickiness ratio of isocitrate is thus ~15 [$\Delta pK = 1.2 = \log(1 + S_r)$].

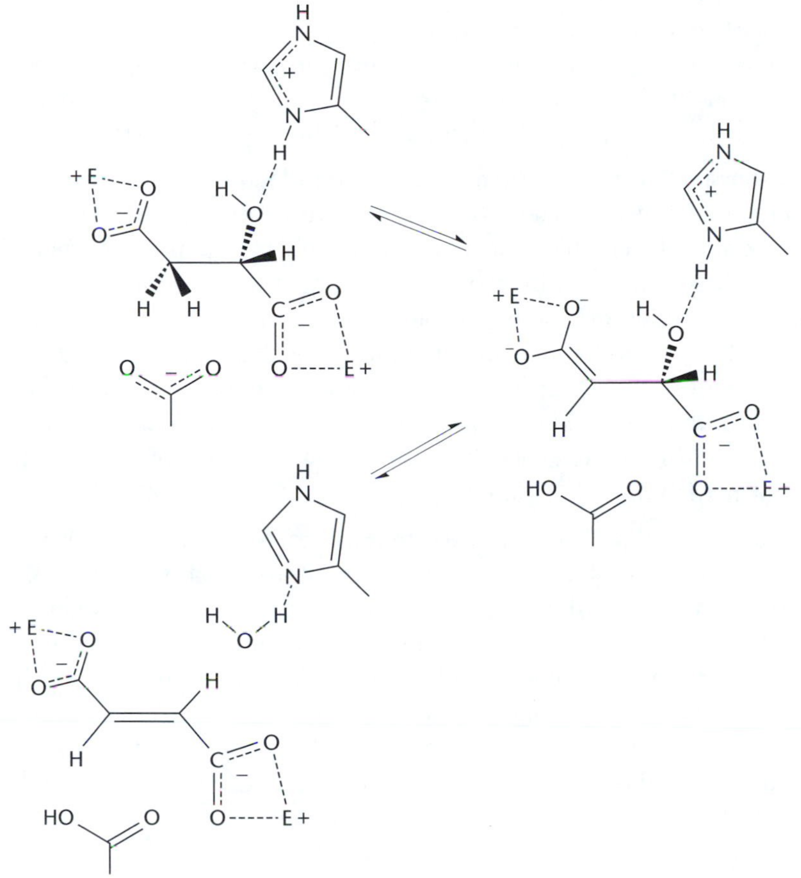

Figure 10-11. Proposed acid–base mechanism for fumarase. Protonated groups on enzyme bind the α- and β-carboxylates of malate. An enzymic carboxylate acts as a general base to accept the 3*R* proton and an imidazole acts as the general acid to donate a proton to leaving hydroxide as fumarate is formed. An aci-acid intermediate is formed as a result of delocalization of electrons to the β-carboxylate.

Identifying the Groups Seen in V/K Profiles

X-ray and NMR structures are a great help not only in locating the active site of an enzyme but also in showing which groups are in a position to act as general acids or bases or to help bind the substrates. One must keep in mind that an X-ray structure usually represents only one conformation of the protein, and unless structures of complexes with substrates and/or inhibitors bound have been determined, with open and closed active sites, one does not have a full picture of the conformation changes that occur. Even when several structures are available, one may be looking at a nonproductive complex and it must be remembered that an inhibitor may bind differently from the substrate (that may be why it is an inhibitor!).

Once candidates for specific roles in the mechanism are identified, one can make active-site-directed mutations and check the activity and pH profiles for changes. Lack of any appreciable change rules out an important role for a particular group, but mere loss of activity may result from structural changes rather than a specific role in catalysis. Substitution of residues involved in general acid–base catalysis may decrease activity by up to 5 orders of magnitude (the change will depend on the enzyme), and the p*K* of the catalytic group should no longer appear in the *V*/*K* profile (ME *249*, 91). Removal of an involved residue may unmask a role for protons or hydroxide, in which case activity should show increases of a factor of 10 per pH unit as pH is lowered or raised.

There are two ways other than X-ray or NMR to try to identify a catalytic group whose p*K* is seen in a *V*/*K* profile. The first is temperature variation of the p*K*. Phosphate and carboxyl groups show near-zero values of ΔH_{ion}, while histidine, cysteine, and tyrosine have values of ~6 kcal/mol, and amino groups and water coordinated to a metal ion have values of ~12 kcal/mol. One determines the pH profiles for K_i or *V*/*K* at different temperatures and plots p*K* versus 1/*T*. The slope is $\Delta H_{ion}/2.3R$. The p*K* values are higher at lower temperatures, with Δp*K* between 0 °C and 25 °C being 0.4 and 0.8 pH unit for ΔH_{ion} values of 6 and 12 kcal/mol, respectively. A potential limitation of this method is that if conformation changes accompany ionization (that is, the p*K* differs in the different conformations of the enzyme), a portion of the ΔH of the conformation change (often 15–25 kcal/mol) will add to ΔH_{ion} and complicate interpretations. The observed value limits the choices to groups having the same or lower ΔH_{ion} values. Thus a value of ~6 kcal/mol could be His, Cys, or Tyr or a carboxyl group whose ionization is partly coupled to a conformation change, but this value rules out Lys or metal-bound water.

In the case of the pigeon liver NADP-malic enzyme, the V/K_{malate} pH–rate profile is bell-shaped with p*K* values of 6 and 8, while the *V* profile gives p*K* values of 4.6 and 9.2. The ΔH_{ion} values of the p*K*s in the *V* profile were 22.0 ± 4.8 kcal/mol (p*K* = 4.6) and 9.3 ± 3 kcal/mol (p*K* = 9.2). Clearly the protonation at pH 4.6 is coupled to a conformational change!

The second method involves solvent perturbation. The p*K* values of neutral acids that ionize to a proton and a negative ion (carboxylate, phosphate, thiolate, tyrosinate, metal-bound hydroxide) are elevated by adding organic solvents, while the p*K* values of nitrogen bases that are cationic before ionization to a proton and a neutral group (imidazole, amine, and guanidinium) are largely unchanged. One can usually add up to 30–50% of a solvent such as dimethyl sulfoxide (DMSO) or *N*,*N*-dimethylformamide (DMF), although formamide, dioxane, and various alcohols have been used, without killing enzyme activity. The effect is not one of changing dielectric constant but rather of decreasing the mole fraction of water.

There are two ways to carry out such an experiment. First, one can measure pH values after solvent addition, which allows one to use any convenient buffer, but one must be certain that the pH meter is reading accurately in the solvent mixture. One then compares *V/K* profiles run with and without solvent. The other method involves measurement of pH values before solvent addition. One covers the pH range of interest with two buffer systems, one with a neutral acid buffer and the other with a cationic acid buffer. In the absence of solvent, one should obtain the same *V/K* profile with both buffers. One then determines the profiles in the presence of solvent in the two buffer systems, measuring the pH values before solvent addition.

The p*K* of a cationic buffer is not much affected by the solvent, so one expects no change in pH values on addition of solvent. In the *V/K* profile, the p*K* values of neutral acid groups will increase and those of cationic acid groups will not change. The p*K* of a neutral acid buffer is elevated by the solvent, so the pH values after solvent addition are higher. This causes little change in the apparent p*K* of a neutral acid group, since its p*K* is elevated along with that of the buffer. A cationic group, on the other hand, will show an apparent p*K* lower than in the absence of solvent, since its p*K* does not change as that of the buffer increases. Comparison of the changes in both cationic and neutral acid buffers allows ready determination of the nature of the groups responsible for the *V/K* profile. A neutral acid group shows no change in p*K* in the neutral acid buffer but an increase in the cationic one after the solvent is added. A cationic acid group shows no change in p*K* in the cationic acid buffer and a decrease in a neutral acid buffer.

A limitation of the solvent perturbation method is that the group must be exposed to the solvent for the method to work. Thus only the p*K*s of the substrate, and of those groups exposed in the enzyme form it combines with, will show a solvent perturbation. An example is yeast hexokinase. The p*K* of the aspartate that acts as a general base is seen in the V/K_{MgATP} profile, but it does not show solvent perturbation because this group is buried under glucose in the E–glucose complex and not exposed to the solvent.

pH Dependence of V_{max}

V is determined by the rate constants for all unimolecular steps in the mechanism when substrate concentrations are infinite, so that bimolecular steps do not limit the rate. Thus isomerization of an EA or EQ complex and all steps between the addition of the last substrate to add and product release, as well as the release of all products, may limit *V*. With alanine dehydrogenase, isomerization of E–NAD is the major rate-limiting step in reaction of alanine and NAD (B *20*, 5650). For many other dehydrogenases, however, release of the product NADH is rate-limiting.

The profile of log *V* can have shapes similar to those discussed above for *V*/*K*, but the p*K*s are those of enzyme–reactant complexes. As noted above for fumarase, covering up a neutral acid group by a substrate elevates its p*K* by removing it from aqueous solution as long as it does not participate in specific hydrogen bonding. Groups that form specific hydrogen bonds will have their p*K*s shifted up to 2 pH units, as was the case with the putative His in fumarase, and these effects may oppose each other. Thus a carboxylate functional group that forms a specific hydrogen bond may have its p*K* lowered only by one pH unit, since the removal from solvent opposes the effect of hydrogen bonding. Substrate p*K*s are often not visible in *V* profiles, because the properly ionized form is involved in specific hydrogen bonding. Examples are phosphate esters, which typically bind only as dianions, and carboxylate groups.

p*K*s in *V* profiles are often displaced because the pH-dependent step is not originally rate-limiting. *V* starts to decrease only when the pH-dependent step has slowed sufficiently to become rate-limiting, and thus the outward displacement on the pH profile equals the log of the original ratio of rates of pH-dependent and rate-limiting steps. The p*K* values for malic enzyme are displaced outwards by 1.3 pH units in log *V* relative to those in log V/K_{malate} because release of NADPH is 20-fold slower than catalysis and release of CO_2 (B *16*, 576). Figure 10-12 shows this effect, where (1) is the pH profile of the catalytic step which is pH-dependent, (2) is the rate of second product release, which is pH independent, and (3) is the observed log *V* profile, where the apparent p*K*s are perturbed to one pH unit below and above pK_1 and pK_2, respectively.

V Profile with a Sticky Substrate and Proton

If we change mechanism 10-23 to add a step after the first irreversible one:

$$\mathrm{EA} \xrightarrow{k_3} () \xrightarrow{k_9} \qquad (10\text{-}49)$$

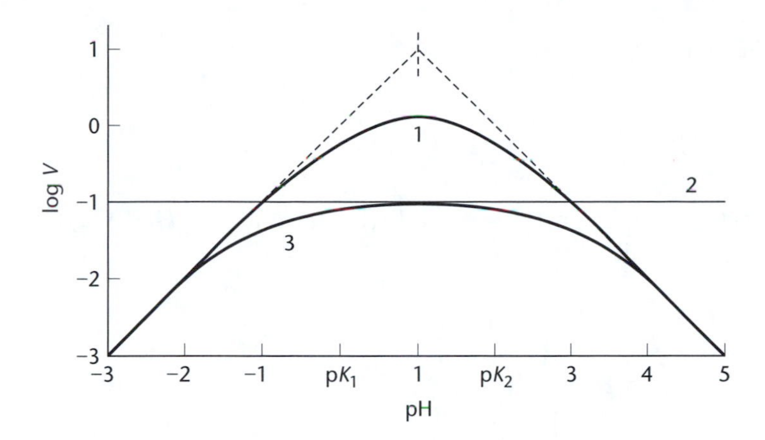

Figure 10-12. pH dependence of *V* illustrating rate-limiting product dissociation: (1) pH dependence of the catalytic step; (2) pH dependence of the rate of product release; (3) observed pH dependence of *V*. The pH dependence of the catalytic step is not observed until its rate approaches that of product release.

the equation for $V/\mathbf{E_t}$, assuming rapid equilibration of E and EH, but slower interconversion of EA and EAH, is

$$\frac{V}{\mathbf{E_t}} = \frac{\left(\dfrac{k_3k_9}{k_3+k_9}\right)\left(1+\dfrac{\mathbf{H}}{K_\alpha}\right)}{1+\dfrac{(k_3k_5k_8+k_5k_8k_9+k_3k_8k_9+k_2k_8k_9+k_2k_5k_9)\left(\dfrac{\mathbf{H}}{K_2}\right)+k_5k_8k_9\left(\dfrac{\mathbf{H}}{K_2}\right)^2}{k_2(k_5+k_8)(k_3+k_9)}} \tag{10-50}$$

where

$$K_2 = \frac{k_5}{k_6} = \frac{K_1k_1k_8}{k_2k_7} \tag{10-51}$$

and

$$K_\alpha = \left(\frac{k_2}{k_8}\right)\left(1+\frac{k_8}{k_5}\right)K_2 = \left(\frac{k_1}{k_7}\right)\left(1+\frac{k_8}{k_5}\right)K_1 \tag{10-52}$$

The asymptotes to this curve cross at

$$\text{app p}K = \text{p}K_2 - \log\left(1 + \frac{k_3}{k_9}\right) \tag{10-53}$$

Equation 10-50 will factor to equation 10-54 if the $k_3k_8k_9$ term in the $(\mathbf{H}/K_2)$ part of the denominator is very small.

$$\frac{V}{\mathbf{E_t}} = \frac{\left(\dfrac{k_3k_9}{k_3 + k_9}\right)}{1 + \left(\dfrac{k_9}{k_3 + k_9}\right)\left(\dfrac{\mathbf{H}}{K_2}\right)} \tag{10-54}$$

This occurs if the substrate is not sticky ($k_2 \gg k_3$, so that the $k_3k_8k_9$ term is much smaller than the $k_2k_8k_9$ one). Equation 10-50 will also factor to equation 10-54 if $k_5 \gg [k_3k_9/(k_3 + k_9)]$ (that is, k_5 is much greater than the smaller of k_3 and k_9. The $k_3k_8k_9$ term is then much smaller than the $k_3k_5k_8$ term if $k_5 \gg k_9$ or the $k_5k_8k_9$ term if $k_5 \gg k_3$).

If both the substrate and the proton in EAH are sticky ($k_3 > k_2$ and $k_5 < [k_3k_9/(k_3 + k_9)]$, the curve can have a flattening or hollow like Figure 10-7. For a pronounced hollow, k_2 and k_8 should be an order of magnitude greater than k_5, and k_3 an order of magnitude greater than k_2 and k_8.

If $\text{p}K_2$ is above 7.3 and a step is added to mechanism 10-31 as indicated by equation 10-49, the equation for $V/\mathbf{E_t}$ is different from equation 10-50 in that K_α in the numerator is given by equation 10-32, and the denominator is

$$1 + \frac{\left[\dfrac{k_3k_8}{k_2 + k_6} + k_9(k_2k_6 + k_2k_8 + k_6k_8)\right]\left(\dfrac{\mathbf{H}}{K_2}\right) + (k_2 + k_3 + k_6)k_8k_9\left(\dfrac{\mathbf{H}}{K_2}\right)^2}{k_2k_6(k_3 + k_9)} \tag{10-55}$$

The asymptotes of the curve cross at

$$\text{app p}K = \text{p}K_2 - \log\left[\frac{\left(1 + \dfrac{k_3}{k_9}\right)}{\left(1 + \dfrac{k_3}{(k_2 + k_6)}\right)}\right] \tag{10-56}$$

The equation for $V/\mathbf{E_t}$ will factor to the simple form:

$$\frac{V}{\mathbf{E_t}} = \frac{\left(\dfrac{k_3 k_9}{k_3 + k_9}\right)}{\left(1 + \left[\dfrac{1 + \left(\dfrac{k_3}{k_2 + k_6}\right)}{\left(1 + \dfrac{k_3}{k_9}\right)}\right]\right)\left(\dfrac{\mathbf{H}}{K_2}\right)} \qquad (10\text{-}57)$$

if either $k_2 \gg k_3$ (nonsticky substrate) or $k_6 \gg k_3$ (nonsticky proton). If $k_3 > (k_2 + k_6)$, however, the curve will be flattened or show a hollow as in Figure 10-7.

When EAH is the active form and EA is not (mechanisms 10-36 or 10-42), adding a step as in equation 10-49 gives rate equations like equation 10-50 (if $pK_2 > 7.3$) or 10-55 (if $pK_2 < 7.3$), except that $K_2/\mathbf{H}$ replaces $\mathbf{H}/K_2$. Equation 10-55 or 10-56 applies, except that the log terms have positive signs. Similar criteria apply for factoring to simpler equations or showing a hollow in the curve.

pH Dependence of V_{max} in a Ping Pong Mechanism

The pH dependence of V/K for the individual half reactions in a Ping Pong mechanism will adhere to the mechanisms discussed above on page 336. The maximum rate, however, can contain contributions from both half-reactions, and both half-reactions will likely be pH-dependent. Consider mechanism 10-58, assuming all proton transfer steps in rapid equilibrium:

$$\begin{array}{ccccccccc}
\mathrm{EH} & \underset{k_8}{\overset{k_7\mathbf{A}}{\rightleftharpoons}} & \mathrm{EHA} & & \mathrm{FH} & \underset{k_{14}}{\overset{k_{13}\mathbf{B}}{\rightleftharpoons}} & \mathrm{FHB} & & \\
\mathbf{H}/K_1 \updownarrow & & \updownarrow \mathbf{H}/K_2 & & \mathbf{H}/K_3 \updownarrow & & \updownarrow \mathbf{H}/K_4 & & \\
\mathrm{E} & \underset{k_2}{\overset{k_1\mathbf{A}}{\rightleftharpoons}} & \mathrm{EA} & \xrightarrow{k_9} & \mathrm{F} & \underset{k_{12}}{\overset{k_{11}\mathbf{B}}{\rightleftharpoons}} & \mathrm{FB} & \xrightarrow{k_{15}} &
\end{array} \qquad (10\text{-}58)$$

In mechanism 10-58, both half-reactions are shown adhering to the full mechanism above, that is, random addition of the reactant and proton to enzyme. The maximum rate, obtained at saturating A and B, gives the following:

$$\begin{array}{ccccc}
\mathrm{EHA} & & \mathrm{FHB} & & \\
\mathbf{H}/K_2 \updownarrow & & \mathbf{H}/K_4 \updownarrow & & \\
\mathrm{EA} & \xrightarrow{k_9} & \mathrm{FB} & \xrightarrow{k_{15}} &
\end{array} \qquad (10\text{-}59)$$

and the expression for V is given in equation 10-60:

$$\frac{V}{\mathbf{E}_t} = \frac{k_9 k_{15}}{k_9\left(1 + \frac{\mathbf{H}}{K_4}\right) + k_{15}\left(1 + \frac{\mathbf{H}}{K_2}\right)} \tag{10-60}$$

If the first half-reaction limits the overall rate, $k_9 < k_{15}$, the equation reduces to $k_9/(1 + \mathbf{H}/K_2)$ and the pH dependence of V is determined by the protonation state of EA. If the second half-reaction limits the overall rate, $k_{15} < k_9$, the equation reduces to $k_{15}/(1 + \mathbf{H}/K_4)$ and the pH dependence of V is determined by the protonation state of FB. If, on the other hand, both half-reactions contribute to the overall rate, the full equation applies and the pH dependence will be determined by a weighted average (based on which is slower) of the two half-reactions.

Metal Ion Binding

pK_i Profiles for Metal Ions

Metal ions bind to various ligands in enzyme active sites such as carboxyl, imidazole, or sulfhydryl groups, which have p*K*s in the accessible pH range. Protonation of a ligand either prevents metal ion binding or greatly decreases the strength of binding. Thus the pK_i profile for a metal ion decreases below the p*K* of the ligand but may level off (Figure 10-3).

There will usually be one or more water molecules remaining in the coordination sphere of a metal ion when it binds to the enzyme, and these will typically ionize to OH^- at a pH that is accessible. This will tighten metal ion binding, if the p*K* for this ionization is lower than that for water bound to the free metal ion, which is usually the case. The pK_i profile will thus increase above the p*K* of water in the E–metal ion complex and level out again at the p*K* of bound water in the free metal ion (this may be above the observable pH range).

Pyruvate kinase catalyzes the decarboxylation of oxaloacetate with a kinetic mechanism that is equilibrium ordered, with Mg^{2+} adding first and oxaloacetate second (B *24*, 5870). The pK_i profile for Mg^{2+} drops below a p*K* of 6.9, which presumably results from protonation of a ligand (Glu271 or Asp295), and increases above a p*K* of 9.2, which corresponds to Mg^{2+}-coordinated water. The p*K* of 6.9 does not appear in either the V/K_{OAA} or V profiles, since the p*K*s of the Mg^{2+} ligands will be displaced to very low pH by binding of the metal ion. Both profiles decrease above the p*K* of 9.2, however. The decrease in the V/K_{OAA} profile shows that oxaloacetate cannot complete the reaction when coordinated water is ionized to OH^-. The decrease in V at this pH shows that the E–Mg^{2+}–oxaloacetate complex is mainly second-sphere, since otherwise the pK of 9.2 would not appear in the V profile. Thus we can conclude that (1) an inner-sphere complex of Mg^{2+} and oxaloacetate is necessary for decarboxylation, (2) only a small proportion of the E–Mg^{2+}–oxaloacetate complex is an inner-sphere one, and (3) oxaloacetate will not form an inner-sphere complex when OH^- is coordinated to Mg^{2+}. Presumably oxaloacetate must replace water in the coordination sphere but cannot replace OH^-, which is typical behavior for ligand substitution.

Analysis of pK_i profiles for metal ions is a powerful tool and deserves more use than has been the case in the past. Together with structural evidence from X-ray, this analysis goes a long way to establish what actually happens in an enzyme active site.

pH Dependence of Isotope Effects

The magnitude of the isotope effects on V and V/K will vary depending on the contribution of the isotope-sensitive step to overall rate limitation at saturating and limiting reactant concentration, respectively. As discussed above, the rate limitation of the chemical steps can change as the pH changes, and the pH dependence of isotope effects may thus provide information on the relative contribution of the chemical steps and other steps along the reaction pathway, that is, the kinetic mechanism of the enzyme, the magnitude of the intrinsic isotope effect, and in some instances the chemical mechanism of the reaction. Below we will consider two general models for the pH dependence of isotope effects. The first assumes the pH-dependent step(s) is(are) isotope-sensitive, while the second assumes they are not. For a more complete treatment see the review article by Cook (1991, page 231–235).

pH-Dependent Step is Sensitive to Isotopic Substitution

Consider the following model for a bireactant mechanism at saturating concentrations of A:

$$\begin{array}{ccccccc} \mathrm{EAH} & \underset{k_8}{\overset{k_7\mathbf{B}}{\rightleftharpoons}} & \mathrm{EHAB} & & & & \\ \mathbf{H}/K_1 \updownarrow & & \updownarrow \mathbf{H}/K_2 & & & & \\ \mathrm{EA} & \underset{k_4}{\overset{k_3\mathbf{B}}{\rightleftharpoons}} & \mathrm{EAB} & \xrightarrow{k_9} \mathrm{P} & \mathrm{EQ} & \xrightarrow{k_{11}} \mathrm{Q} & \mathrm{E} \end{array} \tag{10-61}$$

In mechanism 10-61, k_9 represents the chemical step(s), where k_9 is not a microscopic rate constant but a net rate constant for conversion of the EAB complex to EQ and P. The rate constants k_3 and k_7 are for binding of B to EA and EAH, respectively, while k_4 and k_8 are for release of B from EAB and EHAB, respectively, and k_{11} is for release of the last product, Q. K_1 and K_2 are acid dissociation constants for EAH and EHAB, respectively, and **H** is hydrogen ion concentration. Mechanism 10-61 indicates a decrease in activity as the pH decreases, but the treatment presented below will be identical if the EHAB form of the enzyme is the active one and activity decreases as the pH increases, or if two groups, one protonated and another unprotonated, were required for activity and the activity decreases as the pH decreases below the p*K* of one group or above the p*K* of the second group.

At pH values above pK_1 and pK_2, that is, $\mathbf{H} \ll K_1, K_2$, equations 10-62 and 10-63 are obtained for $V/\mathbf{E_t}$, $V/K_b\mathbf{E_t}$, ${}^{D}V$, and ${}^{D}(V/K_b)$. The *V*/*K* measured is for the second reactant to bind in an ordered mechanism, for either reactant at saturating concentration of the other in a random mechanism, or for the first reactant in a Ping Pong mechanism. (The expression for *V* and ${}^{D}V$ will differ in the case of a Ping Pong kinetic mechanism and will be discussed below.) An isotope effect will always be observed on *V*/*K* as long as the chemical step contributes to rate limitation (see discussion of kinetic mechanism in Chapter 9).

$$\frac{V}{\mathbf{E_t}} = \frac{k_9 k_{11}}{k_9 + k_{11}} \qquad \frac{V}{K_b \mathbf{E_t}} = \frac{k_3 k_9}{k_4 + k_9} \tag{10-62}$$

$${}^{D}V = \frac{{}^{D}k_9 + k_9/k_{11}}{1 + k_9/k_{11}} \qquad {}^{D}\left(\frac{V}{K_b}\right) = \frac{{}^{D}k_9 + k_9/k_4}{1 + k_9/k_4} \tag{10-63}$$

As discussed in Chapter 9, the net rate constant k_9 is common to all of the mechanisms that will be discussed below and includes any reverse commitment or internal forward one that might be expressed. Thus, only the forward external commitment (c_f) k_9/k_4, and the catalytic ratio (c_{Vf}) k_9/k_{11} will determine differences in the isotope effects (equations 10-62 and 10-63).

There are several possibilities for the pH dependence of kinetic parameters and isotope effects based on mechanism 10-61, and each gives different results for at least one of the parameters in equations 10-62 and 10-63. Three of the possibilities are obtained as a result of some of the steps in mechanism 10-61 being absent.

Random Addition of Proton and Substrate to Enzyme. If all rate and equilibrium constants in mechanism 10-61 are finite, both $V/\mathbf{E_t}$ ($k_9K_2/\mathbf{H}$) and

$V/K_b\mathbf{E_t}$ ($k_3k_9K_1/k_4\mathbf{H}$) will decrease by a factor of 10 per pH unit at pH values below pK_2 and pK_1. Under these conditions, the following expressions are obtained for ^DV and $^D(V/K_b)$:

$$^DV={}^Dk_9 \qquad {}^D\left(\frac{V}{K_b}\right)={}^Dk_9 \tag{10-64}$$

The largest observable isotope effect on both V and V/K_b will be observed at low pH as a result of the decrease in the concentration of the correctly protonated forms of EA and EAB. The measured value will not be the intrinsic isotope effect unless the chemical step is completely limiting for the overall reaction under these conditions. The isotope effects, if not equal to one another at high pH, $k_4 \neq k_{11}$, will become equal at low pH, that is, approximately 1 pH unit below the pK values. (If the chemical step limits at all pH values, the isotope effects on V and V/K_b will be equal to one another at all pH values.) Given a value for $^Dk_9$, one can estimate values for c_f and c_{Vf}, and thus from their ratio, k_{11}/k_4.

Data for the pigeon liver NADP-malic enzyme, which catalyzes the divalent metal ion dependent oxidative decarboxylation of L-malate to give pyruvate, CO_2, and NADPH, adhere to this mechanism (B *16*, 576). V_{max} in the direction of malate oxidative decarboxylation decreases at low and high pH with pK values of 4.5 and 9.5, while V/K_{malate} decreases at low and high pH with pK values of 6 and 8. The $^D(V/K_{malate})$ is pH-independent and equal to 1.5. In contrast, ^DV is unity at pH 7 but increases to 1.4 and 1.29 at pH 4 and 9.5, respectively (B *16*, 576). The V isotope effect becomes nearly equal to that on V/K at pH 4, which is below the pK of 4.5. The V isotope effect does not become equal to the one on V/K at high pH but was determined at the pK value of 9.5 rather than one pH unit above the pK. At the pK value, the prediction is that the isotope effect should be halfway between the minimum seen at lower pH and the maximum at pH values above the pK. The isotope effect on V/K is pH-independent and suggests that k_9/k_4 is equal to zero for L-malate over the entire pH range (there is still an internal forward as well as reverse commitment (B *20*, 1790)). With the exception of low and high pH, where the catalytic pathway limits and an isotope effect on V is observed, the absence of an isotope effect on V suggests that NADPH release limits the reaction at neutral pH values.

Dead-End Protonation of Enzyme. If reactants bind only to the correctly protonated form of the enzyme, and vice versa, k_7, k_8, and K_2 are absent in mechanism 10-61. Once reactants are bound to enzyme, the protonation state of all reactant and enzyme functional groups is locked in optimum protonation state for catalysis. The expressions for $V/\mathbf{E_t}$ (equation 10-62) and ${}^{D}V$ (equation 10-63) will be pH-independent since EAB cannot be protonated. The pH independence of V distinguishes this mechanism from the full mechanism with random addition of substrate and proton to enzyme. The expression for $V/K_b\mathbf{E_t}$ at low pH ($k_3k_9K_1/[(k_4+k_9)\mathbf{H}]$) differs from that of the full mechanism. Reactant competes with the proton at low reactant concentration. As a result, although V/K is pH-dependent, ${}^{D}(V/K_b)$ will be independent of pH and given by equation 10-63.

The NAD-malic enzyme from *Ascaris suum* exhibits a mechanism in which reactants bind to only the correctly protonated enzyme form (B 25, 227). The kinetic parameters V/K_{malate}, V/K_{NAD}, and V decrease below a pK of about 5. The primary deuterium isotope effect on the V/K values for malate and NAD is 1.5 and is pH-independent, but ${}^{D}V$ decreases from 1.6 at neutral pH to a value of 1 at low pH. The decrease in ${}^{D}V$ suggests a non-isotope-sensitive step is becoming slower at low pH, leading to the decrease in the isotope effect. [This step was shown to be isomerization of the E–NAD complex (B 32, 1928).] However, with the slow alternative substrate thio-NAD, V/K_{malate} decreases below a pK of about 5 but V does not decrease at low pH, suggesting substrate binds to only the correctly protonated form of the enzyme. The ${}^{D}V$ value of 1.9 for thio-NAD is pH-independent, as expected for a mechanism involving dead-end protonation of the enzyme.

Dead-End Protonation of Enzyme and Enzyme–Reactant Complex. If reactants bind only to the correctly protonated form of the enzyme, B can compete with a proton for the EA complex. However, the EAB complex can be protonated to give dead-end EHAB, that is, k_7 and k_8, are absent in mechanism 10-61. For this mechanism, the expressions for $V/K_b\mathbf{E_t}$ and ${}^{D}(V/K_b)$ are identical to those obtained for dead-end protonation of enzyme, see above, and for the same reasons. $V/\mathbf{E_t}$ and ${}^{D}V$ will be pH-dependent and equal to $k_9k_4/\mathbf{H}$ and ${}^{D}k_9$, respectively, at low pH, because of the dead-end protonation of EAB. This case may be distinguished from the others by the pH independence of ${}^{D}(V/K_b)$ with ${}^{D}V \neq {}^{D}V/K_b$ at low pH. If the substrate is not sticky, that is, k_9/k_4 tends to zero, it cannot be distinguished from the full mechanism. It is difficult to imagine why substrate would be prevented from binding to protonated enzyme, but possible. Only a single example of an enzyme reaction exhibiting this behavior has been reported and is discussed below.

Ketopantoate reductase catalyzes the NADPH-dependent reduction of ketopantoate. The V/K_{NADPH} pH-rate profile decreases at low and high pH and exhibits pK values of 6.2 and 8.7. $V/K_{\text{ketopantoate}}$ is pH-independent at low pH but decreases above a single pK of 8.1, while the V pH–rate profile decreases above a single pK of pH 8.4. The pK at low pH in the V/K_{NADPH} profile was attributed to the 2′-phosphate of NADPH, which indicates preferential binding of the dianionic form of the phosphate monoester of NADPH. A general acid mechanism has been proposed for the reductase in which an enzyme residue is required to be protonated to donate a proton to the carbonyl oxygen as it is reduced to the hydroxypantoate product (B *39*, 3708). ${}^{D}(V/K_{\text{ketopantoate}})$ is 1.8 and ${}^{D}(V/K_{\text{NADPH}})$ is 1.5, and both are pH-independent from pH 6.5 to 9.5. However, ${}^{D}V$ increases from a value of about 1.1 at pH 6.5 to a maximum value of 2.5 at pH 9.6. The isotope effect on ${}^{D}V$ appears to increase above a pK similar to the pK in the log V profile. The data suggest that deprotonation of the E–NADPH–ketopantoate complex locks the substrate on the enzyme. The authors suggest the general acid that protonates the carbonyl of ketopantoate as it is reduced is a lysine. Once the lysine is deprotonated, it could covalently interact with the substrate to form an imine and lock the substrate on the enzyme. This suggestion could be confirmed by reduction of the imine formed at high pH. Formation of an unprotonated dead-end enzyme–substrate complex will show the pH dependence of the isotope effects displayed by ketopantoate reductase where ${}^{D}V$ increases above the value of the ${}^{D}V/K$ isotope effects.

Dead-End Formation of Protonated Enzyme–Reactant Complex. If the EA complex can be protonated and substrate can bind to the EAH complex but locks the proton on enzyme, a dead-end complex is formed. For this case the equilibrium constant K_2 is absent. (In actuality the rate processes that contribute to K_2 are very slow, that is, it is a kinetic phenomenon.) The following expressions are obtained for the parameters V, V/K_b, ${}^{D}V$, and ${}^{D}(V/K_b)$ at low pH:

$$\frac{V}{\mathbf{E_t}} = \frac{k_3k_9k_8K_1}{(k_4+k_9)k_7\mathbf{H}} \qquad \frac{V}{K_b\mathbf{E_t}} = \frac{k_3k_9K_1}{(k_4+k_9)\mathbf{H}} \tag{10-65}$$

$${}^{D}V = \frac{{}^{D}k_9 + k_9/k_4}{1+k_9/k_4} \qquad {}^{D}\left(\frac{V}{K_b}\right) = \frac{{}^{D}k_9 + k_9/k_4}{1+k_9/k_4} \tag{10-66}$$

Once B is bound, the partitioning of EAB is unaffected by pH, so that ${}^{D}(V/K_b)$ is pH independent, although V/K_b does depend on pH. This case differs from dead-end protonation of E alone (see above) in the behavior of V, which is pH dependent as a result of the ability of B to bind to EAH to give a dead-end

EAHB complex. The concentration of B is never saturating for the EA to EAB pathway, and conditions are those for V/K. To date, no examples of this case have been reported.

^{D}V in a Ping Pong Mechanism. The expressions for ^{D}V on the basis of mechanism 10-59 and equation 10-60 above, at high and low pH, are given in equation 10-67 if k_9 is the only isotope-sensitive step:

$$^{D}V_{\text{high pH}} = \frac{^{D}k_9 + \dfrac{k_9}{k_{15}}}{1 + \dfrac{k_9}{k_{15}}} \qquad ^{D}V_{\text{low pH}} = \frac{^{D}k_9 + \left(\dfrac{k_9 K_2}{k_{15} K_4}\right)}{1 + \left(\dfrac{k_9 K_2}{k_{15} K_4}\right)} \tag{10-67}$$

If the first half-reaction limits the reaction, ^{D}V is equal to $^{D}k_9$ and is pH-independent unless K_2/K_4 is greater than or equal to k_{15}/k_9.

pH- and Isotope-Sensitive Steps Differ

Chemistry can occur with proton transfer either preceding or following a second chemical step. In this case the behavior differs from that discussed for the concerted cases discussed above. Consider the following general mechanism:

$$\mathrm{E} \underset{k_2}{\overset{k_1\mathbf{A}}{\rightleftharpoons}} \mathrm{EA} \underset{k_4\mathbf{H}}{\overset{k_3}{\rightleftharpoons}} \mathrm{EX} \underset{k_6}{\overset{k_5}{\rightleftharpoons}} \mathrm{EQ} \xrightarrow{k_7} \mathrm{E} \tag{10-68}$$

In mechanism 10-68, all rate constants have their usual meaning, and k_5 and k_6 are isotope-sensitive. Two chemical steps are depicted, a proton transfer, k_3 and $k_4\mathbf{H}$, generating an intermediate EX, and a second chemical step to give the final product EQ, k_5 and k_6. There will also, of course, be pH dependence as shown above in mechanism 61 as a result of protonation of E and EA, but for simplicity, these pH dependencies are not considered in this mechanism. Expressions for kinetic parameters V and V/K_a are shown in equations 10-69 and 10-70:

$$\frac{V}{\mathbf{E_t}} = \frac{\dfrac{k_5}{\left(1 + \dfrac{\mathbf{H}}{K_3}\right)}}{\left(1 + \dfrac{k_6}{k_7}\right) + \dfrac{k_5\left(\dfrac{1}{k_3} + \dfrac{1}{k_7}\right)}{\left(1 + \dfrac{\mathbf{H}}{K_3}\right)}} \tag{10-69}$$

$$\frac{V}{K_a\mathbf{E}_t} = \frac{\dfrac{k_1k_3k_5}{k_2k_4\mathbf{H}}}{1 + \dfrac{k_6}{k_7} + \dfrac{k_5}{k_4\mathbf{H}}\left(1 + \dfrac{k_3}{k_2}\right)} \tag{10-70}$$

In equation 10-69, K_3 is equal to k_3/k_4, the acid dissociation constant for the pH-dependent step. V and V/K_a exhibit pH dependence as a result of enzyme being maintained in the EA form. From these equations, expressions for the isotope effects on V and V/K_a are generated:

$$^{D}V = \frac{^{D}k_5 + k_5\left(\dfrac{1/k_3 + 1/k_7}{1 + \mathbf{H}/K_3}\right) + \left(\dfrac{k_6}{k_7}\right)^{D}K_{eq}}{1 + k_5\left(\dfrac{1/k_3 + 1/k_7}{1 + \mathbf{H}/K_3}\right) + \dfrac{k_6}{k_7}} \tag{10-71}$$

$$^{D}\left(\frac{V}{K_a}\right) = \frac{^{D}k_5 + \left(\dfrac{k_5}{k_4\mathbf{H}}\right)\left(1 + \dfrac{k_3}{k_2}\right) + \left(\dfrac{k_6}{k_7}\right)^{D}K_{eq}}{1 + \left(\dfrac{k_5}{k_4\mathbf{H}}\right)\left(1 + \dfrac{k_3}{k_2}\right) + \dfrac{k_6}{k_7}} \tag{10-72}$$

The isotope effect on V will be highest at pH values below pK_3 as c_{Vf}, $(k_5[(1/k_3 + 1/k_7)/(1 + \mathbf{H}/K_3)]$, approaches zero and will go to a constant value at high pH depending on the values of c_{Vf} $[k_5(1/k_3 + 1/k_7)]$ and c_r (k_6/k_7). The pH dependence of $^{D}(V/K)$ is diagnostic for this mechanism. Its value is predicted to be identical to that for ^{D}V at low pH as c_f, $(k_5/k_4\mathbf{H})(1 + k_3/k_2)$, tends to zero and will decrease to a value of 1 as the pH increases and c_f tends to infinity. The decrease to unity can occur in a pH range where there is no change in the value of the kinetic parameter and results from the loss of the proton produced in the proton transfer step, committing the reaction to go toward products.

In the reverse reaction direction c_r is identical to c_f in the forward reaction direction, and the isotope effect on V/K_q will decrease at high pH to a value equal to $^{D}K_{eq}$. The prototypical example of this behavior is exhibited by horse liver alcohol dehydrogenase, which catalyzes the NAD-dependent oxidation of a number of primary and secondary alcohols. The V and V/K profiles for cyclohexanol decrease at low pH with pK values of 6.2 and 7.1, respectively (B *20*, 1805). In the direction of ketone reduction, V decreases above a single pK of 8.4 and $V/K_{cyclohexanone}$ decreases with a limiting slope of 2 above pK values of 8.8 and 9.7. $^{D}(V/K_{cyclohexanol})$ is 2.5 and pH-independent below pH 8.5, where

$V/K_{cyclohexanol}$ is pH-dependent, but decreases to a value of 1 at high pH where it is not. $^{D}(V/K_{cyclohexanone})$, in the direction of ketone reduction, is 2.2 and pH-independent below pH 8.5 and decreases to a value of 0.85 at high pH. The equilibrium isotope effect has been determined to be 0.85 (B *19*, 4853). The p*K* values determined for the isotope effect data in both reaction directions are both 9.4. The isotope effect on *V* is 1.15 in both reaction directions at pH 8.5 and decreases to a value of 1 at high pH. The data suggest that a non-isotope-dependent pH-dependent step precedes hydride transfer from alcohol to NAD. The reaction proceeds via an alkoxide mechanism in which inner-sphere coordination to the active-site Zn allows transfer of the alcoholic proton via an active-site proton relay consisting of serine-48, the 2′-hydroxyl of NAD, and histidine-51 (B *29*, 4289; B *44*, 12797). Since one side of His51 faces solvent, it can lose a proton at high pH, thus committing the alkoxide to undergo hydride transfer. In the reverse reaction direction at high pH, hydride transfer to the ketone produces alkoxide, which cannot be released from the enzyme until a proton is available to protonate His51. Thus the hydride transfer step comes to equilibrium and one sees the equilibrium isotope effect of 0.85.

APPENDICES

These pages are provided so that the reader can photocopy them and use them as work sheets. To use them, first label the enzyme forms on the diagram and in the distribution equations. Then in the basic diagram add concentration factors to the rate constants for bimolecular steps. Thus if substrate A adds in the step with rate constant k^1, write "A" after k^1. Then in the rest of the work sheet write "A" after each term containing k^1. Repeat this for all bimolecular steps. If any rate constants are zero (irreversible step), cross out all terms containing them. The denominator of the equations is then the sum of the remaining terms on the work sheet. The rate equation is obtained from the net flux through any convenient step (rate constant in forward direction times the enzyme form concentration given by the appropriate distribution equation, minus the same product in the reverse direction).

Appendix 1 King and Altman Patterns and Distribution Equations

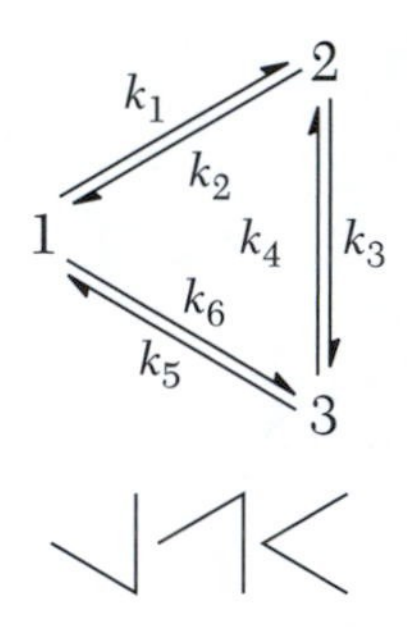

Distribution Equations

$$\frac{1}{\mathbf{E_t}}(\text{denom}) = k_3k_5 \quad + k_2k_4 \quad + k_2k_5$$

$$\frac{2}{\mathbf{E_t}}(\text{denom}) = k_4k_6 \quad + k_1k_4 \quad + k_1k_5$$

$$\frac{3}{\mathbf{E_t}}(\text{denom}) = k_3k_6 \quad + k_1k_3 \quad + k_2k_6$$

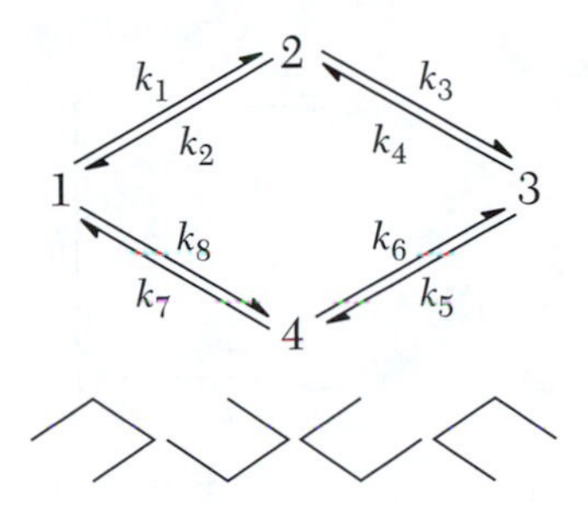

Distribution Equations

$$\frac{1}{\mathbf{E_t}}(\text{denom}) = k_2k_4k_6 \quad + k_3k_5k_7 \quad + k_2k_5k_7 \quad + k_2k_4k_7$$

$$\frac{2}{\mathbf{E_t}}(\text{denom}) = k_1k_4k_6 \quad + k_4k_6k_8 \quad + k_1k_5k_7 \quad + k_1k_4k_7$$

$$\frac{3}{\mathbf{E_t}}(\text{denom}) = k_1k_3k_6 \quad + k_3k_6k_8 \quad + k_2k_6k_8 \quad + k_1k_3k_7$$

$$\frac{4}{\mathbf{E_t}}(\text{denom}) = k_1k_3k_5 \quad + k_3k_5k_8 \quad + k_2k_5k_8 \quad + k_2k_4k_8$$

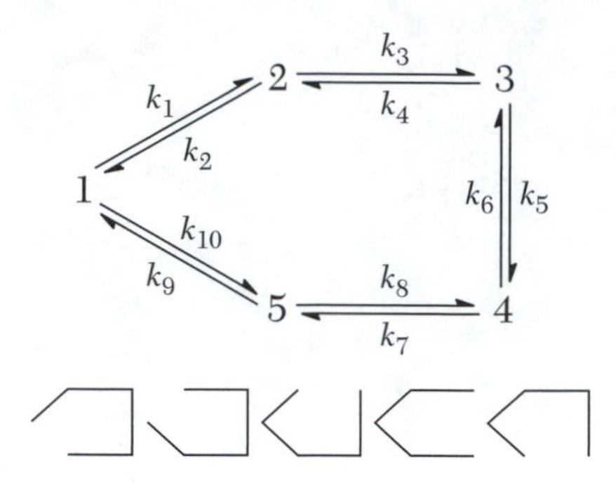

Distribution Equations

$$\frac{1}{\mathbf{E_t}}(\text{denom}) = k_2k_4k_6k_8 + k_3k_5k_7k_9 + k_2k_5k_7k_9 + k_2k_4k_7k_9 + k_2k_4k_6k_9$$

$$\frac{2}{\mathbf{E_t}}(\text{denom}) = k_1k_4k_6k_8 + k_4k_6k_8k_{10} + k_1k_5k_7k_9 + k_1k_4k_7k_9 + k_1k_4k_6k_9$$

$$\frac{3}{\mathbf{E_t}}(\text{denom}) = k_1k_3k_6k_8 + k_3k_6k_8k_{10} + k_2k_6k_8k_{10} + k_1k_3k_7k_9 + k_1k_3k_6k_9$$

$$\frac{4}{\mathbf{E_t}}(\text{denom}) = k_1k_3k_5k_8 + k_3k_5k_8k_{10} + k_2k_5k_8k_{10} + k_2k_4k_8k_{10} + k_1k_3k_5k_9$$

$$\frac{5}{\mathbf{E_t}}(\text{denom}) = k_1k_3k_5k_7 + k_3k_5k_7k_{10} + k_2k_5k_7k_{10} + k_2k_4k_7k_{10} + k_2k_4k_6k_{10}$$

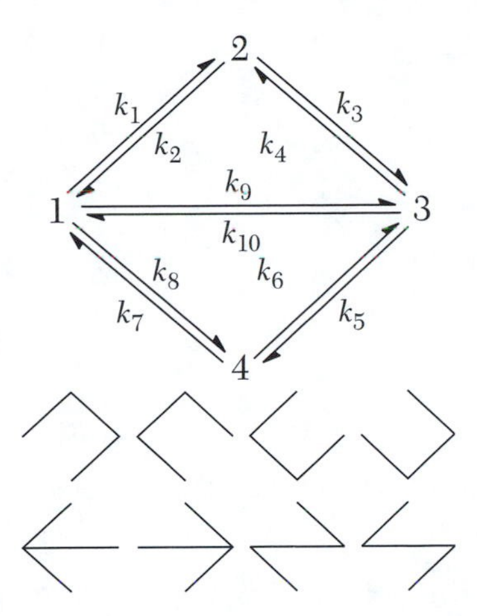

Distribution Equations

$$\begin{aligned}\frac{1}{\mathbf{E_t}}(\text{denom}) = {} & k_2k_4k_6 + k_2k_4k_7 + k_2k_5k_7 + k_3k_5k_7 \\ & + k_2k_7k_{10} + k_3k_6k_{10} + k_3k_7k_{10} + k_2k_6k_{10}\end{aligned}$$

$$\begin{aligned}\frac{2}{\mathbf{E_t}}(\text{denom}) = {} & k_1k_4k_6 + k_1k_4k_7 + k_1k_5k_7 + k_4k_6k_8 \\ & + k_1k_7k_{10} + k_4k_6k_9 + k_4k_7k_9 + k_1k_6k_{10}\end{aligned}$$

$$\begin{aligned}\frac{3}{\mathbf{E_t}}(\text{denom}) = {} & k_1k_3k_6 + k_1k_3k_7 + k_2k_6k_8 + k_3k_6k_8 \\ & + k_2k_7k_9 + k_3k_6k_9 + k_3k_7k_9 + k_2k_6k_9\end{aligned}$$

$$\begin{aligned}\frac{4}{\mathbf{E_t}}(\text{denom}) = {} & k_1k_3k_5 + k_2k_4k_8 + k_2k_5k_8 + k_3k_5k_8 \\ & + k_2k_8k_{10} + k_3k_5k_9 + k_3k_8k_{10} + k_2k_5k_9\end{aligned}$$

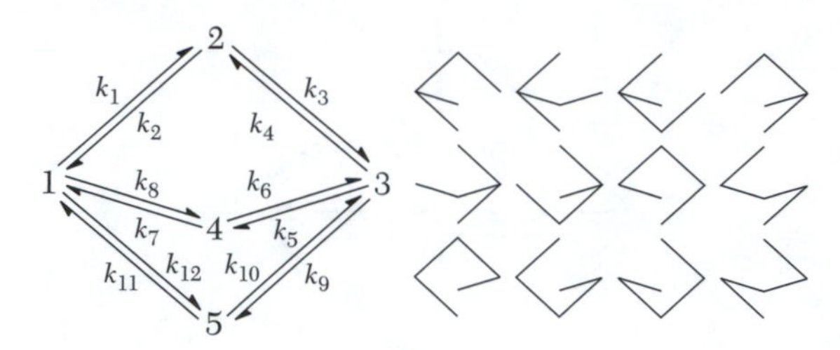

Distribution Equations

$$\begin{aligned}
(1/\mathbf{E_t})(\Delta) = {} & k_2k_4k_7k_{11} + k_2k_5k_7k_{11} + k_2k_7k_9k_{11} + k_2k_4k_6k_{10} \\
& + k_3k_5k_7k_{10} + k_3k_6k_9k_{11} + k_2k_4k_7k_{10} + k_2k_5k_7k_{10} \\
& + k_2k_4k_6k_{11} + k_2k_6k_9k_{11} + k_3k_7k_9k_{11} + k_3k_5k_7k_{11}
\end{aligned}$$

$$\begin{aligned}
(2/\mathbf{E_t})(\Delta) = {} & k_1k_4k_7k_{11} + k_1k_5k_7k_{11} + k_1k_7k_9k_{11} + k_1k_4k_6k_{10} \\
& + k_4k_6k_8k_{10} + k_4k_6k_{10}k_{12} + k_1k_4k_7k_{10} + k_1k_5k_7k_{10} \\
& + k_1k_4k_6k_{11} + k_1k_6k_9k_{11} + k_4k_7k_{10}k_{12} + k_4k_6k_8k_{11}
\end{aligned}$$

$$\begin{aligned}
(3/\mathbf{E_t})(\Delta) = {} & k_1k_3k_7k_{11} + k_2k_6k_8k_{11} + k_2k_7k_{10}k_{12} + k_1k_3k_6k_{10} \\
& + k_3k_6k_8k_{10} + k_3k_6k_{10}k_{12} + k_1k_3k_7k_{10} + k_2k_6k_8k_{10} \\
& + k_1k_3k_6k_{11} + k_2k_6k_{10}k_{12} + k_3k_7k_{10}k_{12} + k_3k_6k_8k_{11}
\end{aligned}$$

$$\begin{aligned}
(4/\mathbf{E_t})(\Delta) = {} & k_2k_4k_8k_{11} + k_2k_5k_8k_{11} + k_2k_8k_9k_{11} + k_1k_3k_5k_{10} \\
& + k_3k_5k_8k_{10} + k_3k_5k_{10}k_{12} + k_2k_4k_8k_{10} + k_2k_5k_8k_{10} \\
& + k_1k_3k_5k_{11} + k_2k_5k_{10}k_{12} + k_3k_8k_9k_{11} + k_3k_5k_8k_{11}
\end{aligned}$$

$$\begin{aligned}
(5/\mathbf{E_t})(\Delta) = {} & k_2k_4k_7k_{12} + k_2k_5k_7k_{12} + k_2k_7k_9k_{12} + k_1k_3k_6k_9 \\
& + k_3k_6k_8k_9 + k_3k_6k_9k_{12} + k_1k_3k_7k_9 + k_2k_6k_8k_9 \\
& + k_2k_4k_6k_{12} + k_2k_6k_9k_{12} + k_3k_7k_9k_{12} + k_3k_5k_7k_{12}
\end{aligned}$$

Δ = denominator

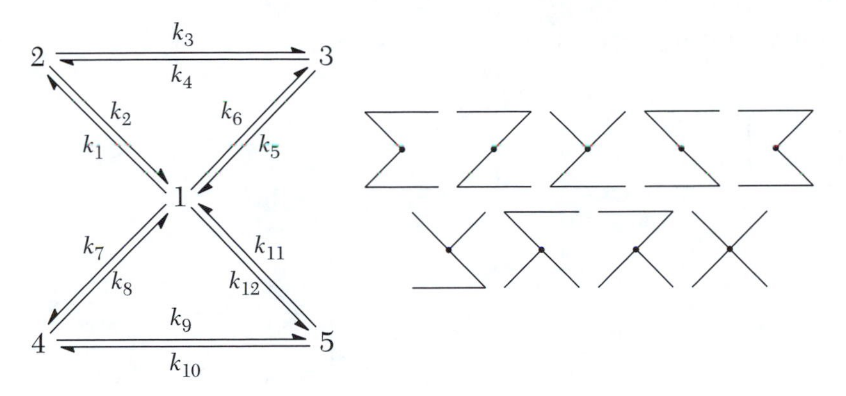

Distribution Equations

$$\begin{aligned}(1/\mathbf{E_t})(\Delta) = {} & k_2k_4k_8k_{10} + k_3k_5k_8k_{10} + k_2k_5k_8k_{10} + k_2k_4k_9k_{11} \\ & + k_3k_5k_9k_{11} + k_2k_5k_9k_{11} + k_2k_4k_8k_{11} + k_3k_5k_8k_{11} \\ & + k_2k_5k_8k_{11}\end{aligned}$$

$$\begin{aligned}(2/\mathbf{E_t})(\Delta) = {} & k_1k_4k_8k_{10} + k_4k_6k_8k_{10} + k_1k_5k_8k_{10} + k_1k_4k_9k_{11} \\ & + k_4k_6k_9k_{11} + k_1k_5k_9k_{11} + k_1k_4k_8k_{11} + k_4k_6k_8k_{11} \\ & + k_1k_5k_8k_{11}\end{aligned}$$

$$\begin{aligned}(3/\mathbf{E_t})(\Delta) = {} & k_1k_3k_8k_{10} + k_3k_6k_8k_{10} + k_2k_6k_8k_{10} + k_1k_3k_9k_{11} \\ & + k_3k_6k_9k_{11} + k_2k_6k_9k_{11} + k_1k_3k_8k_{11} + k_3k_6k_8k_{11} \\ & + k_2k_6k_8k_{11}\end{aligned}$$

$$\begin{aligned}(4/\mathbf{E_t})(\Delta) = {} & k_2k_4k_7k_{10} + k_3k_5k_7k_{10} + k_2k_5k_7k_{10} + k_2k_4k_{10}k_{12} \\ & + k_3k_5k_{10}k_{12} + k_2k_5k_{10}k_{12} + k_2k_4k_7k_{11} + k_3k_5k_7k_{11} \\ & + k_2k_5k_7k_{11}\end{aligned}$$

$$\begin{aligned}(5/\mathbf{E_t})(\Delta) = {} & k_2k_4k_7k_9 + k_3k_5k_7k_9 + k_2k_5k_7k_9 + k_2k_4k_9k_{12} \\ & + k_3k_5k_9k_{12} + k_2k_5k_9k_{12} + k_2k_4k_8k_{12} + k_3k_5k_8k_{12} \\ & + k_2k_5k_8k_{12}\end{aligned}$$

Δ = denominator

Other King–Altman patterns available

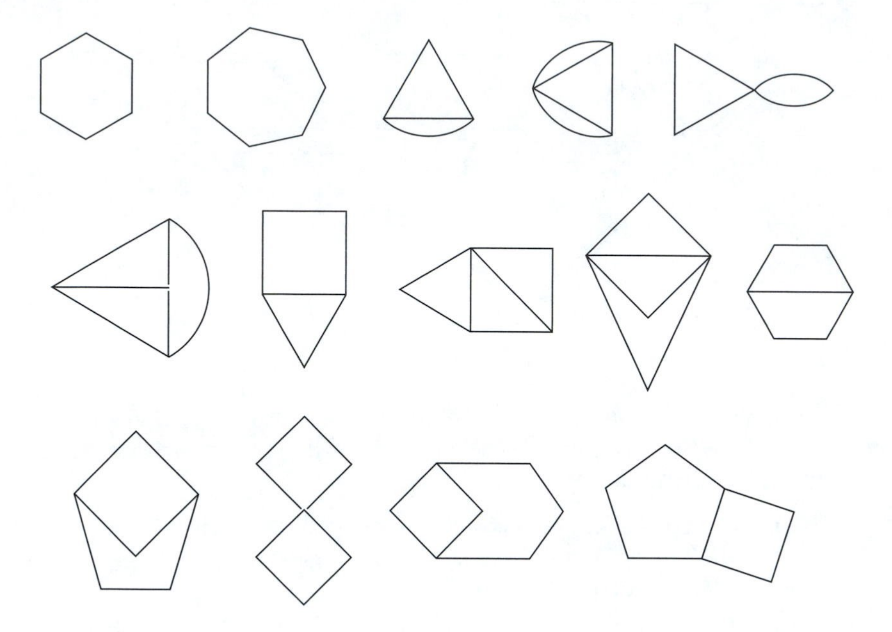

Appendix 2 Rate Equations, Definitions of Kinetic and Inhibition Constants, Haldanes, Distribution Equations, and Rate Constant Calculations for a Number of Multireactant Mechanisms

To convert equations written in terms of kinetic constants for use in the forward direction, divide numerator and denominator by V_2. For the reverse direction, divide by V_1/K_{eq}. Substitution from a Haldane will be needed for some terms to remove V_2K_{eq}/V_1 or its reciprocal.

Ordered Uni Bi

$$\mathbf{E} \underset{k_2}{\overset{k_1\mathbf{A}}{\rightleftharpoons}} (\mathbf{EA\text{–}EPQ}) \underset{k_4\mathbf{P}}{\overset{k_3}{\rightleftharpoons}} \mathbf{EQ} \underset{k_6\mathbf{Q}}{\overset{k_5}{\rightleftharpoons}} \mathbf{E}$$

$$v = \frac{(k_1k_3k_5\mathbf{A} - k_2k_4k_6\mathbf{PQ})\mathbf{E_t}}{(k_2+k_3)k_5 + k_1(k_3+k_5)\mathbf{A} + k_2k_6\mathbf{P} + (k_2+k_3)k_6\mathbf{Q} + k_1k_4\mathbf{AP} + k_4k_6\mathbf{PQ}}$$

Definitions:

$$V_1 = \frac{\text{num}_1}{\text{coef A}} \quad V_2 = \frac{\text{num}_2}{\text{coef PQ}} \quad K_a = \frac{\text{const}}{\text{coef A}} \quad K_p = \frac{\text{coef Q}}{\text{coef PQ}} \quad K_q = \frac{\text{coef P}}{\text{coef PQ}}$$

$$K_{ia} = \frac{\text{coef P}}{\text{coef AP}} = \frac{k_2}{k_1} \quad K_{ip} = \frac{\text{coef A}}{\text{coef AP}} \quad K_{iq} = \frac{\text{const}}{\text{coef Q}} = \frac{k_5}{k_6}$$

$$v = \frac{V_1V_2\left(\mathbf{A} - \frac{\mathbf{PQ}}{K_{eq}}\right)}{K_aV_2 + V_2\mathbf{A} + \frac{K_qV_1\mathbf{P}}{K_{eq}} + \frac{K_pV_1\mathbf{Q}}{K_{eq}} + \frac{V_1\mathbf{PQ}}{K_{eq}} + \frac{V_2\mathbf{AP}}{K_{ip}}}$$

Haldanes:

$$K_{eq} = \frac{V_1K_{ip}K_q}{V_2K_{ia}} = \frac{V_1K_pK_{iq}}{V_2K_a}$$

Distribution equations:

$$\frac{\mathbf{E}}{\mathbf{E_t}} = \frac{K_aV_2 + \frac{K_qV_1\mathbf{P}}{K_{eq}}}{\text{denominator of rate equation}}$$

$$\frac{(\mathbf{EA-EPQ})}{\mathbf{E_t}} = \frac{\left(V_2 - \frac{K_qV_1}{K_{iq}}\right)\mathbf{A} + \frac{V_2\mathbf{AP}}{K_{ip}} + \frac{V_1\mathbf{PQ}}{K_{eq}}}{\text{denominator of rate equation}}$$

$$\frac{\mathbf{EQ}}{\mathbf{E_t}} = \frac{\dfrac{K_p V_1 \mathbf{Q}}{K_{eq}} + \dfrac{K_q V_1 \mathbf{A}}{K_{iq}}}{\text{denominator of rate equation}}$$

Rate constants:

$$k_1 = \frac{V_2/\mathbf{E_t}}{K_{ia}} \quad k_2 = \frac{V_2}{\mathbf{E_t}} \quad k_3 = \frac{1}{\left(\dfrac{\mathbf{E_t}}{V_1} - \dfrac{1}{k_5}\right)} \quad k_4 = \frac{(k_2 + k_3)}{K_p}$$

$$k_5 = \left(\frac{V_2}{\mathbf{E_t}}\right)\left(\frac{K_{iq}}{K_q}\right) \quad k_6 = \frac{V_2/\mathbf{E_t}}{K_q}$$

Effect of isomerizations:

(**EA–EPQ**): Calculations of k_1, k_2, k_3, and k_4 are invalid.

EQ: Calculations of all rate constants are invalid. Only $\mathbf{E/E_t}$ can be calculated.

Rapid Equilibrium Random Uni Bi (E, EP, and EQ in rapid equilibrium)

$$\mathbf{E} \underset{k_2}{\overset{k_1\mathbf{A}}{\rightleftharpoons}} (\mathbf{EA\text{–}EPQ}) \underset{k_4\mathbf{P}}{\overset{k_3}{\rightleftharpoons}} \mathbf{EQ} \overset{K_{iq}/\mathbf{Q}}{\rightleftharpoons} \mathbf{E} \qquad (\mathbf{EA\text{–}EPQ}) \underset{k_6\mathbf{Q}}{\overset{k_5}{\rightleftharpoons}} \mathbf{EP} \overset{K_{ip}/\mathbf{P}}{\rightleftharpoons} \mathbf{E}$$

$$v = \frac{\left[k_1(k_3 + k_5)\mathbf{A} - k_2\left(\dfrac{k_4}{K_{iq}} + \dfrac{k_6}{K_{ip}}\right)\mathbf{PQ}\right]\mathbf{E_t}}{(k_2 + k_3 + k_5) + k_1\mathbf{A} + \dfrac{(k_2 + k_3 + k_5)\mathbf{P}}{K_{ip}} + \dfrac{(k_2 + k_3 + k_5)\mathbf{Q}}{K_{iq}} + \left(\dfrac{k_4}{K_{iq}} + \dfrac{k_6}{K_{ip}}\right)\mathbf{PQ}}$$

Definitions:

$$V_1 = \frac{\text{num}_1}{\text{coef A}} \quad V_2 = \frac{\text{num}_2}{\text{coef PQ}} \quad K_a = \frac{\text{const}}{\text{coef A}} \quad K_p = \frac{\text{coef Q}}{\text{coef PQ}} \quad K_q = \frac{\text{coef P}}{\text{coef PQ}}$$

$$K_{ip} = \frac{\text{const}}{\text{coef P}} \quad K_{iq} = \frac{\text{const}}{\text{coef Q}}$$

$$v = \frac{V_1\left(\mathbf{A} - \dfrac{\mathbf{PQ}}{K_{eq}}\right)}{K_a + \mathbf{A} + \dfrac{K_a\mathbf{P}}{K_{ip}} + \dfrac{K_a\mathbf{Q}}{K_{iq}} + \dfrac{K_a\mathbf{PQ}}{K_p K_{iq}}}$$

Haldanes:

$$K_{eq} = \frac{V_1 K_{ip} K_q}{V_2 K_a} = \frac{V_1 K_p K_{iq}}{V_2 K_a} \qquad K_{ip}K_q = K_p K_{iq}$$

Distribution equations:

$$\frac{\mathbf{E}}{\mathbf{E_t}} = \frac{K_a}{\text{denominator of rate equation}}$$

$$\frac{(\mathbf{EA}-\mathbf{EPQ})}{\mathbf{E_t}} = \frac{\mathbf{A} + \dfrac{K_a\mathbf{PQ}}{K_pK_{iq}}}{\text{denominator of rate equation}}$$

$$\frac{\mathbf{EP}}{\mathbf{E_t}} = \frac{\dfrac{K_a\mathbf{P}}{K_{ip}}}{\text{denominator of rate equation}}$$

$$\frac{\mathbf{EQ}}{\mathbf{E_t}} = \frac{\dfrac{K_a\mathbf{Q}}{K_{iq}}}{\text{denominator of rate equation}}$$

If **EA** and **EPQ** are treated as separate complexes, the **A** term represents **EA** and the **PQ** term represents **EPQ**. If all steps other than the **EA–EPQ** interconversion are in rapid equilibrium, the rate equation is the same, but all Michaelis constants are dissociation constants and the $V/\mathbf{E_t}$ values are the forward and reverse rate constants for **EA–EPQ** interconversion.

Rate constants:

$$k_2 = \frac{V_2}{\mathbf{E_t}} \qquad \text{Not valid if } \mathbf{EA} \text{ and } \mathbf{EPQ} \text{ are not combined.}$$

If only **E** and **EP** are in rapid equilibrium (that is, there is an **EQ** complex present in the steady state in the forward direction), the denominator is the same as that for the Ordered Uni Bi mechanism with added $\mathbf{P}^2$ and $\mathbf{P}^2\mathbf{Q}$ terms. If there are no rapid equilibria (both **EP** and **EQ** occur in the steady state), there are also $\mathbf{Q}^2$, $\mathbf{PQ}^2$, **AQ**, and **APQ** terms in the denominator of the rate equation.

Ordered Bi Bi

$$\mathbf{E} \underset{k_2}{\overset{k_1\mathbf{A}}{\rightleftharpoons}} \mathbf{EA} \underset{k_4}{\overset{k_3\mathbf{B}}{\rightleftharpoons}} (\mathbf{EAB\text{–}EPQ}) \underset{k_6\mathbf{P}}{\overset{k_5}{\rightleftharpoons}} \mathbf{EQ} \underset{k_8\mathbf{Q}}{\overset{k_7}{\rightleftharpoons}} \mathbf{E}$$

$$v = \frac{(k_1k_3k_5k_7\mathbf{AB} - k_2k_4k_6k_8\mathbf{PQ})\mathbf{E_t}}{\begin{array}{l} k_2(k_4+k_5)k_7 + k_1(k_4+k_5)k_7\mathbf{A} + k_3k_5k_7\mathbf{B} + k_1k_3(k_5+k_7)\mathbf{AB} \\ + k_2(k_4+k_5)k_8\mathbf{Q} + k_2k_4k_6\mathbf{P} + (k_2+k_4)k_6k_8\mathbf{PQ} + k_1k_4k_6\mathbf{AP} \\ + k_3k_5k_8\mathbf{BQ} + k_1k_3k_6\mathbf{ABP} + k_3k_6k_8\mathbf{BPQ} \end{array}}$$

Definitions:

$$V_1 = \frac{\text{num}_1}{\text{coef AB}} \quad V_2 = \frac{\text{num}_2}{\text{coef PQ}} \quad K_a = \frac{\text{coef B}}{\text{coef AB}} \quad K_b = \frac{\text{coef A}}{\text{coef AB}} \quad K_p = \frac{\text{coef Q}}{\text{coef PQ}}$$

$$K_q = \frac{\text{coef P}}{\text{coef PQ}} \quad K_{ia} = \frac{\text{const}}{\text{coef A}} = \frac{\text{coef P}}{\text{coef AP}} = \frac{k_2}{k_1} \quad K_{ib} = \frac{\text{coef PQ}}{\text{coef BPQ}} \quad K_{ip} = \frac{\text{coef AB}}{\text{coef ABP}}$$

$$K_{iq} = \frac{\text{const}}{\text{coef Q}} = \frac{\text{coef B}}{\text{coef BQ}} = \frac{k_7}{k_8}$$

$$v = \frac{V_1 V_2\left(\mathbf{AB} - \dfrac{\mathbf{PQ}}{K_{eq}}\right)}{K_{ia}K_bV_2 + K_bV_2\mathbf{A} + K_aV_2\mathbf{B} + V_2\mathbf{AB} + \dfrac{K_qV_1\mathbf{P}}{K_{eq}} + \dfrac{K_pV_1\mathbf{Q}}{K_{eq}} + \dfrac{V_1\mathbf{PQ}}{K_{eq}} + \dfrac{K_qV_1\mathbf{AP}}{K_{ia}K_{eq}} + \dfrac{K_aV_2\mathbf{BQ}}{K_{iq}} + \dfrac{V_2\mathbf{ABP}}{K_{ip}} + \dfrac{V_1\mathbf{BPQ}}{K_{ib}K_{eq}}}$$

Haldanes:

$$K_{eq} = \frac{V_1K_pK_{iq}}{V_2K_{ia}K_b} = \left(\frac{V_1}{V_2}\right)^2 \frac{K_{ip}K_q}{K_aK_{ib}}$$

Distribution equations:

$$\frac{\mathbf{E}}{\mathbf{E_t}} = \frac{K_{ia}K_bV_2 + \dfrac{K_qV_1\mathbf{P}}{K_{eq}} + K_aV_2\mathbf{B}}{\text{denominator of rate equation}}$$

$$\frac{\mathbf{EA}}{\mathbf{E_t}} = \frac{K_bV_2\mathbf{A} + \dfrac{K_qV_1\mathbf{AP}}{K_{ia}K_{eq}} + \dfrac{K_aV_2\mathbf{PQ}}{K_{ia}K_{eq}}}{\text{denominator of rate equation}}$$

$$\frac{(\mathbf{EA}-\mathbf{EPQ})}{\mathbf{E_t}} = \frac{\left(V_2 - \dfrac{K_qV_1}{K_{iq}}\right)\mathbf{AB} + \left(\dfrac{V_1}{K_{eq}} - \dfrac{K_aV_2}{K_{ia}K_{eq}}\right)\mathbf{PQ} + \dfrac{V_2\mathbf{ABP}}{K_{ip}} + \dfrac{V_1\mathbf{BPQ}}{K_{ib}K_{eq}}}{\text{denominator of rate equation}}$$

$$\frac{\mathbf{EQ}}{\mathbf{E_t}} = \frac{\dfrac{K_pV_1\mathbf{Q}}{K_{eq}} + \dfrac{K_qV_1\mathbf{AB}}{K_{iq}} + \dfrac{K_aV_2\mathbf{BQ}}{K_{iq}}}{\text{denominator of rate equation}}$$

Rate constants:

$$k_1 = \frac{V_1}{K_a\mathbf{E_t}} \quad k_2 = \frac{V_1K_{ia}}{K_a\mathbf{E_t}} \quad k_3 = \left(\frac{V_1}{K_a\mathbf{E_t}}\right)\left(1+\frac{k_4}{k_5}\right) \quad \frac{1}{k_4} = \frac{\mathbf{E_t}}{V_2} - \frac{1}{k_2} \quad \frac{1}{k_5} = \frac{\mathbf{E_t}}{V_1} - \frac{1}{k_7}$$

$$k_6 = \left(\frac{V_2}{K_p\mathbf{E_t}}\right)\left(1+\frac{k_5}{k_4}\right) \quad k_7 = \left(\frac{V_2K_{iq}}{K_q\mathbf{E_t}}\right) \quad k_8 = \frac{V_2}{K_q\mathbf{E_t}}$$

Effect of isomerizations:

(**EAB–EPQ**): Distributions are valid; k_1, k_2, k_7, and k_8 can still be calculated.

EA: $\mathbf{E/E_t}$, $\mathbf{EQ/E_t}$, k_7, and k_8 can still be calculated.

EQ: $\mathbf{E/E_t}$, $\mathbf{EA/E_t}$, k_1, and k_2 can still be calculated.

EA and **EQ**: Only $\mathbf{E/E_t}$ can be calculated.

Theorell–Chance Mechanism

$$\mathbf{E} \underset{k_2}{\overset{k_1\mathbf{A}}{\rightleftharpoons}} \mathbf{EA} \underset{k_4\mathbf{P}}{\overset{k_3\mathbf{B}}{\rightleftharpoons}} \mathbf{EQ} \underset{k_6\mathbf{Q}}{\overset{k_5}{\rightleftharpoons}} \mathbf{E}$$

$$v = \frac{(k_1k_3k_5\mathbf{AB} - k_2k_4k_6\mathbf{PQ})\mathbf{E_t}}{k_2k_5 + k_1k_5\mathbf{A} + k_3k_5\mathbf{B} + k_1k_3\mathbf{AB} + k_2k_6\mathbf{Q} + k_4k_6\mathbf{PQ} + k_2k_4\mathbf{P} + k_1k_4\mathbf{AP} + k_3k_6\mathbf{BQ}}$$

Definitions:

$$V_1 = \frac{\text{num}_1}{\text{coef AB}} \quad V_2 = \frac{\text{num}_2}{\text{coef PQ}} \quad K_a = \frac{\text{coef B}}{\text{coef AB}} \quad K_b = \frac{\text{coef A}}{\text{coef AB}} \quad K_p = \frac{\text{coef Q}}{\text{coef PQ}}$$

$$K_q = \frac{\text{coef P}}{\text{coef PQ}} \quad K_{ia} = \frac{\text{const}}{\text{coef A}} = \frac{\text{coef P}}{\text{coef AP}} = \frac{k_2}{k_1} \quad K_{ib} = \frac{\text{coef Q}}{\text{coef BQ}} \quad K_{ip} = \frac{\text{coef A}}{\text{coef AP}}$$

$$K_{iq} = \frac{\text{const}}{\text{coef Q}} = \frac{\text{coef B}}{\text{coef BQ}} = \frac{k_5}{k_6}$$

$$v = \frac{V_1V_2\left(\mathbf{AB} - \frac{\mathbf{PQ}}{K_{eq}}\right)}{K_{ia}K_bV_2 + K_bV_2\mathbf{A} + K_aV_2\mathbf{B} + V_2\mathbf{AB} + \frac{K_qV_1\mathbf{P}}{K_{eq}} + \frac{K_pV_1\mathbf{Q}}{K_{eq}} + \frac{V_1\mathbf{PQ}}{K_{eq}} + \frac{K_qV_1\mathbf{AP}}{K_{ia}K_{eq}} + \frac{K_aV_2\mathbf{BQ}}{K_{iq}}}$$

Haldanes:

$$K_{eq} = \frac{V_1 K_p K_{iq}}{V_2 K_{ia} K_b} = \frac{V_1 K_{ip} K_q}{V_2 K_a K_{ib}} = \frac{V_1 K_p K_{iq}}{V_2 K_a K_{ib}} = \frac{V_1 K_{ip} K_q}{V_2 K_{ia} K_b}$$

There are 12 more Haldanes, which apply for the simple mechanism with six rate constants, but they are not valid when **EA** or **EQ** isomerizes.

Distribution equations:

$$\frac{\mathbf{E}}{\mathbf{E_t}} = \frac{K_{ia} K_b V_2 + \dfrac{K_q V_1 \mathbf{P}}{K_{eq}} + K_a V_2 \mathbf{B}}{\text{denominator of rate equation}}$$

$$\frac{\mathbf{EA}}{\mathbf{E_t}} = \frac{K_b V_2 \mathbf{A} + \dfrac{K_q V_1 \mathbf{AP}}{K_{ia} K_{eq}} + \dfrac{V_1 \mathbf{PQ}}{K_{eq}}}{\text{denominator of rate equation}}$$

$$\frac{\mathbf{EQ}}{\mathbf{E_t}} = \frac{\dfrac{K_p V_1 \mathbf{Q}}{K_{eq}} + V_2 \mathbf{AB} + \dfrac{K_a V_2 \mathbf{BQ}}{K_{iq}}}{\text{denominator of rate equation}}$$

Rate constants:

$$k_1 = \frac{V_1}{K_a \mathbf{E_t}} \quad k_2 = \frac{V_1 K_{ia}}{K_a \mathbf{E_t}} \quad k_3 = \frac{V_1}{K_b \mathbf{E_t}} \quad k_4 = \frac{V_2}{K_p \mathbf{E_t}} \quad k_5 = \frac{V_2 K_{iq}}{K_q \mathbf{E_t}} \quad k_6 = \frac{V_2}{K_q \mathbf{E_t}}$$

Effect of isomerizations:

EA: k_4, k_5, and k_6 can still be calculated. In addition, the rate constant for conversion of **EA** to **E′A** in the forward direction is given by $1/k = \mathbf{E_t}/V_1 - 1/k_5$. The equation $\mathbf{E}/\mathbf{E_t}$ is unchanged, but

$$\frac{(\mathbf{EA} + \mathbf{E'A})}{\mathbf{E_t}} = \frac{K_b V_2 \mathbf{A} + \dfrac{K_q V_1 \mathbf{AP}}{K_{ia} K_{eq}} + \dfrac{V_1 \mathbf{PQ}}{K_{eq}} + \left(V_2 - \dfrac{K_q V_1}{K_{iq}}\right)\mathbf{AB}}{\text{denominator of rate equation}}$$

$$\frac{\mathbf{EQ}}{\mathbf{E_t}} = \frac{\dfrac{K_q V_1 \mathbf{Q}}{K_{eq}} + \dfrac{K_q V_1 \mathbf{AB}}{K_{iq}} + \dfrac{K_a V_2 \mathbf{BQ}}{K_{iq}}}{\text{denominator of rate equation}}$$

EQ: k_1, k_2, and k_3 can still be calculated. In addition, the rate constant for conversion of **EQ** to **EQ′** in the reverse direction is given by $1/k = \mathbf{E_t}/V_2 - 1/k_2$.

The equation for $\mathbf{E}/\mathbf{E_t}$ is unchanged, but

$$\frac{\mathbf{EA}}{\mathbf{E_t}} = \frac{K_b V_2 \mathbf{A} + \dfrac{K_q V_1 \mathbf{AP}}{K_{ia} K_{eq}} + \dfrac{K_a V_2 \mathbf{PQ}}{K_{ia} K_{eq}}}{\text{denominator of rate equation}}$$

$$\frac{(\mathbf{EQ} + \mathbf{E'Q})}{\mathbf{E_t}} = \frac{\dfrac{K_p V_1 \mathbf{Q}}{K_{eq}} + V_2 \mathbf{AB} + \dfrac{K_a V_2 \mathbf{BQ}}{K_{iq}} + \left[\dfrac{V_1 - K_a V_2 / K_{ia}}{K_{eq}}\right]\mathbf{PQ}}{\text{denominator of rate equation}}$$

EA and **EQ**: Only $\mathbf{E}/\mathbf{E_t}$ can be calculated.

Rapid Equilibrium Random Bi Bi (E, EA, EB, EP, EQ in rapid equilibrium)

$$(\mathbf{E} \underset{}{\overset{\mathbf{A}/K_{ia}}{\rightleftharpoons}} \mathbf{EA}) \underset{k_4}{\overset{k_3\mathbf{B}}{\rightleftharpoons}} \mathbf{EAB{-}EPQ} \underset{k_{10}\mathbf{P}}{\overset{k_9}{\rightleftharpoons}} (\mathbf{EQ} \overset{K_{iq}/\mathbf{Q}}{\rightleftharpoons} \mathbf{E})$$

$$(\mathbf{E} \underset{}{\overset{\mathbf{B}/K_{ib}}{\rightleftharpoons}} \mathbf{EB}) \underset{k_7}{\overset{k_8\mathbf{A}}{\rightleftharpoons}} \mathbf{EAB{-}EPQ} \underset{k_{14}\mathbf{Q}}{\overset{k_{13}}{\rightleftharpoons}} (\mathbf{EP} \overset{K_{ip}/\mathbf{P}}{\rightleftharpoons} \mathbf{E})$$

$$v = \frac{\left[(k_9 + k_{13})\left(\dfrac{k_3}{K_{ia}} + \dfrac{k_8}{K_{ib}}\right)\mathbf{AB} - (k_4 + k_7)\left(\dfrac{k_{10}}{K_{iq}} + \dfrac{k_{14}}{K_{ip}}\right)\mathbf{PQ}\right]\mathbf{E_t}}{(k_4 + k_7 + k_9 + k_{13})\left(1 + \dfrac{\mathbf{A}}{K_{ia}} + \dfrac{\mathbf{B}}{K_{ib}} + \dfrac{\mathbf{P}}{K_{ip}} + \dfrac{\mathbf{Q}}{K_{iq}}\right) + \left(\dfrac{k_3}{K_{ia}} + \dfrac{k_8}{K_{ib}}\right)\mathbf{AB} + \left(\dfrac{k_{10}}{K_{iq}} + \dfrac{k_{14}}{K_{ip}}\right)\mathbf{PQ}}$$

Definitions:

$$V_1 = \frac{\text{num}_1}{\text{coef AB}} \quad V_2 = \frac{\text{num}_2}{\text{coef PQ}} \quad K_a = \frac{\text{coef B}}{\text{coef AB}} \quad K_b = \frac{\text{coef A}}{\text{coef AB}} \quad K_p = \frac{\text{coef Q}}{\text{coef PQ}}$$

$$K_q = \frac{\text{coef P}}{\text{coef PQ}} \quad K_{ia} = \frac{\text{const}}{\text{coef A}} \quad K_{ib} = \frac{\text{const}}{\text{coef B}} \quad K_{ip} = \frac{\text{const}}{\text{coef P}} \quad K_{iq} = \frac{\text{const}}{\text{coef Q}}$$

$$v = \frac{V_1 V_2\left(\mathbf{AB} - \dfrac{\mathbf{PQ}}{K_{eq}}\right)}{K_{ia} K_b V_2 + K_b V_2 \mathbf{A} + K_a V_2 \mathbf{B} + V_2 \mathbf{AB} + \dfrac{K_q V_1 \mathbf{P}}{K_{eq}} + \dfrac{K_p V_1 \mathbf{Q}}{K_{eq}} + \dfrac{V_1 \mathbf{PQ}}{K_{eq}}}$$

Haldanes:

$$K_{eq} = \frac{V_1 K_p K_{iq}}{V_2 K_{ia} K_b} = \frac{V_1 K_{ip} K_q}{V_2 K_a K_{ib}} = \frac{V_1 K_p K_{iq}}{V_2 K_a K_{ib}} = \frac{V_1 K_{ip} K_q}{V_2 K_{ia} K_b}$$

Distribution equations:

$$\frac{\mathbf{E}}{\mathbf{E_t}} = \frac{K_{ia} K_b V_2}{\text{denominator of rate equation}}$$

$$\frac{\mathbf{EA}}{\mathbf{E_t}} = \frac{K_b V_2 \mathbf{A}}{\text{denominator of rate equation}}$$

$$\frac{\mathbf{EB}}{\mathbf{E_t}} = \frac{K_a V_2 \mathbf{B}}{\text{denominator of rate equation}}$$

$$\frac{(\mathbf{EAB} - \mathbf{EPQ})}{\mathbf{E_t}} = \frac{V_2 \mathbf{AB} + \dfrac{V_1 \mathbf{PQ}}{K_{eq}}}{\text{denominator of rate equation}}$$

$$\frac{\mathbf{EP}}{\mathbf{E_t}} = \frac{\dfrac{K_q V_1 \mathbf{P}}{K_{eq}}}{\text{denominator of rate equation}}$$

$$\frac{\mathbf{EQ}}{\mathbf{E_t}} = \frac{\dfrac{K_p V_1 \mathbf{Q}}{K_{eq}}}{\text{denominator of rate equation}}$$

Rate constants:

No rate constants can be determined. If all steps other than the **EAB–EPQ** interconversion are in rapid equilibrium, the rate equation is the same, but all Michaelis constants are dissociation constants and the $V/\mathbf{E_t}$ values are the forward and reverse rate constants for **EA–EPQ** interconversion.

If only the **EA**, **E**, and **EQ** complexes are in rapid equilibrium, the denominator contains additional terms such as **AP**, **AQ**, **BP**, **ABQ**, and **APQ** as well as terms containing $\mathbf{A}^2$ or $\mathbf{Q}^2$. There are no terms containing $\mathbf{B}^2$ or $\mathbf{P}^2$, however.

Ping Pong Bi Bi

$$\mathbf{E} \underset{k_2}{\overset{k_1\mathbf{A}}{\rightleftharpoons}} (\mathbf{EA{-}FP}) \underset{k_4\mathbf{P}}{\overset{k_3}{\rightleftharpoons}} \mathbf{F} \underset{k_6}{\overset{k_5\mathbf{B}}{\rightleftharpoons}} (\mathbf{FB{-}EQ}) \underset{k_8\mathbf{Q}}{\overset{k_7}{\rightleftharpoons}} \mathbf{E}$$

$$v = \frac{(k_1k_3k_5\mathbf{AB} - k_2k_4k_6\mathbf{PQ})\mathbf{E_t}}{k_1k_3(k_6+k_7)\mathbf{A} + k_5k_7(k_2+k_3)\mathbf{B} + k_1k_5(k_3+k_7)\mathbf{AB} + k_6k_8(k_2+k_3)\mathbf{Q} + k_2k_4(k_6+k_7)\mathbf{P} + k_4k_8(k_2+k_6)\mathbf{PQ} + k_1k_4(k_6+k_7)\mathbf{AP} + k_5k_8(k_2+k_3)\mathbf{BQ}}$$

Definitions:

$$V_1 = \frac{\text{num}_1}{\text{coef AB}} \quad V_2 = \frac{\text{num}_2}{\text{coef PQ}} \quad K_a = \frac{\text{coef B}}{\text{coef AB}} \quad K_b = \frac{\text{coef A}}{\text{coef AB}} \quad K_p = \frac{\text{coef Q}}{\text{coef PQ}}$$

$$K_q = \frac{\text{coef P}}{\text{coef PQ}} \quad K_{ia} = \frac{\text{coef P}}{\text{coef AP}} = \frac{k_2}{k_1} \quad K_{ib} = \frac{\text{coef Q}}{\text{coef BQ}} = \frac{k_6}{k_5} \quad K_{ip} = \frac{\text{coef A}}{\text{coef AP}} = \frac{k_3}{k_4}$$

$$K_{iq} = \frac{\text{coef B}}{\text{coef BQ}} = \frac{k_7}{k_8}$$

$$v = \frac{V_1V_2\left(\mathbf{AB} - \frac{\mathbf{PQ}}{K_{eq}}\right)}{K_bV_2\mathbf{A} + K_aV_2\mathbf{B} + V_2\mathbf{AB} + \frac{K_qV_1\mathbf{P}}{K_{eq}} + \frac{K_pV_1\mathbf{Q}}{K_{eq}} + \frac{V_1\mathbf{PQ}}{K_{eq}} + \frac{K_bV_2\mathbf{AP}}{K_{ip}} + \frac{K_pV_1\mathbf{BQ}}{K_{ib}K_{eq}}}$$

Haldanes:

$$K_{eq} = \frac{K_{ip}K_{iq}}{K_{ia}K_{ib}} = \frac{V_1K_{ip}K_q}{V_2K_{ia}K_b} = \frac{V_1K_pK_{iq}}{V_2K_aK_{ib}} = \left(\frac{V_1}{V_2}\right)^2\frac{K_pK_q}{K_aK_b}$$

Distribution equations:

$$\frac{\mathbf{E}}{\mathbf{E_t}} = \frac{\frac{K_qV_1\mathbf{P}}{K_{eq}} + K_aV_2\mathbf{B}}{\text{denominator of rate equation}}$$

$$\frac{\mathbf{F}}{\mathbf{E_t}} = \frac{\frac{K_pV_1\mathbf{Q}}{K_{eq}} + K_bV_2\mathbf{A}}{\text{denominator of rate equation}}$$

The distribution between the central complexes cannot be determined.

Rate constants: None can be calculated.

Effects of isomerizations of central complexes: No change.

Ordered Ter Bi

$$\mathbf{E} \underset{k_2}{\overset{k_1\mathbf{A}}{\rightleftharpoons}} \mathbf{EA} \underset{k_4}{\overset{k_3\mathbf{B}}{\rightleftharpoons}} \mathbf{EAB} \underset{k_6}{\overset{k_5\mathbf{C}}{\rightleftharpoons}} (\mathbf{EABC\text{–}EPQ}) \underset{k_8\mathbf{P}}{\overset{k_7}{\rightleftharpoons}} \mathbf{EQ} \underset{k_{10}\mathbf{Q}}{\overset{k_9}{\rightleftharpoons}} \mathbf{E}$$

$$v = \frac{(k_1k_3k_5k_7k_9\mathbf{ABC} - k_2k_4k_6k_8k_{10}\mathbf{PQ})\mathbf{E_t}}{\begin{array}{l} k_2k_4(k_6+k_7)k_9 + k_1k_4(k_6+k_7)k_9\mathbf{A} + k_2k_5k_7k_9\mathbf{C} + k_1k_3(k_6+k_7)k_9\mathbf{AB} \\ + k_1k_5k_7k_9\mathbf{AC} + k_3k_5k_7k_9\mathbf{BC} + k_1k_3k_5(k_7+k_9)\mathbf{ABC} + k_2k_4(k_6+k_7)k_{10}\mathbf{Q} \\ + k_2k_4k_6k_8\mathbf{P} + (k_2k_4 + k_4k_6 + k_2k_6)k_8k_{10}\mathbf{PQ} + k_1k_4k_6k_8\mathbf{AP} + k_2k_5k_7k_{10}\mathbf{CQ} \\ + k_1k_3k_6k_8\mathbf{ABP} + k_3k_5k_7k_{10}\mathbf{BCQ} + k_3k_6k_8k_{10}\mathbf{BPQ} + k_2k_5k_8k_{10}\mathbf{CPQ} \\ + k_1k_3k_5k_8\mathbf{ABCP} + k_3k_5k_8k_{10}\mathbf{BCPQ} \end{array}}$$

Definitions:

$$V_1 = \frac{\text{num}_1}{\text{coef ABC}} \quad V_2 = \frac{\text{num}_2}{\text{coef PQ}} \quad K_a = \frac{\text{coef BC}}{\text{coef ABC}} \quad K_b = \frac{\text{coef AC}}{\text{coef ABC}}$$

$$K_c = \frac{\text{coef AB}}{\text{coef ABC}}$$

$$K_p = \frac{\text{coef Q}}{\text{coef PQ}} \quad K_q = \frac{\text{coef P}}{\text{coef PQ}} \quad K_{ia} = \frac{\text{const}}{\text{coef A}} = \frac{\text{coef P}}{\text{coef AP}} = \frac{\text{coef C}}{\text{coef AC}} = \frac{k_2}{k_1}$$

$$K_{ib} = \frac{\text{coef A}}{\text{coef AB}} = \frac{\text{coef AP}}{\text{coef ABP}} = \frac{k_4}{k_3} \quad K_{ic} = \frac{\text{coef BPQ}}{\text{coef BCPQ}} = \frac{\text{coef ABP}}{\text{coef ABCP}} = \frac{k_6}{k_5}$$

$$K_{ip} = \frac{\text{coef CQ}}{\text{coef CPQ}} = \frac{\text{coef BCQ}}{\text{coef BCPQ}} = \frac{k_7}{k_8} \quad K_{iq} = \frac{\text{const}}{\text{coef Q}} = \frac{\text{coef C}}{\text{coef CQ}} = \frac{\text{coef BC}}{\text{coef BCQ}} = \frac{k_9}{k_{10}}$$

$$v = \frac{V_1V_2\left(\mathbf{ABC} - \dfrac{\mathbf{PQ}}{K_{eq}}\right)}{\begin{array}{l} K_{ia}K_{ib}K_cV_2 + K_{ib}K_cV_2\mathbf{A} + K_{ia}K_bV_2\mathbf{C} + K_cV_2\mathbf{AB} + K_bV_2\mathbf{AC} + K_aV_2\mathbf{BC} \\ + V_2\mathbf{ABC} + \dfrac{K_qV_1\mathbf{P}}{K_{eq}} + \dfrac{K_pV_1\mathbf{Q}}{K_{eq}} + \dfrac{V_1\mathbf{PQ}}{K_{eq}} + \dfrac{K_bV_1\mathbf{AP}}{K_{ia}K_{eq}} + \dfrac{K_{ia}K_bV_2\mathbf{CQ}}{K_{iq}} + \dfrac{K_qV_1\mathbf{ABP}}{K_{ia}K_{ib}K_{eq}} \\ + \dfrac{K_aV_2\mathbf{BCQ}}{K_{iq}} + \dfrac{K_aK_{ic}V_2\mathbf{BPQ}}{K_{ip}K_{iq}} + \dfrac{K_{ia}K_bV_2\mathbf{CPQ}}{K_{ip}K_{iq}} + \dfrac{K_qV_1\mathbf{ABCP}}{K_{ia}K_{ib}K_{ic}K_{eq}} + \dfrac{K_aV_2\mathbf{BCPQ}}{K_{ip}K_{iq}} \end{array}}$$

Haldanes:

$$K_{eq} = \frac{K_{ip}K_{iq}}{K_{ia}K_{ib}K_{ic}} = \frac{V_1K_pK_{iq}}{V_2K_{ia}K_{ib}K_c}$$

Distribution equations:

$$\frac{\mathbf{E}}{\mathbf{E_t}} = \frac{K_{ia}K_{ib}K_cV_2 + \dfrac{K_qV_1\mathbf{P}}{K_{eq}} + K_{ia}K_bV_2\mathbf{C} + K_aV_2\mathbf{BC}}{\text{denominator of rate equation}}$$

$$\frac{\mathbf{EA}}{\mathbf{E_t}} = \frac{K_{ib}K_cV_2\mathbf{A} + K_bV_2\mathbf{AC} + \dfrac{K_qV_1\mathbf{AP}}{K_{ia}K_{eq}} + \dfrac{K_aV_2\mathbf{PQ}}{K_{ia}K_{eq}}}{\text{denominator of rate equation}}$$

$$\frac{\mathbf{EAB}}{\mathbf{E_t}} = \frac{K_cV_2\mathbf{AB} + \dfrac{K_aK_{ic}V_2\mathbf{BPQ}}{K_{ip}K_{iq}} + \dfrac{K_qV_1\mathbf{ABP}}{K_{ia}K_{ib}K_{eq}} + \dfrac{K_bV_2\mathbf{PQ}}{K_{ib}K_{eq}}}{\text{denominator of rate equation}}$$

$$\frac{(\mathbf{EABC}-\mathbf{EPQ})}{\mathbf{E_t}} = \frac{\left(V_2 - \dfrac{K_qV_1}{K_{iq}}\right)\mathbf{ABC} + \left[\dfrac{V_1}{K_{eq}} - \left(\dfrac{V_2}{K_{eq}}\right)\left(\dfrac{K_a}{K_{ia}} + \dfrac{K_b}{K_{ib}}\right)\right]\mathbf{PQ} + \dfrac{K_{ia}K_bV_2\mathbf{CPQ}}{K_{ip}K_{iq}} + \dfrac{K_qV_1\mathbf{ABCP}}{K_{ia}K_{ib}K_{ic}K_{eq}} + \dfrac{K_aV_2\mathbf{BCPQ}}{K_{ip}K_{iq}}}{\text{denominator of rate equation}}$$

$$\frac{\mathbf{EQ}}{\mathbf{E_t}} = \frac{\dfrac{K_pV_1\mathbf{Q}}{K_{eq}} + \dfrac{K_qV_1\mathbf{ABC}}{K_{iq}} + \dfrac{K_{ia}K_bV_2\mathbf{CQ}}{K_{iq}} + \dfrac{K_aV_2\mathbf{BCQ}}{K_{iq}}}{\text{denominator of rate equation}}$$

Rate constants:

$$k_1 = \frac{V_1}{K_a\mathbf{E_t}} \quad k_2 = \frac{V_1K_{ia}}{K_a\mathbf{E_t}} \quad k_3 = \frac{V_1}{K_b\mathbf{E_t}} \quad k_4 = \frac{K_{ib}V_1}{K_b\mathbf{E_t}} \quad k_5 = \left(\frac{V_1}{K_c\mathbf{E_t}}\right)\left(1 + \frac{k_6}{k_7}\right)$$

$$\frac{1}{k_6} = \frac{\mathbf{E_t}}{V_2} - \frac{1}{k_2} - \frac{1}{k_4} \quad \frac{1}{k_7} = \frac{\mathbf{E_t}}{V_1} - \frac{1}{k_9} \quad k_8 = \left(\frac{V_2}{K_p\mathbf{E_t}}\right)\left(1 + \frac{k_7}{k_6}\right) \quad k_9 = \frac{K_{iq}V_2}{K_q\mathbf{E_t}}$$

$$k_{10} = \frac{V_2}{K_q\mathbf{E_t}}$$

Effect of isomerizations:

(**EABC–EPQ**): Distributions are valid; calculations of k_5, k_6, k_7, and k_8 are invalid.

EA: **E/E$_t$**, **EAB/E$_t$**, **EQ/E$_t$**, k_4, k_9, and k_{10} can still be calculated.

EAB: **E/E$_t$**, **EA/E$_t$**, **EQ/E$_t$**, k_1, k_2, k_9, and k_{10} can still be calculated.

EQ: **E/E$_t$**, **EA/E$_t$**, **EAB/E$_t$**, k_1, k_2, k_3, and k_4 can still be calculated.

Bi Uni Uni Uni Ping Pong

$$\mathbf{E} \underset{k_2}{\overset{k_1\mathbf{A}}{\rightleftharpoons}} \mathbf{EA} \underset{k_4}{\overset{k_3\mathbf{B}}{\rightleftharpoons}} (\mathbf{EAB}\text{–}\mathbf{FP}) \underset{k_6\mathbf{P}}{\overset{k_5\mathbf{C}}{\rightleftharpoons}} \mathbf{F} \underset{k_8}{\overset{k_7\mathbf{C}}{\rightleftharpoons}} (\mathbf{FC}\text{–}\mathbf{EQ}) \underset{k_{10}\mathbf{Q}}{\overset{k_9}{\rightleftharpoons}} \mathbf{E}$$

$$v = \frac{(k_1k_3k_5k_7k_9\mathbf{ABC} - k_2k_4k_6k_8k_{10}\mathbf{PQ})\mathbf{E_t}}{\begin{aligned}&k_2(k_4+k_5)k_7k_9\mathbf{C} + k_1k_3k_5(k_8+k_9)\mathbf{AB} + k_1(k_4+k_5)k_7k_9\mathbf{AC} + k_3k_5k_7k_9\mathbf{BC}\\ &+ k_1k_3k_7(k_5+k_9)\mathbf{ABC} + k_2(k_4+k_5)k_8k_{10}\mathbf{Q} + k_2k_4k_6(k_8+k_9)\mathbf{P}\\ &+ (k_2k_4+k_4k_8+k_2k_8)k_6k_{10}\mathbf{PQ} + k_1k_4k_6(k_8+k_9)\mathbf{AP} + k_2(k_4+k_5)k_7k_{10}\mathbf{CQ}\\ &+ k_3k_5k_8k_{10}\mathbf{BQ} + k_1k_3k_6(k_8+k_9)\mathbf{ABP} + k_3k_5k_7k_{10}\mathbf{BCQ} + k_3k_6k_8k_{10}\mathbf{BPQ}\end{aligned}}$$

Definitions:

$$V_1 = \frac{\text{num}_1}{\text{coef ABC}} \quad V_2 = \frac{\text{num}_2}{\text{coef PQ}} \quad K_\text{a} = \frac{\text{coef BC}}{\text{coef ABC}} \quad K_\text{b} = \frac{\text{coef AC}}{\text{coef ABC}}$$

$$K_\text{c} = \frac{\text{coef AB}}{\text{coef ABC}} \quad K_\text{p} = \frac{\text{coef Q}}{\text{coef PQ}} \quad K_\text{q} = \frac{\text{coef P}}{\text{coef PQ}} \quad K_\text{ia} = \frac{\text{coef P}}{\text{coef AP}} = \frac{\text{coef C}}{\text{coef AC}} = \frac{k_2}{k_1}$$

$$K_\text{ib} = \frac{\text{coef AP}}{\text{coef ABP}} = \frac{k_4}{k_3} \quad K_\text{ic} = \frac{\text{coef Q}}{\text{coef CQ}} = \frac{\text{coef BQ}}{\text{coef BCQ}} = \frac{k_8}{k_7}$$

$$K_\text{ip} = \frac{\text{coef BQ}}{\text{coef BPQ}} = \frac{\text{coef AB}}{\text{coef ABP}} = \frac{k_5}{k_6} \quad K_\text{iq} = \frac{\text{coef C}}{\text{coef CQ}} = \frac{\text{coef BC}}{\text{coef BCQ}} = \frac{k_9}{k_{10}}$$

$$v = \frac{V_1V_2\left(\mathbf{ABC} - \dfrac{\mathbf{PQ}}{K_\text{eq}}\right)}{\begin{aligned}&K_\text{ia}K_\text{b}V_2\mathbf{C} + K_\text{c}V_2\mathbf{AB} + K_\text{b}V_2\mathbf{AC} + K_\text{a}V_2\mathbf{BC} + V_2\mathbf{ABC} + \frac{K_\text{q}V_1\mathbf{P}}{K_\text{eq}} + \frac{K_\text{p}V_1\mathbf{Q}}{K_\text{eq}}\\ &+ \frac{V_1\mathbf{PQ}}{K_\text{eq}} + \frac{K_\text{q}V_1\mathbf{AP}}{K_\text{ia}K_\text{eq}} + \frac{K_\text{a}K_\text{ic}V_2\mathbf{BQ}}{K_\text{iq}} + \frac{K_\text{ia}K_\text{b}V_2\mathbf{CQ}}{K_\text{iq}} + \frac{K_\text{q}V_1\mathbf{ABP}}{K_\text{ia}K_\text{ib}K_\text{eq}}\\ &+ \frac{K_\text{a}V_2\mathbf{BCQ}}{K_\text{iq}} + \frac{K_\text{a}K_\text{ic}V_2\mathbf{BPQ}}{K_\text{ip}K_\text{iq}}\end{aligned}}$$

Haldanes:

$$K_\text{eq} = \frac{K_\text{ip}K_\text{iq}}{K_\text{ia}K_\text{ib}K_\text{ic}} = \frac{V_1K_\text{p}K_\text{iq}}{V_2K_\text{ia}K_\text{b}K_\text{ic}} = \frac{V_1K_\text{ip}K_\text{q}}{V_2K_\text{ia}K_\text{ib}K_\text{c}} = \left(\frac{V_1}{V_2}\right)^2 \frac{K_\text{p}K_\text{q}}{K_\text{ia}K_\text{b}K_\text{c}}$$

Distribution equations:

$$\frac{\mathbf{E}}{\mathbf{E_t}} = \frac{\dfrac{K_\text{q}V_1\mathbf{P}}{K_\text{eq}} + K_\text{ia}K_\text{b}V_2\mathbf{C} + K_\text{a}V_2\mathbf{BC}}{\text{denominator of rate equation}}$$

$$\frac{\mathbf{EA}}{\mathbf{E_t}} = \frac{K_\text{b}V_2\mathbf{AC} + \dfrac{K_\text{q}V_1\mathbf{AP}}{K_\text{ia}K_\text{eq}} + \dfrac{K_\text{a}V_2\mathbf{PQ}}{K_\text{ia}K_\text{eq}}}{\text{denominator of rate equation}}$$

$$\frac{\mathbf{F}}{\mathbf{E_t}} = \frac{K_c V_2 \mathbf{AB} + \dfrac{K_p V_1 \mathbf{Q}}{K_{eq}} + \dfrac{K_a K_{ic} V_2 \mathbf{BQ}}{K_{iq}}}{\text{denominator of rate equation}}$$

The distribution between the central complexes cannot be determined.

Rate constants:

$$k_1 = \frac{V_1}{K_a \mathbf{E_t}} \qquad k_2 = \frac{V_1 K_{ia}}{K_a \mathbf{E_t}}$$

The other rate constants cannot be determined.

Effect of isomerizations:

(**EAB–FP**) and/or (**FC–EQ**): No change.

EA: Only $\mathbf{E/E_t}$ and $\mathbf{F/E_t}$ may be calculated.

Ordered Ter Ter

$$\mathbf{E} \underset{k_2}{\overset{k_1\mathbf{A}}{\rightleftharpoons}} \mathbf{EA} \underset{k_4}{\overset{k_3\mathbf{B}}{\rightleftharpoons}} \mathbf{EAB} \underset{k_6}{\overset{k_5\mathbf{C}}{\rightleftharpoons}} (\mathbf{EABC\text{–}EPQR}) \underset{k_8\mathbf{P}}{\overset{k_7}{\rightleftharpoons}} \mathbf{EQR} \underset{k_{10}\mathbf{Q}}{\overset{k_9}{\rightleftharpoons}} \mathbf{ER} \underset{k_{12}\mathbf{R}}{\overset{k_{11}}{\rightleftharpoons}} \mathbf{E}$$

$$v = \frac{(k_1k_3k_5k_7k_9k_{11}\mathbf{ABC} - k_2k_4k_6k_8k_{10}k_{12}\mathbf{PQR})\mathbf{E_t}}{\begin{aligned} & k_2k_4(k_6+k_7)k_9k_{11} + k_1k_4(k_6+k_7)k_9k_{11}\mathbf{A} + k_2k_5k_7k_9k_{11}\mathbf{C} \\ & + k_1k_3(k_6+k_7)k_9k_{11}\mathbf{AB} + k_1k_5k_7k_9k_{11}\mathbf{AC} + k_3k_5k_7k_9k_{11}\mathbf{BC} \\ & + k_1k_3k_5(k_7k_9 + k_7k_{11} + k_9k_{11})\mathbf{ABC} + k_2k_4(k_6+k_7)k_9k_{12}\mathbf{R} + k_2k_4k_6k_8k_{11}\mathbf{P} \\ & + k_2k_4k_6k_8k_{10}\mathbf{PQ} + k_2k_4k_6k_8k_{12}\mathbf{PR} + k_2k_4(k_6+k_7)k_{10}k_{12}\mathbf{QR} \\ & + (k_2k_4 + k_4k_6 + k_2k_6)k_8k_{10}k_{12}\mathbf{PQR} + k_1k_4k_6k_8k_{11}\mathbf{AP} + k_2k_5k_7k_9k_{12}\mathbf{CR} \\ & + k_1k_3k_6k_8k_{11}\mathbf{ABP} + k_1k_4k_6k_8k_{10}\mathbf{APQ} + k_3k_5k_7k_9k_{12}\mathbf{BCR} + k_2k_5k_7k_{10}k_{12}\mathbf{CQR} \\ & + k_1k_3k_5k_8k_{11}\mathbf{ABCP} + k_1k_3k_5k_7k_{10}\mathbf{ABCQ} + k_1k_3k_6k_8k_{10}\mathbf{ABPQ} \\ & + k_3k_5k_7k_{10}k_{12}\mathbf{BCQR} + k_3k_6k_8k_{10}k_{12}\mathbf{BPQR} + k_2k_5k_8k_{10}k_{12}\mathbf{CPQR} \\ & + k_1k_3k_5k_8k_{10}\mathbf{ABCPQ} + k_3k_5k_8k_{10}k_{12}\mathbf{BCPQR} \end{aligned}}$$

Definitions:

$$V_1 = \frac{\text{num}_1}{\text{coef ABC}} \qquad V_2 = \frac{\text{num}_2}{\text{coef PQR}} \qquad K_a = \frac{\text{coef BC}}{\text{coef ABC}} \qquad K_b = \frac{\text{coef AC}}{\text{coef ABC}}$$

$$K_c = \frac{\text{coef AB}}{\text{coef ABC}} \qquad K_p = \frac{\text{coef QR}}{\text{coef PQR}} \qquad K_q = \frac{\text{coef PR}}{\text{coef PQR}} \qquad K_r = \frac{\text{coef PQ}}{\text{coef PQR}}$$

$$K_{ia} = \frac{\text{const}}{\text{coef A}} = \frac{\text{coef P}}{\text{coef AP}} = \frac{\text{coef C}}{\text{coef AC}} = \frac{\text{coef PQ}}{\text{coef APQ}} = \frac{k_2}{k_1}$$

$$K_{ib} = \frac{\text{coef A}}{\text{coef AB}} = \frac{\text{coef AP}}{\text{coef ABP}} = \frac{\text{coef APQ}}{\text{coef ABPQ}} = \frac{k_4}{k_3}$$

$$K_{ic} = \frac{\text{coef ABP}}{\text{coef ABCP}} = \frac{\text{coef ABPQ}}{\text{coef ABCPQ}} = \frac{\text{coef BPQR}}{\text{coef BCPQR}} = \frac{k_6}{k_5}$$

$$K_{ip} = \frac{\text{coef CQR}}{\text{coef CPQR}} = \frac{\text{coef ABCQ}}{\text{coef ABCPQ}} = \frac{\text{coef BCQR}}{\text{coef BCPQR}} = \frac{k_7}{k_8}$$

$$K_{iq} = \frac{\text{coef R}}{\text{coef QR}} = \frac{\text{coef CR}}{\text{coef CQR}} = \frac{\text{coef BCR}}{\text{coef BCQR}} = \frac{k_9}{k_{10}}$$

$$K_{ir} = \frac{\text{const}}{\text{coef R}} = \frac{\text{coef P}}{\text{coef PR}} = \frac{\text{coef C}}{\text{coef CR}} = \frac{\text{coef BC}}{\text{coef BCR}} = \frac{k_{11}}{k_{12}}$$

$$v = \frac{V_1V_2\left(\mathbf{ABC} - \frac{\mathbf{PQR}}{K_{eq}}\right)}{\begin{aligned}&K_{ia}K_{ib}K_cV_2 + K_{ib}K_cV_2\mathbf{A} + K_{ia}K_bV_2\mathbf{C} + K_cV_2\mathbf{AB} + K_bV_2\mathbf{AC} + K_aV_2\mathbf{BC}\\ &+ V_2\mathbf{ABC} + \frac{K_{ir}K_qV_1\mathbf{P}}{K_{eq}} + \frac{K_{iq}K_pV_1\mathbf{R}}{K_{eq}} + \frac{K_rV_1\mathbf{PQ}}{K_{eq}} + \frac{K_qV_1\mathbf{PR}}{K_{eq}} + \frac{K_pV_1\mathbf{QR}}{K_{eq}}\\ &+ \frac{V_1\mathbf{PQR}}{K_{eq}} + \frac{K_qK_{ir}V_1\mathbf{AP}}{K_{ia}K_{eq}} + \frac{K_{ia}K_bV_2\mathbf{CR}}{K_{ir}} + \frac{K_qK_{ir}V_1\mathbf{ABP}}{K_{ia}K_{ib}K_{eq}} + \frac{K_rV_1\mathbf{APQ}}{K_{ia}K_{eq}}\\ &+ \frac{K_aV_2\mathbf{BCR}}{K_{ir}} + \frac{K_{ia}K_bV_2\mathbf{CQR}}{K_{iq}K_{ir}} + \frac{K_qK_{ir}V_1\mathbf{ABCP}}{K_{ia}K_{ib}K_{ic}K_{eq}} + \frac{K_aV_2\mathbf{BCQR}}{K_{ir}K_{iq}}\\ &+ \frac{K_{ip}K_rV_1\mathbf{ABCQ}}{K_{ia}K_{ib}K_{ic}K_{eq}} + \frac{K_rV_1\mathbf{ABPQ}}{K_{ia}K_{ib}K_{eq}} + \frac{K_aK_{ic}V_2\mathbf{BPQR}}{K_{ip}K_{iq}K_{ir}} + \frac{K_{ia}K_bV_2\mathbf{CPQR}}{K_{ip}K_{iq}K_{ir}}\\ &+ \frac{K_rV_1\mathbf{ABCPQ}}{K_{ia}K_{ib}K_{ic}K_{eq}} + \frac{K_aV_2\mathbf{BCPQR}}{K_{ip}K_{iq}K_{ir}}\end{aligned}}$$

Haldanes:

$$K_{eq} = \frac{K_{ip}K_{iq}K_{ir}}{K_{ia}K_{ib}K_{ic}} = \frac{V_1K_pK_{iq}K_{ir}}{V_2K_{ia}K_{ib}K_c}$$

Distribution equations:

$$\frac{\mathbf{E}}{\mathbf{E_t}} = \frac{K_{ia}K_{ib}K_cV_2 + \frac{K_{ir}K_qV_1\mathbf{P}}{K_{eq}} + K_{ia}K_bV_2\mathbf{C} + K_aV_2\mathbf{BC} + \frac{K_rV_1\mathbf{PQ}}{K_{eq}}}{\text{denominator of rate equation}}$$

$$\frac{\mathbf{EA}}{\mathbf{E_t}} = \frac{K_{ib}K_cV_2\mathbf{A} + K_bV_2\mathbf{AC} + \frac{K_qK_{ir}V_1\mathbf{AP}}{K_{ia}K_{eq}} + \frac{K_rV_1\mathbf{APQ}}{K_{ia}K_{eq}} + \frac{K_aV_2\mathbf{PQR}}{K_{ia}K_{eq}}}{\text{denominator of rate equation}}$$

$$\frac{\mathbf{EAB}}{\mathbf{E_t}} = \frac{K_c V_2 \mathbf{AB} + \frac{K_b V_2 \mathbf{PQR}}{K_{ib} K_{eq}} + \frac{K_q K_{ir} V_1 \mathbf{ABP}}{K_{ia} K_{ib} K_{eq}} + \frac{K_r V_1 \mathbf{ABPQ}}{K_{ia} K_{ib} K_{eq}} + \frac{K_a K_{ic} V_2 \mathbf{BPQR}}{K_{ip} K_{iq} K_{ir}}}{\text{denominator of rate equation}}$$

$$\frac{(\mathbf{EABC}-\mathbf{EPQR})}{\mathbf{E_t}} = \frac{\begin{aligned}&\left(V_2 - V_1\left[\frac{K_q}{K_{iq}} + \frac{K_r}{K_{ir}}\right]\right)\mathbf{ABC} + \frac{K_q K_{ir} V_1 \mathbf{ABCP}}{K_{ia} K_{ib} K_{ic} K_{eq}}\\ &+\left[\frac{V_1}{K_{eq}} - \left(\frac{V_2}{K_{eq}}\right)\left(\frac{K_a}{K_{ia}} + \frac{K_b}{K_{ib}}\right)\right]\mathbf{PQR} + \frac{K_{ia} K_b V_2 \mathbf{CPQR}}{K_{ip} K_{iq} K_{ir}}\\ &+\frac{K_r V_1 \mathbf{ABCPQ}}{K_{ia} K_{ib} K_{ic} K_{eq}} + \frac{K_a V_2 \mathbf{BCPQR}}{K_{ip} K_{iq} K_{ir}}\end{aligned}}{\text{denominator of rate equation}}$$

$$\frac{\mathbf{EQR}}{\mathbf{E_t}} = \frac{\frac{K_p V_1 \mathbf{QR}}{K_{eq}} + \frac{K_q V_1 \mathbf{ABC}}{K_{iq}} + \frac{K_{ia} K_b V_2 \mathbf{CQR}}{K_{iq} K_{ir}} + \frac{K_a V_2 \mathbf{BCQR}}{K_{iq} K_{ir}} + \frac{K_{ip} K_r V_1 \mathbf{ABCQ}}{K_{ia} K_{ib} K_{ic} K_{eq}}}{\text{denominator of rate equation}}$$

$$\frac{\mathbf{ER}}{\mathbf{E_t}} = \frac{\frac{K_p K_{iq} V_1 \mathbf{R}}{K_{eq}} + \frac{K_r V_1 \mathbf{ABC}}{K_{ir}} + \frac{K_q V_1 \mathbf{PR}}{K_{eq}} + \frac{K_a V_2 \mathbf{BCR}}{K_{ir}} + \frac{K_{ia} K_b V_2 \mathbf{CR}}{K_{ir}}}{\text{denominator of rate equation}}$$

Rate constants:

$$k_1 = \frac{V_1}{K_a \mathbf{E_t}} \quad k_2 = \frac{V_1 K_{ia}}{K_a \mathbf{E_t}} \quad k_3 = \frac{V_1}{K_b \mathbf{E_t}} \quad k_4 = \frac{K_{ib} V_1}{K_b \mathbf{E_t}} \quad k_5 = \left(\frac{V_1}{K_c \mathbf{E_t}}\right)\left(1 + \frac{k_6}{k_7}\right)$$

$$\frac{1}{k_6} = \frac{\mathbf{E_t}}{V_2} - \frac{1}{k_2} - \frac{1}{k_4} \quad \frac{1}{k_7} = \frac{\mathbf{E_t}}{V_1} - \frac{1}{k_9} - \frac{1}{k_{11}} \quad k_8 = \left(\frac{V_2}{K_p \mathbf{E_t}}\right)\left(1 + \frac{k_7}{k_6}\right) \quad k_9 = \frac{K_{iq} V_2}{K_q \mathbf{E_t}}$$

$$k_{10} = \frac{V_2}{K_q \mathbf{E_t}} \quad k_{11} = \frac{V_2 K_{ir}}{K_r \mathbf{E_t}} \quad k_{12} = \frac{V_2}{K_r \mathbf{E_t}}$$

Effect of isomerizations:

(**EABC–EPQR**): Distributions are valid; calculations of k_5, k_6, k_7, and k_8 are invalid.

EA: Calculations of **EA/E**$_\mathbf{t}$, **EAB/E**$_\mathbf{t}$, and (**EABC–EPQR**)/**E**$_\mathbf{t}$ are invalid. k_{11}–k_{14} can still be calculated, by use of equations for k_9–k_{12}.

EAB: Calculations of **EA/E**$_\mathbf{t}$, **EAB/E**$_\mathbf{t}$, and (**EABC–EPQR**)/**E**$_\mathbf{t}$ are invalid. k_1, k_2, and k_{11}–k_{14} can still be calculated, by use of equations for k_9–k_{12} to obtain k_{11}–k_{14}.

EQR: $\mathbf{E/E_t}$, $\mathbf{EA/E_t}$, $\mathbf{EAB/E_t}$, k_1, k_2, k_3, k_4, k_{13}, and k_{14} can still be calculated, by use of equations for k_{11} and k_{12} for the latter two.

ER: $\mathbf{E/E_t}$, $\mathbf{EA/E_t}$, $\mathbf{EAB/E_t}$, k_1, k_2, k_3, and k_4 can still be calculated.

Bi Uni Uni Bi Ping Pong

$$\mathbf{E} \underset{k_2}{\overset{k_1\mathbf{A}}{\rightleftharpoons}} \mathbf{EA} \underset{k_4}{\overset{k_3\mathbf{B}}{\rightleftharpoons}} (\mathbf{EAB{-}FP}) \underset{k_6\mathbf{P}}{\overset{k_5}{\rightleftharpoons}} \mathbf{F} \underset{k_8}{\overset{k_7\mathbf{C}}{\rightleftharpoons}} (\mathbf{FC{-}EQR}) \underset{k_{10}\mathbf{Q}}{\overset{k_9}{\rightleftharpoons}} \mathbf{ER} \underset{k_{12}\mathbf{R}}{\overset{k_{11}}{\rightleftharpoons}} \mathbf{E}$$

$$v = \frac{(k_1k_3k_5k_7k_9k_{11}\mathbf{ABC} - k_2k_4k_6k_8k_{10}k_{12}\mathbf{PQR})\mathbf{E_t}}{\begin{aligned} & k_2(k_4+k_5)k_7k_9k_{11}\mathbf{C} + k_1k_3k_5(k_8+k_9)k_{11}\mathbf{AB} + k_1(k_4+k_5)k_7k_9k_{11}\mathbf{AC} \\ & + k_3k_5k_7k_9k_{11}\mathbf{BC} + k_1k_3k_7(k_5k_9+k_5k_{11}+k_9k_{11})\mathbf{ABC} + k_2k_4k_6(k_8+k_9)k_{11}\mathbf{P} \\ & + k_2k_4k_6k_8k_{10}\mathbf{PQ} + k_2k_4k_6(k_8+k_9)k_{12}\mathbf{PR} + k_2(k_4+k_5)k_8k_{10}k_{12}\mathbf{QR} \\ & + (k_2k_4+k_4k_8+k_2k_8)k_6k_{10}k_{12}\mathbf{PQR} + k_1k_4k_6(k_8+k_9)k_{11}\mathbf{AP} \\ & + k_2(k_4+k_5)k_7k_9k_{12}\mathbf{CR} + k_1k_3k_6(k_8+k_9)k_{11}\mathbf{ABP} + k_1k_4k_6k_8k_{10}\mathbf{APQ} \\ & + k_3k_5k_7k_9k_{12}\mathbf{BCR} + k_3k_5k_8k_{10}k_{12}\mathbf{BQR} + k_2(k_4+k_5)k_7k_{10}k_{12}\mathbf{CQR} \\ & + k_1k_3k_5k_8k_{10}\mathbf{ABQ} + k_1k_3k_5k_7k_{10}\mathbf{ABCQ} + k_1k_3k_6k_8k_{10}\mathbf{ABPQ} \\ & + k_3k_5k_7k_{10}k_{12}\mathbf{BCQR} + k_3k_6k_8k_{10}k_{12}\mathbf{BPQR} \end{aligned}}$$

Definitions:

$$V_1 = \frac{\text{num}_1}{\text{coef ABC}} \quad V_2 = \frac{\text{num}_2}{\text{coef PQR}} \quad K_a = \frac{\text{coef BC}}{\text{coef ABC}} \quad K_b = \frac{\text{coef AC}}{\text{coef ABC}}$$

$$K_c = \frac{\text{coef AB}}{\text{coef ABC}} \quad K_p = \frac{\text{coef QR}}{\text{coef PQR}} \quad K_q = \frac{\text{coef PR}}{\text{coef PQR}} \quad K_r = \frac{\text{coef PQ}}{\text{coef PQR}}$$

$$K_{ib} = \frac{\text{coef AP}}{\text{coef ABP}} = \frac{\text{coef APQ}}{\text{coef ABPQ}} = \frac{k_4}{k_3} \quad K_{ia} = \frac{\text{coef P}}{\text{coef AP}} = \frac{\text{coef C}}{\text{coef AC}} = \frac{\text{coef PQ}}{\text{coef APQ}} = \frac{k_2}{k_1}$$

$$K_{ip} = \frac{\text{coef AB}}{\text{coef ABP}} = \frac{\text{coef ABQ}}{\text{coef ABPQ}} = \frac{k_5}{k_6}$$

$$K_{ic} = \frac{\text{coef QR}}{\text{coef CQR}} = \frac{\text{coef ABQ}}{\text{coef ABCQ}} = \frac{\text{coef BQR}}{\text{coef BCQR}} = \frac{k_8}{k_7}$$

$$K_{iq} = \frac{\text{coef CR}}{\text{coef CQR}} = \frac{\text{coef BCR}}{\text{coef BCQR}} = \frac{k_9}{k_{10}}$$

$$K_{ir} = \frac{\text{coef P}}{\text{coef PR}} = \frac{\text{coef C}}{\text{coef CR}} = \frac{\text{coef BC}}{\text{coef BCR}} = \frac{k_{11}}{k_{12}}$$

$$v = \frac{V_1 V_2\left(\mathbf{ABC} - \frac{\mathbf{PQR}}{K_{eq}}\right)}{K_{ia}K_b V_2\mathbf{C} + K_c V_2\mathbf{AB} + K_b V_2\mathbf{AC} + K_a V_2\mathbf{BC} + V_2\mathbf{ABC} + \frac{K_{ir}K_q V_1\mathbf{P}}{K_{eq}} + \frac{K_r V_1\mathbf{PQ}}{K_{eq}} + \frac{K_q V_1\mathbf{PR}}{K_{eq}} + \frac{K_p V_1\mathbf{QR}}{K_{eq}} + \frac{V_1\mathbf{PQR}}{K_{eq}} + \frac{K_q K_{ir} V_1\mathbf{AP}}{K_{ia}K_{eq}} + \frac{K_{ia}K_b V_2\mathbf{CR}}{K_{ir}} + \frac{K_q K_{ir} V_1\mathbf{ABP}}{K_{ia}K_{ib}K_{eq}} + \frac{K_{ip}K_r V_1\mathbf{ABQ}}{K_{ia}K_{ib}K_{eq}} + \frac{K_a K_{ic} V_2\mathbf{BQR}}{K_{iq}K_{ir}} + \frac{K_r V_1\mathbf{APQ}}{K_{ia}K_{eq}} + \frac{K_a V_2\mathbf{BCR}}{K_{ir}} + \frac{K_{ia}K_b V_2\mathbf{CQR}}{K_{iq}K_{ir}} + \frac{K_a V_2\mathbf{BCQR}}{K_{ir}K_{iq}} + \frac{K_{ip}K_r V_1\mathbf{ABCQ}}{K_{ia}K_{ib}K_{ic}K_{eq}} + \frac{K_r V_1\mathbf{ABPQ}}{K_{ia}K_{ib}K_{eq}} + \frac{K_a K_{ic} V_2\mathbf{BPQR}}{K_{ip}K_{iq}K_{ir}}}$$

Haldanes:

$$K_{eq} = \frac{K_{ip}K_{iq}K_{ir}}{K_{ia}K_{ib}K_{ic}} = \frac{V_1 K_{ip}K_q K_{ir}}{V_2 K_{ia}K_{ib}K_c} = \frac{V_1 K_p K_{iq}K_{ir}}{V_2 K_{ia}K_b K_{ic}} = \left(\frac{V_1}{V_2}\right)^2 \frac{K_p K_q K_{ir}}{K_{ia}K_b K_c}$$

Distribution equations:

$$\frac{\mathbf{E}}{\mathbf{E_t}} = \frac{\frac{K_{ir}K_q V_1\mathbf{P}}{K_{eq}} + K_{ia}K_b V_2\mathbf{C} + K_a V_2\mathbf{BC} + \frac{K_r V_1\mathbf{PQ}}{K_{eq}}}{\text{denominator of rate equation}}$$

$$\frac{\mathbf{EA}}{\mathbf{E_t}} = \frac{K_b V_2\mathbf{AC} + \frac{K_{ir}K_q V_1\mathbf{AP}}{K_{ia}K_{eq}} + \frac{K_r V_1\mathbf{APQ}}{K_{ia}K_{eq}} + \frac{K_a V_2\mathbf{PQR}}{K_{ia}K_{eq}}}{\text{denominator of rate equation}}$$

$$\frac{\mathbf{F}}{\mathbf{E_t}} = \frac{K_c V_2\mathbf{AB} + \frac{K_p V_1\mathbf{QR}}{K_{eq}} + \frac{K_{ip}K_r V_1\mathbf{ABQ}}{K_{ia}K_{ib}K_{eq}} + \frac{K_a K_{ic} V_2\mathbf{BQR}}{K_{iq}K_{ir}}}{\text{denominator of rate equation}}$$

$$\frac{\mathbf{ER}}{\mathbf{E_t}} = \frac{\frac{K_r V_1\mathbf{ABC}}{K_{ir}} + \frac{K_q V_1\mathbf{PR}}{K_{eq}} + \frac{K_a V_2\mathbf{BCR}}{K_{ir}} + \frac{K_{ia}K_b V_2\mathbf{CR}}{K_{ir}}}{\text{denominator of rate equation}}$$

The distribution between the central complexes cannot be determined.

Rate constants:

$$k_1 = \frac{V_1}{K_a\mathbf{E_t}} \qquad k_2 = \frac{V_1 K_{ia}}{K_a\mathbf{E_t}} \qquad k_{11} = \frac{V_2 K_{ir}}{K_r\mathbf{E_t}} \qquad k_{12} = \frac{V_2}{K_r\mathbf{E_t}}$$

Other rate constants cannot be determined.

Effect of isomerizations:

(**EAB** + **FP**) and/or (**FC** + **EQR**): No change.

EA: Calculations of k_1, k_2, and $\mathbf{EA/E_t}$ are invalid.

ER: Calculations of k_{11}, k_{12}, and $\mathbf{ER/E_t}$ are invalid.

Bi Bi Uni Uni Ping Pong

$$\mathbf{E} \underset{k_2}{\overset{k_1\mathbf{A}}{\rightleftharpoons}} \mathbf{EA} \underset{k_4}{\overset{k_3\mathbf{B}}{\rightleftharpoons}} (\mathbf{EAB{-}FPQ}) \underset{k_6\mathbf{P}}{\overset{k_5}{\rightleftharpoons}} \mathbf{FQ} \underset{k_8\mathbf{Q}}{\overset{k_7}{\rightleftharpoons}} \mathbf{F} \underset{k_{10}}{\overset{k_9\mathbf{C}}{\rightleftharpoons}} (\mathbf{FC{-}ER}) \underset{k_{12}\mathbf{R}}{\overset{k_{11}}{\rightleftharpoons}} \mathbf{E}$$

$$v = \frac{(k_1k_3k_5k_7k_9k_{11}\mathbf{ABC} - k_2k_4k_6k_8k_{10}k_{12}\mathbf{PQR})\mathbf{E_t}}{\begin{aligned} &k_2(k_4+k_5)k_7k_9k_{11}\mathbf{C} + k_1k_3k_5k_7(k_{10}+k_{11})\mathbf{AB} + k_1(k_4+k_5)k_7k_9k_{11}\mathbf{AC} \\ &+ k_3k_5k_7k_9k_{11}\mathbf{BC} + k_1k_3k_9(k_5k_7+k_5k_{11}+k_7k_{11})\mathbf{ABC} \\ &+ k_2(k_4+k_5)k_7k_{10}k_{12}\mathbf{R} + k_2k_4k_6k_8(k_{10}+k_{11})\mathbf{PQ} + k_2k_4k_6k_{10}k_{12}\mathbf{PR} \\ &+ k_2(k_4+k_5)k_8k_{10}k_{12}\mathbf{QR} + (k_2k_4+k_4k_{10}+k_2k_{10})k_6k_8k_{12}\mathbf{PQR} \\ &+ k_2k_4k_6k_9k_{11}\mathbf{CP} + k_2(k_4+k_5)k_7k_9k_{12}\mathbf{CR} + k_3k_5k_7k_{10}k_{12}\mathbf{BR} \\ &+ k_1k_4k_6k_9k_{11}\mathbf{ACP} + k_1k_4k_6k_8(k_{10}+k_{11})\mathbf{APQ} + k_3k_5k_7k_9k_{12}\mathbf{BCR} \\ &+ k_3k_5k_8k_{10}k_{12}\mathbf{BQR} + k_2k_4k_6k_9k_{12}\mathbf{CPR} + k_1k_3k_5k_8(k_{10}+k_{11})\mathbf{ABQ} \\ &+ k_1k_3k_6k_9k_{11}\mathbf{ABCP} + k_1k_3k_6k_8(k_{10}+k_{11})\mathbf{ABPQ} + k_3k_6k_8k_{10}k_{12}\mathbf{BPQR} \end{aligned}}$$

Definitions:

$$V_1 = \frac{\text{num}_1}{\text{coef ABC}} \quad V_2 = \frac{\text{num}_2}{\text{coef PQR}} \quad K_\text{a} = \frac{\text{coef BC}}{\text{coef ABC}} \quad K_\text{b} = \frac{\text{coef AC}}{\text{coef ABC}}$$

$$K_\text{c} = \frac{\text{coef AB}}{\text{coef ABC}} \quad K_\text{p} = \frac{\text{coef QR}}{\text{coef PQR}} \quad K_\text{q} = \frac{\text{coef PR}}{\text{coef PQR}} \quad K_\text{r} = \frac{\text{coef PQ}}{\text{coef PQR}}$$

$$K_\text{ib} = \frac{\text{coef ACP}}{\text{coef ABCP}} = \frac{\text{coef APQ}}{\text{coef ABPQ}} = \frac{k_4}{k_3}$$

$$K_\text{ia} = \frac{\text{coef CP}}{\text{coef ACP}} = \frac{\text{coef C}}{\text{coef AC}} = \frac{\text{coef PQ}}{\text{coef APQ}} = \frac{k_2}{k_1}$$

$$K_\text{ip} = \frac{\text{coef BQR}}{\text{coef BPQR}} = \frac{\text{coef ABQ}}{\text{coef ABPQ}} = \frac{k_5}{k_6}$$

$$K_\text{ic} = \frac{\text{coef BR}}{\text{coef BCR}} = \frac{\text{coef PR}}{\text{coef CPR}} = \frac{\text{coef R}}{\text{coef CR}} = \frac{k_{10}}{k_9}$$

$$K_{iq} = \frac{\text{coef AB}}{\text{coef ABQ}} = \frac{\text{coef BR}}{\text{coef BQR}} = \frac{k_7}{k_8}$$

$$K_{ir} = \frac{\text{coef CP}}{\text{coef CPR}} = \frac{\text{coef C}}{\text{coef CR}} = \frac{\text{coef BC}}{\text{coef BCR}} = \frac{k_{11}}{k_{12}}$$

$$v = \frac{V_1 V_2\left(\mathbf{ABC} - \frac{\mathbf{PQR}}{K_{eq}}\right)}{\begin{array}{l} K_{ia}K_bV_2\mathbf{C} + K_cV_2\mathbf{AB} + K_bV_2\mathbf{AC} + K_aV_2\mathbf{BC} + V_2\mathbf{ABC} + \frac{K_{iq}K_pV_1\mathbf{R}}{K_{eq}} \\ + \frac{K_rV_1\mathbf{PQ}}{K_{eq}} + \frac{K_qV_1\mathbf{PR}}{K_{eq}} + \frac{K_pV_1\mathbf{QR}}{K_{eq}} + \frac{V_1\mathbf{PQR}}{K_{eq}} + \frac{K_aK_{ic}V_2\mathbf{BR}}{K_{ir}} + \frac{K_qK_{ir}V_1\mathbf{CP}}{K_{ic}K_{eq}} \\ + \frac{K_{ia}K_bV_2\mathbf{CR}}{K_{ir}} + \frac{K_qK_{ir}V_1\mathbf{ACP}}{K_{ia}K_{ic}K_{eq}} + \frac{K_{ip}K_rV_1\mathbf{ABQ}}{K_{ia}K_{ib}K_{eq}} + \frac{K_aK_{ic}V_2\mathbf{BQR}}{K_{iq}K_{ir}} + \frac{K_rV_1\mathbf{APQ}}{K_{ia}K_{eq}} \\ + \frac{K_aV_2\mathbf{BCR}}{K_{ir}} + \frac{K_qV_1\mathbf{CPR}}{K_{ic}K_{eq}} + \frac{K_{ir}K_qV_1\mathbf{ABCP}}{K_{ia}K_{ib}K_{ic}K_{eq}} + \frac{K_rV_1\mathbf{ABPQ}}{K_{ia}K_{ib}K_{eq}} + \frac{K_aK_{ic}V_2\mathbf{BPQR}}{K_{ip}K_{iq}K_{ir}} \end{array}}$$

Haldanes:

$$K_{eq} = \frac{K_{ip}K_{iq}K_{ir}}{K_{ia}K_{ib}K_{ic}} = \frac{V_1K_{ip}K_{iq}K_r}{V_2K_{ia}K_{ib}K_c} = \frac{V_1K_pK_{iq}K_{ir}}{V_2K_{ia}K_bK_{ic}} = \left(\frac{V_1}{V_2}\right)^2\frac{K_pK_{iq}K_r}{K_{ia}K_bK_c}$$

Distribution equations:

$$\frac{\mathbf{E}}{\mathbf{E_t}} = \frac{\frac{K_{ir}K_qV_1\mathbf{CP}}{K_{ic}K_{eq}} + K_{ia}K_bV_2\mathbf{C} + K_aV_2\mathbf{BC} + \frac{K_rV_1\mathbf{PQ}}{K_{eq}}}{\text{denominator of rate equation}}$$

$$\frac{\mathbf{EA}}{\mathbf{E_t}} = \frac{K_bV_2\mathbf{AC} + \frac{K_{ir}K_qV_1\mathbf{ACP}}{K_{ia}K_{ic}K_{eq}} + \frac{K_rV_1\mathbf{APQ}}{K_{ia}K_{eq}} + \frac{K_aV_2\mathbf{PQR}}{K_{ia}K_{eq}}}{\text{denominator of rate equation}}$$

$$\frac{\mathbf{FQ}}{\mathbf{E_t}} = \frac{\frac{K_qV_1\mathbf{ABC}}{K_{iq}} + \frac{K_pV_1\mathbf{QR}}{K_{eq}} + \frac{K_{ip}K_rV_1\mathbf{ABQ}}{K_{ia}K_{ib}K_{eq}} + \frac{K_aK_{ic}V_2\mathbf{BQR}}{K_{iq}K_{ir}}}{\text{denominator of rate equation}}$$

$$\frac{\mathbf{F}}{\mathbf{E_t}} = \frac{\frac{K_{iq}K_pV_1\mathbf{R}}{K_{eq}} + \frac{K_aK_{ic}V_2\mathbf{BR}}{K_{ir}} + \frac{K_qV_1\mathbf{PR}}{K_{eq}} + K_cV_2\mathbf{AB}}{\text{denominator of rate equation}}$$

The distribution between the central complexes cannot be determined.

Rate constants:

$$k_1 = \frac{V_1}{K_a\mathbf{E_t}} \qquad k_2 = \frac{V_1 K_{ia}}{K_a\mathbf{E_t}} \qquad k_7 = \frac{V_2 K_{iq}}{K_q\mathbf{E_t}} \qquad k_8 = \frac{V_2}{K_q\mathbf{E_t}}$$

Other rate constants cannot be determined.

Effect of isomerizations:

(**EAB** + **FPQ**) and/or (**FC** + **ER**): No change.

EA: Calculations of k_1, k_2, and $\mathbf{EA/E_t}$ are invalid.

FQ: Calculations of k_7, k_8, and $\mathbf{FQ/E_t}$ are invalid.

Hexa-Uni Ping Pong (Uni Uni Uni Uni Uni Uni)

$$\mathbf{E} \underset{k_2}{\overset{k_1\mathbf{A}}{\rightleftharpoons}} (\mathbf{EA{-}FP}) \underset{k_4\mathbf{P}}{\overset{k_3}{\rightleftharpoons}} \mathbf{F} \underset{k_6}{\overset{k_5\mathbf{B}}{\rightleftharpoons}} (\mathbf{FB{-}GQ}) \underset{k_8\mathbf{Q}}{\overset{k_7}{\rightleftharpoons}} \mathbf{G} \underset{k_{10}}{\overset{k_9\mathbf{C}}{\rightleftharpoons}} (\mathbf{GC{-}ER}) \underset{k_{12}\mathbf{R}}{\overset{k_{11}}{\rightleftharpoons}} \mathbf{E}$$

$$v = \frac{(k_1k_3k_5k_7k_9k_{11}\mathbf{ABC} - k_2k_4k_6k_8k_{10}k_{12}\mathbf{PQR})\mathbf{E_t}}{\begin{array}{l} k_1k_3k_5k_7(k_{10}+k_{11})\mathbf{AB} + k_1k_3(k_6+k_7)k_9k_{11}\mathbf{AC} + (k_2+k_3)k_5k_7k_9k_{11}\mathbf{BC} \\ + k_1k_5k_9(k_3k_7+k_7k_{11}+k_3k_{11})\mathbf{ABC} + k_2k_4k_6k_8(k_{10}+k_{11})\mathbf{PQ} \\ + k_2k_4(k_6+k_7)k_{10}k_{12}\mathbf{PR} + (k_2+k_3)k_6k_8k_{10}k_{12}\mathbf{QR} \\ + (k_2k_6+k_6k_{10}+k_2k_{10})k_4k_8k_{12}\mathbf{PQR} + k_1k_3k_6k_8(k_{10}+k_{11})\mathbf{AQ} \\ + (k_2+k_3)k_5k_7k_{10}k_{12}\mathbf{BR} + k_2k_4(k_6+k_7)k_9k_{11}\mathbf{CP} + k_1k_4(k_6+k_7)k_9k_{11}\mathbf{ACP} \\ + k_1k_4k_6k_8(k_{10}+k_{11})\mathbf{APQ} + k_1k_3k_5k_8(k_{10}+k_{11})\mathbf{ABQ} + (k_2+k_3)k_5k_8k_{10}k_{12}\mathbf{BQR} \\ + (k_2+k_3)k_5k_7k_9k_{12}\mathbf{BCR} + k_2k_4(k_6+k_7)k_9k_{12}\mathbf{CPR} \end{array}}$$

Definitions:

$$V_1 = \frac{\text{num}_1}{\text{coef ABC}} \qquad V_2 = \frac{\text{num}_2}{\text{coef PQR}} \qquad K_a = \frac{\text{coef BC}}{\text{coef ABC}} \qquad K_b = \frac{\text{coef AC}}{\text{coef ABC}}$$

$$K_c = \frac{\text{coef AB}}{\text{coef ABC}} \qquad K_p = \frac{\text{coef QR}}{\text{coef PQR}} \qquad K_q = \frac{\text{coef PR}}{\text{coef PQR}} \qquad K_r = \frac{\text{coef PQ}}{\text{coef PQR}}$$

$$K_{ia} = \frac{\text{coef CP}}{\text{coef ACP}} = \frac{\text{coef PQ}}{\text{coef APQ}} = \frac{k_2}{k_1} \qquad K_{ib} = \frac{\text{coef QR}}{\text{coef BQR}} = \frac{\text{coef AQ}}{\text{coef ABQ}} = \frac{k_6}{k_5}$$

$$K_{ic} = \frac{\text{coef BR}}{\text{coef BCR}} = \frac{\text{coef PR}}{\text{coef CPR}} = \frac{k_{10}}{k_9} \qquad K_{ip} = \frac{\text{coef AC}}{\text{coef ACP}} = \frac{\text{coef AQ}}{\text{coef APQ}} = \frac{k_3}{k_4}$$

$$K_{iq} = \frac{\text{coef AB}}{\text{coef ABQ}} = \frac{\text{coef BR}}{\text{coef BQR}} = \frac{k_7}{k_8} \qquad K_{ir} = \frac{\text{coef CP}}{\text{coef CPR}} = \frac{\text{coef BC}}{\text{coef BCR}} = \frac{k_{11}}{k_{12}}$$

$$v = \frac{V_1V_2\left(\mathbf{ABC} - \dfrac{\mathbf{PQR}}{K_{eq}}\right)}{\begin{array}{l} K_cV_2\mathbf{AB} + K_bV_2\mathbf{AC} + K_aV_2\mathbf{BC} + V_2\mathbf{ABC} + \dfrac{K_rV_1\mathbf{PQ}}{K_{eq}} + \dfrac{K_qV_1\mathbf{PR}}{K_{eq}} + \dfrac{K_pV_1\mathbf{QR}}{K_{eq}} \\ + \dfrac{V_1\mathbf{PQR}}{K_{eq}} + \dfrac{K_{ip}K_rV_1\mathbf{AQ}}{K_{ia}K_{eq}} + \dfrac{K_{ia}K_bV_2\mathbf{CP}}{K_{ip}} + \dfrac{K_aK_{ic}V_2\mathbf{BR}}{K_{ir}} + \dfrac{K_qK_{ir}V_1\mathbf{ACP}}{K_{ia}K_{ic}K_{eq}} \\ + \dfrac{K_rV_1\mathbf{APQ}}{K_{ia}K_{eq}} + \dfrac{K_aV_2\mathbf{BCR}}{K_{ir}} + \dfrac{K_{ia}K_bV_2\mathbf{CPR}}{K_{ip}K_{ir}} + \dfrac{K_pV_1\mathbf{BQR}}{K_{ib}K_{eq}} + \dfrac{K_cV_2\mathbf{ABQ}}{K_{iq}} \end{array}}$$

$$K_{eq} = \frac{K_{ip}K_{iq}K_{ir}}{K_{ia}K_{ib}K_{ic}} = \frac{V_1K_{ip}K_{iq}K_r}{V_2K_{ia}K_{ib}K_c} = \frac{V_1K_pK_{iq}K_{ir}}{V_2K_aK_{ib}K_{ic}} = \frac{V_1K_{ip}K_qK_{ir}}{V_2K_{ia}K_bK_{ic}}$$

Haldanes:

$$= \left(\frac{V_1}{V_2}\right)^2 \frac{K_pK_qK_{ir}}{K_aK_bK_{ic}} = \left(\frac{V_1}{V_2}\right)^2 \frac{K_pK_{iq}K_r}{K_aK_{ib}K_c} = \left(\frac{V_1}{V_2}\right)^2 \frac{K_{ip}K_qK_r}{K_{ia}K_bK_c}$$

$$= \left(\frac{V_1}{V_2}\right)^3 \frac{K_pK_qK_r}{K_aK_bK_c}$$

Distribution equations:

$$\frac{\mathbf{E}}{\mathbf{E_t}} = \frac{K_aV_2\mathbf{BC} + \dfrac{K_{ia}K_bV_2\mathbf{CP}}{K_{ip}} + \dfrac{K_rV_1\mathbf{PQ}}{K_{eq}}}{\text{denominator of rate equation}}$$

$$\frac{\mathbf{F}}{\mathbf{E_t}} = \frac{K_bV_2\mathbf{AC} + \dfrac{K_{ip}K_rV_1\mathbf{AQ}}{K_{ia}K_{eq}} + \dfrac{K_pV_1\mathbf{QR}}{K_{eq}}}{\text{denominator of rate equation}}$$

$$\frac{\mathbf{G}}{\mathbf{E_t}} = \frac{K_cV_2\mathbf{AB} + \dfrac{K_aK_{ic}V_2\mathbf{BR}}{K_{ir}} + \dfrac{K_qV_1\mathbf{PR}}{K_{eq}}}{\text{denominator of rate equation}}$$

The distribution between the central complexes cannot be determined.

No rate constants can be calculated.

Isomerizations of the central complexes have no effect.

INDEX

acetate kinase
  inhibition by acetaldehyde hydrate 305
  isotopic exchange 228
O-acetylserine sulfhydrylase 211–212
  competitive substrate inhibition 181
  mixed product and dead-end inhibition 166
activation
  substrate 77
aconitase
  slow binding inhibition 199
adenosine deaminase
  viscosity effect 251
adenylate kinase
  competitive substrate inhibition 181
alanine dehydrogenase
  pH profiles 342
alcohol dehydrogenase, equine
  alternate product inhibition 152
  alternate substrate inhibition 191
  double inhibition 195
  isotope effects 279, 285, 295
  isotopic exchange 233
  mixed product and dead-end inhibition 164
  pH profiles 365–366
  Theorell-Chance mechanism 74
  transition state 308–309
alcohol dehydrogenase, yeast
  isotope effects 295
  partial substrate inhibition 183
alternative substrate
  anhydro sugars 101
  ATPβS 103
  exchange inert metal nucleotides 101
  kinetic mechanism by (Northrop) 104
allosterism 112
  transmission of signal 116
5-aminolevulinate synthase
  isotopic exchange 234
argininosuccinate lyase
  PIX 241–242
asparagine synthetase
  isotopic exchange 230
aspartate aminotransferase
  competitive substrate inhibition 180

aspartate aminotransferase (*Cont.*):
  isotope effects 289
  Ping Pong kinetic mechanism 81
  multiple combination of dead-end inhibitors 172
  temperature jump 220
aspartate transcarbamoylase
  allosteric regulation 112
  isotope effects 299
  mechanism 299–301
assay
  analysis of time course 32
  continuous 19
  coupled chemical 32
  coupled enzyme 25
    lag time 26
    optimization 28
  fixed-time 24
  ATPase
    aldehyde-induced 305

buffers
  effect on activity 326
  for pH profiles 327

cAMP-dependent protein kinase
  pH profiles 326
carbamoyl-P
  mechanism of nonenzymatic breakdown 300–301
carbamoyl-P synthetase
  PIX 242
carbonic anhydrase
  countertransport 239–240
chorismate mutase
  isotope effects 301
  mechanism 301
chymotrypsin
  viscosity effect 250
cooperativity
  positive 108
  negative 11
countertransport of label 238–240
creatine kinase
  equilibrium ordered kinetic mechanism 71
  pH profiles 326, 340, 344
crossover analysis 88

data processing 56
dihydrofolate reductase
  tight binding inhibition 199
distribution equations, common mechanisms 375

enolase
  metal ion kinetics 107
  pH profiles 336
  slow binding inhibition 197

formate dehydrogenase
  transition state 307
  binding isotope effect 307–308, 320
fractionation factors 269–270
fructokinase 101
  isotope partitioning 247
fumarase
  aldehyde-induced ATPase 305
  pH profiles 349–350
  mechanism 351
  oversaturation, countertransport of label 239

galactose-1-P uridylyltransferase
  PIX 243
glucose-6-P dehydrogenase
  isotope effects 278–279
glucose-6-phosphatase
  isotopic exchange 234
glutamate dehydrogenase
  induced substrate inhibition 185
  isotope effects 297
  pre-steady state 216
  regulatory mechanism 115

glutamate dehydrogenase (*Cont.*):
terreactant kinetic mechanism 97
transient state isotope effects 323
glutamate mutase
transition state 309–310
glutamine synthetase
isotope partitioning 244
PIX 240
glyceraldehyde-3-phosphate dehydrogenase
uncompetitive substrate inhibition 178
glycerokinase
aldehyde induced ATPase 305

Haldane relationship
Ping Pong mechanisms 100
sequential mechanisms 98
hexokinase
aldehyde-induced ATPase 305
binding isotope effects 321
induced substrate inhibition 185
iso mechanisms 235
isotope partitioning 244, 247–248
pH profile 353
slow binding inhibition 198
Hill plot 112

initial velocity patterns
primary plot 62
reading the plot 64
secondary plot 63
tertiary plots 95
inhibition
alternate product 150
alternate substrate
Ping Pong mechanisms 186
sequential mechanisms 187
competitive 122
dead-end
Bi Bi mechanisms 154
double 192
induced substrate in ordered mechanism 184

inhibition (*Cont.*):
noncompetitive 124
patterns
rules for prediction of product inhibition 132
rules for prediction of dead-end inhibition 157
partial substrate 77, 182
practical considerations 204
product
Bi Bi mechanism
ordered
steady state 134
Theorell-Chance 142
equilibrium 143
Ping Pong
classical 147
two site 150
random
rapid equilibrium with EBQ 144
steady state 147
Uni Bi mechanism
steady state ordered 128
rapid equilibrium random 130
slow binding 196
effect of reversible inhibitor on 199
slow tight binding 203
substrate
complete 173
Ping Pong with FA and/or EB 178
steady state ordered 175
tight binding 199
uncompetitive 125
inhibitor
combination with multiple enzyme forms
mixed product and dead-end 161
ordered with EAP 162
Ping Pong with EP 165
multiple combination of dead-end
ordered 166
rapid equilibrium random 170
Ping Pong 171

initial velocity studies
absence of inhibitors
Bi Bi mechanisms
sequential
ordered 68
equilibrium 69, 75
steady-state 72
Theorell-Chance 73
random
rapid equilibrium 75
steady state 76
resemble Ping Pong 85
Bi Bi Ping Pong
classical 80
two site 82
resemble sequential 86
terreactant
Ping Pong 90
sequential 91
detecting denominator terms 94
Uni Bi steady state ordered 67
Uni Bi rapid equilibrium random 62
practical considerations 117
rules for sequential mechanisms 91
isocitrate dehydrogenase
double inhibition 195
isotope effects 272
isotopic exchange 233
pH profiles 350
isocitrate lyase
slow binding inhibition 199
isotope partitioning 244
iso-mechanism 4, 235–238
isotope effects
as a probe of chemical mechanism 297–305
as a probe of regulatory mechanism 295–297
as a probe of transition state structure 306
acyl transfer 310–313
alcohol dehydrogenase 308–309

isotope effects (*Cont.*):
formate dehydrogenase 307–308
glutamate mutase 309–310
glycosyl transferases 318–320
phosphoryl transfer 313–318
binding isotope effects 320–321
calculation of K_d from ^DV and $^D(V/K)$ 273–274
determination of intrinsic isotope effects 275–279
multiple isotope effect method 277–279
Northrop's method 276
equation for isotope effects
on V/K 271–272
on V_{max} 273
measurement
direct comparison 254–257
equilibrium perturbation 257–260
internal competition 260–264
remote label method 264–267
equilibrium 267–270
intermediate partitioning 280
multiple isotope effects in stepwise mechanisms 280
nomenclature 254
on more than one step 275
on V_{max} in Ping Pong mechanism 364
product dependence 290–295
ordered mechanism 291
Ping Pong mechanism 294–295
random mechanism 292–293
substrate dependence 282–290
Ping Pong mechanism 287–289
ordered mechanism 284–285
random mechanism 285–287
terreactant mechanisms 289–292
transient-state kinetic isotope effects 321–323
isotopic exchange 226–235
at equilibrium 227–234
Ping Pong mechanisms 227–230
sequential mechanisms 230–234

isotopic exchange (*Cont.*):
  measurement of 226–227
  not at equilibrium 234–235
irreversible step 122

kanamycin nucleotidyltransferase
  transition state 316
ketopantoate reductase
  pH dependence of isotope effects 362–363
kinetics
  chemical
    first order 9
    pseudo-first order 11
    saturation 12
    second order 11
  pre-steady state
    burst in time course 213–216
    enzyme, reactant at comparable levels 216–218
    first order reactions
      consecutive 207–212
      irreversible 206
      parallel 212
      reversible 206–207
    methods 220–223
      rapid quench 222
      relaxation methods 223
      stopped flow 221
    temperature jump 218–219
      aspartate aminotransferase 220
      metmyoglobin ligands 220
kinetic mechanism, definition
  sequential 3
  Ping Pong 3
King-Altman patterns for photocopying 367
K_m
  definition 8, 15
  dependence on substrate concentration 78
kinetic parameters
  definitions, common mechanisms 375
  temperature dependence 12

β-lactamase
  viscosity effect 250

lactate dehydrogenase
  binding isotope effects 320
  kinetic isotope effects 295
  measurement of isotope effects 256
leucine amino peptidase
  binding isotope effect 320
Lineweaver-Burk plot 17

mandelate racemase
  oversaturation 238
malate dehydrogenase
  uncompetitive substrate inhibition 177
metal ions, kinetics 105
Michaelis-Menten equation 14

NAD malic enzyme
  double inhibition 195
  isotope effects 286
  mechanism 298
  pH dependence of isotope effects 362
NADP malic enzyme
  isotope effects 257, 277, 280–282
  mechanism 298
  pH profiles 352, 354
  pH dependence of isotope effects 361
  viscosity effect 251
metmyoglobin
  ligand binding by temperature jump 220
m-nitrobenzyl-P diesters
  as remote label 316
  transition state, ribonuclease 316
  transition state, kanamycin nucleotidyltransferase 316
p-nitrophenyl acetate
  as remote label 310
  transition states, non-enzymatic reactions 311–312
  transition states, enzymatic reactions 311, 313

p-nitrophenyl-P
as remote label 313
transition states, enzymatic and non-enzymatic 314–315
p-nitrophenyl-P diesters
isotope effects, snake venom diesterase 317
mechanism, snake venom diesterase 317
nucleoside hydrolase
transition state 318
nucleoside diphosphate kinase
competitive substrate inhibition 181
isotopic exchange 229

OMP decarboxylase
isotope effects 304
mechanism 304
overall rate equation 15
oversaturation 235–238
oxalate decarboxylase
isotope effects 302
mechanism 302

PEP carboxylase
binding isotope effects 310
induced substrate inhibition 186
pH
effect on enzyme stability 325
pH dependence
dissociation constant 328–336
binding less at high than at low pH 332–333
binding less at low than at high pH 330–332
binding lost at high pH 330
binding lost at low pH 329
binding lost at low and high pH 334–335
metal ion binding 358
two groups ionize at low or high pH 333–334
pH dependence (*Cont.*):
isotope effects 359–366
pH-dependent step isotope dependent 359–364
dead-end protonation of enzyme 361–362
dead-end protonation of E and EA 362–363
dead-end formation of EAH 363–364
random addition of proton and substrate 360–361
pH and isotope-sensitive steps differ 364–366
V/K 336–353
activity lost at high pH 343–346
dead-end protonation of free enzyme 343
E and EA can be deprotonated 343–345
sticky substrate 344–345
pK below 7.3 345–346
activity lost at low pH 336–342
dead-end protonation of free enzyme 337
E and EA can be protonated 337–342
pK above 7.3 340
sticky substrate 338–340
activity lost at both low and high pH 346–350
reverse protonation 349–350
sticky substrate 348–350
identity of groups 351–353
temperature dependence of pK 352
solvent perturbation of pK 352–353
two pK's at low or high pH 350
V_{max} 354–358
sticky substrate and proton 354–357
Ping Pong mechanism 357–358
phosphodiesterase, snake venom
mechanism 317

phosphofructokinase
aldehyde-induced ATPase 305
anomeric specificity 101
3-P-glycerate kinase
inhibition by D-glyceraldehyde-3-P 306
phosphatase, potato acid
alternate product inhbition 152
6-phosphogluconate dehydrogenase
isotope effects 287
mechanism 298
pH profiles 336
phosphotriester hydrolysis
transition states 317
phosphotriesterase
viscosity effect 250
positional isotopic exchange (PIX) 240–243
induced by presence of product 242
prephenate dehydrogenase
allosteric regulation 113
isotope effects 278
mechanism 298
proline racemase
oversaturation 235–238
countertransport of label 238
purine nucleoside phosphorylase
transition state 320
pyrophosphate-phosphofructokinase
multiple combination of dead-end inhibitor 171
pyruvate carboxylase
isotopic exchange 230
pyruvate kinase
binding isotope effect 320
pH profiles 358–359

rate constants
calculation, common mechanisms 375
macroscopic 7
microscopic 5
shorthand notation 55
rate equation, derivation
algebraic 36
definition of kinetic constants 40
Cha's method 46
distribution equations, common mechanisms 367–396
isotope effects 54
isotope exchange 52
King-Altman
alternative pathway 39
linear pathway 37
patterns for derivation of equations 367–374
net rate constants 43
rate equations, common mechanisms 375
regulation, kinetic mechanism 115
ribonuclease
transition state 316
L-ribulose-5-P 4-epimerase
isotope effects 302
mechanism 302

SAM synthetase
transition state 320
shikimate dehydrogenase
steady state ordered kinetic mechanism 73
shorthand notation
reactants 1
enzyme forms
stable 2
transient 2
kinetic mechanism 6
steady-state assumption 14
stepwise vs. concerted mechanisms 298
stickiness factors
determination of by isotope effects 339
determination of by isotope partitioning 243–251
succinate dehydrogenase
dead-end inhibition 153

thymidylate synthase
  induced substrate inhibition 185
triosephosphate isomerase
  pH profiles 337

UDP-glucose pyrophosphorylase
  PIX 243

Viscosity
  effects on V/K 249–251
V_{max}, estimation of 83

yeast osmoregulatory system 210